Asia-Pacific Interdisciplinary Research Conference 2011

(AP-IRC 2011)

Journal of Physics: Conference Series Volume 352

Toyohashi, Aichi, Japan
17-18 November 2011

Editors:

Adarsh Sandhu
Hiroshi Okada
Toru Maekawa
Ken Okano

ISBN: 978-1-62276-769-4
ISSN: 1742-6588

Printed from e-media with permission by:

Curran Associates, Inc.
57 Morehouse Lane
Red Hook, NY 12571

Printed by Curran Associates, Inc. (2013)

For permission requests, please contact the Institute of Physics
at the address below.

Institute of Physics
Dirac House, Temple Back
Bristol BS1 6BE UK

Phone: 44 1 17 929 7481
Fax: 44 1 17 920 0979

techtracking@iop.org

Additional copies of this publication are available from:

Curran Associates, Inc.
57 Morehouse Lane
Red Hook, NY 12571 USA
Phone: 845-758-0400
Fax: 845-758-2634
Email: curran@proceedings.com
Web: www.proceedings.com

TABLE OF CONTENTS

012001 Behavior of Multi-component Magnetic Colloidal Systems in Tunable Magnetic Fields and Applications in Biosensing 1
Y Yang, Z Li, P J Ko, A Sandhu

012002 Optical Transmittance for Analysis of the Dynamics of Self-Assembled Rotating Chains of Superparamagnetic Micro- and Nano Beads in Solution 18
P J Ko, T Takamura, A Sandhu

012003 Electrophoretic Deposition of High Quality Transparent Conductive Graphene Films on Insulating Glass Substrates 28
R Ishikawa, P J Ko, Y Kurokawa, M Konagai, A Sandhu

012004 CMOS On-Chip Optoelectronic Neural Interface Device with Integrated Light Source for Optogenetics 35
Y Sawadsaringkarn, H Kimura, Y Maezawa, A Nakajima, T Kobayashi, K Sasagawa, T Noda, T Tokuda, J Ohta

012005 Optimization of Sputtering Condition of IrOx Thin Film Stimulation Electrode for Retinal Prosthesis Application 43
Y Pan, T Noda, K Sasagawa, T Tokuda, H Kanda, T Fujikado, J Ohta

012006 Ferroelectric and Ferromagnetic Nanofibers: Synthesis, Properties and Applications 51
A Jalalian, A M Grishin, S X Dou

012007 Luminescent Magneto-Optical Photonic Crystals 60
A M Grishin, S I Khartsev

012008 Synthesis of Gd_2O_3 Nanoparticles for MRI Contrast Agents 70
N Sakai, L Zhu, A Kurokawa, H Takeuchi, S Yano, T Yanoh, N Wada, S Taira, Y Hosokai, A Usui, Y Machida, H Saito, Y Ichiyanagi

012009 High Temperature Hall Sensors using AlGaN/GaN HEMT Structures 76
S Koide, H Takahashi, A Abderrahmane, I Shibasaki, A Sandhu

012010 Effects of Proton Irradiation on the Magnetoelectric Properties of 2DEG AlGaN/GaN Micro-Hall Sensors 80
H Okada, A Abderrahmane, S Koide, H Takahashi, S Sato, T Ohshima, A Sandhu

012011 Microorganism Mediated Synthesis of Reduced Graphene Oxide Films 85
Y Tanizawa, Y Okamoto, K Tsuzuki, Y Nagao, N Yoshida, R Tero, S Iwasa, A Hiraishi, Y Suda, H Takikawa, R Numano, H Okada, R Ishikawa, A Sandhu

012012 Electronic Structure and Magnetic Properties of $^{31}P@C_{60}$-SWCNT as Peapods 90
A Suzuki, T Oku

012013 A Solution of Time Dependent Schrodinger Equation by Quantum Walk 98
H Sekino, M Kawahata, S Hamada

012014 A New Implementation of Dynamic Polarizability Evaluation Using a Multi-resolution Multi-wavelet Basis Set 105
H Sekino, Y Yokoi, R J Harrison

012015 Detecting Ferrite Nanobeads for Sentinel Lymph Node Mapping with a Highly Sensitive Hall Differential Magnetic Field Sensor 110
M Abe, T Ueda, T Masaki, Y Kitamoto, N Matsushita, H Handa

012016 Reduced Graphene Oxide as the Support for Lipid Bilayer Membrane 114
K Tsuzuki, Y Okamoto, S Iwasa, R Ishikawa, A Sandhu, R Tero

012017 Fabrication of Supported Lipid Bilayer on Graphene Oxide 121
Y Okamoto, K Tsuzuki, S Iwasa, R Ishikawa, A Sandhu, R Tero

012018 Fabrication and Characterization of Poly[diphenylsilane]-based Solar Cells 127
M Iwase, T Oku, A Suzuki, T Akiyama, K Tokumitsu, M Yamada, M Nakamura

012019 Fabrication and Characterization of Polysilane/C_{60} Thin Film Solar Cells 132
J Nakagawa, T Oku, A Suzuki, T Akiyama, K Tokumitsu, M Yamada, M Nakamura

012020 Low Temperature Synthesis of FePt Alloy Nanoparticles by Polyol Process 137
K Fukuda, S Fujieda, K Shinoda, S Suzuki, B Jeyadevan

012021 Synthesis of Multiferroic $DyFeO_3$ Nanoparticles and Study of Their Magnetic Properties 143
L Zhu, N Sakai, T Yanoh, S Yano, N Wada, H Takeuchi, A Kurokawa, Y Ichiyanagi

012022 Fabrication and Characterization of CuO/ZnO Solar Cells 148
H Kidowaki, T Oku, T Akiyama

012023 Microstructure Analysis of Spherical Silicon Solar Cells Coated with Anti-reflection Films 153
Y Ono, T Oku, T Akiyama, Y Kanamori, Y Ohnishi, Y Ohtani, M Murozono

012024 Microstructures and Photovoltaic Properties of Fullerene-based Organic-inorganic Hybrid Solar Cells 158
T Oku, A Takeda, A Nagata, K Fujimoto, T Akiyama, A Suzuki
012025 Design of Trench Gate GaN Power MOSFET using Al_2O_3 Gate Oxide 167
E G Kang, Y T Kim
012026 Fabrication and Properties of Thermoelectric Oxide Thick Films Deposited with Aerosol Deposition Method 175
Y Nakamura, Y Matsufuji, M Inoue
012027 Development of Cell Culture Microdevice Actuated by Piezoelectric Thin Films for Delivering Mechanical Vibratory Stimuli to Cells 183
Y Yamada, G Umegaki, T Kawashima, M Nagai, T Shibata, T Masuzawa, T Kimura, A Kishida
012028 Aqueous Synthesis and Transmission Electron Microscopy Observation of Seed-grown Spherical Ferrite Nanoparticles 190
T Tanaka, M Tada, T Nakagawa, M Abe
012029 Circular Dichroism in Optical Second Harmonic Generated in Reflection from Chiral G-shaped Metamaterials 199
E Mamonov, I Kolmychek, T Murzina, A Maydykovsky, O Aktsipetrov, V Valev, T Verbiest, A Silhanek, V Moshchalkov
012030 Flow Regulator Powered by the Stalk of Vorticella Convarallia 204
N Matsumoto, M Nagai, Y Hayasaka, M Oguri, T Kawashima, T Shibata
012031 Biofouling of Cr-Nickel Spray Coated Films on Steel Surfaces 211
K Yoshida, H Kanematsu, D Kuroda, H Ikigai, T Kogo, S Yokoyama
012032 Electrochemical Properties of Arc-black and Carbon Nano-balloon As Electrochemical Capacitor Electrodes 217
T Sato, Y Suda, H Uruno, H Takikawa, H Tanoue, H Ue, N Aoyagi, T Okawa, K Shimizu
012033 Remote Sensing of Radiation Dose Rate by Customizing an Autonomous Robot 225
T Kobayashi, M Nakahara, K Morisato, T Takashina, H Kanematsu
012034 Optimisation of Flux Concentrator-Coil Design for Application in Magnetic Hyperthermia 231
M Kallumadil, M Abe, T Ueda, H Handa
012035 Development of High Sensitive Magnetic Contaminant Detection System Using an HTS-rf-SQUID Covered with HTS Thin Films 239
Y Hatsukade, R Kurosawa, Y Uchida, S Tanaka
012036 Detection System for Metallic Contaminants by Eight-Channel SQUIDs 246
Y Kitamura, Y Hatsukade, S Tanaka, T Ohtani, S Suzuki
012037 Image Identification Based on Color and Luminance Information by Using an Optical Correlator 252
H Kuboyama, K Moriyama, S Arai, M Fukuda, M Kato, T Kawaguchi, S Yamamoto, M Inoue
012038 Observation of Biological Tissues Using Common Path Optical Coherence Tomography with Gold Coated Conical Tip Lens Fiber 268
K Taguchi, J Sugiyama, M Totsuka, S Imanaka
012039 Three-Dimensional Optical Trapping for Cell Isolation Using Tapered Fiber Probe by Dynamic Chemical Etching 274
K Taguchi, J Okada, Y Nomura, K Tamura
012040 Optical Manipulation of Symbiotic Chlorella in Paramecium Bursaria Using a Fiber Axicon Microlens 278
K Taguchi, S Hirota, H Nakayama, D Kunugihara, Y Mihara
012041 Development of a Cell Microarray Chip for Detection of Circulating Tumor Cells 284
S Yamamura, S Yatsushiro, K Abe, Y Baba, M Kataoka
012042 High-affinity RNA Aptamers to C-reactive Protein (CRP): Newly Developed Pre-elution Methods for Aptamer Selection 288
N Orito, S Umekage, K Sato, S Kawauchi, H Tanaka, E Sakai, T Tanaka, Y Kikuchi
012043 Crystal Structure of an L-Carnitine Complex with Pyrogallol[4]arene 301
I Fujisawa, D Takeuchi, Y Kitamura, R Okamoto, K Aoki
012044 Quantitative Analysis of Plasma Interleiukin-6 by Immunoassay on Microchip 307
K Abe, Y Hashimoto, S Yatsushiro, S Yamamura, M Tanaka, T Ooie, Y Baba, M Kataoka
012045 Emotion Telepresence: Emotion Augmentation through Affective Haptics and Visual Stimuli 312
D Tsetserukou, A Neviarouskaya
012046 A Neurological Enigma: The Inborn Numerical Competence of Humans and Animals 317
H J Gross
012047 The "Ouzo Effect", Recent Developments and Application to Therapeutic Drug Carrying 323
R Botet

012048 Biofouling of Various Metal Oxides in Marine Environment 333
T Kougo, D Kuroda, N Wada, H Ikegai, H Kanematsu
012049 Applicability of Carbonated Electric Arc Furnace Slag to Mortar 338
S Yokoyama, R Arisawa, M N N Hisyamudin, K Murakami, A Maegawa, M Izaki
012050 Elution of Zinc in Dust Discharged from Electric Arc Furnace in Carbonic Acid Solution 347
S Yokoyama, T Sasaki, J Sasano, M Izaki
012051 Influence of Amount of Oxidizing Slag Discharged from Stainless Steelmaking Process of Electric Arc Furnace on Elution Behavior Into Fresh Water 354
S Yokoyama, T Shimomura, M N N Hisyamudin, T Takahashi, M Izaki
012052 Ammonia-oxidizing Activity and Microbial Community Structure in Acid Tea (Camellia Sinensis) Orchard Soil 362
K Okamura, A Takanashi, T Yamada, A Hiraishi
012053 Extraction and Refining of Essential Oil from Australian Tea Tree, Melaleuca Alterfornia, and the Antimicrobial Activity in Cosmetic Products 367
Q Huynh, T D Phan, V Q Q Thieu, S T Tran, S H Do
012054 Production of D-lactic Acid from Sugarcane Bagasse Using Steam-explosion 374
C Sasaki, R Okumura, A Asakawa, C Asada, Y Nakamura
Author Index

Journal of Physics: Conference Series **352** (2012) 011001 doi:10.1088/1742-6596/352/1/011001

The Asia-Pacific Interdisciplinary Research Conference 2011 (AP-IRC 2011)

PREFACE

AP-IRC 2011

The Asia-Pacific Interdisciplinary Research Conference 2011

Organized by Electronics-Inspired Interdisciplinary Research Institute (EIIRIS), Toyohashi University of Technology, Japan

Scientists, engineers, entrepreneurs and policymakers gather at the first truly interdisciplinary conference held in Asia-Pacific (http://www.apirc.jp).

The inaugural Asia-Pacific Interdisciplinary Research Conference 2011 (AP-IRC 2011) was held at Toyohashi University of Technology (Toyohashi Tech) on 17–18 November 2011. The conference is a forum for enhancing mutual understanding between scientists, engineers, policymakers and experts from a wide spectrum of pure and applied sciences, to resolve the daunting global issues facing mankind.

The conference attracted approximately 300 participants including delegates from France, Germany, India, Indonesia, Korea, Malaysia, Russia, Sweden, United Kingdom, USA and Vietnam. AP-IRC 2011 was chaired by Dr Yoshiyuki Sakaki, President of Toyohashi Tech, who opened the proceedings by stressing the importance of an interdisciplinary approach to research, to resolve global scientific and technical issues. Recalling his own experience as the leader of Japan's efforts in the Human Genome Project, Sakaki also encouraged participants to make an effort to try to understand the sometimes difficult concepts and terminology of other areas of research.

The presentations at AP-IRC 2011 were divided into three focus sessions: innovative mechano-magneto-electronic systems, life sciences, and green science and technology. A total of 174 papers were presented over the two-day conference including eight by invited speakers. Highlights of AP-IRC 2011 included a first-hand account of the damage caused by the massive earthquake in March 2011 to experimental facilities at Tohoku University by Masayoshi Esashi; the fascinating world of bees and the inborn numerical competence of humans and animals by Hans J Gross; research on robots and cognition-enabled technical systems at Technische Universität München by Sandra Hirche; the history of events leading to the invention of the world's strongest NdFeB permanent magnet by Masato Sagawa; a novel method for the synthesis of graphene using bacteria extracted from a riverside in Toyohashi by Toyohashi Tech scientists; and ambitious plans to

Journal of Physics: Conference Series **352** (2012) 011001 doi:10.1088/1742-6596/352/1/011001

harvest energy by laying massive numbers of solar cells in North Africa as part of the "Sahara Solar Breeder (SSB) Plan" for a global clean-energy superhighway, described by Hideomi Koinuma.

In addition to the technical sessions, the conference banquet included a short session during which the invited speakers described notable trends in research and policy in their part of the world. The short speeches led to animated discussions between the delegates, particularly the young scientists and graduate students, who were able to talk directly with veteran researchers for a first-hand view of the issues raised during the day's presentations.

In closing the conference, Professor Makoto Ishida, co-chair of the conference and vice-president of Toyohashi Tech announced that this conference will be held annually at the same time each year, with AP-IRC 2012 scheduled for 15–16 November 2012 at the Irago Sea-Park & Spa Hotel in Aichi Prefecture, Japan.

Journal of Physics: Conference Series **352** (2012) 011001 doi:10.1088/1742-6596/352/1/011001

COMMITTEES

Chairmen

Chairman

Yoshiyuki Sakaki, *President, Toyohashi University of Technology*

Vice chairs

Yasuyoshi Inagaki, *Executive Trustee, Vice-president, Toyohashi Tech*

Kiyokatsu Jinno, *Executive Trustee, Vice-president, Toyohashi Tech*

Toshiaki Tsuji, *Executive Trustee, Toyohashi Tech*

Makoto Ishida, *Vice-president, Toyohashi Tech*

Editorial Committee

Editor

Adarsh Sandhu, *Electronics-Inspired Interdisciplinary Research Institute (EIIRIS), Toyohashi Tech*

Associate Editors

Hiroshi Okada, *Electronics-Inspired Interdisciplinary Research Institute (EIIRIS), Toyohashi Tech*

Toru Maekawa, *Bio-Nano Electronics Research Centre, Toyo University, Japan*

Ken Okano, *Graduate School of Natural Science, International Christian University, Japan*

Special Advisory Committee

Shiro Hiruta, *Senior Advisor, Asahi Kasei Corporation, Japan*

Takashi Mimura, *Fellow, Fujitsu Laboratories Ltd., Japan*

Quentin Pankhurst, *Director of Research, The Royal Institution of Great Britain, UK*

Masato Sagawa, *Representative, Intermetallics Co., Ltd., Japan*

Hiroyuki Sakaki, *President, Toyota Technological Institute, Japan*

Takao Suzuki, *Director of Center for Materials for Information Technology, University of Alabama, USA*

Mathias Uhlén, *Professor, Division of Proteomics, Royal Institute of Technology, Sweden*

Shiro Usui, *Professor, Brain Science Institute, RIKEN, Japan*

Hiroyuki Watanabe, *Senior Technical Executive, Toyota Motor Corporation, Japan*

Steering Committee

Makoto Ishida (Chairman)

Shin-ichi Aoki

Masaki Aono

Mitsuo Fukuda
Takanobu Inoue
Shinichi Itsuno
Noriyoshi Kakuta
Mihoko Kato
Shozo Kawamura
Yo Kikuchi
Yoshiaki Kiso
Kenzo Kitamura
Shigeru Kuriyama
Yuzuru Miyata
Hirofumi Takikawa
Saburo Tanaka

International Advisory Committee

Pulickel M. Ajayan, *Rice University, USA*
Jeyadevan Balachandran, *The University of Shiga Prefecture, Japan*
Matty Caymax, *IMEC, Belgium*
Ching-Ray Chang, *National Taiwan University, Taiwan*
Ho Jung Chang, *Dankook University, Korea*
Sudhir Chandra, *Indian Institute of Technology Delhi, India*
Masayoshi Esashi, *Tohoku University, Japan*
Chao-Ming Fu, *National Taiwan University, Taiwan*
Hiroyuki Fujita, *University of Tokyo, Japan*
Toshio Fukuda, *Nagoya University, Japan*
Arindam Ghosh, *Indian Institute of Science, India*
Alexander B. Granovski, *Moscow State University, Russia*
Alexander Grishin, *Royal Institute of Technology, Sweden*
Hans J. Gross, *Würzburg University, Germany*
Dirk Grundler, *Technische Universität München, Germany*
Hiroshi Handa, *Tokyo Institute of Technology, Japan*
Shin-won Kang, *Kyungpook National University, Korea*
Haruhisa Kawasaki, *Gifu University, Japan*
Kwang Sun Kim, *Korea University of Technology and Education, Korea*
Yamuna Krishnan, *Tata Institute for Fundamental Research, India*
D. Sakthi Kumar, *Toyo University, Japan*
SangHoon Lee, *Korea University, Korea*

Journal of Physics: Conference Series **352** (2012) 011001 doi:10.1088/1742-6596/352/1/011001

Miguel Levy, *Michigan Technological University, USA*
G.P.Li, *University of California, Irvine, USA*
Jing-Feng Li, *Tsinghua University, China*
Sang Ho Lim, *Korea University, Korea*
Liwei Lin, *University of California Berkeley, USA*
Mikhail F. Limonov, *Ioffe Physical Technical Institute, Russia*
Shuang-Jiang Liu, *Chinese Academy of Sciences, China*
Justin Llandro, *University of Cambridge, UK*
Masud Mansuripur, *The University of Arizona, USA*
Ahmet Oral, *Sabanci University, Turkey*
Caroline Ross, *Massachusetts Institute of Technology, USA*
Vladimir M. Shalaev, *Purdue University, USA*
André Touboul, *University of Bordeaux, France*
Chen S. Tsai, *University of California, Irvine, USA*

Local Organizing Committee

Makoto Ishida (Chairman)
Yo Kikuchi
Daisuke Akai
Kunihiko Hara
Yasoo Harada
Tsuneo Hashizume
Hiromu Ishii
Shigeki Nakauchi
Hiroshi Okada
Adarsh Sandhu
Kazuaki Sawada
Ichiro Shibasaki
Takayuki Shibata
Tetsuo Shinoda
Akihiro Wakahara

Program Committee

Akira Hiraishi (Chair, Green Science and Technology)
Mitsuteru Inoue (Chair, Innovative Mechano-Magneto-Electronic Systems)
Yo Kikuchi (Chair, Life Sciences)
Seiji Iwasa

Organizing Institutes

Electronics-Inspired Interdisciplinary Research Institute (EIIRIS), Toyohashi University of Technology.

Websites

AP-IRC Home page: http://www.apirc.jp
Toyohashi University of Technology: http://www.tut.ac.jp/english
Electronics-Inspired Interdisciplinary Research Institute (EIIRIS) at Toyohashi Tech: http://www.eiiris.tut.ac.jp

All papers published in this volume of *Journal of Physics: Conference Series* have been peer reviewed through processes administered by the proceedings Editors. Reviews were conducted by expert referees to the professional and scientific standards expected of a proceedings journal published by IOP Publishing.

Behavior of multi-component magnetic colloidal systems in tunable magnetic fields and applications in biosensing

Ye Yang [1,2], Zhengcao Li [1], Pil Ju Ko [2] and Adarsh Sandhu [*,2,3]

[1] Advanced Materials Laboratory, Department of Materials Science and Engineering, Tsinghua University, Beijing 100084, People's Republic of China

[2] Department of Electrical and Electronic Engineering, Tokyo Institute of Technology, 2-12-1 O-okayama, Meguro-ku, Tokyo 152-8552, Japan

[3] Electronics-Inspired Interdisciplinary Research Institute (EIIRIS), Toyohashi University of Technology, 1-1 Hibarigaoka, Tempaku, Toyohashi 441-8580, Japan

*Email: sandhu.a.aa@m.titech.ac.jp Phone: +81-3-5734-2807, Fax: +81-3-5734-2807

Abstract. A system consisting of multiple-component beads, such as superparamagnetic beads, nonmagnetic beads and magnetorheological (MR) fluid, can display some very amazing and special properties when subjected to an external magnetic field because the MR fluid can act on both types of beads synchronously as a magnetic medium. Some novel structures and phenomena were discovered and are discussed in our work, including 'ring-structures', 'small-ring' and 'ring-chains' in static or rotational magnetic fields. If both probe and target molecules are attached consisting of functionalized superparamagnetic beads and non-magnetic beads, respectively, the ring-structure could be maintained due to biomolecular bonding, even after removing the external magnetic field. Using these remnant rings, we raised two protocols for biosensing: a two-dimensional biosensor using a magnetic self-assembled colloidal ring-structure, and an improved magneto-optical transmittance (MT) method. In the former protocol, we define the small nonmagnetic particles as "petals" because the whole structure looks like a flower. It was proved that the number of remnant ring petals was a function of the concentration of the target molecules', with a concentration range from 0.0768 ng/mL ~ 3.8419 ng/mL, making it a promising technology for applications involving biosensing. In the latter protocol, the use of larger individual units made the magnetic particle chain longer, which was considered to be a promising way of improving the sensitivity of the MT method.

1. Introduction

The concept of micro total analysis systems was introduced into science in 1990. Parallel to the boom in microfluidic systems, nanomaterials and nanoparticles have become a hot topic in research. Magnetic micro/nano-particles can display several special properties involving magnetic actuation, which forms the theoretical basis for a wide range of research work. Therefore, there are a series of new biomedical applications such as MRI contrast enhancement [1, 2-4], hyperthermia [5, 6, 7], drug targeting [8, 9], magnetic separation [10, 11, 12], magnetic washing, magneto-optical biosensing [13, 14], cell biophysics [15, 16], and biological engineering [17]. Almost all of these applications are based on the phenomenon whereby the particles can be magnetically probed and manipulated using permanent magnets or electromagnets. Because of this, three of the most important advantages of using these beads for analytical applications are known to be: (1) the possibility to reduce sample sizes and reagents to minute quantities; (2) relatively fast reaction times; (3) a large surface-to volume ratio, which is considered to be one of the most effective ways to obtain high sensitivity. Specifically, applications of magnetic nanoparticles for point-of-care medical treatments include functionalized fluorescent magnetic beads for ultrafast medical diagnosis, biosensing using Hall biosensors, the detection of sub-200 nm magnetic particles by magnetically-induced self-assembly, and label-less homogeneous biosensing by magneto-optical scattering. [18]

A system consisting of paramagnetic and nonmagnetic particles in a magnetorheological (MR) fluid can display some very amazing and special properties. Because the MR fluid acts as a magnetic medium, it acts on both types of beads synchronously. Using such a multiple-component system, colloidal superstructures with multi-pole symmetry can be manufactured, such as "Saturn's rings", flowers, linear quadrupoles (poles) and mixed multipole arrangements ('two tone'). [19] Kwan H. Li and Benjamin B. Yellen [20] also conducted some research into the relationship between the ferrofluid concentration and the separation distance of particles within the ring structure.

In our study, some novel phenomena were observed in the kinds of multiple-component magnetic colloidal systems that exist in a tunable magnetic field, such as the formation of rings, small-rings, ring-chains, and we observed their behavior in a rotating magnetic field. These dynamical studies are expected to become the basis for much relevant research in the future.

Also, in our work, superparamagnetic beads and nonmagnetic beads were attached to probe molecules (streptavidin) and target molecules (biotin) respectively, and the ring-structures could be maintained even after removing the external magnetic field because of bio-interactions.

Since the shape of the ring-structure that we created is just like a flower, we define the small nonmagnetic beads around it as "petals". We proved that the average number of petals in the ring was a function (a linear function at low concentration) of the concentration of biotin, which means that this procedure can be applied in biosensing. Since the whole procedure was carried out in a two-dimensional system, this approach offers a means of cutting down reagent consumption and waste generation, which is important for biosensing. Furthermore, the driving force generated by the external magnetic field enables the whole procedure to only take a few minutes, which makes this method very promising for applications requiring rapid diagnosis.

In addition, these remnant rings were also used to improve the sensitivity of the magneto-optical transmittance (MT) technique. With paramagnetic particles replaced by rings as the basic unit, the magnetic chains became longer. Combined with the working principle of the MT method, this change is expected to lead to promising improvements.

1.1 Formation of Ring-structure

Randall M. Erb et al. used a theoretical model to describe the formation of these multi-component superstructures. [19] Their physical model employs a continuum approach to approximate the local fluid magnetization, which is considered to be justifiable when the ferrofluid particles are much smaller than the other colloidal particles in the fluid. In our experiments, the MR fluid consisted of a colloid of nanoparticles mixed with micro-sized superparamagnetic and nonmagnetic beads. The effective dipole moment of a linearly-magnetizable spherical particle immersed in a magnetizable carrier fluid is given by

$$\mathbf{m}_i = 4\pi a^3 \frac{\chi_i-\chi_f}{\chi_f+2\chi_i+3}\mathbf{H}_{\text{ext}} \qquad (1)$$

where a is the radius of the particle, $\mathbf{H}_{\text{ext}}$ is the magnetic field at the location of the particle center, and χ_f and χ_i are the magnetic susceptibilities of the ferrofluid and the particle, respectively. If we can control the susceptibility of the ferrofluid to satisfy the relationship $\chi_{NM} < \chi_f < \chi_{PM}$, then, according to the above equation, we can predict that paramagnetic particles for which $\chi_i = \chi_{PM} > \chi_f$ will behave as point dipoles aligned parallel to the applied field, whereas the non-magnetic particles with $\chi_i = \chi_{NM} \approx 0$ will behave as point dipoles aligned antiparallel to the field. Here, in addition, compared with a ferrofluid with a typical susceptibility value of between 0.1-1.0, the susceptibility of non-magnetic beads is only on the order of 10^{-6}, so the approximation $\chi_{NM} \approx 0$ is reasonable.

Equilibrium particle configurations can be explained from the expression for the magnetostatic potential energy of a dipole m in a locally-applied field **H**, given by

$$U_i(\mathbf{r}) = -2\pi a^3 \mu_0 \frac{\chi_i-\chi_f}{\chi_f+2\chi_i+3}|\mathbf{H}(\mathbf{r})|^2 \qquad (2)$$

where μ_0 is the magnetic permeability of free space. For a non-magnetic particle ($\chi_{NM} < \chi_f$), according to (2), its potential energy must be strictly positive, which causes the particle to move towards regions of minimum magnetic field, $|H(r)|=|H_{min}|$. In contrast to non-magnetic particles, for paramagnetic particles, with $\chi_{PM} > \chi_f$, the potential energy is strictly negative, which causes the particle to move towards regions of maximum magnetic field, $|H(r)|=|H_{max}|$. Hence, in mixed suspensions of two types of particles, the magnetic force drives non-magnetic particle towards the 'equator' of the paramagnetic particle, where the total magnetic field is equal to the dipolar field of the paramagnetic particle subtracted from the

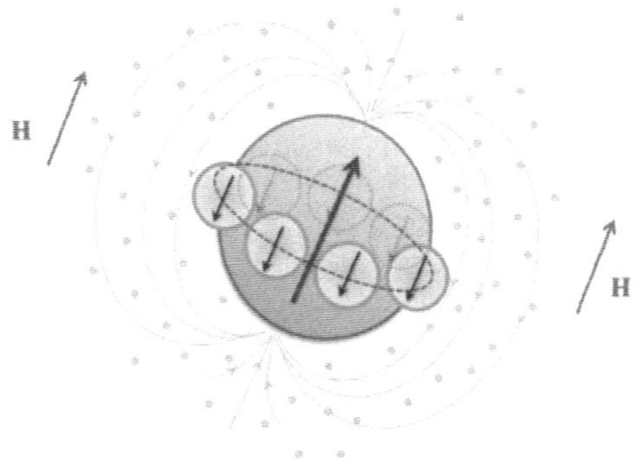

Figure 1 The yellow, blue and small green balls represent paramagnetic beads, nonmagnetic beads and MR nanoparticles respectively. When an external magnetic field is applied, at the "equator" of a paramagnetic particle, the total magnetic field equals the dipolar field of the paramagnetic particle subtracted from the external field, which is a minimum according to equation (2), so the magnetic force drives nonmagnetic particles to this area.

external field. In a similar way, the paramagnetic particles move towards the equator of the non-magnetic particles, where the total magnetic field equals the dipolar field of the non-magnetic particle added to the external field. Fig.1 shows a schematic diagram for this principle.

In fact, the size-ratio between the paramagnetic particles and the non-magnetic particles is a very important parameter to determine the configuration of the ring-structure. When the system consists of large paramagnetic particles and small non-magnetic particles, the larger paramagnetic particles will act as a core and will be surrounded by pairs of smaller non-magnetic particles around their equatorial areas. When the non-magnetic particles are larger than the paramagnetic ones, the situation will be just the opposite.

1.2 Streptavidin-Biotin interaction

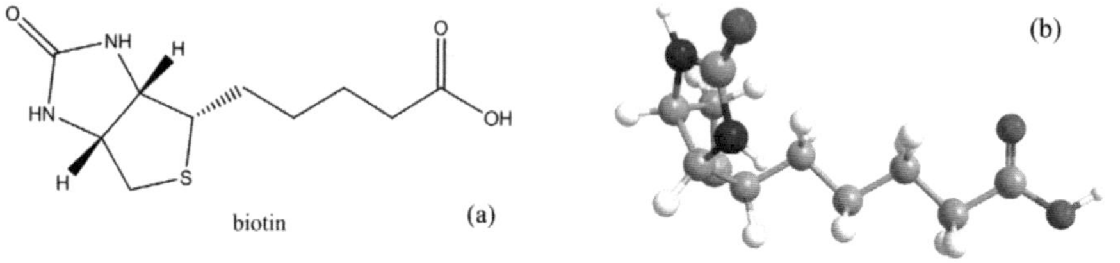

Figure 2 Structural formula (a) and 3D molecule model (b) of Biotin. The gray, white, blue, red and yellow balls represent C, H, N, O and S atoms respectively.

Streptavidin is a tetrameric protein that binds very tightly to the small molecule, biotin. The biotin-streptavidin system is the strongest noncovalent biological interaction known, having a dissociation constant K(d) of about ~10-15mol/L, and the complexes are also extremely stable over a wide range of temperature and pH. The strength and specificity of the interaction has led to it being one of the most widely used affinity pairs in molecular, immunological, and cellular assays.

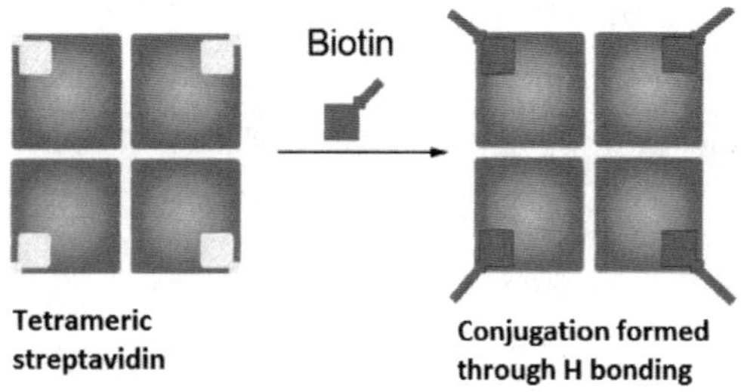

Figure 3 Hydrogen bonding in the streptavidin-biotin binding site. The four blue squares represent the four parts of a tetrameric streptavidin protein molecule. The red squares in each part are Biotin molecules. Hence, every tetrameric streptavidin protein molecule can bond with four Biotin molecules at most.

The molecular structure of Biotin is shown in Fig. 2, both in a structural formula and a 3D model. Fig. 3 shows how streptavidin and biotin connect with each other.

1.3 Working principle of the MT technique

As we know, paramagnetic beads tend to form chains in an external magnetic field, and if the external magnetic field rotates, the chains will rotate as the magnetic field rotates in a stationary frequency. Sang Yoon Park et al. [13, 14, 21] applied this phenomenon in biosensing combined with a light scattering technique.

A rotating external field (H_{ex}) applied to an aqueous solution containing biotinylated superparamagnetic beads (SBs) produced linear chains of SBs rotating in phase with H_{ex} due to magnetically induced self-assembly. At constant H_{ex}, the addition of avidin to the solution led to the formation of longer SB-chains than those formed without the presence of avidin. This change in length of the SB-chains was revealed by increases in the amplitude of the oscillating optical transmittance signal of the magnetic colloid solution. Monitoring changes in the amplitude of the optical transmittance of the solution enabled quantitative determination of the concentration of avidin that was added to the solution. The rotating chains acted as biomolecular probes and micromagnetic mixers, enabling the detection of biomolecular recognition in a very short time.

Fig. 4 shows a schematic diagram of the MT measurement set-ups that were utilized. In this figure, the angle between the incident beam and the direction of the external magnetic field is defined. Fig. 5 shows how the MT technique can be used for biosensing. First of all, the light that is incident on the SBSs is scattered, and the intensity of the transmitted light depends on the angle (θ) between the SBC direction and the light propagation path, which we refer to as the 'magneto-optical transmittance (MT) effect'. As for the rotation of the external magnetic field, the optical transmittance signal curve is sinusoidal. When θ=0°, the value of the optical transmittance signal reaches a maximum (peak value); while when θ=90°, the value of this signal reaches a minimum (valley value). The average length of a rotating SBC at a given hydrodynamics force (FH) depends only on the magnetic dipolar force between the beads (Fm). After the attachment of a biomolecule, a new avidin biotin conjugation force (FABC) is introduced, which results in stronger and longer biotinylated SBCs. Furthermore, this leads to a greater amplitude of the oscillating signal of the MT effect. The longer the chains, the larger the peak-to-valley ratio of the signal we can obtain.

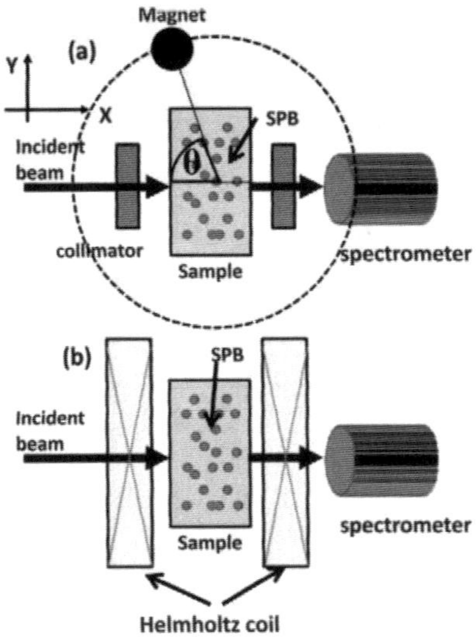

Figure 4 [21] Schematic illustration of the measurement setups utilizing (a) rotatable permanent magnet and (b) Helmholtz magnet. The size of SPBs and the wavelength of incident light were 1.5 μm and 630 nm, respectively.

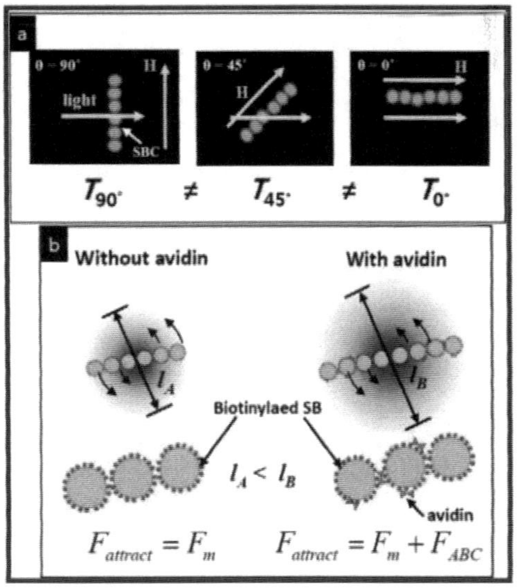

Figure 5 [14] Schematic illustration of the process of magneto-optical biomolecular detection by rotating superparamagnetic bead chains in solution. (a) Different transmission situation with different θ. (b) Addition of biomolecule change the length of the SBs-chain

2. Results and Analysis

2.1 Behavior of a multiple-component magnetic colloid system in a tunable field

2.1.1 Formation of a ring-structure

Superparamagnetic beads with diameters of 2.8-μm *35μL*(Invitrogen Dynal AS, Oslo. Norway), nonmagnetic beads with diameters of 1.0-μm *20μL*(Micromod Partikel Technologie GmbH), and MR fluid (Sigma-hc. Co.) with a volume fraction of 1.2% *50μL* were mixed to form a solution. The initial concentrations of superparamagnetic beads and nonmagnetic beads were 6.7×10^{8} particles/mL and 9.3×10^{10} particles/ mL respectively. The sample was treated with supersonic vibration for 3 min to cause the different components to mix sufficiently. An external magnetic field was then applied using an electric solenoid, and the magnetic field intensity could be adjusted by changing the current through the solenoid.

The configuration of the ring structure depends not only on the concentration of the MR fluid, [19] but also on the magnetic field intensity and the concentration ratio between the individual components.

We also monitored the ring-structure formation process for different concentrations of MR fluid while slowly increasing the intensity of the magnetic field.

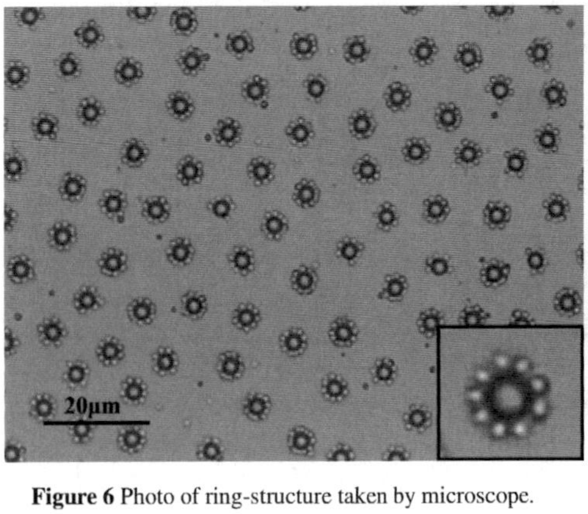

Figure 6 Photo of ring-structure taken by microscope.

Fig. 6 shows a perfect ring structure. In this case, the magnetic field was generated by a permanent magnet at a magnetic field of ~260Oe. In this structure, paramagnetic beads with larger sizes act as cores, and are surrounded by nonmagnetic beads. Every "flower" has 8~12 petals.

We also monitor the ring-structure formation process by varying the concentration of the MR fluid and slowly increasing the intensity of the magnetic field. Four different stages are defined: incompact, semi-formed, integral and uniformity. A schematic diagram is shown in Fig. 7.

Table 1. Magnetic field intensity needed for different stages.

Volume fraction of MR Fluid	Magnetic field intensity needed (Oe)			
	Incompact	Semi-formed	Integral	Uniformity
0.3%	29	34	50	53
1.2%	13	18	26	47
2.2%	8-10	13	18	34

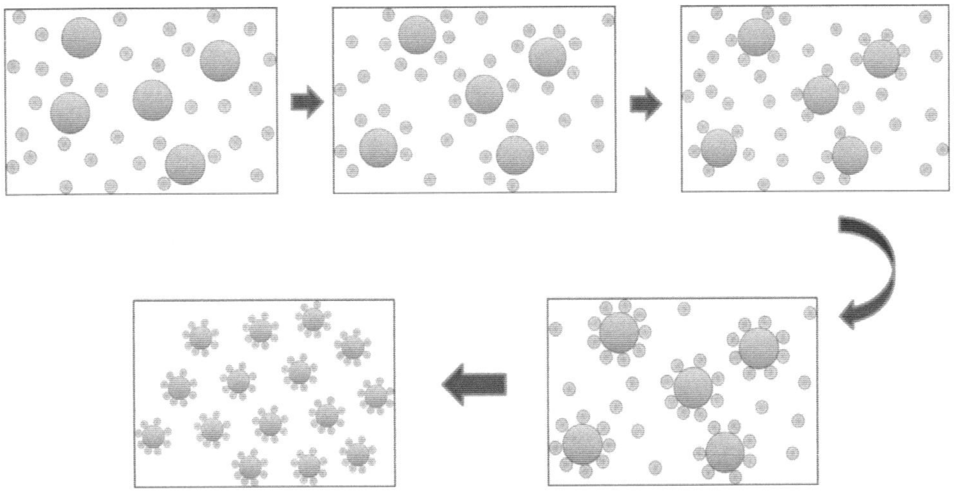

Figure 7 As the intensity of the magnetic field gradually increases, the ring-structure formation process goes through four main stages. This schematic diagram shows these in order. (1) With no magnetic field, the system is totally orderless; (2) Incompact stage: small nonmagnetic beads start to approach core paramagnetic beads; (3) Semi-formed: some (but not fully, about 3~4) small beads have attached onto the surfaces of the core beads; (4) Integral: perfect individual ring-structures form, every core particle is fully surrounded by small particles; (5) Uniformity: the distribution of individual ring-structures in the whole solution system becomes uniform.

Table 1 shows the magnetic field intensities that are needed to form the different stages with different MR fluid concentrations. As the concentration of the MR fluid increases, lower magnetic field intensity is required to form every stage.

2.1.2 Formation of a small-ring

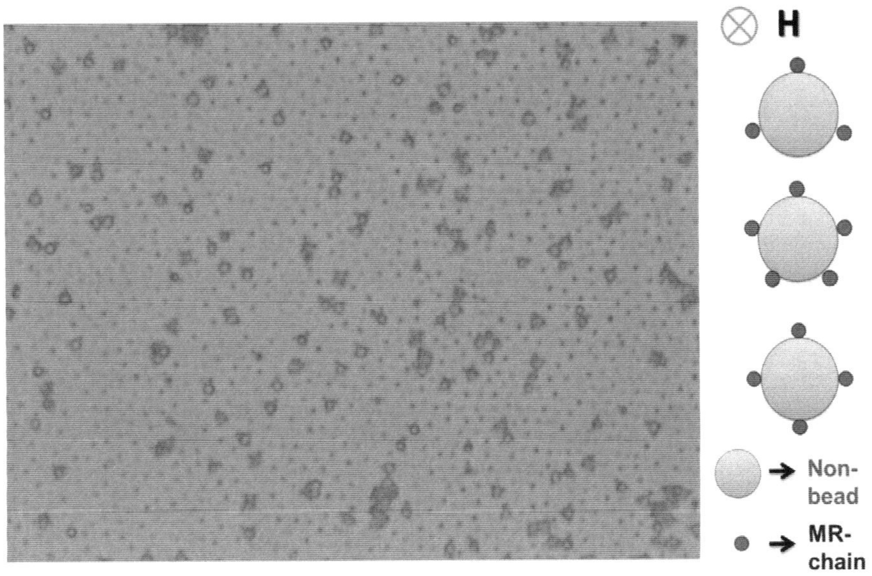

Figure 8 Photo of small-ring. There are many basic units for the formation of small-ring in this photo. A magnified schematic diagram is shown on the right side.

We made a solution with nonmagnetic-1.0μm beads 20μL and MR fluid (with a volume fraction Cv=1.2%) 100μL. Then we exposed this sample in an external vertical magnetic field, obtaining structures as in the following pictures. Fig. 8 shows the basic unit for the formation of a small-ring. The small dark spots that suround the nonmagnetic beads are actually the chain-like clusters of MR nanoparticles. (We cannot see discrete MR nanoparticles with an optical microscope because of the limitation of the amplification factor.) And as Fig. 8 shows, every basic unit for the small-ring structure is composed of one nonmagnetic bead as a core that is surrounded by 3-5 MR nanoparticle-chains.

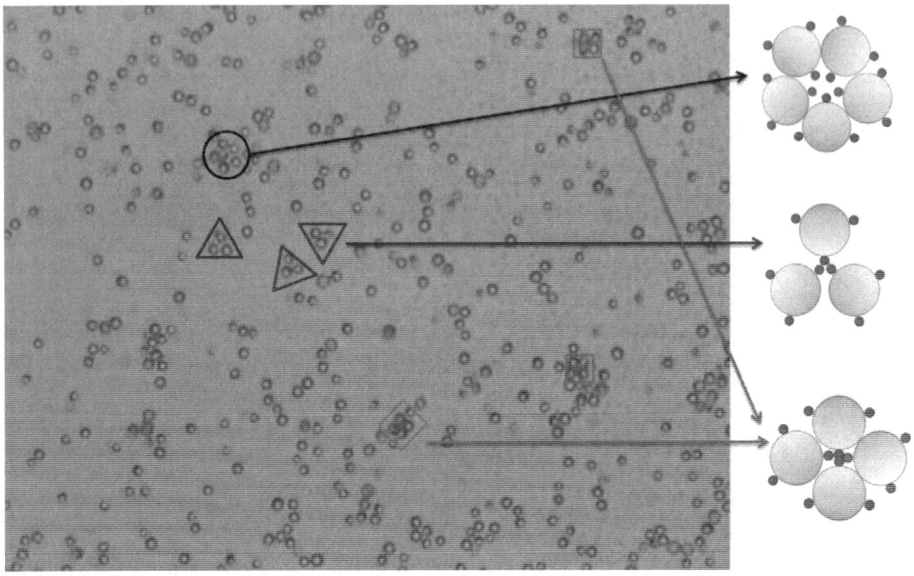

Figure 9 Photo of small-ring. Every small-ring is composed of 3-5 nonmagnetic particles and a 'pistil' made of several MR nanoparticle-chains.

Fig. 9 emphatically shows small-ring structures. Every small-ring is composed of 3-5 nonmagnetic particles and a 'pistil' made of several MR nanoparticle-chains.

Based on our observation, we raised a theory conjecture to explain the formation of small-ring. As we mentioned above, the dark spots in this system are actually MR nanoparticle chains seen from a top view. These MR-chains infibulate nonmagnetic particles just like we use chopsticks to infibulate dishes. In the next, because of the lateral attraction between the MR-chains, these chains tend to bring nonmagnetic particles together. Fig. 10 is a schematic diagram of this process.

2.1.3 Formation of ring-chains

We all know paramagnetic beads form chains in a horizontal magnetic field (that is, relative to the observation angle) and a multiple-component system forms ring-structures in a magnetic field perpendicular to that in the former statement. What would happen if there are magnetic field components in both directions? We carried out an experiment in an aligned magnetic field using a multiple-component system consisting of paramagnetic beads, nonmagnetic beads and MR fluid.

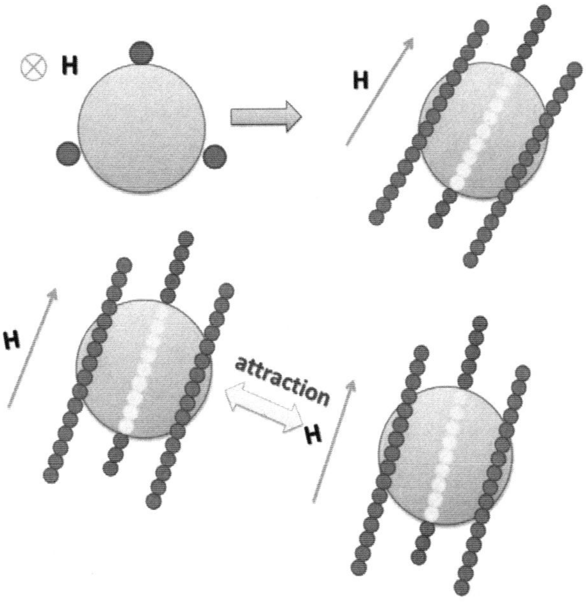

Figure 10 Formation process of a small-ring structure.

A simple device that can offer an aligned magnetic field was set up as shown in Fig. 11. The angle of inclination **α** is defined as the angle between the sample plane and the plane of the magnet. In this part, our study is non-quantitative, so **α** is just for a qualitative description.

It turns out that with an aligned magnetic field, chain-structure and the ring-structure form at the same time, which we call 'ring-chains'. Two experiments were carried out with $\alpha=30^{\circ}$ and $\alpha=70^{\circ}$ respectively. Photos of these are shown in Fig. 12 (a) and (b).

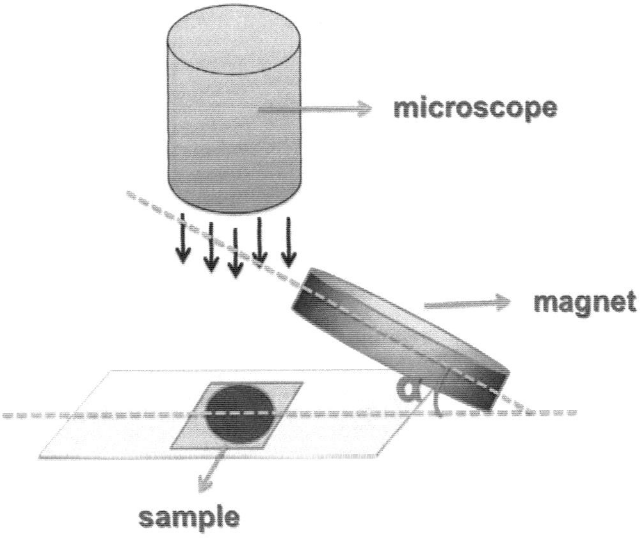

Figure 11 A schematic diagram of a sample device that can offer an aligned magnetic field. The angle of inclination is defined as the angle between the sample plane and the plane of the magnet

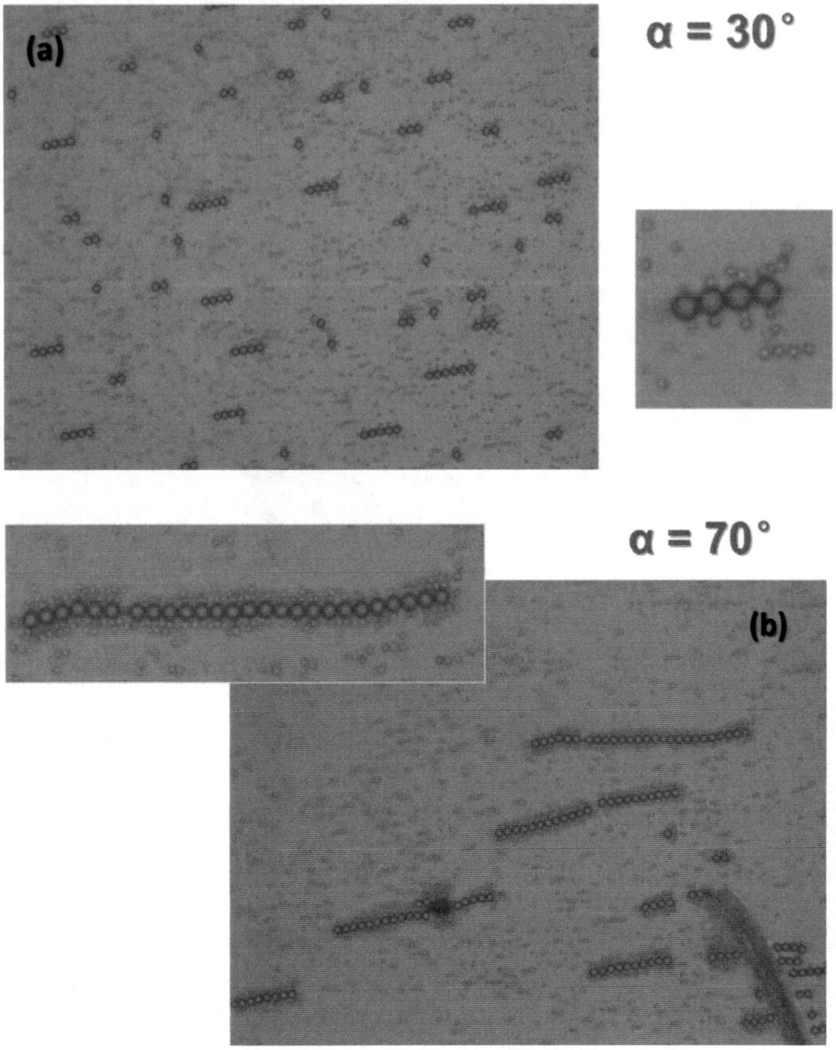

Figure 12 Ring-chains form due to the aligned magnetic field. (a) and (b) are photos in which α=30° and α=70° respectively.

We can see from the photos that the shape of a 'ring-chain' is like a scolopendra. On one hand, paramagnetic beads align in a row according to the direction of the total external magnetic field; on the other hand, nonmagnetic particles revolve around the core particles on the equator plane which is vertical to the total external magnetic field. When α becomes larger, the horizontal component of the external magnetic field becomes higher. Furthermore, the chains become longer.

2.1.4 Ring-chain in a rotating magnetic field

The magnet in this experiment is aligned such that it will generate a relatively large horizontal magnetic field component. In this way, from the visual angle of the microscope, we observe something really interesting, which is the axial rotation of a ring-chain.

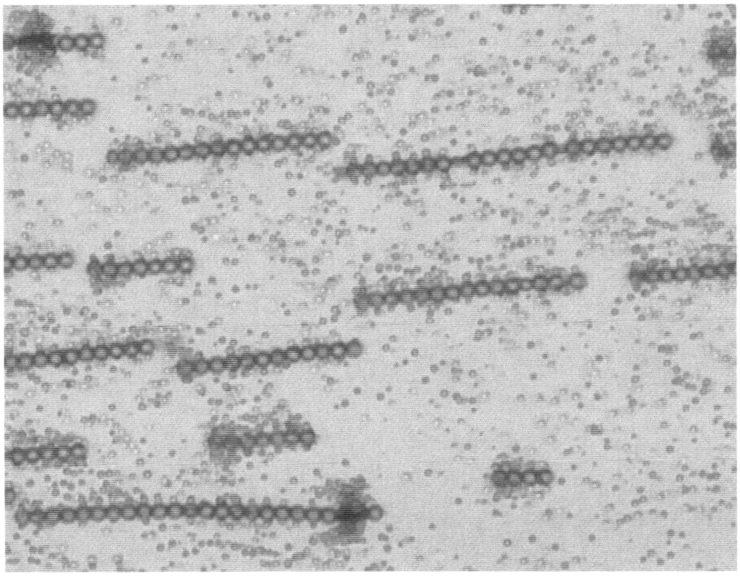

Video 1 Rotation and swing movement of ring-chains.

Video 1 shows this phenomenon which is considered to be the first such discovery in this field of study. We were very surprised to see the ring structure rotate on its axis.

In order to explain this phenomenon, we have to start with force and torque analyses of the ring-chain (See schematic diagram Fig. 13). As we know, the magnetic field generated by a magnet can be deconvolved into two parts: an *x-y* plane component and a *z*-component. For simplicity, we just pre-digest the x-y component as the x-component. $H_{x\text{-mag}}$ offers a force moment in the x-direction, which tends to force the chain to align parallel to the x-direction, and its magnitude can be described as:

$$M_{x\text{-mag}} \propto H_{x\text{-mag}} \cdot l \cdot \sin\theta$$

At the same time, the rotating magnetic field H_{rot} offers a torque to the chain, the magnitude of which can be described as:

$$M_{rot} \propto H_{rot} \cdot l$$

When θ=0°, only M_{rot} acts on the ring-chain, so the chain rotates along the direction of the rotating magnetic field. As θ increases, $M_{x\text{-mag}}$ becomes larger and tends to "drag" the chain harder. When it comes to $M_{x\text{-mag}} = M_{rot}$, the chain is dragged back to its horizontal station. With further increases in θ (when θ>90°), $|\sin\theta|$ becomes smaller and the ring-chain swings to the other side. So far, we can explain the swing movement of the ring-chains that were shown in Video 1.

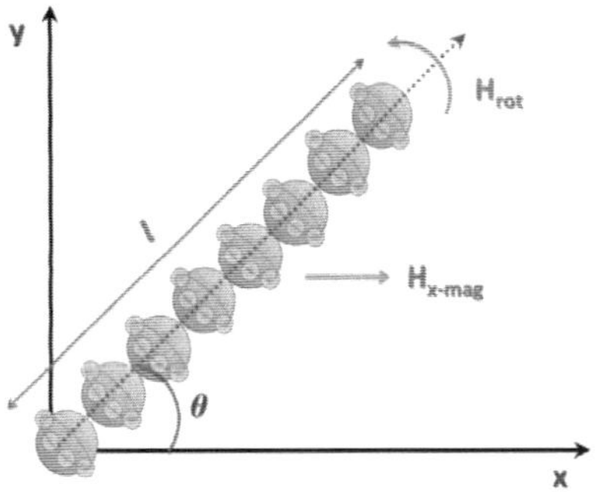

Figure. 13 Schematic diagram (*x-y* plane) to explain swing movement of ring-chains.

Considering the magnetic field generated by a magnet in the *z* direction $H_{z\text{-mag}}$, the ring-chain cannot lie on the x-y plane, as discussed above, so the ring-chains will 'cock up'. To sum up all the factors, the ring-chains tend to describe an 'egg-whisk-like' movement. Hence, a hydrodynamical hindrance acts on the nonmagnetic particles, and this situation can be compared with the situation when a windmill rotates; the nonmagnetic particles are just like the blades of a windmill.

In the video, it is very clear that in one ring-chain, the ring-structure in the center almost doesn't rotate at all, while the ring-structure at the end of the ring-chain rotates most vigorously. This is because every ring-chain swings around its center; if we define the linear velocity as v, the angular velocity as ω and the distance between a ring-structure and the center as r, we have

$$v = \omega \cdot r$$

From this equation, we can see that, the further a ring-structure is from the center, the larger linear velocity it can obtain. Furthermore, according to related hydromechanics theory, the hydrodynamical hindrance is in direct proportion to the linear velocity of movement.

2.2 *Protocols for biosensing*

2.2.1 Two dimensional detection system using a magnetically self-assembled 'ring'

This protocol consists of two parts: the experimental group and the reference group. The experimental procedure goes as follows: (1) Immobilize the target molecules onto the surfaces of the nonmagnetic beads. (2) Mix together paramagnetic particles with probe molecules attached to their surfaces, the former non-magnetic particles, and the MR fluid to obtain a multi-component suspension in solution. (3) Transfer a droplet from the mixture solution onto a glass slide and cover it with a coverslip. (4) Apply an external vertical magnetic field to the two-dimensional plane for a couple of minutes. (5) Remove the external magnetic field and leave it to stand for a while. (6) Take photos of the remnant structure, then count and calculate the statistical average number of ring petals. The entire procedure is shown in Fig. 14.

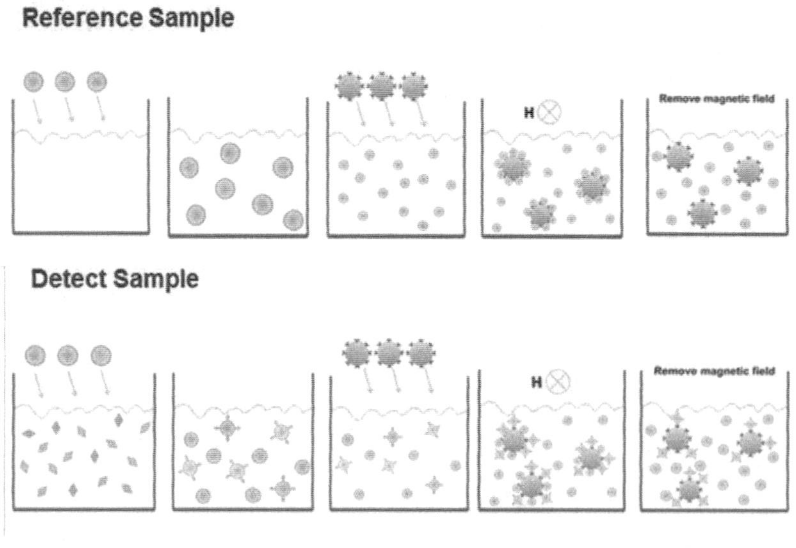

Figure 14 In both the reference and the detection samples, five steps are taken: (1) Immobilize target molecules (the first two pictures); (2) Mix multiple components; (3) Apply external magnetic field; (4) Remove external magnetic field; (5) Observe remnant structures.

For the reference group, there are no target molecules in the initial solution in step (1). The remaining manipulations are the same as those in the experimental group. Under ideal circumstances, the average number for the reference group should be zero because there are no bio-interactions at all.

By varying the concentration of the target molecules in step (2), we can obtain a series of statistical results and generalize the rule.

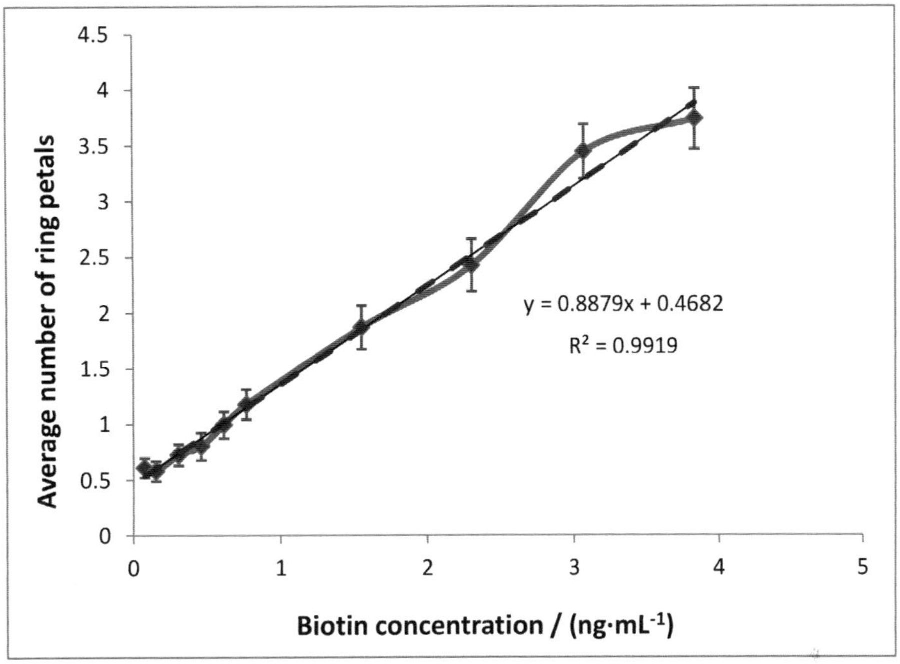

Figure 15 Average number of ring petals is a function of Biotin concentration

Looking into the statistical results, it turns out that the average number of ring petals is a function of the concentration of Biotin (shown as Fig. 15). As the concentration of the biotin increases, the average number of petals on the ring increases monotonously. In the 'low-concentration' region, the function is linear (as shown by the purple line). Because of observations such as these, this protocol is considered to be a promising approach for biosensing.

2.2.2 Improvement of the MT method

In *Section 2.3*, we have already introduced the working theory of the MT (magneto-optical transmittance) method used for biosensing. The most important parameter in this method is the difference in length between the chains in the initial sample system and that in the biomolecules decorating the detection system, because this length difference can be transferred to the vibration amplitude of the sinusoidal curve in the optical transmittance measurements. One of the most effective ways of improving the sensitivity of this method is to increase the length difference as much as possible under the same sample conditions.

Hence, we considered combining our ring-structure with the MT method to improve its sensitivity. According to the statement above, the essence of this protocol is to increase the length difference for the 'ring-structure' as a basic unit in a chain instead of a paramagnetic particle.

Most of the basic manipulations in this experiment are the same with the steps in *Section 3.2.1.* After the remnant rings were formed, a rotating magnetic field was applied for MT detection. The expected result is shown in Fig. 16.

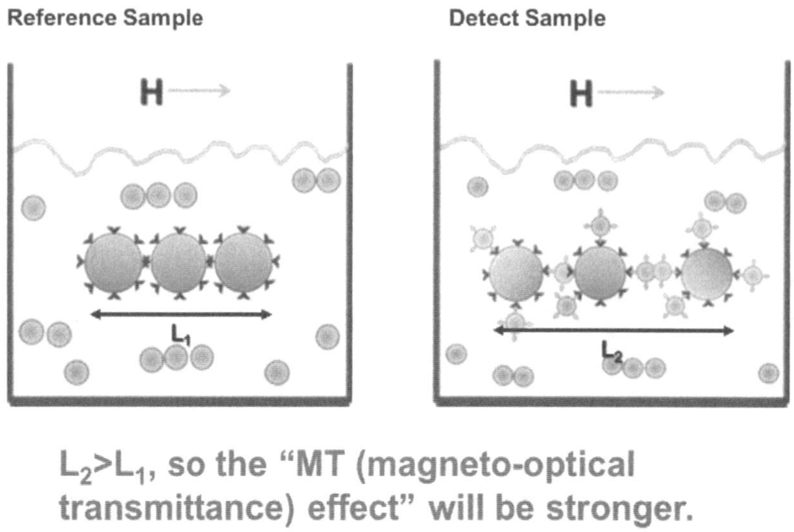

Figure 16 The quintessential idea for MT-improvement protocol: longer length difference results in stronger MT effect.

We carried out two experiments with a reference sample and a detection sample in a rotating magnetic field separately. Both the reference sample system and the detection system consisted of streptavidin-coated paramagnetic-2.8μm particles and MR fluid (C_v=1.2%); the difference lies in whether there is biotin molecules decorating the nonmagnetic particles, which determines whether there is a bio-interaction between the core-particles and the petal-particles. The formula for the two samples are given in Table 2.

Table 2. Formula for reference sample and detect sample.

Reference sample	*Strep-para-2.8μm / μL*	*Non-1.0μm / μL*	*MR fluid (C_v=1.2%) / μL*
	10	10	20
Detection sample	*Strep-para-2.8μm /μL*	*Biotin-non-0.8μm / μL*	*MR fluid (C_v=1.2%) / μL*
	10	10	20

Fig. 17 shows the result of this pair of experiments.

Obviously, because the unit-cell of the chain changed from paramagnetic particles to the ring structure, the size of the units became larger, so the length of chains became longer. It is expected that this phenomenon can help to enhance the "MT effect" (magneto-optical transmittance) and improve its sensitivity.

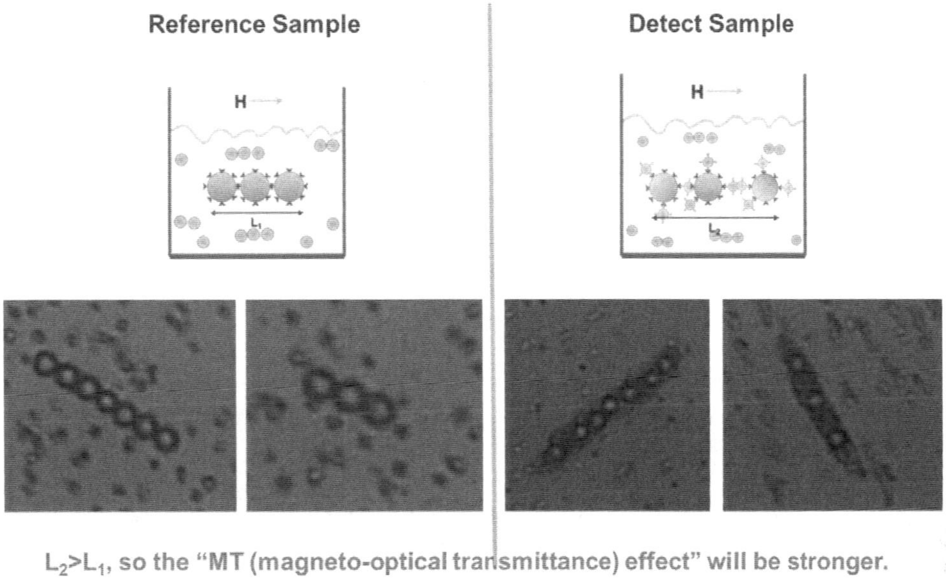

Figure 17 Results of a pair of experiments for MT improvement

3. Conclusion

We introduced our work concerning the study of the dynamics of colloidal solution consisting of multi-component magnetic beads in a magnetic field and its potential applications in biosensing.

The behavior of multiple-component systems built up from paramagnetic, ferromagnetic and nonmagnetic beads were investigated both theoretically and by experiments. Some novel behavior of new structures, such as the rotation of 'ring-chains' in a rotating magnetic field and the formation of 'small-rings' were observed for the first time. We also proposed a theory to explain these novel phenomena. These fundamental theoretical studies are expected to contribute to the industrial application of related fields in the future.

We also tried to exploit these basic studies in biosensing applications ourselves. Two protocols based on ring-structures were raised: a two-dimensional detecting system using magnetically self-assembled 'rings', and improvements in the MT (magneto-optical transmittance) method.

As a result, two-dimensional biosensing using a magnetic self-assembled colloidal ring-structure is considered to be the most promising method for use in biosensing. A suspension solution containing multiple-components including paramagnetic particles, non-magnetic particles and MR fluid was exposed under an external vertical magnetic field for a few minutes. Due to the magnetic force, a ring-structure was formed. If the particles are coated with functional molecules such as streptavidin or biotin, the rings can still remain, even after removing the magnetic field. The average number of remnant ring petals is a function of the concentration of target molecules (biotin). In this way, this technique can be applied to two-dimensional biosensing.

Randall M. Erb et al. (2009) calculated that the size limitation that can be manipulated using this technique is on the order of 60 nm in diameter [19], which means that the use of a higher magnification microscope would allow this protocol to be used to detect smaller carrier particles.

Nowadays, the most commonly used biosensing methods include Enzyme Linked Immunosorbent Assay (ELISA), Gel-shift assay, Quartz crystal microbalance (QCM), Surface plasmon resonance (SPR) and so on. The former two are both time consuming, very cumbersome and require labeling; the latter two offer high quantification but are so expensive that they cannot be applied in portable point-of-care treatment systems. [22]

Compared to these methods, our protocol is much cheaper, and with the support of pattern recognition and statistics software, the detection time can be cut down to just a few minutes. In addition, combined with a portable microscope of the type that can already be purchased on the market, this two-dimensional biosensing technique is expected to be a promising method for diagnosis in remote areas or in fieldwork.

References

[1] Lawaczeck R, Bauer H, Frenzel T, Hasegawa M, Ito Y, Kito K, Miwa N, Tsutsui H, Volger H and Weinmann HJ 1997 Magnetic iron oxide particles coated with carboxydextran for parenteral administration and liver contrasting—pre-clinical profile of SH U555A, Acta Radiologica 38 584–97

[2] Kim D K, Zhang Y, Kehr J, Klason T, Bjelke B and Muhammed M 2001(a) Characterization and MRI study of surfactant-coated superparamagnetic nanoparticles administered into the rat brain, J Magnetism Magn Mater 225 256–61

[3] Kim D K, Zhang Y, Voit W, Kao K V, Kehr J, Bjelke B and Mu-hammed M 2001(b) Superparamagnetic iron oxide nanoparticles for bio-medical applications, Scripta Materialia 44 1713–17

[4] Kim D K, Mikhaylova M, Wang F H, Kehr J, Bjelke B, Zhang Y, Tsakalakos T and Muhammed M 2003 Starch-coated super-paramagnetic nanoparticles as MR contrast agents, Chem Mater 15 4343–51

[5] Lin J C and Wang Y J 1987 Interstitial microwave antennas for thermal therapy, Int J Hyperthermia 3 37–47

[6] Ikeda N, Hayashida O, Kameda H, Ito H and Matsuda T 1994 Experimental study on thermal damage to normal dog brain, Int J Hyperthermia 10 553–61

[7] Jordan A et al 2001 Presentation of a new magnetic field therapy system for the treatment of human solid tumors with magnetic fluid hyperthermia, J Magnetism Magn Mater 225 118–26

[8] Lee H, Kim J H, Kim H K, Kim J Y and Kwon S H 2010 Colour-barcoded magnetic microparticles for multiplexed bioassays, Nature Materials 9 745-49

[9] Alexiou C 2001 Target tumor therapy with "magnetic drug targeting": therapeutic efficacy and biokinetic study of ferrofluid bound mitoxantrone, Clin Cancer Res 7 257

[10] Kondo A, Kamura H and Higashitani K 1994 Development and application of thermosensitive magnetic immunomicrospheres for antibody purification, Appl Microbiol Biotechnol 41 99–105

[11] Koneracka M, Kopcansky P, Timko M and Ramchand C N 2002 Direct binding procedure of proteins and enzymes to fine magnetic particles, J Magnetism Magn Mater 252

409–11
[12] Kourilov V and Steinitz M 2002 Magnetic-bead enzyme-linked immunosorbent assay verifies adsorption of ligand and epitope accessibility, Anal Biochem 311 166–70
[13] Park S Y, Ko P J, Handa H and Sandhu A 2010 Exploring magneto-optical properties of biofunctionalized magnetic chains for developing label-free immunoassays, Journal of Applied Physics 107 09B324(1-3)
[14] Park S Y, Handa H and Sandhu A 2010 Magneto-optical biosensing platform based on light scattering from self-assembled chains of functionalized rotating magnetic beads, Nano Lett. 10 446-51
[15] Wilhelm C, Gazeau F, Roger J, Pons J N and Bacri J C 2002 Interaction of anionic superparamagnetic nanoparticles with cells: kinetic analyses of membrane adsorption and subsequent internalization, Langmuir 18 8148–55
[16] Tiwari A, Punshon G, Kidane A, Hamilton G and Seifalian A M 2003 Magnetic beads (DynabeadTM) toxicity to endothelial cells at high bead concentration: implication for tissue engineering of vascular prosthesis, Cell Biol Toxicol 19 265–72
[17] Krebs M D, Erb R M, Yellen B B, Samanta B, Bajaj A, Rotello V M and Alsberg E 2009 Formation of ordered cellular structures in suspension via label-free negative magnetophoresis, Nano Letters 9 1812-17
[18] Sandhu A, Handa H and Masanori A 2010 Synthesis and applications of magnetic nanoparticles for biorecognition and point of care medical diagnostics, Nanotechnology 21 442001(22pp)
[19] Erb R M, Son H S, Samanta B, Rotello V M and Yellen B B 2009 Magnetic assembly of colloidal superstructures with multipole symmetry, Nature 457 999-1002
[20] Li K H and Yellen B B 2010 Magnetically tunable self-assembly of colloidal rings, Applied Physics Letters 97 083105(1-3)
[21] S. Y. Park, H. Handa and A. Sandhu High speed magneto-optical valve: Rapid control of the optical transmittance of aqueous solutions by the magnetically-induced self-assembly of superparamagnetic particle chains, Journal of Applied Physics 105 07B526
[22] Morimoto Y 2011 Applications of self-assembly of magnetic particles for biorecognition and point-of-care medical diagnostics, Doctoral Dissertation Tokyo Institute of Technology

Optical Transmittance for Analysis of the Dynamics of Self-Assembled Rotating Chains of Superparamagnetic Micro- and Nano Beads in Solution

Pil Ju Ko[1,2], Tsukasa Takamura[1,2], and Adarsh Sandhu[1,3]
[1]Department of Electrical and Electronic Engineering, Tokyo Institute of Technology, 2-12-1 O-okayama, Meguro-ku, Tokyo 152-8552, Japan
[2]Tokyo Tech Global COE Program on Evolving Education and Research Center for Spatio-Temporal Biological Network, Tokyo, 152-8552, Japan
[3]Electronics-Inspired Interdisciplinary Research Institute (EIIRIS), Toyohashi University of Technology, Hibarigaoka, Tempaku-cho, Toyohashi, 441-8580, Japan

E-mail: ko.p.aa@m.titech.ac.jp

Abstract. Optically monitoring biosensing procedures based on the dynamics of an aqueous solution containing functionalized superparamagnetic beads in the application of the external rotating magnetic field has been developed for a rapid, highly sensitive, and inexpensive bioassay. Typically, the dynamics of micrometer diameter beads is observed by conventional optical microscopes. For greater affinity to biomolecules, there is a demand which necessitates the use of nanometer sized superparamagnetic beads, comparable size to actual biomolecules. However, a limited amount of work for monitoring the dynamics of nanometer sized beads has been performed thus far due to the maximum resolution of microscopes. Here, we propose a novel protocol enabling us monitor the dynamics of nanometer-diameter beads via change in the optical transmittance.

1. Introduction

Biosensing protocols based on functionalized superparamagnetic beads (SPBs) have widely been studied for developing inexpensive, rapid, and highly-sensitive biodiagnosis of disorder [1] [2], even testing the use of illegal drag [3]. Several sorts of biosensing protocols utilizing the functionalized SPBs have been introduced over past decades. The optical, magnetic, chemical, and mechanical properties depend mostly on the particle shape, size, composition, and more importantly the applied external magnetic field. Among many, an optical mean to measure the intensity of fluorescence to determine the biorecognition of interest has particularly been utilized [4]. However, the optical method requires expensive equipments, such as a fluorometer and a fluorescence microtiter plate reader, thus it is limited only in affluent research centers and hospitals [5]. For that reason, it finds difficult the protocol to become prevailed. In order to familiarize a biosensing protocol, it must be inexpensive, highly sensitive, and more importantly easy to use [6].

Many researchers have attempted to realize the requisites with using conventional optical microscopes. However, this mean has not been able to response to a recent trend due to its limitation, extremely challenging to observe magnetic beads smaller than intrinsic maximum resolution of a microscope [7]. The recent trend puts great focus on the use of the functionalized SPBs less than 300-nm-diameter comparable size to actual biomolecules for greater affinity to target biomolecules [8].

Recently, there has been considerable interest in magneto-optic devices which combine magnetic and optical phenomena. The optical transmittance of an aqueous solution containing SPBs under an applied external magnetic field is one of the essential areas of magneto-optical effects. Some reports show that the optical transmittance of the magnetic fluids can be modulated by varying the external magnetic field [9-12]. In addition to magnetic field effect, the wavelength effect on the transmission of the magnetic fluids was also reported [13]. Piecing together published reports, the optical transmittance is greatly dependent on the external magnetic field and its direction, the wavelength of the light, and the orientation of the magnetically induced superparamagnetic bead chains.

Here we show a potential application of a magneto optical method to optical biosensing protocols. Our biosensing new protocol have already been published elsewhere[14]. We need to more research of the effect of bead size, composition, and configuration on the transmittance of the light passing through a column filled with an aqueous solution containing a kind of superparamagnetic bead has been investigated as a function of the applied external magnetic field and frequency. The beads we used were 2.8-μm-diameter SPBs (consisting nanometer sized magnetic beads, such as Fe_3O_4, typically 5-10 nm in diameter, are embedded into polystyrene matrix) and 250-nm-diameter SPBs (composing of crosslinked nanometer sized magnetic beads with dextran). Our procedure enables to monitor the dynamics of sub-300-nm-diameter SPBs in application of the rotating external magnetic field, although the dynamics of such beads is not well observable under an optical microscope.

2. Experimental

The experiments were performed using two suspensions of super-paramagnetic particles obtained diluting 0.5% and 0.25% in weight suspensions supplied by dynabeads (M-270 carboxylic acid) and nanomag-® D (09-02-252 carboxylic acid). The beads were diluted with deionized water to concentrations of M-270 = 1.0×10^7 particles/ml and 09-02-252 = 1.15×10^9 beads/ml, respectively. We conducted two different sets of experiments to observe dynamics of SPBs using the rotating external magnetic field. In the first set, we used a conventional optical microscope to directly monitor the dynamics of an aqueous solution containing SPBs subjected to the rotating magnetic field, illustrated in figure 1. The rotating external magnetic field was achieved by applying sinusoidal current to the two orthogonally aligned pairs of Helmholtz coils via a two channel function generator referenced to one another at a phase difference of 90 degrees shown figure 1. The function generator was used for controlling both the magnitude and frequency of the rotating external magnetic field. In the second set, we used an experimental setup depicted in figure 2 to observe the change in the optical transmittance due to the linearly aligned chains of SPBs following up the direction of the applied rotating external magnetic field. Unpolarized incident visible light produced by a tungsten halogen bulb was passed through the column filled with an aqueous solution containing SPBs, and the intensity of the transmitted light through the column was measured by a spectrum analyzer which was set across the column from the light source, shown in figure 2. The wavelength of 630 nm highly transmitted through aqueous solutions, and thus it was used for our analysis.

The SPBs we used were optically observable 2.8-μm-diameter SPBs and 250-nm-diameter SPBs, challenging to observe under conventional microscopes. The surface of both micro- and nano- sized SPBs is modified with carboxylic acid groups which help prevent SPBs from aggregating in an aqueous solution due to the static electric force. . The transmitted light was guided to spectrometer by an optical fiber (Ocean Optics USB4000). The images were recorded with a CCD (a change-coupled divice) video camera Sanyo VDC-3825 connected to a S-VHS VCR Panasonic AG 1975. We digitalized single frames on a computer for their later analysis.

Journal of Physics: Conference Series **352** (2012) 012002 doi:10.1088/1742-6596/352/1/012002

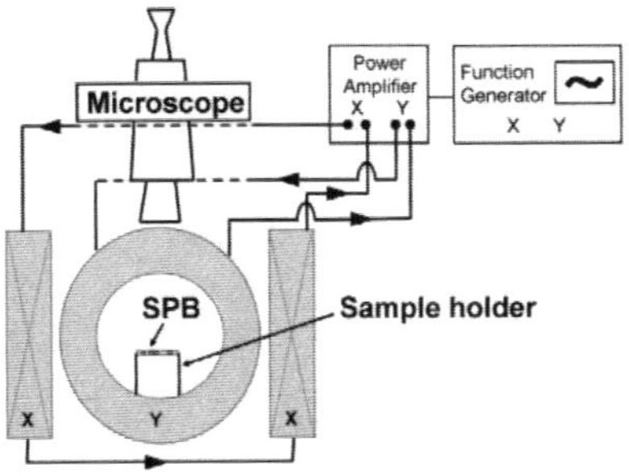

Figure 1. A schematic diagram of experimental setups for monitoring the dynamics of micrometer-diameter SPBs under an optical microscope.

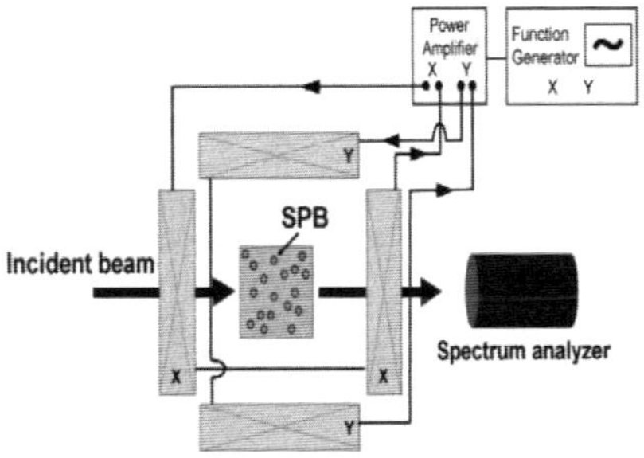

Figure 2 A schematic diagram of transmittance experimental setups utilizing two orthogonal pairs of Helmholtz coils. A column filled with an aqueous solution containing SPBs was located at the center of two pairs of Helmholtz coils. The incident beam passed through the aqueous solution and the intensity of transmitted light was detected by a spectrum analyzer.

3. Results and Discussion

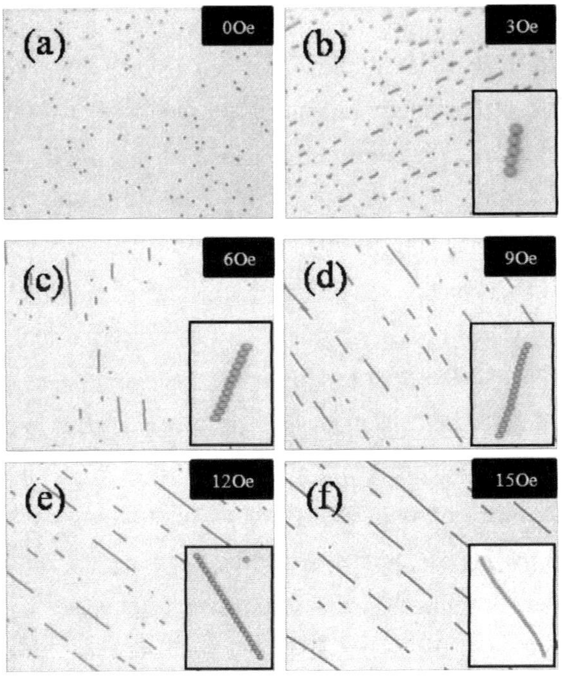

Figure 3. Optical microscope images after 300 s, showing that the 2.8-μm-diameter in diameter SPBs formed column like chains along the direction of the external magnetic field, and their length varies with the magnitude of the external magnetic field.: (a) H_{ext} = 0 Oe, (b) H_{ext} = 3 Oe, (c) H_{ext} = 6 Oe, (d) H_{ext} = 9 Oe, and (e) H_{ext} = 12 Oe, and (f) H_{ext} = 15 Oe.

Figure 3 shows optical microscope images of variation of the length of the linear aligned chain of the 2.8-μm-diameter SPBs with the magnitude of the applied external magnetic field at 0.01 Hz. The longer chains were achieved at higher magnitude of the external magnetic field. The result was also backed up by reciprocal square root of Mason number, ratio of viscous and magnetic forces acting on SPBs under the applied rotating external magnetic field [15, 16].

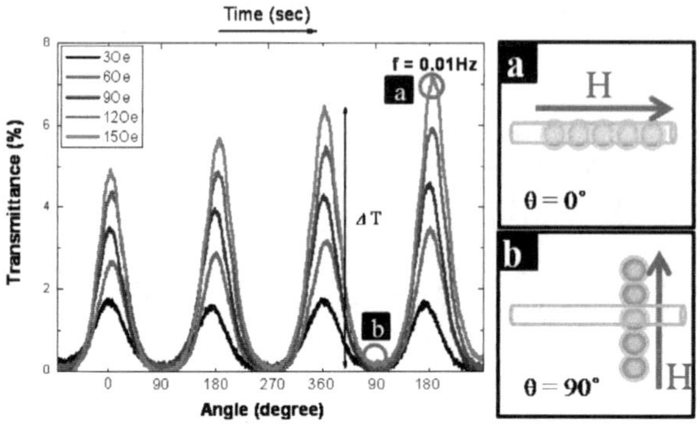

Figure 4. Variation of the optical transmittance of 2.8-μm-diameter super-paramagnetic particles with the angle between the external magnetic field up to 15 Oe and path of incident light. The light incident on the SPBs is scattered and the intensity of transmitted light depends on the angle (θ).

The figure 4 shows the variation of the optical transmittance as a function of the external magnetic field up to 15 Oe with angle (θ). The maximum and minimum optical transmittance for magnitude of the external magnetic occurred along parallel directions (θ = 180n degrees, n = 0, 1, 2...) and othogonal directions (θ = 180n – 90 degrees, n = 1, 2...) with respect to the incident light, respectively. The intensity of transmitted light tended to be greater at higher magnetic field strength. Based on the optical images shown in figure 3, the length of the magnetically induced chain was greatly dependent on the magnitude of the external magnetic field. Thus, it might be thought that the change in optical transmittance was attributed to the length of the chain. As well as the magnitude of the magnetic field, the time, how long the magnetic beads experienced the applied magnetic field, affected the optical transmittance shown in figure 4. 2.8-μm-diameter SPBs are subdivided into polystyrene and magnetite in a ratio of 4 to 1. The approximated diameter of magnetite sphere inside the polystyrene could reach to 1.6-μm-diameter. Visible light with the wavelength of 630 nm, attempting to pass through the magnetite area, was most likely absorbed. Based on the fact, compared to the configuration of Figure 4 (a), the area of magnetic beads blocking the light increased in that of figure 4 (b). Therefore, when the magnetically induced linear chains were aligned parallel to the incident light as shown in figure 4 (a), the visible light went through more and it achieved higher transmittance by geometrical shadowing effects [17, 18].

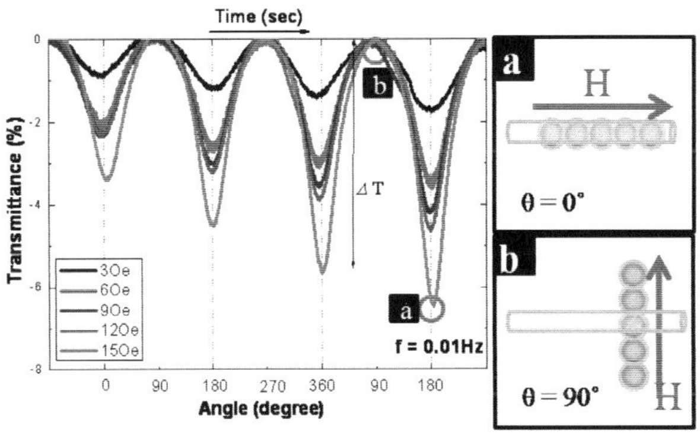

Figure 5. Variation of the optical transmittance of 250-nm-diameter super-paramagnetic particles with the angle between the external magnetic field up to 15 Oe and path of incident light. The light incident on the SBSs is scattered and the intensity of transmitted light depends on the angle (θ).

The figure 5 shows the optical transmittance for 250 nm in diameter SPBs appeared to be similar to that for 2.8-μm-diameter SPBs, but had opposite behavior with respect to the incident light. The maximum and minimum optical transmittance for the external magnetic field occurred along othogonal directions (θ = 180n – 90 degrees, n = 1, 2...) and parallel directions (θ = 180n degrees, n = 0, 1, 2...), respectively, shown in figure 5. The transmittance difference of maximum and minimum, ΔT, for an aqueous solution containing 250-nm-diameter SPBs was smaller than that for 2.8-μm-diameter magnetic beads solution. The conceivable reason would be the composition of SPBs we utilized: 2.8-μm-diameter beads are the polymer matrix in which tiny magnetite is embedded, and 250-nm-diameter magnetic beads which are composed of clustered magnetite (Fe_3O_4). The visible light is hardly absorbed by the transparent polymer, resulting in the high optical transmittance for 2.8-μm-diameter magnetic beads. Composition of 250-nm-diameter SPBs is considerably different: More than 80% are made of magnetite and the great number of small beads is crosslinked with dextran. Notice that when the wavelength of light is very similar to the size of single magnetic bead, it tends to strongly scatter forward, namely Mie scattering [19, 20]. When the orientation of the magnetic beads is othogonal to the incident light as shown in figure 5 (b), it satisfies the condition and thus the wavelength of 630 nm transmitted well. On the other hand, when the configuration of bead chain became parallel to the incident light, the chain length exceeded the wavelength. The light would be blocked by the chain and thus there was almost no transmitted light in the light path. This resulted in making the optical transmittance lowest, at the angle θ=180n – 90 degrees, where n is integer as depicted in figure 5. The difference of transmittance tendency under the rotating magnetic field for 2.8-μm-diameter and 250 nm in diameter magnetic beads would be mainly due to the dependence on the amount of light being scattered and absorbed by the beads. Thus, composition and size of magnetic beads could be the cause of the opposite behavior of the optical transmittance. The above statement is not yet good enough to explain why the lower optical transmittance was achieved at higher magnitude of the external magnetic field. Under the external magnetic field, fluctuations due to Brownian motion in magnetically induced chains gave rise to chain-chain interaction and the formation of thicker column like chains. Needless to say, the column like chains tended to have a bigger cross sectional area under the higher external magnetic field, resulting in increasing the blocking area of the incident light.

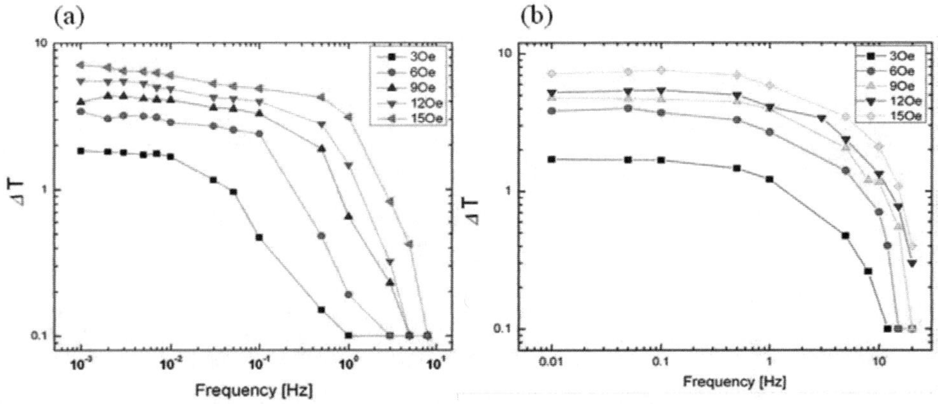

Figure 6. Variation of transmittance difference between maximum and minimum, ΔT, as a function of frequency and the external magnetic field up to 15 Oe. (a) 2.8-μm-diameter SPBs and, (b) 250-nm-diameter SPBs.

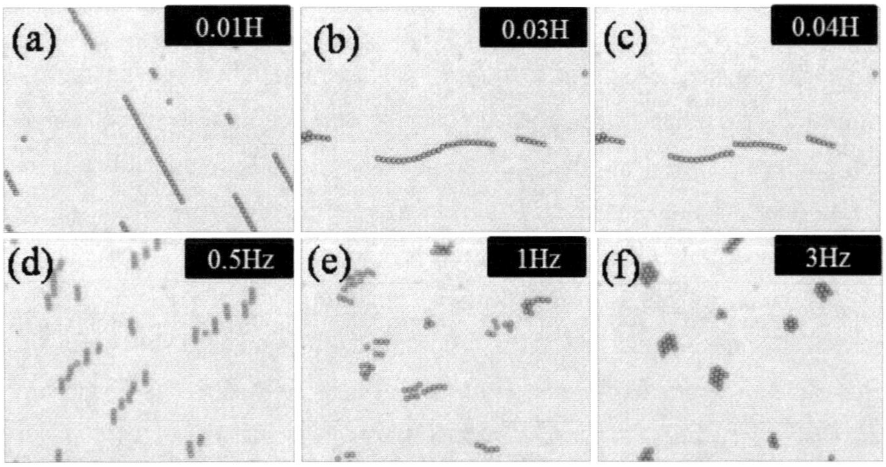

Figure 7. The variation of magnetically induced linear chain with frequency up to 3 Hz. The chain tended to become shorter when higher frequency was used.

Although there was a distinction of transmittance behavior between two kinds of beads, ΔT as a function of frequency made no difference between sorts of beads as shown in figure 6. Fig. 6 (a) and (b) shows after 300sec, the dependence of the maxima and minima of length of the optical transmittance of 2.8-μm-diameter and 250-nm-diameter beads were observed at for various rotation frequencies (*f*) and different external magnetic fields (Oe). Notably, as *f* was increased, ΔT decreased slowly and then rapidly decreased and eventually reach zero. These results show that the average length of chains begin to reduce with increasing *f*, and evantually the chains become dispersed singlet particles due to hydrodynamic forces which oppose the forces due to the applied magnetic torque. As the external magnetic field increased the length of the ΔT with chain structural length increased to comply with (see Fig 7), and transmittance fluctuations become an important component of the observed dynamic of rotating chain of magnetic particles. It is one of the greatest advantages in our method. ΔT as a function of frequency does not discriminate the size and composition of beads. As seen in the figure 6, ΔT was remarkable and remained almost constant until reaching to the frequency threshold. The frequency threshold was higher under the stronger external magnetic field. Once the frequency reached threshold, ΔT dramatically dropped and eventually reached to zero. At this moment, it was thought that none of magnetically induced chain could be remained as a linear chain. Based on the optical images in figure 7, the length of magnetically induced chains became shorter as frequency increased. Ideally, by knowing the frequency at ΔT = 0, we could estimate dipole-dipole interaction of the magnetic beads we used. A considerable application of this mean to bioassay would be to determine the biorecognition forces among biomolecuels.

4. Conclusion

We have investigated the effect of the composition of magnetic bead on optical transmission of light passing through an aqueous solution containing the SPBs as a function of the external magnetic field and the orientation of magnetically induced bead chains. For the solution containing 2.8-μm-diameter SPBs, the transmission increased as the orientation of the chains became parallel to the direction of the incident light. On the contrary, the transmission declined for the aqueous solution containing 250-nm-diameter SPBs under the same condition. Despite the fact that transmission behaviors made distinction, when focusing on the difference between maximum and minimum transmission, ΔT, as function of frequency, we noticed that ΔT reached to zero at higher frequency. Based on results, our procedure enables to monitor the dynamics of sub-300-nm-diameter SPBs without using any expensive, difficult, and unwieldy experimental equipments, although the dynamics of such beads is not well observable under an optical microscope. Thus, our protocol has a great potential to apply into the bioassay field.

References

[1] S. Y. Park, H. Handa, and A. Sandhu: J. Appl. Phys. **105** (2009) 07B526.

[2] Y. Morimoto, M. Abe, M. Hatakayama, H. Handa, and A. Sandhu: IEEE Trans. Magn. **45** (2009) 2871-2874.

[3] A. Sandhu: Biosensing, Nature Nanotechnol. **2** (2007) 746–748.

[4] O. S. Wolfbeis: Anal. Chem. **80** (2008) 4269–4283.

[5] I. L. Medintz, G. P. Anderson, M. E. Lassman, E. R. Goldman, L. A. Bettencourt, and J. M. Mauro: Anal. Chem. **76** (2004) 5620-5629.

[6] K. R. Rogers: Biosens. Bioelectron. **10** (1995) 533–541.

[7] S. W. Hell: Nat. Method **6** (2009) 24-32.

[8] S. X. Wang, and G. Li: IEEE Trans. Magn. 44 (2008) 1687-1720.

[9] J. Baudry, C. Rouzeau, C. Goubault, C. Robic, L. Cohen-Tannoudji, A. Koenig, E. Bertrand, and J. Bibette: Proc. Natl. Acad. Sci. U.S.A. **103** (2006) 16076 – 16078.

[10] R. Lenke, C. Eisenmann, D. Reinke, and G. Maret: Phys. Rev. E **66** (2002) 056610.

[11] J. E. Martin, K. M. Hill, and C. P. Tigges: Phys. Rev. E **59** (1999) 5676-5691.

[12] E. R. Cintra, J. L. Santos Jr, L. M. Socolovsky, N. Buske, and A. F. Bakuzis: J. Magn. Magn. Mater. **320** (2008) e351–e353.

[13] A. D. McFarland, and R. P. Van Duyne: Nano Lett. **3** (2003) 1057-1062.

[14] S. Y. Park, H. Handa, and A. Sandhu: . Nano Lett. **2** (2010) 446-452.

[15] S. Melle, and James E. Martin: J. Chem. Phys. **118** (2003) 9975-9981.

[16] D. J. Klingenberg, J. C. Ulicny, and M. A. Golden: J. Rheol. **51** (2007) 883-893.

[17] S. Y. Yang, Y. P. Chiu, B. Y. Jeang, H. E. Horng, C.-Y. Horng, C.-Y. Hong, and H. C. Yang: Appl. Phys. Lett. **79** (2001) 2372.

[18] J. E. Martin, K. M. Hill, and C. P. Tigges: Phys. Rev. **E 59** (1999) 5676.

[19] C. Bohren, and D. Huffmann: Absorption and scattering of light by small particles; John Wiley & Sons: New York, (1983).

[20] G. Mie: Ann. Phys. (Leipzig): **25** (1908) 377.

Electrophoretic deposition of high quality transparent conductive graphene films on insulating glass substrates

R Ishikawa[1,2], P J Ko[1,2], Y Kurokawa[3], M Konagai[3,4], and A Sandhu[1,2,5]

[1] Department of Electrical and Electronic Engineering, Tokyo Institute of Technology
[2] G-COE Program on Evolving Education and Research Center for Spatio-Temporal Biological Network, Tokyo Institute of Technology
[3] Department of Physical Electronics, Tokyo Institute of Technology
[4] Photovoltaics Research Center (PVREC), Tokyo Institute of Technology
[5] Electronics-Inspired Interdisciplinary Research Institute (EIIRIS), Toyohashi University of Technology

E-mail: ishikawa.r.ab@m.titech.ac.jp

Abstract. Graphene is a promising material for transparent conductive films (TCFs) because of its high electrical conductivity and excellent optical transparency over a wide spectral range. We have previously reported on an inexpensive means of producing graphene-based TCFs using chemically derived graphene flakes. However, the deposition of chemically derived graphene can yield poor stacking of graphene flakes, which degrades the electrical conductivity of the resulting films. Here, we describe an alternative approach for producing large areas of TCFs based on electrophoretic deposition of graphene onto glass substrates using charged graphene oxide flakes. This method enabled the deposition of highly stacked graphene films onto insulating glass substrates with potential for TCFs.

1. Introduction

Transparent conductive films (TCFs) have a wide range of important applications, including flat displays, light emitting diodes, and solar cells [1]. Graphene is one of the most attractive alternative materials for TCFs because of its high electrical conductivity and optical transparency over a wide range of wavelengths. Graphene is an atomically thin layer of carbon with superior electronic properties that has been a subject of intense study since its discovery in 2004 [2]. A major hurdle in the incorporation of graphene into electronics is the exfoliation of graphite into individual sheets on a large-scale and in a controllable way. Recently, the epitaxial growth of wafer-scale graphene on silicon carbide and copper substrates by chemical vapor deposition has been demonstrated [3, 4]. However, these methods require complicated and time consuming processes to transfer the graphene onto more desirable substrates, and uniform deposition over large-areas is proving to be challenging.

Electrophoretic deposition (EPD) is an economical and versatile deposition technique that is used for deposition of coatings and films, such as phosphors for displays [5]. EPD has many attractive features for the preparation of thin films from suspensions, such as high deposition rates and throughput, good uniformity, large-area deposition, and precisely controllable thickness of the resulting films. For example, EPD has been used to produce highly packed microstructural carbon nanotube (CNT) films exhibiting excellent electron-field-emission characteristics from colloidal CNT suspensions [6]. Furthermore, Hasan et al have used EPD for the deposition of large-area graphene oxide (GO) films on stainless steel [7].

A governing principle of EPD indicates that it should not be possible to deposit films on nonconductive materials, such as insulating glass. Here, we describe an innovative EPD-based method for depositing graphene films on insulating glass and other non-conductive surfaces. Our method enabled the deposition of highly stacked graphene films onto insulating glass substrates without the need for sophisticated transfer processes. We successfully deposited graphene films showing lower sheet resistance compared with conventional deposition methods.

2. Experimental

2.1. Preparation of graphene oxide dispersed in solution

Chemically derived graphene was synthesized by the modified Hummer's method [8], a well-known approach for producing monolayered graphene via liquid-phase exfoliation of graphite oxide. Natural graphite powder (SEC Carbon SNO-30) was washed in H_2SO_4 and $K_2S_2O_8$, and oxidized in $KMnO_4$ and H_2SO_4. After centrifugation, the resulting graphite oxide was exfoliated into graphene oxide (GO) by ultra-sonication (100 W, 30 min, 60 °C). Next, a GO aqueous dispersion was produced by centrifugation and dialysis to neutralize the pH of the solution. The surface charge (zeta potential) of obtained GO flakes in neutral water dispersion was examined with an OTSUKA ELECTRONICS ELSZ-1. Zeta potential measurement revealed that GO indicated highly negative charge (ζ_{GO} = -32 mV at pH7) due to ionization of the carboxylic acid and phenolic hydroxyl groups that are known to exist on the GO flakes. This negative characteristic plays an important role in electrophoretic deposition and leads to a stable dispersion without any aggregation for a few months.

The presence of monolayer GO flakes in the resulting GO aqueous solution was verified by atomic force microscopy, Raman spectroscopy, where peak shifts and the shape of the peak of the second-order two phonons at 2700 cm^{-1}, referred to as the 2D band. The size of the reduced graphene flakes was up to 50 x 50 μm^2[9].

2.2. Setup of the electrophoretic deposition (EPD) system

The target substrate faced the Pt plate counter electrode to form a parallel-plate geometry with a separation of 3 to 20 mm. The entire electrode assembly was placed on a custom-built dip-coater (Oriental Motor Co. Ltd., DRS42SA2G-04KA). The set of electrodes were slowly submerged into the GO aqueous solution for concentrations of 0.1, 1.0, and 10 wt%). Electrophoretic deposition was carried out in constant voltage mode at voltages in the range 1 to 30 V using a programmable voltage source (ADVANTEST R6144) at room temperature and atmosphere. After electrophoretic deposition for 1 to 30 min, the graphene-deposited substrates were withdrawn from the solutions at a rate of 1 mm min^{-1} using the dip-coater.

2.3. Electrophoretic deposition on SiO_2 / Si substrates

We used our EPD equipment deposit graphene films on insulating SiO_2 (90 nm) grown on highly doped, degenerate Si substrates. The substrates were first cleaned by sonication in acetone and distilled water, and O_2 plasma treatment immediately before being submerged into GO aqueous solution. Insulating SiO_2 faced the Pt plate counter electrode and a conductive lead was attached to the backside of the Si substrate. After electrophoretic deposition, the substrates were dried by a nitrogen N_2 gun and baked at 100 °C for 5 minutes. The variation of the morphology of the resulting graphene films by due to changes in deposition parameters was monitored by scanning electron microscope (JEOL JSM-7001F).

2.4. Electrophoretic deposited graphene films on glass substrates

To directly deposit graphene-based TCFs on insulating glass substrates, we tested the possibility of using a new EPD procedure as shown in figure 1. In this procedure, we attached conductive Cu tape (or other metallic films) onto the backside of glass substrates (Corning 7059) and then the front sides of substrates were exposed to O_2 plasma. A constant voltage was applied between backside Cu tape and Pt plate electrode during electrochemical deposition. Finally, the substrate was removed from the solution and the Cu tape peeled off (or etched off) and the substrates were dried and baked at 100 °C.

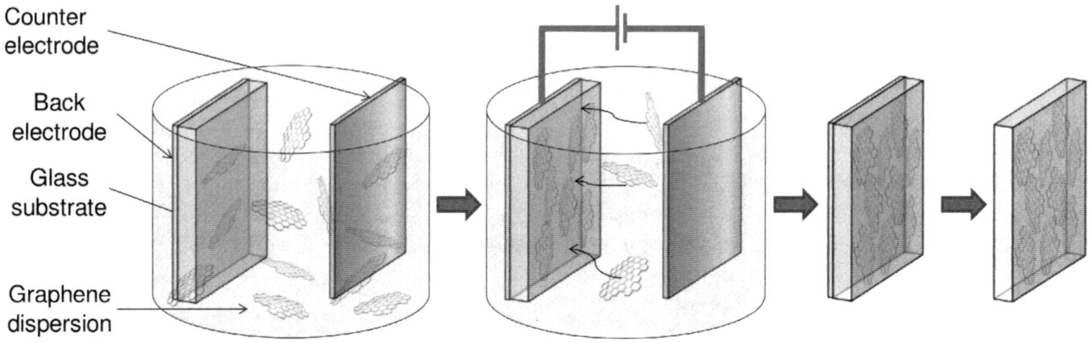

Figure 1. A schematic diagram of our innovative EPD procedure.

Chemical analysis of the resulting graphene films was carried out using X-ray photoelectron spectroscopy (XPS, ULVAC PHI5000, Al K_α). The optical and electrical properties of graphene films were characterized by UV-visible spectrophotometer and van der Pauw measurements.

3. Results and Discussion

3.1. Optimization of deposition parameters

We optimized the electrophoretic deposition parameters (distance between electrodes, applied voltage, concentration of GO solution, and deposition time) by monitoring the morphology of graphene films deposited on SiO_2/Si substrates. Under an applied voltage, the negatively charged graphene oxide sheets migrated toward the positive Si electrode, and were subsequently orderly deposited onto the surface of the positive electrode. The electrophoretic mobility was estimated 2.5 μm V^{-1} s^{-1} from following Helmholtz-Smoluchowski formula;

$$u = \frac{\varepsilon_0 \varepsilon_{H_2O}}{\eta_{H_2O}} \zeta_{GO}, \tag{1}$$

where ζ_{GO} = -32 mV in water dispersion at room temperature. Although other parameters also affect the deposition rate, this estimation supports the high deposition rate (around 2 nm min^{-1}) of EPD method.

Notably, graphene films were deposited on the counter electrode under an opposite applied voltage. The GO flakes easily aggregated in solution when the electric field between electrodes – determined by the distance between the electrodes and the applied voltage– was sufficiently strong. Figure 2 shows the SEM images of graphene films when concentration of GO and deposition time were varied.

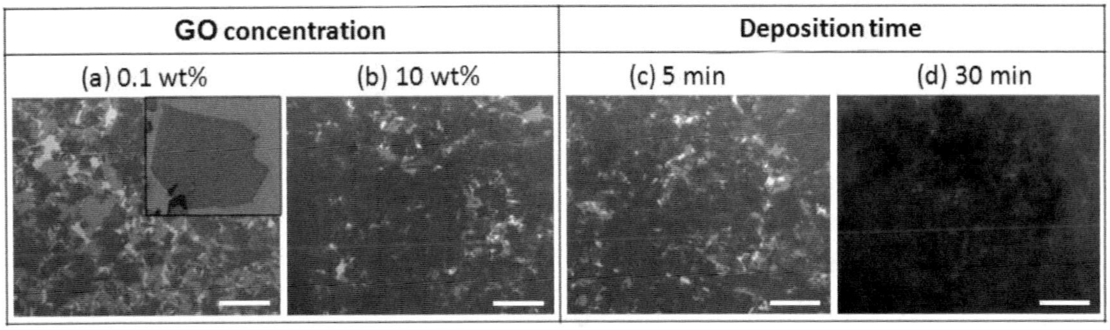

Figure 2. SEM images of EPD-graphene films deposited on SiO_2 / Si substrates for different parameters. The concentrations of the GO solution were (a) 0.1 wt% and (b) 10 wt%. Deposition times were (c) 5 min and (d) 30 min. Scale bars correspond to 100 μm. Inset in (a) shows zoom-in image of a single graphene flake.

We found that the graphene films had a high GO density and uniform morphology. Moreover, we did not observe wrinkles in the deposited GO flakes (see an inset in figure 2a), which are observed in other deposition methods such as filtration and spin-coating [10, 11]. The wrinkle structures are considered to act as scattering centers in graphene-based TCFs as reported in studies on chemical vapour deposition graphene [12-14].

The optimized deposition parameters for producing uniform films of a few layers graphene are summarized in table 1. In the case of deposition on glass, it was required to apply higher voltage around 20 V because the thickness of glass substrate was much thicker than 90 nm.

Table 1. Optimized electrophoretic deposition parameters.

Distance between electrodes	Applied voltage	Concentration of GO	Deposition time
5 mm	10 V	1 wt%	5 min

3.2. Reduction of GO via electrophoretic deposition

The reduction of GO is a critical step for the fabrication of graphene-based TCFs from GO flakes. Therefore, high temperature annealing [15, 16] or chemical reduction by strong reducing agents such as hydrazine [17], and hydroiodic acid [18] is necessary after deposition of GO films. We used XPS measurements to determine the chemical state of graphene films produced by our EPD method. As shown in figure 3, the C1s XPS spectra clearly showed EPD-graphene films to be significantly reduced after deposition without any reduction process. From the semi-quantitative analysis of the XPS results, we estimated the relative amount of oxygen contained in the functional groups for each sample. Peak separation was carried out for all samples after Shirley background was subtracted. The relative ratios of each component consisted of aromatic rings (284.6 eV), C-OH (286.5 eV), C-O-C (287.0 eV), and O=C-OH (288.3 eV) as summarized in table 2. Surprisingly, the relative ratios of oxygen containing functional groups of EPD-graphene films were less than that of GO films reduced by gaseous N_2H_4 and annealing. Thus we conclude that the EPD-graphene films were electrochemically reduced during deposition, but further investigation is necessary to understand the mechanism of reduction process [19, 20].

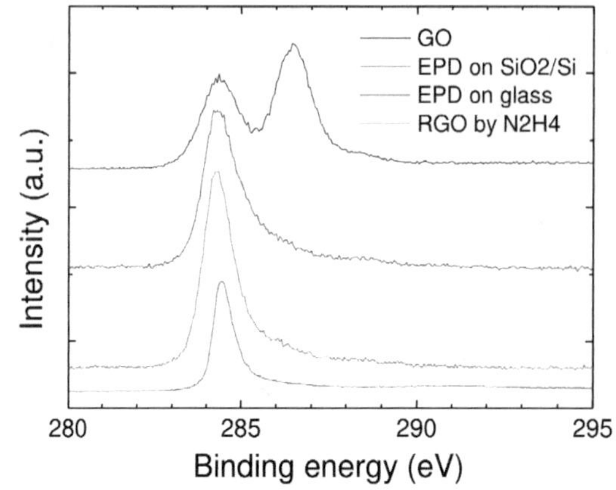

Figure 3. C1s XPS spectra of GO, and EPD-graphene films deposited on SiO_2/Si and glass, and GO films reduced by gaseous N_2H_4 and annealing.

Table 2. Chemical composition of the samples.

	C-C	C-OH	C-O-C	O=C-OH
GO	49.10 %	25.64 %	22.07 %	3.18 %
EPD on SiO_2/Si	81.05 %	6.41 %	5.00 %	7.55 %
EPD on glass	81.60 %	13.52 %	0.26 %	4.61 %
RGO by N_2H_4	73.65 %	19.08 %	0.00 %	7.26 %

3.3. Optical and electrical properties of EPD-graphene films deposited on glass

The optical properties of graphene films deposited by EPD onto glass substrates were characterized by measuring the optical transmittance of the films. They films exhibited a flat optical transmittance spectrum in the visible to infrared region as shown in figure 4. The sheet resistance of the films was approximately $4.59 \times 10^4\ \Omega\ sq^{-1}$ at a transmittance of 83.8 %. Chemically and thermally reduced GO film showed a sheet resistance of $1.13 \times 10^6\ \Omega\ sq^{-1}$ at transmittance of 83.7 %. The lower relative ratios of oxygen containing functional groups of EPD-graphene films might be one of reasons for the lower sheet resistance of the EPD-graphene films. Moreover, high uniformity and an absence of wrinkles may also have resulted in the lower sheet resistance of the EPD-graphene films.

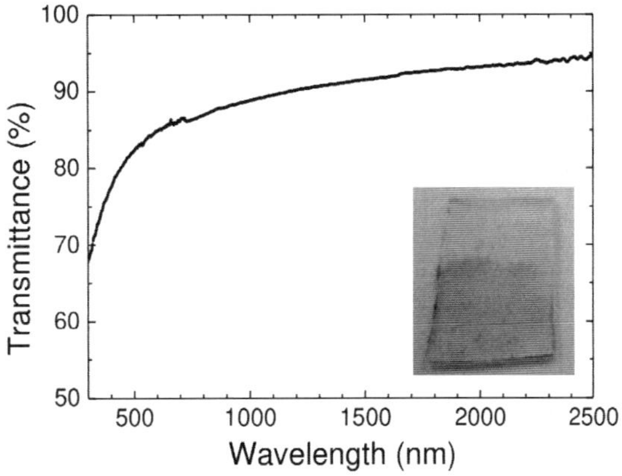

Figure 4. Optical transmittance spectrum of an EPD-graphene film. The inset shows a photograph of EPD-graphene film.

4. Summary

We described an innovative electrophoretic deposition (EPD) based method for depositing graphene films on insulating substrates. Our method enabled the deposition of highly stacked graphene films without requiring a reduction process or sophisticated transfer processes. The graphene films had a lower sheet resistance than that of chemically and thermally reduced GO films because of significant chemical reduction, higher uniformity, and higher stacking. Our EPD method shows promise for the production of large area graphene-based TCFs.

Acknowledgments

This work was conducted as part of the Tokyo Tech Global COE Program on Evolving Education and Research Center for Spatio-Temporal Biological Network, based on a grant from the Ministry of Education, Culture, Sports, Science, and Technology, Japan. The natural graphite powder used in this study was donated by SEC Carbon Ltd.

References

[1] Granqvist C G 2007 *Solar Energy Materials and Solar Cells* **91** 1529

[2] Novoselov K S, Geim A K, Morozov S V, Jiang D, Zhang Y, Dubonos S V, Grigorieva I V and Firsov A A 2004 *Science* **306** 666

[3] Emtsev K V, Bostwick A, Horn K, Jobst J, Kellogg G L, Ley L, McChesney J L, Ohta T, Reshanov S A, Rohrl J, Rotenberg E, Schmid A K, Waldmann D, Weber H B and Seyller T 2009 *Nature Materials* **8** 203

[4] Bae S, Kim H, Lee Y, Xu X F, Park J S, Zheng Y, Balakrishnan J, Lei T, Kim H R, Song Y I, Kim Y J, Kim K S, Ozyilmaz B, Ahn J H, Hong B H and Iijima S 2010 *Nature Nanotechnology* **5** 574

[5] Van der Biest O O, Vandeperre L J 1999 *Ann. Rev. Mater. Sci.* **29** 327

[6] Gao B, Yue G Z, Qiu Q, Cheng Y, Shimoda H, Fleming L and Zhou O 2001 *Adv. Mater.* **13** 1770

[7] Hasan S A, Rigueur J L, Harl R R, Krejci A J, Juan I G, Rogers B R and Dickerson J H 2010 *ACS Nano* **4** 7367

[8] Tung V C, Allen M J, Yang Y and Kaner R B 2009 *Nature Nanotech.* **4** 25

[9] Ishikawa R, Bando M, Morimoto Y and Sandhu A 2010 *Nanoscale Research Letters* **6** 111

[10] Eda G, Lin Y Y, Miller S, Chen C W, Su W F and Chhowalla M 2008 *Applied Physics Letters* **92** 233305
[11] Yamaguchi H, Eda G, Mattevi C, Kim H and Chhowalla M 2010 *Acs Nano* **4** 524
[12] Ahmad M, Han S A, Tien D H, Jung J and Seo Y 2011 *Journal of Applied Physics* **110** 054307
[13] Chae S J, Gunes F, Kim K K, Kim E S, Han G H, Kim S M, Shin H J, Yoon S M, Choi J Y, Park M H, Yang C W, Pribat D and Lee Y H 2009 *Advanced Materials* **21** 2328
[14] Katsnelson M, Geim A 2008 *Phil. Trans. R. Soc. A* **366** 195
[15] Becerril H A, Mao J, Liu Z, Stoltenberg R M, Bao Z and Chen Y 2008 *Acs Nano* **2** 463
[16] Mattevi C, Eda G, Agnoli S, Miller S, Mkhoyan K A, Celik O, Mostrogiovanni D, Granozzi G, Garfunkel E and Chhowalla M 2009 *Advanced Functional Materials* **19** 2577
[17] Eda G, Fanchini G and Chhowalla M 2008 *Nature Nanotechnology* **3** 270
[18] Pei S F, Zhao J P, Du J H, Ren W C and Cheng H M 2010 *Carbon* **48** 4466
[19] Loh K P, Lu J, Yang J X, Wang J Z, Lim A L and Wang S 2009 *Acs Nano* **3** 2367
[20] Su C Y, Lu A Y, Xu Y P, Chen F R, Khlobystov A N and Li L J 2011 *Acs Nano* **5** 2332

CMOS On-Chip Optoelectronic Neural Interface Device with Integrated Light Source for Optogenetics

Y Sawadsaringkarn[1], H Kimura[1], Y Maezawa[1], A Nakajima[1], T Kobayashi[1], K Sasagawa[1], T Noda[1], T Tokuda[1,2] and J Ohta[1]
[1]Graduate School of Materials Science, Nara Institute of Science and Technology, 8916-5 Takayama, Ikoma, Nara 630-0101, Japan
[2]Precursory Research for Embryonic Science and Technology (PREST), Japan Science and Technology Agency (JST), Kawaguchi, Saitama 332-0011, Japan

E-mail: ohta@ms.naist.jp

Abstract. A novel optoelectronic neural interface device is proposed for target applications in optogenetics for neural science. The device consists of a light emitting diode (LED) array implemented on a CMOS image sensor for on-chip local light stimulation. In this study, we designed a suitable CMOS image sensor equipped with on-chip electrodes to drive the LEDs, and developed a device structure and packaging process for LED integration. The prototype device produced an illumination intensity of approximately 1 mW with a driving current of 2.0 mA, which is expected to be sufficient to activate channelrhodopsin (ChR2). We also demonstrated the functions of light stimulation and on-chip imaging using a brain slice from a mouse as a target sample.

1. Introduction

There have been drastic improvements in optogenetics in recent years [1-4]. It is a methodology that realizes optical stimulation of neural cells with the help of genetic modification. It provides a way to stimulate neural cells with good spatial selectivity and smaller invasion. Natural photosensitive protein plays an essential role in optogenetics. The protein works as a photosensitive ion channel on a cell membrane. It can be activated or deactivated by light with a specific wavelength and affects intracellular signaling.

Channelrhodopsin-2 (ChR2) is a photosensitive channel protein commonly used in optogenetics. ChR2 is activated by light with a wavelength of about 470 nm and shows a fast response to the light stimulation. It has been reported that light stimulation using appropriate illumination intensity and a frequency of around 20 – 50 Hz can evoke neural activity [4-6]. It has also been reported that the ChR2 protein does not disturb or change the physiological properties of modified neuron cells [7].

In our previous works, we demonstrated that our CMOS image sensor is capable of observing neural activity using an on-chip fluorescence imaging method [8,9]. In this work, we propose a new CMOS on-chip optoelectronic neural interface device, which integrates an array of light emitting diodes (LEDs) on our CMOS image sensor that are used as an addressable light source for optogenetically modified neural cells.

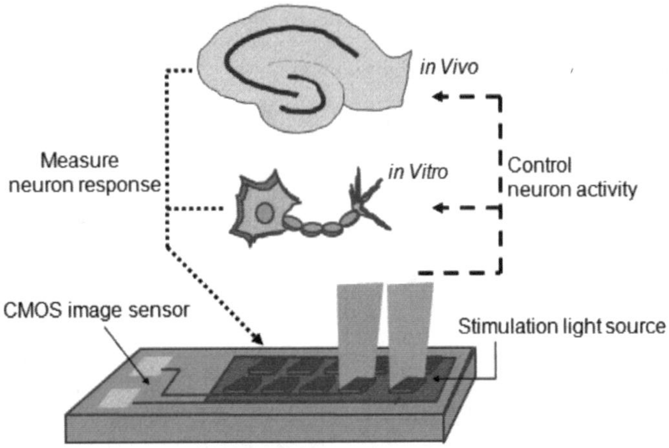

Figure 1. Concept of neural interface device for optogenetics

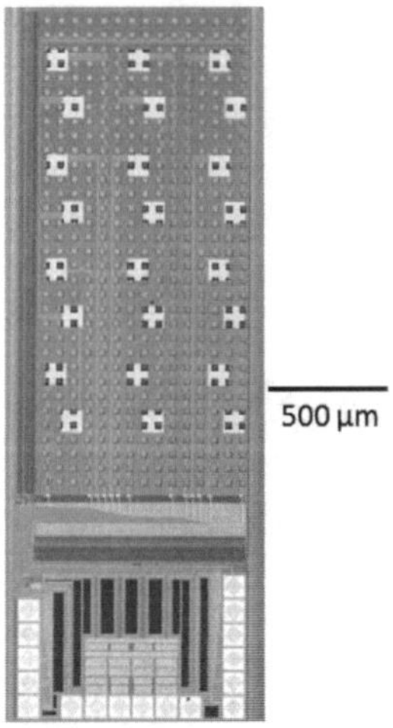

Figure 2. Layout of CMOS image sensor for light-source integration

2. Design of CMOS image sensor chip for optoelectronic neural interface device

The CMOS image sensor is one of the primary parts of the proposed neural interface device. As reported in our previous work, this CMOS image sensor is capable of imaging the brain structure and activity in an on-chip configuration. The architecture of the CMOS image sensor consists of active pixel sensors as the pixel circuitry, analog signal processing circuitry, row and column selectors, and timing control circuitry [9]. The pixel substructure of our sensor uses a three-transistor type active pixel sensor (APS), which consists of a photodiode, select transistor switch, and reset transistor switch [10]. The photodiode is an important part of the APS and is used to change light into an electrical signal. The control and readout circuits for the sensor consist of a row selector at the left side of the pixel array and a column selector at the bottom of the pixel array. The row selector chooses the light sensing rows one-by-one, and the pixels in the selected row are connected to the column circuit via each column signal line. When the column selector is operated, the pixel signal levels in each column are read out consecutively [10-12].

Table I. Specification of CMOS image sensor

Technology	0.35-µm, 2-poly, 4-metal standard CMOS process
Operation voltage	3.3 V
Chip size	1000 µm × 2965 µm
Pixel array size	900 µm × 2010 µm
Pixel Type	3-transistor active pixel sensor
Pixel size	7.5 µm × 7.5 µm
Electrode size	90 µm × 90 µm
Electrode material	Al
Number of LED	12
LED connection	Parallel connection

Journal of Physics: Conference Series **352** (2012) 012004 doi:10.1088/1742-6596/352/1/012004

As the base chip of the proposed optoelectronic neural interface device, we designed a CMOS image sensor with an on-chip current injection electrode array. We used a 0.35-μm, 2-poly, 4-metal standard CMOS process for the fabrication. Figure 2 shows a layout of the CMOS image sensor chip for light source integration. Table I shows the specifications of the CMOS image sensor. As shown in Figure 2, we implemented an array of on-chip electrode pads to operate the LEDs on the sensor chip, with the capability of integrating 3 × 4 LEDs on the sensor. The pitches between the LEDs are 345 μm in the horizontal direction and 435 μm in the vertical direction. All of the pads for the cathodes of the LEDs are connected to the ground line. The LEDs' anode pads are separately connected to connection pads aligned at the edge of the sensor chip for addressable operation. Thus, all of the integrated LEDs in the array are connected in parallel. In addition, the sensor has some area that do not contain pixel. This "empty" area has a size of 22.5 μm × 21.6 μm and was placed in each area of 8 × 8 pixel array (60 μm × 60 μm). This empty area was designed for making a hole on pixel array to deliver light through the sensor chip. However, the structure was not used in this work.

3. Integration of LED array on CMOS image sensor

We chose a commercially available InGaN Blue LED chip with a peak wavelength of 470 nm for integration. The chip size was 280 μm × 305 μm. The specifications of the LED are compatible with the excitation of the ChR2 used in optogenetics. We used a flip-chip bonding technique with gold bumps for the integration of the LED array on the CMOS image sensor.

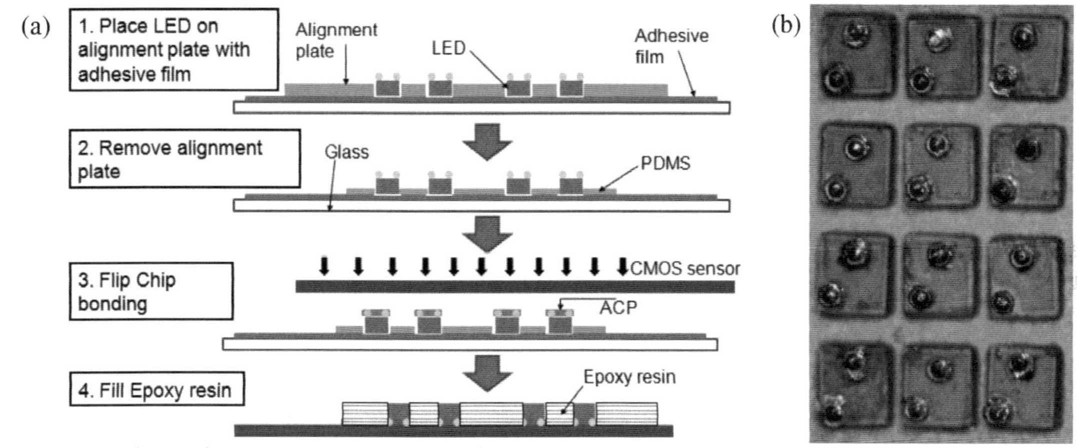

Figure 3. (a) Integration process for multiple discrete LEDs on CMOS base chip, (b) LED array with gold bumps arranged on metal alignment plate

Figure 3(a) shows the process flow for integrating the LED array on the CMOS image sensor. Before we started the process, we formed gold stud bumps on both the anode and cathode pads of the LEDs. As shown in Figure 3(b), we used a metal alignment plate to align the LEDs before the bonding process. The LEDs were tentatively mounted on an adhesive film using the alignment plate as a jig. Then, we removed the metal alignment plate. Before we performed the flip-chip bonding, we covered the exposed adhesive film with polydimethylsiloxane (PDMS) to avoid the fixation of the film on the CMOS image sensor. This PDMS covering layer also helped to fix the position of each LED. We used anisotropic conduction paste (ACP) as an underfill for bonding. We used the following bonding conditions: pressure: 15 N, duration: 60 seconds, and temperature: 140 °C. After we bonded the LEDs, we glued the chip with the LEDs to a printed circuit board using epoxy resin and connected the connection pads of the sensor and the terminals of the printed circuit board with Al wire using a conventional wedge bonder. Then, we covered the surface of the device and the bonding wires with epoxy resin. To expose the backside surface of the LED array, we pressed the epoxy resin with a plate covered with silicone during the cure process. The refractive index of the epoxy resin was approximately 1.40. Since the current injection lines to drive the LEDs are separated each other, we expect crosstalk due to leakage current is negligible. On the other hand, since the gaps between LEDs are filled with transparent epoxy resin, light from one LED partly propagates through the sapphire wafer and epoxy resin. We have not characterized the optical crosstalk and it should be discussed in further works.

4. Characterization

4.1. Measurement setup

Figure 4 schematically shows the experimental setup for device operation. We designed an interface board for computer-controlled device operation and image capturing. The control signals from a PC (5 V digital) were transformed into 3.3 V digital signals by buffers on the interface board and supplied to the CMOS image sensor chip. The signals for controlling the sensor were compatible with those used in our previous works [12]. The CMOS image sensor output 3.3 V analog signals. An analog-to-digital converter (ADC) implemented on the interface board (Analog Devices, AD9225, 0 to 4 V/ 12 bit configuration) transformed the output signals from the CMOS image sensor into 12 bit-parallel digital signals. On the other hand, we used a simple approach to operate the integrated LEDs for light stimulation. The cathode lines of all the integrated LEDs were connected to ground, and the anode lines were separately connected to the terminals on the interface board. DC power supplies with current monitoring capability were used to drive the LEDs. We mounted the optoelectronic neural interface device under a conventional microscope (Olympus BX51WI), which we used to monitor the LED operation and on-chip imaging.

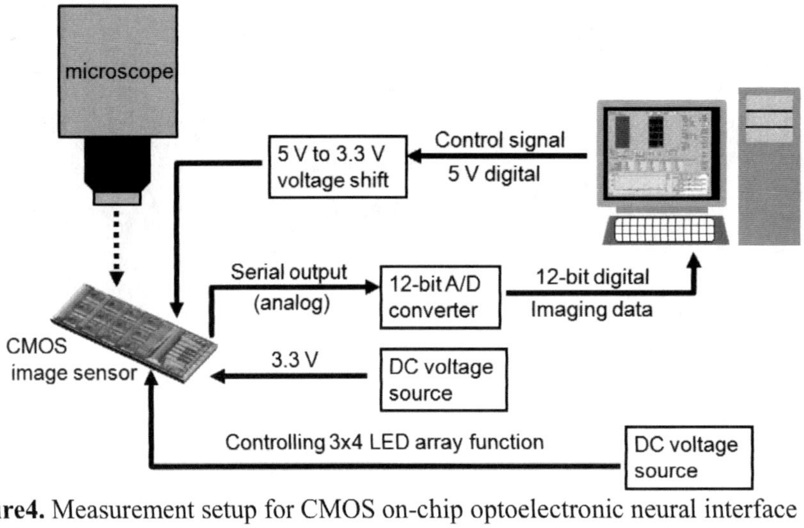

Figure4. Measurement setup for CMOS on-chip optoelectronic neural interface device

4.2. Functional confirmation of LED operation

Figure 5 shows external views of the fabricated device in the following situations: (a) regular observation, (b) with LED illumination, and the images captured by the CMOS image sensor under (c) uniform illumination and (d) a dark situation with one LED operating. As shown in Figure 5(a), the LEDs were aligned on the CMOS image sensor using the flip-chip bonding technique. The LED array was molded with epoxy resin and a flat surface was obtained. Because the LEDs could be operated in parallel, we could perform not only single site stimulation, but also simultaneous multi-site stimulation, as shown in Figure 5(b). This is an advantage of our device compared to the previously reported LED array devices designed for optogenetics [13,14]. Because the LEDs were operated by currents injected from external DC power supplies, we could control the illumination intensity by changing these currents. The current range for LED operation was 0.01-15 mA.

We measured the total illumination power emitted from the LEDs mounted on the chip. We obtained approximately 1 mW with an injection current of 2 mA, which was consistent with the specifications of the LED chip. Based on the literature, which suggests an intensity of 0.1-1 mW/mm^2 as a typical threshold [14-17], we consider the illumination performance of the on-chip LEDs to be sufficient to activate ChR2.

4.3. Characterization of imaging function

The images captured with portion of the CMOS image sensor are shown in Figures 5(c), and (d). Its imaging capability is one of the largest advantages of the present neural interface device, compared to the previously reported LED array device for optogenetics [13,14]. The frame rate was approximately 13 fps. The effective signal range in the image data used to show the intensity was approximately 10 bits. The control software could be used to perform inter-image subtraction, which is a very important function in imaging applications. We could also plot the values measured by selected pixels in real time, which will help with observations of the neural activities visualized by dyes such as Ca^{2+} indicating dyes or voltage sensitive dyes.

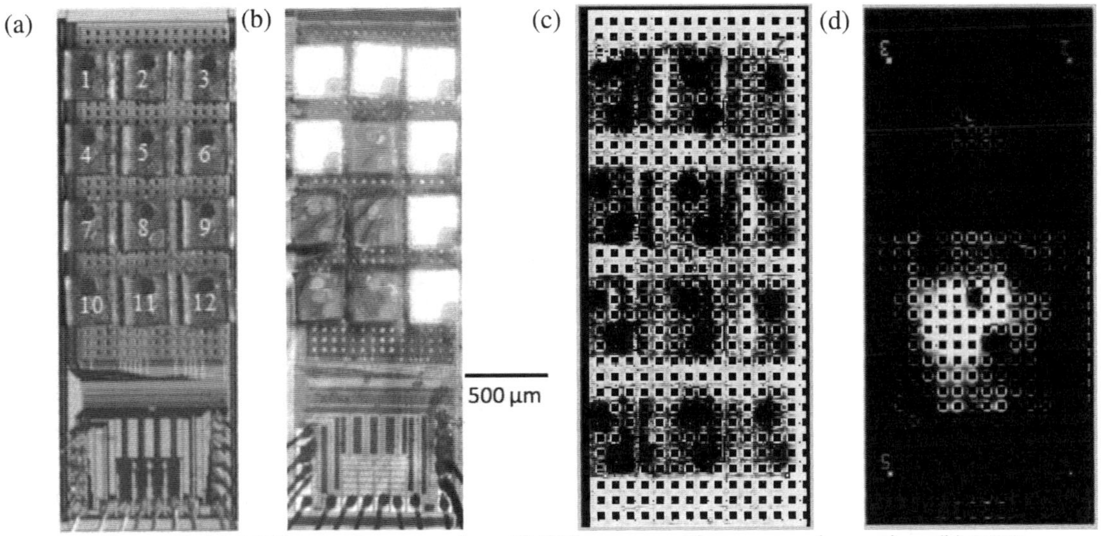

Figure 5. (a) LED array integrated on CMOS sensor with epoxy resin coating, (b) LED array illuminating CMOS sensor, (c) captured image of LED array on CMOS sensor , and (d) captured image of LED illumination

Figure 5(c) is an image taken under uniform illumination. The LED is not operating. Because we have pixel-less areas in each 8 × 8 pixel array, the image has data-less squares. We previously reported this in-array pixel-less design [18]. We can also see the pattern of the anode and cathode pads on the LED array in Figure 5(c). Because we used gold bumps for the flip-chip bonding, we could not avoid these shadows from the electrodes. We are currently developing a bonding technique without large gold bumps to avoid this image deterioration. Figure 5(d) shows an image captured under a dark situation. For Figure 5(d), we operated only one LED in the pixel array (LED number 8 shown in Figure 5(a)). We can observe the light emitted from the operating LED. This result suggests that we can observe and record the operation of the LEDs, as well as the situation of observation targets such as biological cells, brain slices, or the brain itself. It should be mentioned that we can take fluorescence images using the imaging functionality, if we integrate a high-performance optical colour filter with the image sensor [19, 20].

4.4. Functional demonstration: on-chip imaging of brain slice

As a preliminary functional demonstration, we performed an on-chip brain slice imaging experiment.

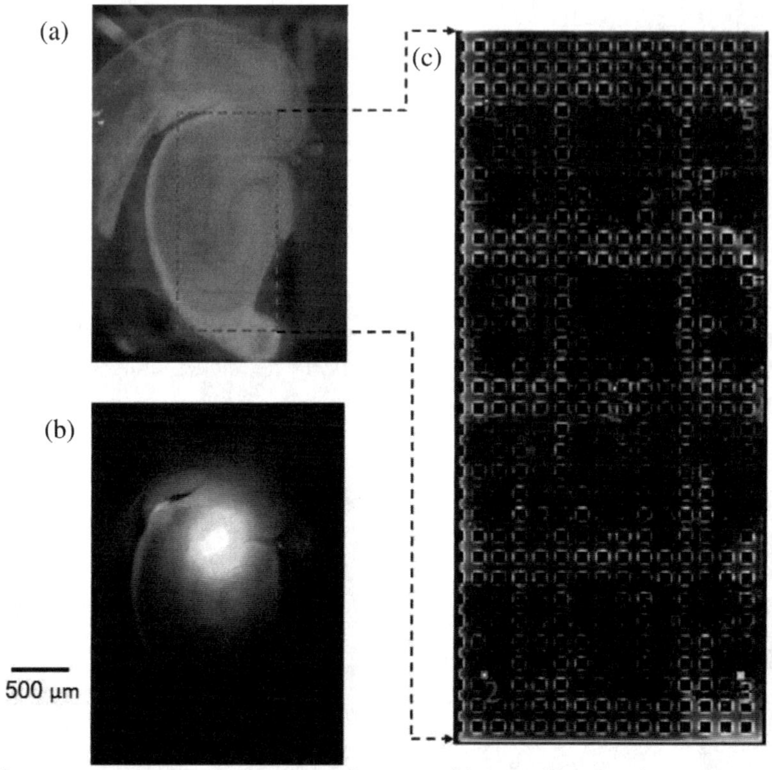

Figure 6. (a) Hippocampus brain slice on CMOS sensor, (b) LED illumination under Hippocampus, (c) captured image of brain slice on CMOS sensor

We used a mouse brain slice from around the Hippocampus area as the measurement target. All of the experiments followed our university's procedure guidelines for animal experiments. The brain slice was fixed and kept in phosphate buffered saline (PBS) at a temperature of 2 °C [21]. The brain slice was placed on a culture dish (μ-dish, 33 mm, low). Prior to the observation, the PBS was drained from the culture dish, and 50% glycerine was dropped onto the slice to provide some moisture. Then, the culture dish was placed on the present optoelectronic neural interface device. Figure 6 shows: (a) the setup of the experiment, (b) demonstration of light emission from one LED, and (c) on-chip image captured under regular illumination (no LED operation). As shown in Figure 6 (b), the light from the LED diffuses within a diameter of approximately 500 μm. This result shows that the current resolution of the light stimulation is not suitable for cell-level stimulation, but region level stimulation. We could only perform region-level stimulation such as CA1, CA2, and CA3. In terms of on-chip imaging function, only a silhouette of the brain slice can be obscurely observed in the captured image, as shown in Figure 6(c). Currently, this imaging function can be used only as an assisting function to monitor the placement of the measurement target.

To improve the performance of the imaging function and realize a neural imaging in on-chip configuration, we need to implement an on-chip optical filter to eliminate excitation light. A filter resist layer will be formed between the CMOS chip and LEDs [22]. The shadow of the contacting electrodes on the LEDs is also an issue to overcome. We are also currently working on introducing a new LED electrodes and bonding materials.

5. Conclusions

We proposed a novel optoelectronic neural interface device for optogenetics. The proposed device is based on the technology of a CMOS image sensor. We integrated an array of GaInN LEDs on the CMOS image sensor with the capability of on-chip current injection. We successfully integrated multiple LEDs on the CMOS image sensor. We demonstrated the light-stimulation functionality of the LED array and performed an on-chip observation of the structure of a mouse's Hippocampus.

The proposed architecture is advantageous because of its flexibility in arranging the number, type, and location of an LED array on a CMOS sensor. We can flexibly design optoelectronic neural interface devices that can fit a variety of target such as culture cells, brain slices, and the brain itself under *in-vivo* situations.

6. Acknowledgment

This work was partially supported by the Japan Science and Technology Agency, Precursory Research for Embryonic Science and Technology (JST-PRESTO).

References

[1] Deisseroth K, Feng G, Majewska A K, Miesenbock G, Ting A and Schnitzer M J 2006 *J. Neurosci.* **26** 10380
[2] Zhang F, Aravanis A M, Adamantidis A, de Lecea L and Deisseroth K 2007 *Nature Reviews Neuroscience* **8** 577
[3] Airan R D, Hu E S, Vijaykumar R, Roy M, Meltzer L A, and Deisseroth K 2007 *Curr. Opin. Neurobiol.* **17** 587
[4] Carter M E and de Lecea L 2011 *Trends in Molecular Medicine* **17** 197
[5] Zhang F, Gradinaru V, Adamantidis A R, Durand R, Airan R D, de Lecea L and Deisseroth K 2010 *Nature Protocol* **5** 439
[6] Diester I, Kaufman M T, Mogri M, Pashaie R, Goo W, Yizhar O, Ramakrishnan C, Deisseroth K and Shenoy K V 2011 *Nature Neuroscience* **14** 387
[7] Kramer R H, Fortin D L and Trauner D 2009 *Curr. Opin. Neurobiol.* **19** 544
[8] Kobayashi T, Tagawa T, Noda T, Sasagawa K, Tokuda T, Hatanaka Y, Tamura H, Ishikawa Y, Shiosaka S and Ohta J 2010 *Jpn. J. Appl. Phys.* **49** 117001
[9] Tagawa A, Higuchi A, Sugiyama T, Sasagawa K, Tokuda T, Tamura H, Hatanaka Y, Shiosaka S and Ohta J 2009 *Jpn. J. Appl. Phys.* **48** 04C195
[10] Ohta J, Tokuda T, Sasagawa K and Noda T 2009 *Sensors* **9** 9073
[11] Ng D C, Nakagawa T, Mizuno T, Tokuda T, Nuoshita M, Tamura H, Ishikawa Y, Shiosaka S, and Ohta J 2008 *IEEE Sensors J* **8** 121
[12] Ng D C, Tamura H, Mizuno T, Tokuda T, Nunoshita M, Ishikawa S and Ohta J 2008 *Sensors & Actuators* **A145-146** 176
[13] Grossman N, Poher V, Grubb M S, Kennedy G T, Nikolic K, McGovern B, Palmini R B, Gong Z, Drakakis E M, Neil M AA, Dawson M D, Burrone J, and Degenaar P 2010 *J. Neural. Eng.* **7** 016004
[14] Huber D, Petreanu L, Ghitani N, Ranade S, Hromadka T, Mainen Z, and Svoboda K 2008 *Nature* **451** 61
[15] Tomita H, Sugano E, Isago H, Hiroi T, Wang Z, Ohta E and Tamai M 2010 *Experimental Eyes Research* **90** 429
[16] Kravitz AVand Kreitzer AC 2011 *Curr. Opin. Neurobiol.* **21** 433
[17] Tomita H, Sugano E, Fukazawa Y, Isago H, Sugiyama Y, Hiroi T, Ishizuka T, Mushiake H, Kato M, Hirabayashi M, Shigemoto R, Yawo H, and Tamai M 2009 *PLoS ONE* **4** e7679
[18] Tokuda T, Tanaka K, Matsuo M, Kagawa K, Nunoshita M, and Ohta J 2007 *Sensors & Actuators* A **135** 315

[19] Tamura H, Ng D C, Tokuda T, Naoki H, Nakagawa T, Mizuno T, Hatanaka Y, Ishikawa Y, Ohta J, and Shiosaka S 2008 *J. Neurosci. Methods.* **173** 114
[20] Tagawa A, Mitani M, Minami H, Noda T, Sasagawa K, Tokuda T, and Ohta J 2010 *Jpn. J. Appl. Phys.* **49** 01AG02-1
[21] Ishikawa Y, Tamura H, and Shiosaka S 2011 *J. Physiol.* **589** 3559
[22] Sasagawa K, Mitani M, Sugiyama T, Noda T, Tokuda T, and Ohta J 2010 *Jpn. J. Appl. Phys.* **49** 04DL03

Journal of Physics: Conference Series **352** (2012) 012005 doi:10.1088/1742-6596/352/1/012005

Optimization of Sputtering Condition of IrOx Thin Film Stimulation Electrode for Retinal Prosthesis Application

Y Pan[1], T Noda[1], K Sasagawa[1], T Tokuda[1], H Kanda[2], T Fujikado[2], and J Ohta[1]
[1] Graduate School of Materials Science, Nara Institute of Science and Technology, 8916-5, Takayama, Ikoma, Nara, 630-0192, Japan
[2] Graduate School of Medicine, Osaka University, 2-15 Yamadaoka, Suita, Osaka 565-0871, Japan

E-mail: ohta@ms.naist.jp

Abstract We have optimized the sputtering condition of the RF sputtering deposition for the fabrication of the iridium oxide thin film used in retinal prosthesis applications. The deposited IrOx thin films were characterized by using the cyclic voltammetry method and the charge delivery capacity was calculated from the integral of the generic CV curve. From the experimental results, the charge delivery capacity of IrOx under the best sputtering condition was improved to more than 50 times that of Pt. We also verified from our *in vivo* experiment results that IrOx has a better charge delivery capacity than that of Pt. The *in vitro* and *in vivo* experimental results also show that IrOx is a promising candidate for retinal prosthesis applications.

1. Introduction

A retinal prosthesis is developed for those who have lost their sight due to degenerative retinal diseases [1]. In these cases, only the photoreceptor cells are damaged, but a large portion of the neural cell on the retina is intact. It is believed that patterned electrical stimulation through a retinal prosthesis device onto the remaining retinal neural cells can partially restore a person's visual sensation.

We proposed a multi-chip architecture-based neural stimulation device for a retinal prosthesis device in our previous research [2-3]. This architecture consists of small unit chips with Pt electrodes that work cooperatively with other unit chips, and the function of the developed retinal prosthesis device was verified from the results of *in vitro* and *in vivo* experiments. With this architecture, a thin and flexible multi-chip retinal stimulation device can be developed. The next stage is to increase the number of electrodes to enhance the spatial resolution of current retinal prostheses. However, the charge delivery capacity (CDC) of current electrode material Pt has been pointed out to be inefficient [4], and therefore, a higher charge delivery capacity material is required for a smaller electrode design in order to increase the number of electrodes in the limited area. Iridium oxide (IrOx) has been reported to hold a high charge delivery capacity and to have resistance to dissolution and corrosion in a saline environment [5-6]. IrOx thin film can be fabricated by electrochemical activating iridium metal (AIROF) [7] using reactive laser ablation [8] or by reactive sputtering from an iridium target (SIROF) [9-10]. IrOx thin film deposited by sputtering is compatible with CMOS process technology, and thus, has the advantage in that it can be multi-electrode array fabricated by using thin film methods, and the CDC of sputtered IrOx thin film can be controlled by using the sputtering condition.

In this work, IrOx thin films were fabricated by using reactive sputtering deposition. Although sputtered IrOx has widely been made by other groups [9-10], little systematic analyses to optimize sputtering conditions have been appeared in the previous research. In order to optimize the sputtering condition to maximize the CDC of IrOx, we fabricated 16 IrOx thin film samples under different sputtering condition combinations, and Pt thin film was also sputter deposited for comparison. The IrOx thin films were characterized by using the cyclic voltammetry (CV) method, and the CDC were calculated from the integral of the CV generic curve and the effect on the CDC for each sputtering parameter was also evaluated. In addition, *in vivo* experiments were also carried out to compare the IrOx and Pt film performances. IrOx and Pt thin films were sputter deposited on a polyimide substrate with eight electrodes and implanted into the eye ball of a cat in the *in vivo* experiment.

2. Fabrication of Electrodes and Electrochemical Evaluation

We fabricated a thin film electrode using a RF reactive sputtering machine (Shibaura Mechatronics, CFS-4ES-231). The sputtering chamber was evacuated to 7 x 10^{-4} Pa before the deposition by first using a rotary pump and then a turbo molecular pump. A 200-nm Ti adhesive layer was deposited on a glass substrate and then an Ar and O_2 gas flow ratio was controlled and introduced to the reactive chamber for IrOx thin film deposition. The CDC of IrOx is strongly related to the surface morphology and composition ratio and can be affected by the sputtering conditions. It is essential to optimize the sputtering conditions, such as the sputtering pressure, RF power, O_2 gas partial pressure, and sputtering time to maximize the CDC of IrOx. In this work, each parameter in the deposition process was set to the High (H) or Low (L) condition, for example the RF power was set at 200 W for the high condition and 50 W for the low condition. Table 1 lists the detailed sputtering conditions for each IrOx sample fabrication. There are a total of 16 IrOx samples that were fabricated under different sputtering condition combinations. In addition, Pt thin film was also deposited by DC sputtering deposition for comparison with IrOx. The deposition condition for Pt was set at 0.5 Pa, a DC power 500 W, an Ar gas flow of 20 sccm and then sputtered for 30 mins.

Table 1. Sputtering conditions for IrOx fabrication

Condition	Pressure [Pa]		RF Power [W]		O_2 Partial Pressure			Time [min]	
						Ar [sccm]	O_2 [sccm]		
1	0.5	(L)	50	(L)	(L)	8	2	30	(L)
2	0.5	(L)	50	(L)	(L)	8	2	60	(H)
3	0.5	(L)	50	(L)	(H)	0	10	30	(L)
4	0.5	(L)	50	(L)	(H)	0	10	60	(H)
5	0.5	(L)	200	(H)	(L)	8	2	30	(L)
6	0.5	(L)	200	(H)	(L)	8	2	60	(H)
7	0.5	(L)	200	(H)	(H)	0	10	30	(L)
8	0.5	(L)	200	(H)	(H)	0	10	60	(H)
9	1.0	(H)	50	(L)	(L)	8	2	30	(L)
10	1.0	(H)	50	(L)	(L)	8	2	60	(H)
11	1.0	(H)	50	(L)	(H)	0	10	30	(L)
12	1.0	(H)	50	(L)	(H)	0	10	60	(H)
13	1.0	(H)	200	(H)	(L)	8	2	30	(L)
14	1.0	(H)	200	(H)	(L)	8	2	60	(H)
15	1.0	(H)	200	(H)	(H)	0	10	30	(L)
16	1.0	(H)	200	(H)	(H)	0	10	60	(H)

The objective of this work is to optimize the charge delivery capacity of sputtered IrOx film for a smaller electrode design and increase the special resolution for a retinal prosthesis device. In order to fabricate mechanically stable amorphous sputtered iridium oxide thin film, an IrOx thin film was deposited at a low temperature without heating the substrate [11]. The electrochemical characteristic was evaluated by using the cyclic voltammetry (CV) method and was measured within the potential limits without causing any irreversible redox reactions. Irreversible processes are avoided because they will generate undesirable toxic species and lead to electrode corrosion. The cyclic voltammetry (CV) measurement method uses an Ag/AgCl reference electrode and a Pt wire counter electrode in 0.01M PBS (Phosphate Buffered Saline). Prior to CV testing, the IrOx film was electrically connected to a lead wire via electrically conductive paste. Figure 1 shows a schematic diagram of the cyclic voltammetry measurement system used in this work. The IrOx film sample was fixed by a jig and the lead wire was electrically connected to the potentiostat. The area of the IrOx defined by the jig for cyclic voltammetry evaluation was 0.5 cm^2. The cyclic voltamogram was taken by being activated using a triangular wave and repetitive potential cycling at a scan rate of 50 mV/s in a potential range from -0.65 to 0.85 V for the IrOx and -0.6 to 1.05 V for the Pt without H_2 and O_2 evolution. The CV generic curve was monitored through a PC and measured when the potential of the IrOx film was stable after being activated for three cycles. The CDC [Q_a] of IrOx can be calculated using equation 1.

$$Q_a = \frac{1}{v}\int_{E_a}^{E_c} |i| dE, \qquad (1)$$

where E is the electrode potential, i is the measured current, v is the scan rate, and E_c and E_a are the cathodic and anodic potential limits, respectively [11]. Therefore, the Q_a for each sample can be calculated by integrating the generic CV curve for a sample evaluation. The thickness of the fabricated IrOx thin film was measured by using a Alpha-step 500 surface profiler.

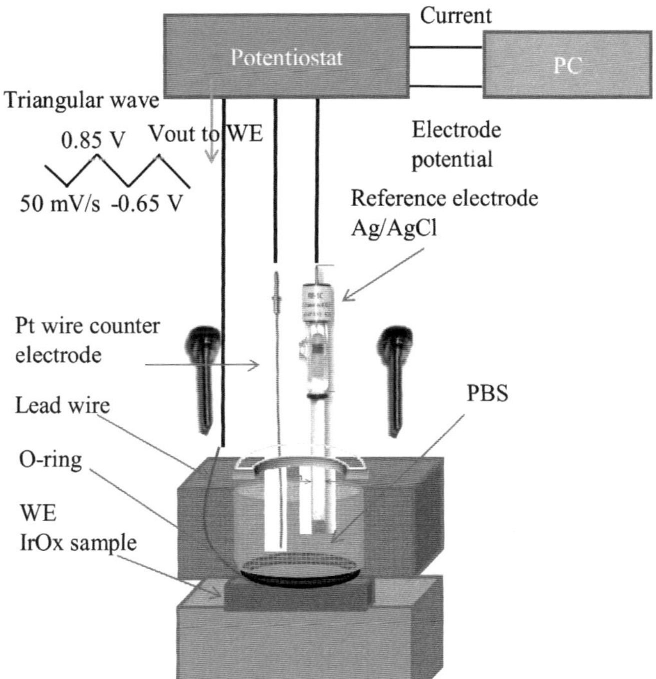

Figure 1. Schematic diagram of cyclic voltammetry measurement system

3. Experimental Results and Discussion

Figure 2 shows the measured thickness of the deposited IrOx thin film, and the thickness of the deposited Pt thin film for comparison is 862 nm. Figure 3 shows the CDC [Q_a] calculated from the integral of the CV generic curve for each IrOx sample, and Pt is shown for comparison. It is apparent that the CDC of the deposited IrOx films is strongly related to the sputtering conditions. The CDC of the sample deposited under conditions 7 and 8 are not available because the deposited film was brittle and easily peeled off. The calculated CDC [Q_a] of the deposited Pt thin film was 3 mC/cm^2. From figures 2 and 3, the CDC of the deposited IrOx thin film has correlation with the thickness (sputtering time) while only sputtering time was varied and the other sputtering parameters were fixed.

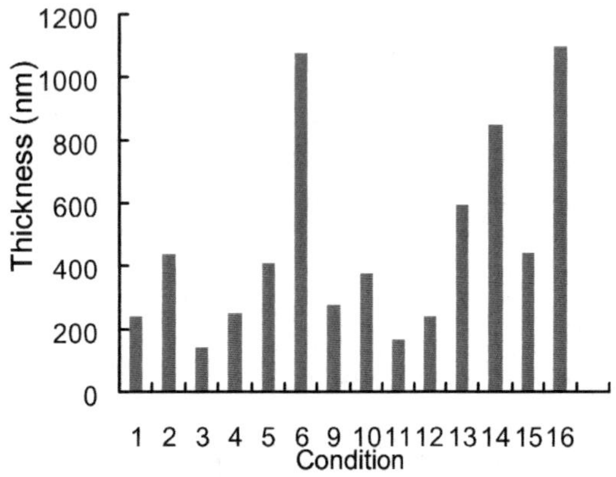

Figure 2. Thickness of deposited IrOx thin film

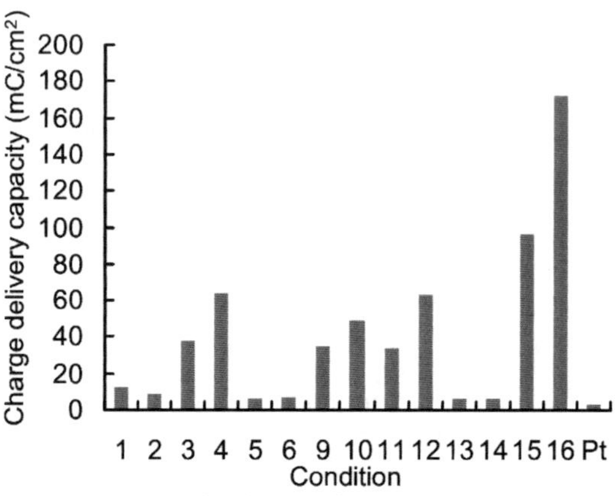

Figure 3. CDC calculated from CV curve

Figure 4 shows the generic CV curve of the deposited IrOx thin film for the best four conditions and compared with Pt. The current peak in each CV curve of the IrOx occurred due to the redox reactions involving the changes in valence of the iridium between Ir^{+3} and Ir^{+4} within the multilayered metal oxide film [11]. The charge delivery capacity of IrOx, because of the multilayer nature, is much greater than that of the monolayer redox reactions on Pt. The larger the closed area of CV curve, the larger the CDC of IrOx thin film. In this work, the CDC of IrOx obtained under the best condition is over 50 times greater than that of Pt thin film.

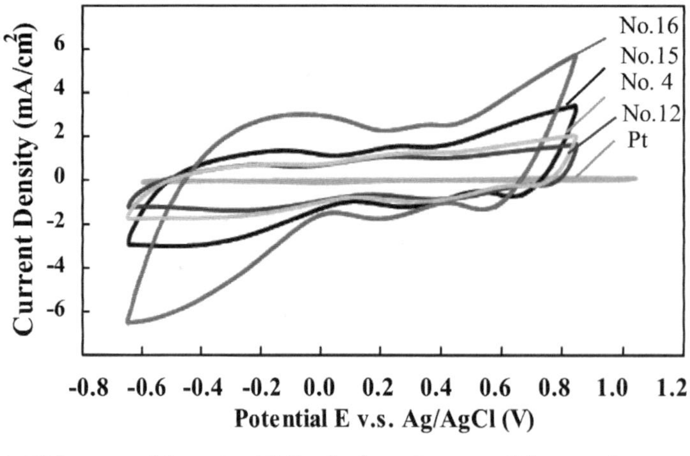

Figure 4. CV curve of deposited IrOx for best four conditions and compared with Pt

From the CV generic curve, the CDC of IrOx obviously varied by changing the sputtering conditions during the deposition process. This is because the composition and morphology of the deposited IrOx thin film can be altered under different sputtering conditions during the deposition process. The current peak in the CV generic curve represents that the iridium changes its oxidation state from Ir^{+} to Ir^{4+} during potential cycling. The possible redox reactions that occurred at these current peaks is described as follows [12]:

$$Ir(OH)\,.3H_2O \longleftrightarrow Ir(OH)_2\,.2H_2O + H^+ + e^-$$
$$Ir(OH)_2\,.\,2H_2O \longleftrightarrow Ir(OH)_3\,.\,H_2O + H^+ + e^-$$
$$Ir(OH)_3\,.\,H_2O \longleftrightarrow Ir(OH)_4 + H^+ + e^-$$

In order to evaluate the effect of each sputtering parameter on CDC, the CDC ratio of the high condition/low condition for each sputtering parameter was calculated. Table 2 lists the representative CDC ratio of the high condition/low condition for each sputtering parameter. We can see from table 2 that the CDC ratio of O_2 partial pressure is the highest among the other sputtering parameters. Therefore O_2 partial pressure plays the most important role in determining the CDC of the sputtered IrOx thin film. The effect of O_2 partial pressure on CDC was also shown in CV curves. For example figure 5 shows the CV generic curve of the IrOx thin film deposited under conditions 14 and 16. Condition 14 was deposited at an oxygen gas flow of 2 sccm and condition 16 was deposited at an oxygen gas flow of 10 sccm. It is clear that the peak current varied with respect to O_2 partial pressure. With the increase in O_2 partial pressure, the iridium atoms were surrounded by more oxide radical and lead to the composition of the sputtered IrOx film deposited at a high O_2 partial pressure was different from that at low O_2 partial pressure.

Although it is not as obvious as that for the O_2 partial pressure, we found that thickness is another factor that affects the CDC of the sputtered IrOx film from the electrochemical experimental results. A thicker sputtered IrOx film and high O_2 partial pressure sputtered IrOx film has a larger CDC probably because the rougher surface, which increases the contact area between the electrode and the electrolyte, and thus, also increases the CDC of the electrode. Figure 6 shows the SEM imaging results for No. 14~No. 16 IrOx samples. From SEM imaging results, No. 16 (sputtered at high O_2 partial pressure and high sputtering time) shows the densely microporous surface lead to the rougher surface than No. 14 (sputtered at low O_2 partial pressure) and No. 15 (thin thickness). However, the pressure in the chamber and the RF power during the deposition process don't have an obvious influence on the CDC from the experimental results.

The CDC of the electrode not only depends on the material but also the electrolyte and stimulation waveforms. However, in this work we only focus on the material. The CDC of the sputtered IrOx film obtained under the best sputtering condition was around 170 mC/cm^2 in this work. However, because the current pulses used *in vivo* are much faster than the scan rates in the CV experiment, only part of the charge injection is available during the real stimulation.

Table 2. Representative CDC ratio of high condition/low condition for each sputtering parameter

Sputtering parameter	RF power	O_2 partial pressure	Sputtering pressure	Sputtering time
CDC ratio condition H/L	0.86	29.06	0.83	1.79

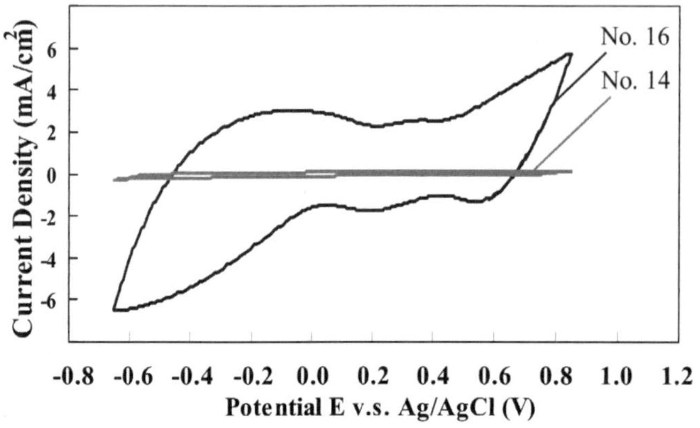

Figure 5. CV curve of IrOx thin film deposited under conditions 14 and 16

(a) (b) (c)

Figure 6. SEM results of (a) No. 14 , (b) No. 15 and (c) No. 16

4. *In Vivo* Evaluation

An *in vivo* experiment was also conducted to compare the performances of the IrOx and Pt electrodes. The IrOx electrode was sputtered at the same condition with No. 16. Figure 7 shows the IrOx and Pt thin film electrodes fabricated on a polyimide substrate for the *in vivo* experiment. The diameter of each electrode was 0.3 mm and the distance between two electrodes was 0.4 mm. The polyimide substrate was implanted into the eye ball of a cat for the *in vivo* experiment. The current controlled retinal stimulations were performed in a Suprachoroidal Transretinal Stimulations (STS) configuration [13-14]. In this method, the retinal prosthesis device was implanted in a sclera pocket and the reference electrode was inserted into the vitreous. Cathodic first charge-balanced biphasic current pulse that was input and the pulse width for the *in vivo* experiment were 0.3 mA and 0.5 ms, respectively, and the electrical potential of the IrOx and Pt electrodes were then measured. The electrical stimulation in cat's eye ball was performed and the responses of retinal cell evoked by the stimulation were observed successfully. Figure 8 shows the electrical potential of the IrOx and Pt electrodes measured in the *in vivo* experiment. We found from the experimental result that IrOx has a lower maximum electrical potential than that of Pt, which means that IrOx has a lower impedance compared with Pt, and therefore, we were also able to verify in this experiment that IrOx has a better charge delivery capacity than that of Pt.

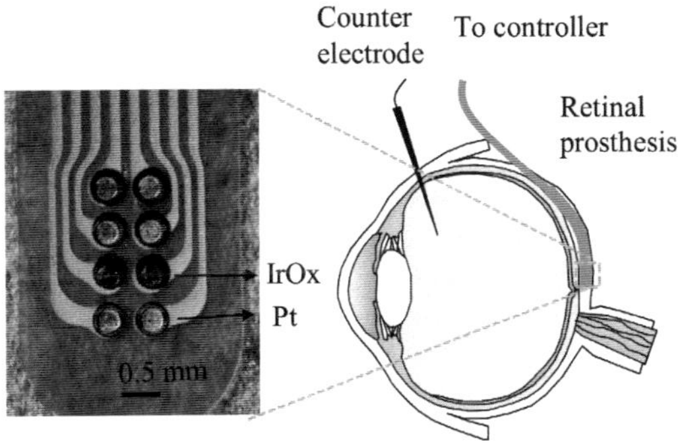

Figure 7. IrOx and Pt thin film electrodes fabricated on polyimide substrate for *in vivo* experiment

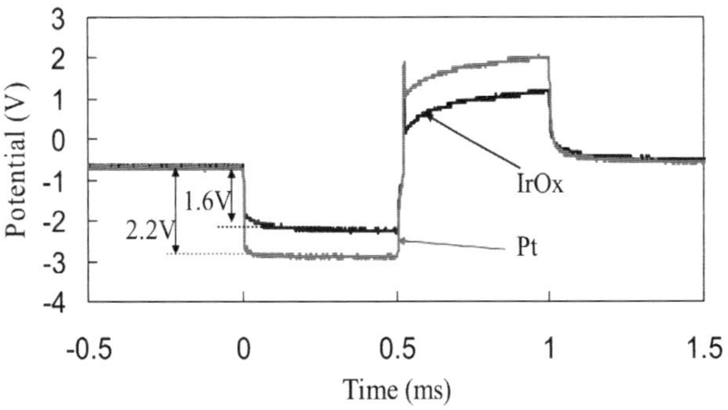

Figure 8. Electrical potential of IrOx and Pt electrodes measured *in vivo* experiment

5. Conclusion

This research focuses on the optimization of the RF sputtering conditions for fabricating IrOx thin film electrodes in retinal prosthesis devices. We found from the electrochemical experimental results when using the cyclic voltammetry method that the O_2 partial pressure and thickness strongly effect the CDC of the sputter deposited IrOx films due to the morphology and the composition of the sputtered IrOx film was altered. The CDC of the sputtered IrOx thin film obtained under the best sputtering condition was over 50 times that of Pt thin film. With the increase in charge delivery capability of the electrode, a smaller electrode design is possible to increase the number of electrodes to 1000 for retinal prosthesis devices. In addition, IrOx has a better CDC than Pt and this was also verified in the results from our *in vivo* experiment. The IrOx electrode fabricated by sputter deposition clearly improved the CDC compared with a Pt one, and thus, is a promising candidate for retinal prosthesis application.

Acknowledgment

This work was supported by the Strategic Research Program for Brain Sciences from the Ministry of Education, Culture, Sports, Science and Technology of Japan.

References

[1] Ohta J, Tokuda T, Kagawa K, Furumiya T, Uehara A, Terazawa Y, Ozawa M, Fujikado T, and Tano Y. 2006 *IEEE Eng. Med. & Biol. Magazine* **25** 47.
[2] Pan Y, Tokuda T, Uehara A, Kagawa K, Ohta J, and Nunoshita M. 2005 *Jpn. J. Appl. Phys.* **44** 2099.
[3] Tokuda T, Hiyama K, Sawamura S, Sasagawa K,Terasawa Y, Nishida K, Kitaguchi Y, Fujikado T, Tano Y, and Ohta J. 2009 *IEEE Trans. Electron Devices* **56** 2577.
[4] Palanker D, Vankou A, Hui P, and Baccus S. 2005 *J. Neural Eng.*, **2** S105-S120.
[5] Tanghe S, Najafi K, and Wise K. 1990 *Sensors Actuators,* B1 464.
[6] Cogan S, Troyk P, Ehrlich J, and Plante T. 2005 *IEEE Trans. Biomed. Eng.*, **52** 1612.
[7] Weiland J and Anderson D. 2000 *IEEE Trans Biomed. Eng.*, **35** 911.
[8] Khakani M and Chaker M. 1998 *Thin Solid Films*, **335** 6.
[9] Wang K, Lin C, and Durand D. 2009 *IEEE Trans Biomed. Eng.*, **56** 6.
[10] Cogan S, Ehrlich J, Plante T, Smirnov A, Shire D, Gingerich M, and Rizzo J. 2009 *Journal of Biomedical Materials Research Part B*: **89B** 353.
[11] Slavcheva E, Vitushinsky R, Mokwa W, and Schnakenberg U. 2004 J*ournal of the Electrochemical Society,* **151** 226.

[12] Mailley S, Hyland M, Mailley P, McLaughlin J, and McAdams E. 2002 *Mater. Sci. Eng.,* **21** 167.
[13] Kanda H, Morimoto T, Fujikado T, Tano Y, Fukuda Y, and Sawai H. 2004 *Invest. Ophthal. & Vis. Sci.*, **45** 560.
[14] Fujikado T, Morimoto T, Kanda H, Kusaka S, Nakauchi K, Ozawa M, Matsushita K, Sakaguchi H, Ikuno Y, Kamei M, and Tano Y. 2007 *Graefes Arch. Clin. Exp. Ophthalmol.*, **245** 1411.

Ferroelectric and Ferromagnetic Nanofibers: Synthesis, Properties and Applications

A Jalalian[1,2], A M Grishin[1*] and S X Dou [2]

[1]Department of Condensed Matter Physics, Royal Institute of Technology, SE-164 40 Kista, Sweden
[2]University of Wollongong, North Wollongong, NSW 2500, Australia

*grishin@kth.se; +4687904176

Key words: Sol-Gel, Electrospinning, Electrical Polarizability, Ferromagnetic Resonance, Biocompatibility.

Abstract. As first time, ferroelectric $(Ba_{0.85}Ca_{0.15})(Ti_{0.9}Zr_{0.1})O_3$ (BCTZ), biocompatible $(Na_{0.5}K_{0.5})NbO_3$ (NKN), and ferromagnetic $Y_3Fe_5O_{12}$ (YIG) nanofibers were synthesized by sol-gel assisted electrospining technique. Non-woven, bead-free 100 µm long fiber mat was formed from close-packed nanocrystals after calcination in air at 800°C (BCTZ and NKN) and 1100°C (YIG). BCTZ fibers are 200 nm in width whereas NKN and YIG fibers have smaller 50-100 nm diameter. X-ray diffraction and Raman spectroscopy show that BCTZ and NKN crystallized in tetragonal and rhombohedral polar phases in the vicinity of morphotropic phase boundary. Room temperature ferromagnetism in YIG fibers is indicated by a well formed *M-H* hysteresis loop with about 70% of remnant magnetization to the saturated value reached at 150 Oe. Broad band microwave spectroscopy revealed strong asymmetry of the shape of ferromagnetic resonance lines that conforms to a three orders of magnitude shape aspect ratio of YIG nanofibers.

1. Introduction

Quasi one dimensional materials including nanofibers, nanowires, nanobelts and nanorods have attracted intensive interest of scientist recently. Ferroelctric and ferromagnetic materials as functional compositions have been employed to invent or improve nanoelectronic, nanophotonic and nanogenerator devices [1-3].

Using piezoelectric nanofibers as the building blocks in nano-generators and nano-sensors makes them promising candidates for energy harvesting [4-6]. X.Chen et.al. designed and fabricated a high efficiency nanogenerator based on Pb(Zr,Ti)O3 nanofibers which produced 1.6V output [6].Recently, a Pb-free piezoelectric ceramics system $Ba(Ti_{0.80}Zr_{0.20})O_3$- x $(Ba_{0.70}Ca_{0.30})TiO_3$ (abbreviated as BZT-x BCT, x= molar fraction of BCT) reported by W.Liu et.al Demonstrated a high piezoelectric coefficient d_{33} about 600 pC/N in Morphotropic Phase Boundary (MPB) region at x= 50 [7]. Regarding to its high piezoelectric performance it would be a promising candidate to use as a piezoelectric segment in generators and sensors.

3D and fibrous structures of the electrospun ferroelectric nanofibers in addition to their response to the external electric field and ultrasound vibrations, enhanced the tissue and neural tissue regeneration

[8,9]. Sodium potassium niobate composition has been studied intensively since it was discovered in 1949 [10] and showed a good potential as a biocompatible ferroelectric material [11-13].
Yttrium iron garnet $Y_3Fe_5O_{12}$ (YIG) as a high frequency passive low loss and possessing a unique magneto-optical property has been used extensively in optical communication and magneto-optical devices [14,15]. The magnetic phenomenon in this material can be employed in different applications including drug delivery and DNA transportation [16].

Electrospinning technique is a sol-gel based, versatile and straightforward method to fabricate the nanofeatures i.e. nanofibers and nanotubes that recently, has been employed intensively to produce oxides and ceramics nanostructures i.e. $Pb(Zr,Ti)O_3$, $BaTiO_3$, $MgFe_2O_4$, TiO_2, $CoFe_2O_4$ nanofiers [17-21].

In this paper we review our recent achievements on synthesis and charachterization of ferroelectric BZT-BCT (BCTZ), biocompatible $(Na_{0.5}K_{0.5})NbO_3$ (NKN), and ferromagnetic $Y_3Fe_5O_{12}$ (YIG) nanofibers were synthesized by the electrospinning technique.

2. Experimental details

Sol-gel assisted electrospinning technique was employed to synthesize the BCTZ, NKN and YIG nanofibers.

2.1 Ferroelectric nanofibers

2.1.1 BCTZ nanofibers BCTZ precursor solution was prepared by disolving barium acetate and calcium acetate in glacial acetic acid at 60 °C followed by mixing the Ti and Zr solution from disolved titanium isopropoxide and zirconium propoxide in ethanol. In order to prepare the BCTZ solution for the electrospining, 350 mg of polyvinylpyrrolidone (PVP) was dissolved in BCTZ precusor solution and stirred for 1 hour in a closed caped glass bottle. The transparent and viscous solution was transferred into a plastic syringe.

Electrospining was carried out by applying 15 kV to the metallic needle of the syringe pump while the solution was fed with 0.5 ml/hour rate. A non-woven and bead-free nanofibers containing hydrolized BCTZ precursor and PVP were collected from the surface of the aluminium foil used as a collector placed at 9 cm below the needle. Crystallized nanofibers were achieved by drying at 100 °C for 1 hour followed by calcination at 800 °C for 1 hour in air.

2.1.2 NKN nanofibers To prepare the NKN precursor solution, potassium acetate and sodium acetate were mixed in 2-methoxyethanol at room temperature and stirred for 1 hour to achieve a clear and transparent solution. Niobium ethoxide was dissolved in acetyl acetone at room temperature and stirred for 1 hour (acetyl acetone was used as a chelating agent) and then added to the Na/K precursor solution and continued stirring. NKN solution was stirred for 12 hours in a closed cap glass ware at room temperature. Polyvinylpyrrolidone (PVP, 0.035g/ml) was added to the precursor mixture to prepare the solution for electrospinning. The solution was transferred to a plastic syringe with a metallic needle. Electrospinning was carried at 1.8 kV/cm electric field strength between the metallic needle and aluminum foil collector located 8cm below the needle. A non-woven and bead-free nanofibers mat was collected from the surface of the collector after drying at 100 °C in nitrogen atmosphere for 12 hours followed by annealing at 800°C for 1 hour in air.

2.2. Ferromagnetic nanofibers

Iron nitrate $Fe(NO_3)_3 \times 9H_2O$ and yttrium acetate $Y(O_2C_2H_3)_3 \times H_2O$ were used as starting materials. The yttrium and iron salts precursors were dissolved in 10 ml solvent contained deionized water: methanol with 2:1 volume ratio, respectively. After stirring for 1 hour at room temperature, a red and transparent solution was achieved. The molar ratio of Fe^{3+}:Y^{3+} was maintained at 5:3. In order to prepare the solution for the electrospining, 0.035gr/ml of polyvinylpyrrolidone polymer was added to the YIG precursor solution and continued stirring for 1 hour. The viscous solution was transferred into a plastic syringe.

The electrospining was carried out by applying 15 kV to the metallic needle of the syringe pump which fed YIG solution at a constant rate of 0.3 ml/h. The as-spun nanofibers containing hydrolyzed YIG precursor and PVP were collected from the surface of the aluminum foil collector placed 10 cm below the tip of the needle and dried at 100 °C for 30 minutes. Synthesis was finalized with the calcination at 1100 °C for 1 hour in air.

Field emission scanning electron microscope (FE-SEM, *ZIESS Ultra* 55) was used to observe the morphology and crystallinity of nanofibers. The images were taken by using 1 kV accelerating voltage at 3 mm work distance without using a conductive coating of tested fibers. Phase content and crystalline structure of the nanofibers were examined by x-ray diffraction (XRD) using *Siemens* D5000 powder diffractometer. Raman spectra of the BCTZ and NKN nanofibers were recorded at room temperature using 514.5 nm line of Ar-laser (*HORIBA Jobin-Yvon*). Scanning capacitance microscope (SCM, *Dimension* 3100, Vecco metrology group) was used to study the polarizability of ferroelectric fibers.

3. Results and discussion

3.1 BCTZ nanofibers

SEM images of nanocrystalline structure of BCTZ nanofibers are shown in Figure 1. Very uniform, non-woven and bead-free mat of BCTZ nanofibers were achieved by sol-gel asisted electrospining technique. Fibers are crystalized in about 200 nm in width and several ten microns in lengh.

Figure 1. FE-SEM image of the BCTZ nanofibers annealed at 800 °C.

Figure 2(a) shows X-ray diffraction (XRD) patterns of electrospun BCTZ nano-fibers annealed at 800 °C for 1 hour. Bragg reflections are indexed according to the PDF reference cards 85-0368 for rhombohedral structure and 80-0601 for the tetragonal structure of the barium titanate [22].

The crystallite size of 18 nm of the BCTZ nanofibers calcined at 800 °C was calculated using Scherrer equation from the full width at half maximum (FWHM) value of the Bragg peak at $2\theta = 31.5$

deg. Figures 1(b) and 1(c) show enlarged profiles of the main Bragg diffraction reflections {110} at 31.5 deg and {200} about 45 degrees in nanofibers fitted by several Lorentzian contour lines. Coexistence of the tetragonal and rhombohedral syngonies confirms nanofibers were crystallized in the vicinity of morphotropic phase boundary (MPB).

Raman spectrum of the electrospun nanofibers calcined at 800 °C is shown in figure 3. Peaks at 260, 306, 520, and 720 cm^{-1} correspond to the $BaTiO_3$ tetragonal phase [23-26] whereas a low intensive band around 181 cm^{-1} is assigned to an A_1(LO) optical mode in rhombohedral phase of crystallized fibers [26].

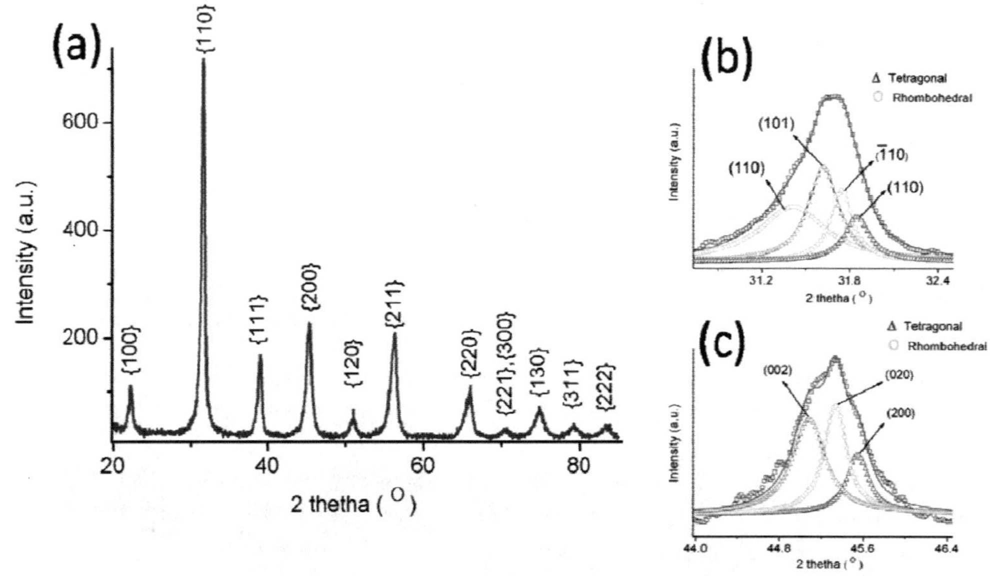

Figure 2. (a) - XRD pattern of nanofibers calcined at 800 °C in air. (b) and (c) – fitting of the main Bragg diffraction reflections {110} and {200} in nanofibers by Lorentzian functions.

Band frequencies around 268 cm^{-1} [A_1(TO)], 290cm^{-1} [B_1,E(TO+LO)], 520cm^{-1} [A_1(TO)], and 720 cm^{-1} [A_1(LO)] which would not be present in the cubic phase are commonly considered as the indication of the polar phases in $BaTiO_3$ crystals [27].

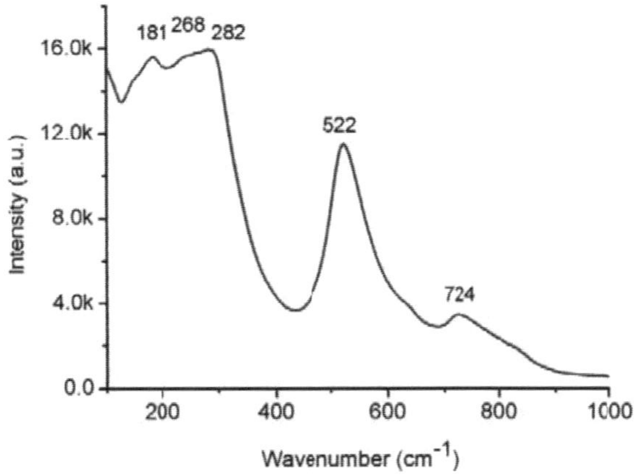

Figure 3. Raman scattering spectrum of $(Ba_{0.85}Ca_{0.15})(Ti_{0.9}Zr_{0.1})O_3$ nanofibers calcined at 800 °C in air.

Scanning capacitance microscopy (SCM) was employed to study ferroelectric properties of $(Ba,Ca)(Ti,Zr)O_3$ nanofibers annealed at 800 °C on the conductive substrate (Pt coated silicon wafer). SCM method enables the measurement of local capacitance at different bias voltages and superimposed ac-signals. Figures 4 (a) and (b) present, respectively, differential capacitance vs. voltage (*dC/dV-V*) and deflection vs. voltage curves recorded contacting the middle of individual $(Ba,Ca)(Ti,Zr)O_3$ nanofiber. *dC/dV-V* plot shows saturated polarization whereas the deflection-*V* indicates the electrostriction effect occurred during bias voltage swiping. Maximum deflection about 11 nm was achieved around 4-5 V where *dC/dV-V* curve reaches the saturation.

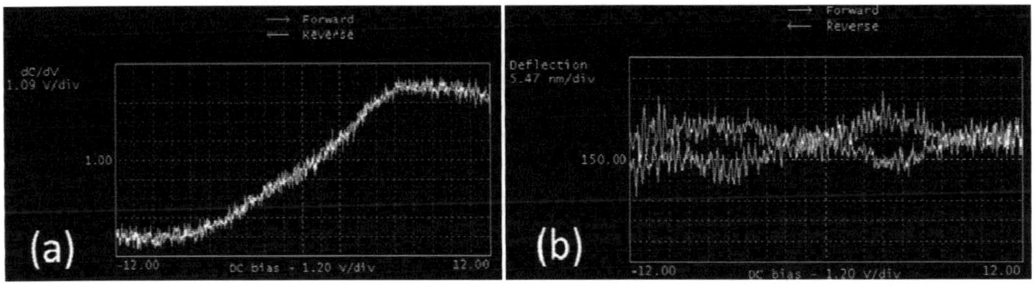

Figure 4. Local measurement of (a) dc/dv versus voltage and (b) deflection versus voltage variation of a BCTZ nanofiber annealed at 800°C from +12 to -12 volts.

3.2. NKN nanofibers

FE-SEM images of as-spun binder-contained fibers after drying at 100 °C in N_2 atmosphere for 12 hours and NKN sample completely calcined at 800°C for 1 hour in air are shown in figures 5(a) and (b), correspondingly. A bead-free and non-woven mat of nanofibers was achieved by controlling the electrospinning and annealing conditions during the synthesis procedure. The smooth surface of the dried binder-contained fibers is due to their amorphous structure. Dried fibers with average diameter about 350 nm experienced a considerable shrinkage of more than 50% and crystallized in the average diameter of 150 nm after annealing at 800 °C for 1 hour. The rough surface of the calcined nanofibers shown in inset of Figure 5(b) indicates the way how NKN plate-like nanocrystals grow on each other and compose nanofibers with several microns in length.

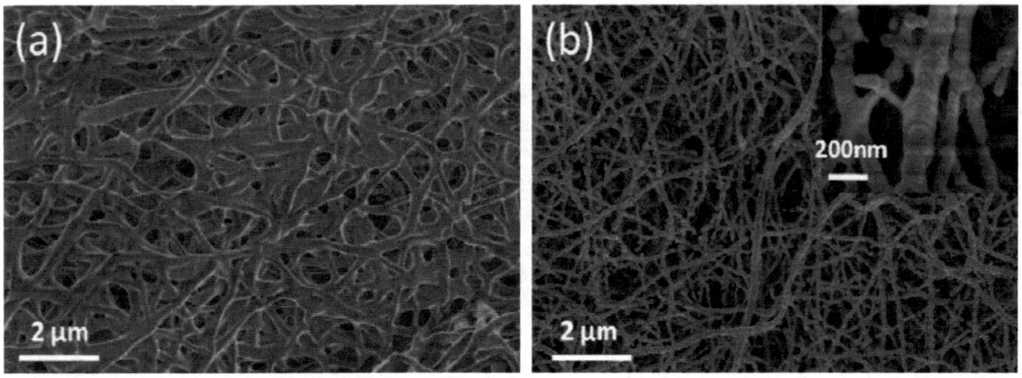

Figure 5. SEM images of electrospun NKN nanofibers: (a) - as-spun binder-contained after drying at 100 °C in N_2 atmosphere, (b) - after calcination at 800 °C in air.

Figure 6(a) shows XRD pattern of the NKN nanofibers calcined at 800°C in air for 1 hour. The pure perovskite-type and highly crystallized structure are observed in annealed sample. The XRD reflections are indexed according to the PDF reference card 77-0038 for monoclinic $(Na_{0.35}K_{0.65})NbO_3$ lattice with Pm (6) symmetry with a ≈ c > b [28]. Crystallite size calculation using Scherrer equation from the FWHM value of the Bragg peak at 2Θ= 31.9 deg, shows that nanofibers crystallized at 800 °C have a size of 42 nm. Enlarged {220} and {440} peaks at 31.9 and 66.6 deg fitted by Lorentzian functions are shown in Figs. 2(b) and 2(c). Absence of any trace of an impurity phase evidences for well homogenized and uniform structure of NKN nanofibers synthesized by electrospinning technique followed by a well controlled heating and cooling regimes.

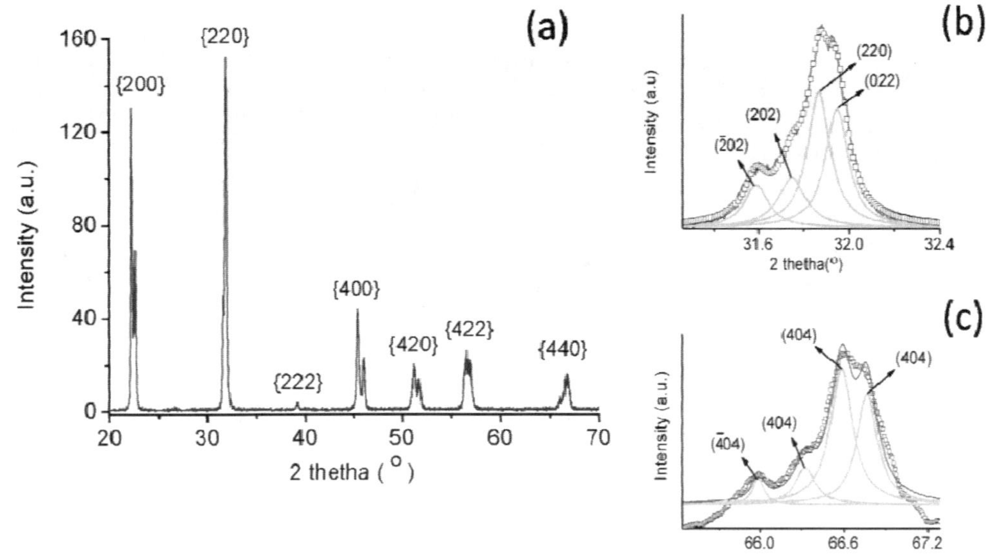

Figure 6. (a) - XRD pattern of the NKN nanofibers calcined at 800°C in air for 1 hour. (b) and (c) – enlarged {220} and {440} Bragg reflections fitted by Lorentzian contour lines.

Raman spectrum of NKN-nanofibers annealed at 800 °C recorded at room temperature is shown in figure 7. Comparing this spectrum with Raman spectra of NKN and $KNbO_3$ ceramics [29-31] shows that internal vibrational modes of the NbO_6 octahedra produce the Raman shift bands at frequencies above 160 cm^{-1}.

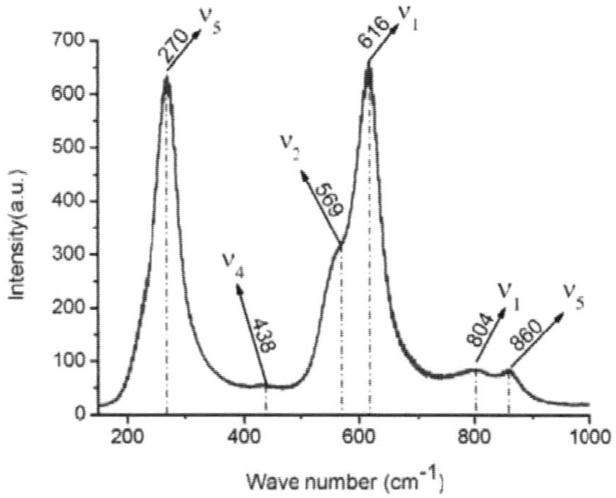

Figure 7. Raman spectrum of the NKN-nanofibers annealed at 800 °C for 1 hour in air.

Stretching mode vibration of the O-Nb-O bonds are represented by strong peaks at 569 (ν_2) and 616 cm^{-1} (ν_1). The bands located in 270 (ν_5) and 438 cm^{-1} (ν_4) are attributed to the bending mode of O-Nb-O bonds. The Raman signals at higher frequencies located at 804 and 860 cm^{-1} are produced by the fundamental vibrational ν_1 (A_{1g}) and ν_5 (F_{1g}) modes of the short Nb=O bonds in NbO_6 octahedra [29-31].

3.2. YIG nanofibers

Morphology and dimension YIG nanofibers calcined at 1100 °C for 1 hour are displayed by FE-SEM images in figure 8. A crystallized YIG nanofiber with 100nm in diameter is shown in inset to Fig. 8.

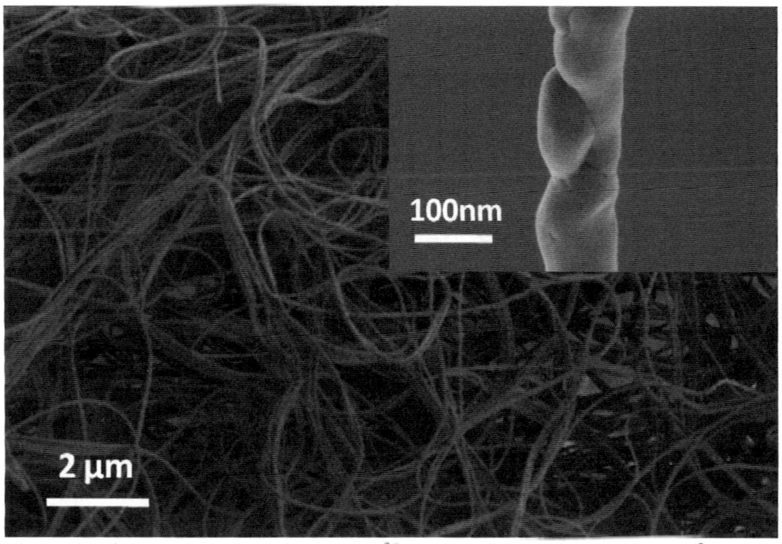

Figure 8. High resolution SEM images of YIG nanofibers annealed at 1100 °C in air.

Figure 9 shows XRD pattern of ferromagnetic YIG nanofibers annealed at 1100 °C. Bragg peaks are indexed according to the standard PDF card 33-0693. Sample was confirmed as almost a single phase yttrium iron garnet $Y_3Fe_5O_{12}$ with the only two noticeable orthoferrite Bragg reflections: the main

$YFeO_3(121)$ at $2\Theta = 32.89$ deg and $YFeO_3(111)$ at $2\Theta = 25.95$ deg [32]. The relative intensity ratios of all the $Y_3Fe_5O_{12}$ reflections are in a good correspondence with $Y_3Fe_5O_{12}$ powder.

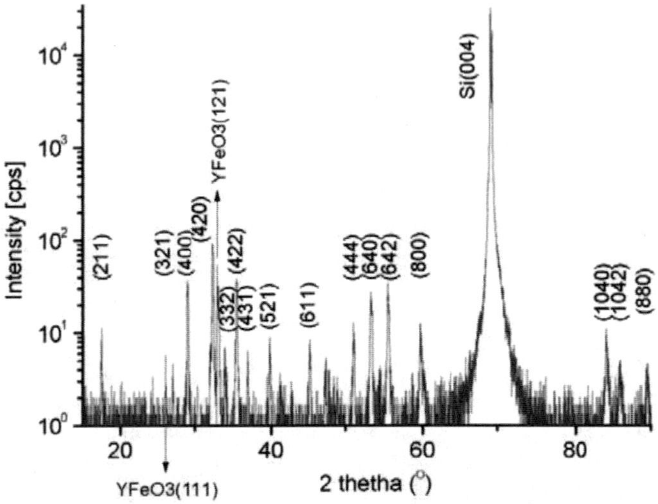

Figure 9. XRD pattern of YIG nanofibers annealed at 1100 °C for 1 hour in air. Fibers are clamped onto single crystal Si(001) wafer.

Room temperature ferromagnetism in YIG nanofibers was revealed by tracing *M-H* hysteresis loops presented in figure 10(a). YIG nanofibers saturate in a weak magnetic field 150 - 200 Oe and possess high remnant magnetization at the level of 68% of saturated value.

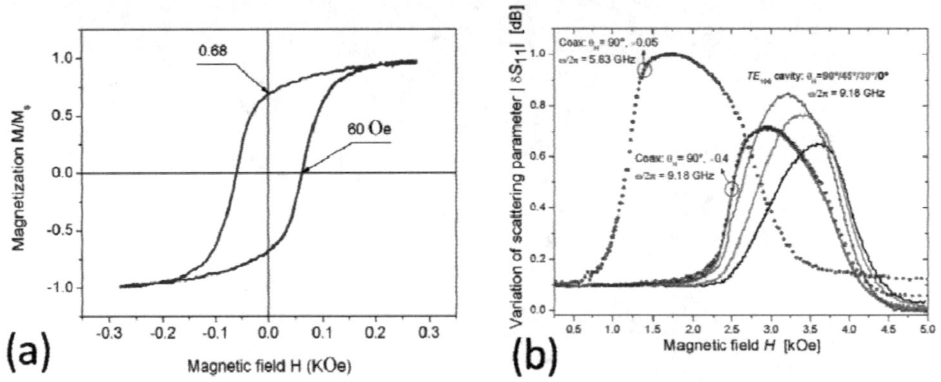

Figure 10. (a) M-H hyteresis curve traced with a rate of 20 Hz in YIG nanofibers. (b) Resonance variation of the reflection scattering parameter | S_{11} | recorded for $Y_3Fe_5O_{12}$ nanofibers.

To detect very weak FMR signals from a diminutive amount of fibers, we employed broad Ferromagnetic Resonance (FMR) spectroscopy. Figure 10(b) displays FMR lines detected at 5.83 and 9.18 GHz. Resonant response reveals strong magnetic anisotropy of YIG nanofibers: non-symmetric FMR lines are inverted when external magnetic field rotates from in-plane (where fibers are aligned) to out-of-plane geometry. This effect conforms to a three orders of magnitude length-to-diameter aspect ratio of YIG nanofibers.

4. Conclusion

- $(Ba_{0.85}Ca_{0.15})(Ti_{0.9}Zr_{0.1})O_3$ (BCTZ) piezoelectric and ferroelectric nanofibers were fabricated successfully at 800 °C with high aspect ratio in morphotropic phase boundary region which can be used as piezoelectric nanogenerator in nano-devices.
- Biocompatible and ferroelectric (Na,K)NbO3 nanofibers were synthesized with 3D structure that make them a promising candidate as scaffolds for engineering, repair, and regeneration of damaged tissue.
- Ferromagnetic yttrium iron garnet (YIG) nanofibers with about 100nm in diameter and several tens of microns in length have been synthesized and showed strong resonance response in a microwave irradiation. YIG ferromagnetic fibers have the potential to use as delivery guides for pharmaceutical drugs, protein complexes, and DNA transportation.

Acknowledgment

This work was supported by the Vetenskapsrådet (Swedish Research Council) through the Advanced Optics and Photonics (ADOPT) Linné center grant and Australian Research Council (ARC) support under program DP0879070.

References

[1] Thomas L, Moriya R, Rettner C and Parkin S S P 2010 *Science* **330** 1810
[2] Martı´na J I, Nogue´s J, Liuc K, Vicente J L and Schuller I K 2003 *J. Magn.Magn.Mater.* **256** 449
[3] Ryan Tu, Zhang L, Nishi Y and Dai H 2007 *Nano Lett.* **7** 2778
[4] Chen C Y, Huang J H, Song J, Zhou Y, Lin L, Huang P C,Y Zhang, Liu C P, He J H and Wang Z L 2011 *ACS NANO* **5** 6707
[5] Que R, Shao M, Wang S, Ma D D D and Lee S T 2011 *Nano lett.* **11** 4870
[6] Chen X, Xu S, Yao N and Shi Y 2010 *Nano lett.* **10** 2133
[7] Liu W and Ren X 2009 *Phys. Rev. Lett.***103** 257602
[8] Weber N, Lee Y S, Shanmugasundaram S, Jaffe M and Arinzeh T L 2010 *Acta Biomater.* **6** 3550
[9] Jianqing F, Huipin Y and Xingdong Z 1997 *Biomaterials* **18** 1531
[10] Matthias B T 1949 *Phys. Rev.* **75** 1771
[11] Hagh N M, Jadidian B and Safari A 2007 *J. Electroceram.* **18** 339
[12] Cho C R and Grishin A M 1999 *Appl. Phys. Lett.* **75** 268
[13] Nilsson K. Lidman J , Ljungstrom K and Kjellman U.S. patent 6,526,984, March 4, 2003
[14] Jalalian A, Kavrik M S, Khartsev S I and Grishin A M 2011 *Appl. Phys. Lett.* **99** 102501
[15] Manuilov S A, Fors R, Khartsev S I and Grishin A M 2009 *J. Appl. Phys.* **105** 033917
[16] Zhang J, Sun W, Bergman L, Rosenholm J M, Lindén M, Wu G, Xu H and Gu H C 2012 *Mater. Lett.* **67** 379
[17] Wang Y and Santiago-Avil´es J J 2004 *Nanotechnology* **15** 32
[18] McCann J T, Chen J I L, Dan L, Zuo-Guang Y and X Younan 2006 *Chem. Phys.Lett.* **424** 162
[19] Maensiri S, Sangmanee M and Wiengmoon A 2009 *Nanoscale Res. Lett.* **4** 221
[20] Lee S H, TekmenC and Sigmunda W M 2005 *Mater. Sci. Eng.* A **398** 77
[21] Lamastra F R, Nanni F, Camilli L, Matassa R, Carbone M and Gusmano G 2010 *Chem. Eng. J.* **162** 430
[22] 2001 JCPDS-International Center for Diffraction Data, Cards No. 85-0368 and 80-0601.
[23] Pizani P S, Mastelaro V R, Santos C O P, Longo E and Varela J A 2008 *Chem. Mater.* **20** 5381
[24] Urban J J, Yun W S, Gu Q and Park H 2002 *J. Am. Chem. Soc.* **124** 1186
[25] Choi Y K, Hoshina T, Takeda H and Tsurumi T 2010 *J. Ceram. Soc. Jpn.* **118** 881
[26] Xiao C J, Jin C Q and Wang X H 2008 *Mater. Chem. Phys.* **111** 209
[27] Souza A E, Silva R A, Santos G T A, Moreira M L, Volanti D P, Teixeira S R and Longo E 2010 *Chem. Phys. Lett.* **488** 54
[28] 2001 JCPDS-International Center for Diffraction Data, Card No. 77-0038.
[29] Trodahl H J, Klein N, Damjanovic D, Setter N, Ludbrook B, Rytz D and Kuball M 2008 *Appl. Phys. Lett.* **93** 262901
[30] Rubio-Marcos F, Romero J J, Martín-Gonzalez M S and Fernández J F 2010 *J. Eur. Ceram. Soc.* **30** 2763
[31] Ge H, Hou Y, Rao X, Zhu M, Wang H, and Yan H 2011 *Appl. Phys. Lett.* **99** 032905
[32] 2001 JCPDS-International Center for Diffraction Data, Cards No. 33-0693 ($Y_3Fe_5O_{12}$) and 39-1489 ($YFeO_3$)

Journal of Physics: Conference Series **352** (2012) 012007 doi:10.1088/1742-6596/352/1/012007

Luminescent Magneto-Optical Photonic Crystals

AM Grishin[1] and SI Khartsev
Department of Condensed Matter Physics, Royal Institute of Technology,
SE-164 40 Stockholm-Kista, SWEDEN

E-mail: grishin@kth.se

Abstract. We compare luminescent properties of several Er-doped garnet films as building blocks in all-garnet heteroepitaxial magneto-optical photonic crystals: $La_3Ga_5O_{12}$, $Gd_3Ga_5O_{12}$, $Y_3Fe_5O_{12}$, $Bi_3Fe_5O_{12}$, and $Bi_{2.97}Er_{0.03}Fe_4Al_{0.5}Ga_{0.5}O_{12}$. Er substituents on the dodecahedral lattice sites do not decrease giant Faraday rotation in $Bi_3Fe_5O_{12}$ garnet; meanwhile providing intense room temperature *C*-band photoluminescence (PL). Fe^{3+} ion works as a sensitizer for Er resulting in fivefold PL enhancement in iron garnets compared to gallium ones. PL lifetime in gallium garnets is in millisecond range reaching 6 ms in $Gd_{2.9}Er_{0.1}Ga_5O_{12}$. The first luminescent one-dimensional heteroepitaxial all-garnet magneto-optical (MO) photonic crystal was composed from diamagnetic $Sm_3Ga_5O_{12}$ and MO-active $Bi_{2.97}Er_{0.03}Al_{0.5}Ga_{0.5}O_{12}$ garnet layers by rf-magnetron sputtering on $Gd_3Ga_5O_{12}(111)$ substrate. Substitution of ferric ions by aluminum and gallium improved transparency and induced perpendicular anisotropy in pure $Bi_3Fe_5O_{12}$. Photonic crystals owned a record high magneto-optical quality and a latching type (magnetic remnant) Faraday rotation (FR). At the resonance wavelength 775 nm, specific FR $\theta_F = -14.1$ deg/μm and MO-quality factor $Q = 99.3$ deg represent the highest MO performance achieved so far. Long-lived near-IR luminescence in Er substituted gallium and iron garnet layers used both as Bragg mirrors and microcavities promises magneto-optical photonic crystals to become an active lasing medium.

Introduction

Newer photonic integrated circuits (PICs) own variety of functions as lasing, modulation, wavelength multiplexing, etc. integrating all optical components on the same semiconducting platform. Optical isolators are irreplaceable part of PICs. Laser sources are sensitive to feedback – a light backscattered from other elements the laser is coupled and from external elements on the transmission line. Optical isolators prevent unwanted feedbacks providing unidirectional propagation of optical signals. Due to spontaneous magnetization breaking time inversion symmetry $t \rightarrow -t$, ferromagnetic materials demonstrate nonreciprocal Faraday effect: rotation of the plane of polarization of the light in a different direction according to the direction of propagation of the light. Rare earth iron garnets are the principal materials for non-reciprocal passive optical devices in telecommunications applications.

Bismuth substituted rare earth iron garnets combine high Faraday rotation and transparency in visible and near infrared light. Completely substituted bismuth iron garnet $Bi_3Fe_5O_{12}$ in the form of a single layer epitaxial film keeps the record value of FR $\theta_F = -8.4°$/μm at 633 nm [1,2]. Further increase of FR was offered by light localization in magneto-optical photonic crystals (MOPCs). Rosenberg's idea [3] to enhance FR by placing MO-material in a resonant cavity has been practically realized in one dimensional (1D) MOPC by Inoue *et al* [4]. Ultimate MO performance was recently achieved in all-

[1] To whom any correspondence should be addressed.

garnet MOPCs: 1D heteroepitaxial $Bi_3Fe_5O_{12}/Y_3Fe_5O_{12}$ [5], $Bi_3Fe_5O_{12}/Gd_3Ga_5O_{12}$ [6], $Bi_3Fe_5O_{12}/Sm_3Ga_5O_{12}$ [7], and $Bi_3Fe_4Al_{0.5}Ga_{0.5}O_{12}/Sm_3Ga_5O_{12}$ [8] crystals. So far, $Bi_3Fe_5O_{12}/Sm_3Ga_5O_{12}$ MOPC with 6 pair quarter-wavelength reflectors in two Bragg mirrors and half-wavelength $Bi_3Fe_5O_{12}$ microcavity in-between demonstrated the highest specific FR $\theta_F = -20.5°/\mu m$ at 750 nm [7]. Even though superior FR, the accompanying high optical absorption hinders practical applications of $Bi_3Fe_5O_{12}$ films. It is still a challenging task to engineer MO crystals which compromise a strong FR with modest optical insertion loss.

Reduction of bismuth concentration and substitution of the Fe^{3+} ion by other trivalent diamagnetic ions strongly suppress light absorption [9]. We employed combined Al, Ga, and Er substitution of, respectively, ferric and bismuth ions to enhance transmissivity and make MOPCs luminescent. In this paper we review our recent results on synthesis and characterization of optical properties of garnet materials and magneto-optical photonic crystals with enhanced transmissivity, remnant Faraday rotation and strong luminescence at room temperature.

Experimental
Erbium-doped ferrimagnetic iron- and diamagnetic gallium-contained garnet films and their multilayers were sintered onto $Gd_3Ga_5O_{12}(111)$ single crystal substrates. $La_3Ga_5O_{12}$, $Gd_3Ga_5O_{12}$, and $Y_3Fe_5O_{12}$ films were prepared by pulsed laser deposition (PLD) whereas $Bi_3Fe_5O_{12}$, $Sm_3Ga_5O_{12}$, and $Bi_3Fe_4Al_{0.5}Ga_{0.5}O_{12}$ films were grown by rf-magnetron sputtering. The details of pulsed laser deposition and sputtering of garnet films have been published elsewhere [5] and [7], correspondingly.

Synthesis and characterization of single layer iron- and gallium contained films should always precede preparation of MOPCs. First, reference epitaxial films were deposited to optimize growth conditions and to determine refractive indices $n(\lambda)$ and deposition rates. Then, MOPCs for the resonance wavelength λ_{res} were grown with corresponding Bragg reflectors having thickness of $\lambda_{res}/4n(\lambda_{res})$ and a microcavity which thickness is a multiple of $\lambda_{res}/2n(\lambda_{res})$.

X-ray diffraction analyses verified morphotropic epitaxial quality of all garnet films [7]. The composition of deposited garnets was controlled by the Rutherford backscattering spectroscopy. Doping with 0.5 at.% of Er (0.1 garnet formula units) corresponds to the following volume concentration of Er^{3+}: 4.21×10^{20} cm^{-3} in $Gd_3Ga_5O_{12}$ and $Y_3Fe_5O_{12}$, 3.97×10^{20} cm^{-3} in bismuth substituted iron garnets, and 3.84×10^{20} cm^{-3} in $La_3Ga_5O_{12}$.

Optical dispersion in garnets within 450 to 1000 nm range was examined using fiber-optic *OceanOptics* PC200 and HR4000 spectrometers. Transmittance vs. λ spectra were normalized to the intensity of the light transmitted through a blank $Gd_3Ga_5O_{12}$ substrate. Photoluminescence (PL) in Er-doped garnets we studied using confocal *LabRam* HR800 Raman microscope with nitrogen cooled InGaAs array (900–1700nm) at 514.5 nm Ar-laser pumping. Decay of the luminescence at room temperature was measured pumping Er-doped films with square pulse modulated 980 nm laser and detecting 1530 nm PL with the Ge uncooled detector shielded with 2 mm thick Si filter.

To compare PL intensities spectra were normalized to films thickness. Such normalization gives reliable relative estimation for transparent gallium garnets meantime underestimates the effectiveness of PL in iron garnets. Incident Ar-laser light in iron garnets is absorbed completely at film surface at the depth ~ $\lambda/(2\pi\sqrt{|\varepsilon|})$. Dielectric permittivity ε grows rapidly at the absorption edge. Therefore, in Er-doped iron garnets green light generates PL in very shallow film surface layer.

Films optical properties
As an example, Figures 1 and 2 present transmission and Faraday rotation spectra in 0.9 μm thick $Bi_{2.9}Er_{0.1}Fe_5O_{12}$ film and in 1.6 μm thick $Bi_{2.97}Er_{0.03}Fe_4Al_{0.5}Ga_{0.5}O_{12}$ film. For all rare earth iron garnets

Faraday rotation has a local maximum close to the absorption edge around 520 nm (19 231 cm^{-1}, 2.38 eV). In $Bi_{2.9}Er_{0.1}Fe_5O_{12}$ specific FR has a peak value of θ_F = - 29.9 deg/μm at 535 nm and equals - 1.63 deg/μm at 980 nm. Aluminum-gallium substitution for iron leads to reduction of saturation magnetization and consequent decrease of FR: peak θ_F = - 12.6 deg/μm at 526 nm and - 0.58 deg/μm at 980 nm. Diamagnetic ions substitution for Fe^{3+} significantly increases transmissivity: $Bi_{2.97}Er_{0.03}Fe_4Al_{0.5}Ga_{0.5}O_{12}$ film has the same transmittance as unsubstituted $Bi_{2.9}Er_{0.1}Fe_5O_{12}$ though its thickness almost twice larger. The absorption coefficient α = ln(1/*T*)/$Bi_{2.97}Er_{0.03}Fe_4Al_{0.5}Ga_{0.5}O_{12}$ film thickness is 0.21 dB/μm at 750 nm against 0.62 dB/μm in pure $Bi_3Fe_5O_{12}$ film.

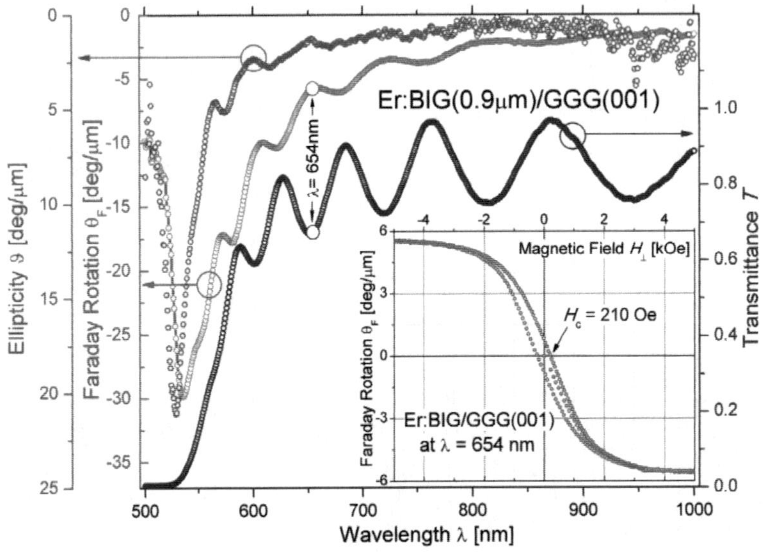

Figure 1 Transmittance *T*(λ), specific FR $\theta_F(\lambda)$, and ellipticity ϑ(λ) spectra for the 0.9 μm thick $Bi_{2.9}Er_{0.1}Fe_5O_{12}$ film recorded in saturating perpendicular magnetic field *H* = 3.5 kOe. Inset shows specific FR θ_F[deg/μm] vs. *H* hysteresis loop traced at the wavelength λ =654 nm. Two circular symbols ○ depict the magnitudes of the transmittance and FR at this wavelength.

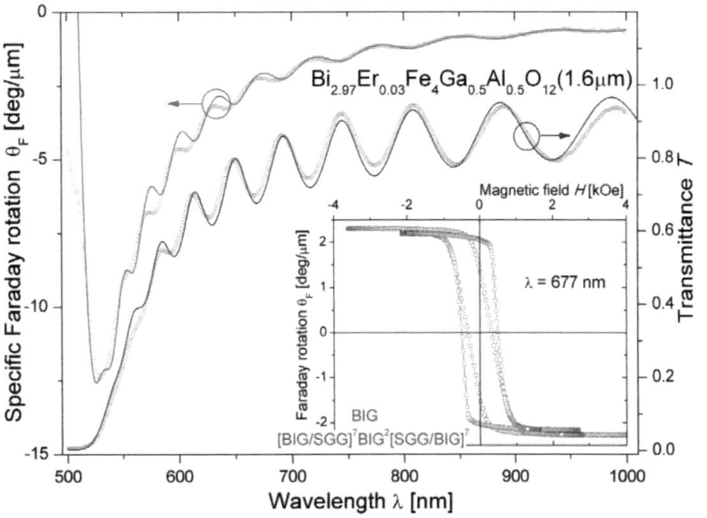

Figure 2 Transmittance T and specific FR θ_F spectra of the reference 1.6 μm thick $Bi_{2.97}Er_{0.03}Fe_4Ga_{0.5}Al_{0.5}O_{12}$ film. Solid lines show spectra modelled with a fitting parameters collected in the Table I. Inset compares FR θ_F vs. perpendicular magnetic field H hysteresis loops traced at λ = 677 nm in the reference $Bi_{2.97}Er_{0.03}Fe_4Ga_{0.5}Al_{0.5}O_{12}$ film and magneto-optical photonic crystal (MOPC) $[BIG/SGG]^7BIG^2[SGG/BIG]^7$ designed for λ_{res} = 775 nm.

The spectra in $Bi_{2.97}Er_{0.03}Fe_4Al_{0.5}Ga_{0.5}O_{12}$ film were fitted (solid lines in Fig. 2) to the theory which accounts three resonance transitions in electric dipole approximation [10]. A complete set of microscopic

Table 1. Material parameters of $Bi_{2.97}Er_{0.03}Al_{0.5}Ga_{0.5}O_{12}$ garnet†

Transition	λ_o[nm]	$\hbar\omega_o$[eV]	$\hbar\omega_p\sqrt{f}$[eV]	$\hbar\Gamma$[eV]	$\hbar\Delta$[eV]
First	300	4.14	8.82	$< 10^{-3}$	$< 10^{-3}$
Second	480	2.59	0.84	0.09	0.13
Third	407	3.05	0.84	0.03	0.17

† resonance transition wavelength λ_o and frequency ω_o, plasma frequency ω_p and oscillator strength f, half linewidth Γ and parameter of spin-orbit coupling Δ.

parameters giving the best fit to the experimental data in $Bi_{2.97}Er_{0.03}Al_{0.5}Ga_{0.5}O_{12}$ garnet is listed in Table 1. Also, useful interpolation Sellmeier-like formulas for the complex refractive indices can be presented in the wavelength range of our interest as follows: $n^2(\lambda) = 1 + \frac{2.75}{1-(128\text{nm}/\lambda)^2}$ for $Sm_3Ga_5O_{12}$ and

$$\left[n(\lambda) - ik(\lambda)\right]^2 = 1 + \frac{4.58}{1-(303\text{nm}/\lambda)^2} + \frac{0.11}{1-(494\text{nm}/\lambda)^2 + i0.08(494\text{nm}/\lambda)}$$ for $Bi_{2.97}Er_{0.03}Al_{0.5}Ga_{0.5}O_{12}$ garnet.

Besides increased transparency, substitution of ferric ions on tetrahedral positions by diamagnetic Al^{3+} and Ga^{3+} ions induces a perpendicular magnetic anisotropy. Insets in Figs. 1 and 2 show θ_F vs. H hysteresis curves recorded under normal incident red laser illumination. $Er{:}Bi_3Fe_5O_{12}$ film is in-plane magnetized - θ_F vs. H loop shows a typical magnetization in the "hard"-axis direction. $Bi_{2.97}Er_{0.03}Al_{0.5}Ga_{0.5}O_{12}$ films demonstrate perpendicular magnetization with characteristic square hysteresis loops and remnant Faraday rotation.

Luminescence in garnet films

As seen in Figs. 1 and 2, the maximum of FR in iron garnets occurs near the absorption edge around 520 nm. Therefore, for new MO-applications we should trade off FR against absorption to achieve superior MO-figure of merit Q [deg] = $2\ \theta_F/\ \alpha$. As a necessitative condition, any method chosen to compensate the absorption of the signal light should not harm the resonant electronic transition (s) responsible for FR. One of the solutions would be to introduce optical gain simultaneously preserving FR.

Er-doped fiber amplifiers, invented in 1987, dominate in commercial WDM signal transmission systems operating in C(1530–1565 nm) and L(1565–1625 nm) bands. Pumping Er-doped fibers with solid state lasers, the inverted electron population at $^4I_{13/2}$ level (see below Fig. 4) can be achieved. As a result, absorption of telecom C and L optical signal is reduced and optical gain occurs when a pumping overcomes a threshold power. Since Er^{3+} ion easily substitutes any rare earth occupying dodecahedral sites in the garnet structure, both MO-active iron garnet and transparent gallium garnet layers could be grown to host luminescent erbium centers.

Figure 3 shows photoluminescence spectra in Er-doped films pumped by green Ar-laser. Radiative intra-4f Er^{3+} transitions appear to be very different in gallium and iron garnets. The latter films show very strong C-band (1530–1565 nm) PL. In Er:$Y_3Fe_5O_{12}$ and Er:$Bi_3Fe_5O_{12}$ at λ_{PL}=1530 nm luminescence, respectively, is 5.2 and 4.1 times stronger than in Er:$La_3Ga_5O_{12}$. On the contrary, at 514.5 nm pumping, intensive luminescence at λ_{PL}=540 and 980 nm caused by radiative transitions from, respectively, $^4S_{3/2}$ and $^4I_{11/2}$ excited Er^{3+} states to the $^4I_{15/2}$ ground state (see the diagram in Fig. 4) was clearly observed only in Er:$La_3Ga_5O_{12}$ garnet. As seen in the left frame of Fig. 3, PL in Er:$Gd_3Ga_5O_{12}$ in this range is weak and broaden; meanwhile it is completely undetectable for Er:$Bi_3Fe_5O_{12}$ and Er:$Y_3Fe_5O_{12}$. Strong absorption in a broad charge transfer band and weaker one at the resonant $^6A_{1g}\rightarrow{}^4T_{1g}$ transition of octahedrally coordinated Fe^{3+} ion completely extinguishes 540 and 980 nm PL in Er-doped iron garnets.

Two effects are responsible for enhanced C-band PL in iron garnets. The first is a strong absorption that pumping green light experiences in iron garnets below 520 nm. The second is the *sensitizing effect*. As seen in Fig. 4, the $^4I_{11/2}$ energy level of Er^{3+} and the $^4T_{1g}$ level of octahedrally coordinated ferric ion are nearly resonant in energy. Fe^{3+} can be excited either by a pump 514.5 nm Ar-laser in its very broad charge transfer absorption band, or by a solid state 980 nm laser which light is absorbed at the narrow discrete band around $^4T_{1g}$ level. There are sixteen octahedrally coordinated Fe ions per one Er atom in the erbium substituted iron garnet unit cell. In PL process, the net Fe^{3+} absorption cross section at 980 nm is 16 times higher than that of Er^{3+}. Therefore, pump radiation at 980 nm is efficiently absorbed by Fe^{3+} and then is transferred to Er^{3+}.

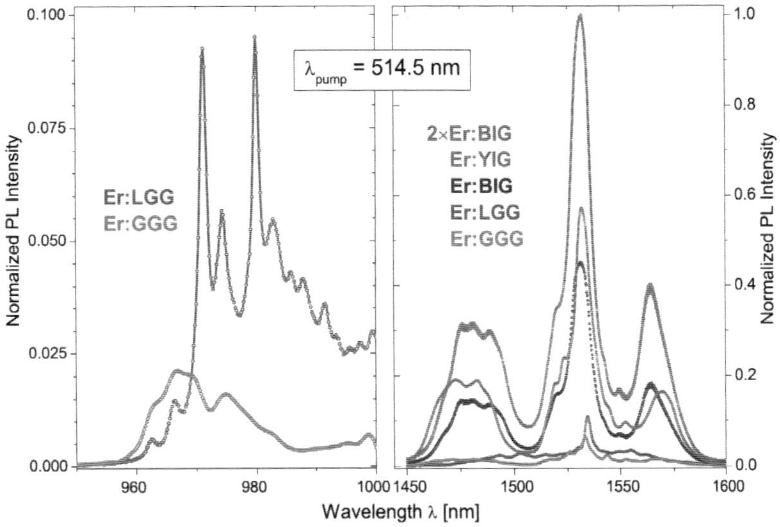

Figure 3 Comparison of PL spectra in Er-doped garnet films pumped by 514.5 nm Ar-laser. Film compositions $Bi_{2.8}Er_{0.2}Fe_5O_{12}$ (2×Er:BIG), $Bi_{2.9}Er_{0.1}Fe_5O_{12}$ (Er:BIG), $Y_{2.9}Er_{0.1}Fe_5O_{12}$ (Er:YIG), $La_{2.9}Er_{0.1}Ga_5O_{12}$ (Er:LGG), and $Gd_{2.9}Er_{0.1}Ga_5O_{12}$ (Er:GGG) are shown as the shorthands. All the spectra have been normalized to the films thickness. There was no noticeable luminescence observed in 2×Er:BIG, Er:BIG, and Er:YIG iron garnet films at 540 and 980 nm.

Table 2. Properties of $Bi_{2.97}Er_{0.03}Al_{0.5}Ga_{0.5}O_{12}/Sm_3Ga_5O_{12}$ photonic crystal and reference films†

Film composition	λ_{res} [nm]	MOPC thickness [nm]	MO-layers thickness [nm]	Resonance bandwidth $\delta\lambda$ [nm]	Resonance FR $\lvert\theta_F\rvert$ [deg/μm]	$\lvert\Theta_F\rvert$ [deg]	Quality factor at λ_{res} $Q = 2\lvert\Theta_F\rvert/\ln(1/T)$ [deg]
$[BIG/SGG]^7/BIG^2/[SGG/BIG]^7$	775	2594	1207	3.8	14.1	17.0	99.3
	at 775 nm		1600		1.2	1.9	57.0
Reference BIG	at 750 nm				1.6	2.6	68.7
	at 640 nm		700		3.3	5.3	42.0
Reference $Bi_3Fe_5O_{12}$	at 750 nm				3.6	2.5	50.0

† $Bi_{2.97}Er_{0.03}Fe_4Al_{0.5}Ga_{0.5}O_{12}$ is notified in the Table 2 with the abbreviation BIG.

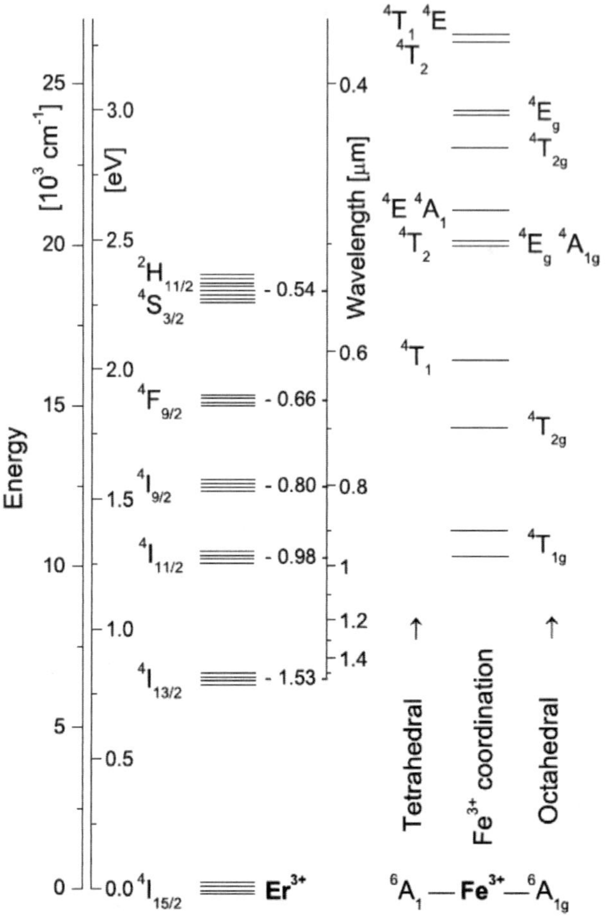

Figure 4 Schematic energy level diagram of Er^{3+} and Fe^{3+} ions. Er^{3+} levels are split due to the Stark effect. Discrete lines show crystal field transition of Fe^{3+} ion lodged on the tetrahedral and octahedral sites.

Data on *C*-band luminescence decay at room temperature are presented in Fig. 5. The excited $^4I_{13/2}$ state lifetime is ranged from 400 μs for Er:$Y_3Fe_5O_{12}$ and 1 ms for Er:$Bi_3Fe_5O_{12}$ to 4 ms for Er:$La_3Ga_5O_{12}$ and almost 6 ms for Er:$Gd_3Ga_5O_{12}$. While pumped by 980 nm laser, the integral *C*-band PL intensity from Er:$La_3Ga_5O_{12}$, Er:$Bi_3Fe_5O_{12}$, and 2×Er:BIG was found to be in the ratio of 0.79:0.49:1 compared to 0.21:0.45:1 at 514 nm pumping (see Fig. 3). It evidences that the 514 nm pumping of Er:$Bi_3Fe_5O_{12}$ films is much more efficient than 980 nm one. We rely this upon incomparable stronger absorption of green light in $Bi_3Fe_5O_{12}$ compared to $La_3Ga_5O_{12}$.

Magneto-optical photonic crystals

Erbium ions can be added both to the garnet layers in Bragg reflectors and/or microcavities. This raises a challenging task to engineer Er-substituted all-garnet MOPCs that combine lasing/amplifying and nonreciprocal optical properties with MO remanence. Luminescent MOPCs promise built-in *intelligence*: ability to simultaneously recognize, process and store optical data, make color filtering, and amplify optical signals.

Figure 6 exemplifies dispersion properties of photonic crystal with 7 $Bi_{2.97}Er_{0.03}Fe_4Al_{0.5}Ga_{0.5}O_{12}$/ $Sm_3Ga_5O_{12}$ reflectors in two Bragg mirrors and half-wavelength $Bi_{2.97}Er_{0.03}Fe_4Al_{0.5}Ga_{0.5}O_{12}$ microcavity in-between designed for 775 nm. MOPC spectra have a stop band structure – the band gap with the transmittance central peak at the resonance wavelength λ_{res} = 775 nm. Table 2 collects the properties of fabricated MOPC: resonance net Θ_F and specific θ_F Faraday rotation (the latter is normalized to the total thickness of $Bi_{2.97}Er_{0.03}Fe_4Al_{0.5}Ga_{0.5}O_{12}$ layers notified in the Table 2 with the abbreviation BIG), resonance bandwidth $\delta\lambda$ as well as MO-quality factor defined at λ_{res} as Q [deg] = $2|\Theta_F|/\ln(1/T)$. Compared to the reference $Bi_{2.97}Er_{0.03}Fe_4Al_{0.5}Ga_{0.5}O_{12}$ film, MOPC demonstrates enhanced MO properties: specific FR θ_F and quality factor Q were increased, correspondingly, by the factor of 12 and 2. As a band pass filter, 775 nm MOPC possesses narrow bandwidth $\delta\lambda$ and a strong light rejection within the stop band characterized by peak-to-valley transmittance ratio as high as 23 dB.

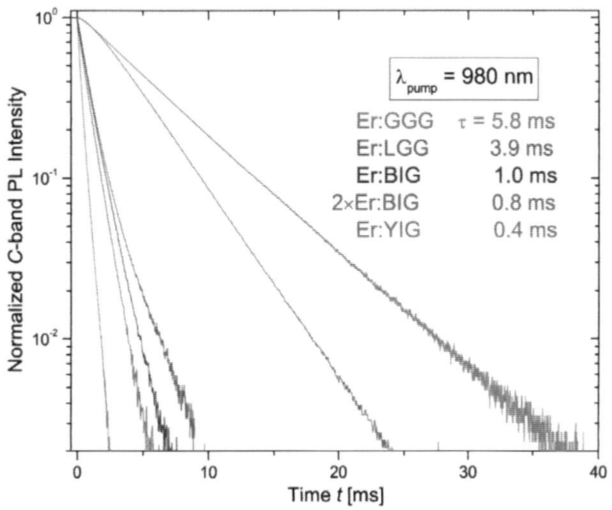

Figure 5 Normalized *C*-band (1530-1565 nm) photoluminescence in the time domain at 980 nm laser pumping.

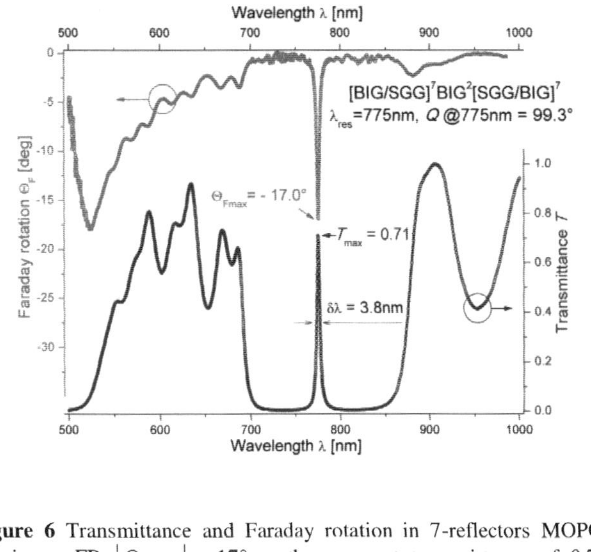

Figure 6 Transmittance and Faraday rotation in 7-reflectors MOPC. Maximum FR $|\Theta_{Fmax}|$ = 17° at the resonant transmittance of 0.71 corresponds to the MO-quality factor $Q = 2|\Theta_F|/\ln(1/T)$ = 99.3° at λ_{res} = 775 nm.

Single layer $Bi_{2.97}Er_{0.03}Fe_4Al_{0.5}Ga_{0.5}O_{12}$ film and $Bi_{2.97}Er_{0.03}Fe_4Al_{0.5}Ga_{0.5}O_{12}/Sm_3Ga_5O_{12}$ MOPC have out-of-plane magnetization. As it is seen in inset to Fig. 2, squareness of magnetization loop increased for MOPC and remnant (latched) Faraday rotation appeared as large as 95% of the saturated value. Strengthening of perpendicular anisotropy in MOPC compare to a reference $Bi_{2.97}Er_{0.03}Fe_4Al_{0.5}Ga_{0.5}O_{12}$ film, we rely upon the in-plane compressive strain and corresponding out-of-plane rhombohedral distortions of $Bi_{2.97}Er_{0.03}Fe_4Al_{0.5}Ga_{0.5}O_{12}$ unit cell. In a single layer $Bi_{2.97}Er_{0.03}Fe_4Al_{0.5}Ga_{0.5}O_{12}$ film, a strain induced by a large film-to-substrate lattice mismatch releases through a nucleation of copious misfit dislocations leading to cracks in thicker films. In multilayered garnet films, finite thickness of layers hinders a nucleation of misfit dislocations. As a result, mismatch strain is accumulated while a magnetostriction converts it to the uniaxial magnetic anisotropy.

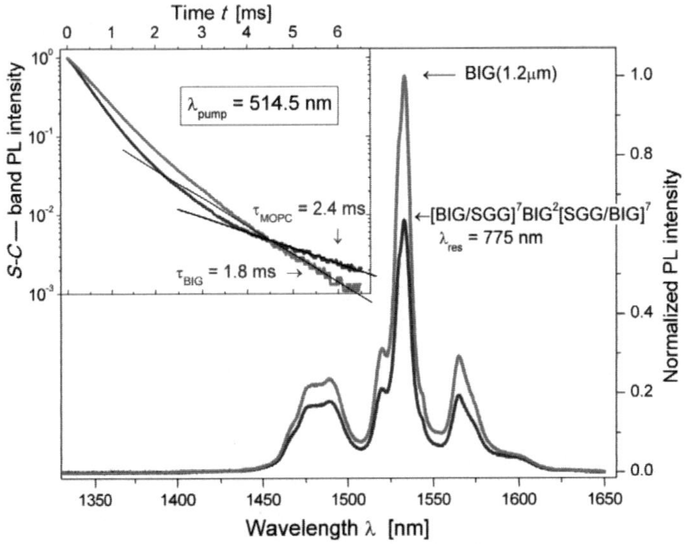

Figure 7 Photoluminescence spectra in Er-doped films pumped by 514.5 nm Ar–laser. Reference 1.2 μm thick $Bi_{2.97}Er_{0.03}Fe_4Ga_{0.5}Al_{0.5}O_{12}$ film is compared to 7-reflectors MOPC designed for λ_{res} = 775 nm. Inset shows a decay of C – band photoluminescence in MOPC and a single layer BIG film with the same as in MOPC equivalent thickness of 1.2 μm.

Normalized C –band PL spectra in 7-reflectors MOPC and single layer $Bi_{2.97}Er_{0.03}Fe_4Al_{0.5}Ga_{0.5}O_{12}$ film of the same effective thickness are compared in Fig. 7. Both samples have sharp emission peaks related to the main $^4I_{13/2} \rightarrow {}^4I_{15/2}$ laser transition. Fine structure of the spectrum reflects Stark splitting of $^4I_{13/2}$ and $^4I_{15/2}$ manifolds of dodecahedral coordinated Er^{3+} ions. Peak PL intensity in MOPC decreased compare to a single layer $Bi_{2.97}Er_{0.03}Fe_4Al_{0.5}Ga_{0.5}O_{12}$ film by a factor of 1.6. This is related to a larger total thickness of the MOPC (2.6 against 1.2 μm).

C–band PL decay curves in 7-reflectors MOPC and reference $Bi_{2.97}Er_{0.03}Fe_4Al_{0.5}Ga_{0.5}O_{12}$ film are compared in inset to Fig. 7. Both experimental time dependencies are fitted to the exponential decay with two lifetimes. The main PL component in a single layer $Bi_{2.97}Er_{0.03}Fe_4Al_{0.5}Ga_{0.5}O_{12}$ and MOPC has a magnitude of 96% and 97% and is quenched at 700 μs and 600 μs, respectively. Long-life component survives within 1.8 ms in Er:$Bi_{2.97}Er_{0.03}Fe_4Al_{0.5}Ga_{0.5}O_{12}$ film and 2.4 ms in MOPC. Longer lifetime in MOPC we rely upon the Er^{3+} excitations in distorted dodecahedral complexes localized at $Bi_{2.97}Er_{0.03}Fe_4Al_{0.5}Ga_{0.5}O_{12}/Sm_3Ga_5O_{12}$ interfaces.

Conclusions

Long-lived near-infrared PL of Er^{3+} ions at 1530 nm related to the main $^4I_{13/2} \rightarrow ^4I_{15/2}$ laser transition occurs in Er-doped rare earths/bismuth iron and gallium garnet epitaxial films. Intensive PL at room temperature suggests that a number of different pumping and lasing schemes can be realized through the proper design of Er-doped $Bi_3Fe_5O_{12}/RE_3Ga_5O_{12}$ magneto-optical photonic crystals. Optical gain combined with high FR can be achieved localizing light either in optically dense Er:$Bi_3Fe_5O_{12}$ cavity or in Er:$Bi_3Fe_5O_{12}$ mirrors in crystals with transparent Er:$RE_3Ga_5O_{12}$ cavities.

Al and Ga substitution of ferric ions in $Bi_3Fe_5O_{12}$ garnet reduces optical absorption by a factor of 3 and induces perpendicular magnetic anisotropy. Photonic crystals earn a record high magneto-optical quality and remnant Faraday rotation. At $\lambda_{res} = 775$ nm, specific FR $\theta_F = -14.1$ deg/μm and MO-quality factor $Q = 99.3$ deg exhibit the highest to date MOPC performance. Owning gain [11], luminescent all-garnet heteroepitaxial photonic crystals promise great potential for MO memory, light guiding, filtering and switching, exceptional dispersion, nonreciprocal properties as well as integration with semiconductor platforms.

Acknowledgment

Authors wishing to acknowledge financial support from the Swedish Research Council (Vetenskapsrådet) through the Advanced Optics and Photonics (ADOPT) Linné center grant.

References

[1] Okuda T, Koshizuka N, Hayashi K, Satoh K, Taniguchi H, and Yamamoto H 1988 *IEEE Translation Journal on Magnetics in Japan* **3** 483
[2] Adachi N, Denysenkov V P, Khartsev S I, Grishin A M, and Okuda T 2000 *J. Appl. Phys.* **88** 2734
[3] Rosenberg R, Rubinstein C B, and Herriot D R 1964 *Appl. Opt.* **3** 1079
[4] Inoue M, Arai K, Fujii T, Abe M 1998 *J. Appl. Phys.* **83** 6768
[5] Kahl S, Grishin A M 2004 *Appl. Phys. Lett.* **84** 1438
[6] Khartsev S I and Grishin A M 2005 *Appl. Phys. Lett.* **87** 122504
[7] Khartsev S I and Grishin A M 2007 *J. Appl. Phys.* **101** 053906; Grishin A M, Khartsev S I, and Kawasaki H *Appl. Phys. Lett.* **90** 191113
[8] Khartsev S I and Grishin A M 2011 *Optics Letters* **36** 2806
[9] Scott G B, Lacklison D E, Page J L, and Hewett J 1976 *Appl. Phys.* **9** 71; Scott G B and Lacklison D E 1976 *IEEE Trans. Magn.* **MAG-12** 292
[10] Dzibrou D O and Grishin A M 2009 *J. Appl. Phys.* **106** 043901
[11] Grishin A M 2011 *Appl. Phys. Lett.* **97** 061116

Synthesis of Gd_2O_3 nanoparticles for MRI contrast agents

N Sakai[1], L Zhu[1], A Kurokawa[1], H Takeuchi[1], S Yano[1], T Yanoh[1], N Wada[1], S. Taira[2], Y Hosokai[3], A Usui[3], Y Machida[3], H Saito[3] and Y Ichiyanagi[1]*

[1]Graduate School of Engineering, Yokohama National University, 79-5 Tokiwadai, Hodogaya, Yokohama, Kanagawa 240-8501, Japan +81-(0)45-339-4185

[2]Japan Advanced Institute of Science and Technology School of Material Science, 1-1 Asahidai, Nomi, Ishikawa 923-1292, Japan

[3]Graduate School of Medicine, Tohoku University, 2-1 Seiryo-machi, Aoba, Sendai, Miyagi 980-8575, Japan

*E-mail: yuko@ynu.ac.jp

Abstract. Gd_2O_3 nanoparticles were synthesized by using our original method for application as MRI contrast agents. The sample diameters were controlled in the range 18–66 nm by adjusting the annealing temperature between 773 and 1273 K in air or in an Ar atmosphere. Magnetization measurements were performed at 300 K, and the effective magnetic moment of each sample was calculated. They showed paramagnetism at 300 K and they had large effective magnetic moment μ_{eff} of 7.15–8.05 μ_B. MRI measurements were performed in 0.8 wt% agarose solution, and the Gd_2O_3 nanoparticles were found to work as effective T_1-shortening MRI contrast agents.

1. Introduction

Gadolinium in the form of a chelate complex is used as a T_1-shortening MRI (Magnetic Resonance Imaging) contrast agent [1]. MRI is a common medical imaging technique that allows visualization through measurement of the magnetic relaxation of protons in the human body. MRI contrast agents accelerate this magnetic relaxation through the magnetic interactions between the protons and agents, and define the contrast of the MR images obtained. Gd^{3+} ions have a large calculated magnetic moment of 7.94 μ_B, which gives a large magnetic interaction and strongly accelerates the magnetic relaxation of the proton. Efficient MRI contrast agents can alleviate the burden on patients through the shortening of the diagnosis time and the reduction in administering agents.

Metal oxide nanoparticles and their hybrid materials with organic compounds are anticipated to find use in the field of multifunctionalized nanomedicine [2], with possible applications in drug-delivery systems (DDSs) [3], magnetic hyperthermia [4], and as medical markers. Through the employment of these techniques, gadolinium oxide nanoparticles could be applied as new types of multifunctional MRI contrast agents [5].

Previously, one of the authors produced magnetic nanoparticles through a unique method and reported their magnetic properties [6][7]. In this study, Gd_2O_3 materials were produced in order to apply to MRI contrast agents. Their diameters were controlled by annealing temperature, and magnetization measurements were performed for each sample. Same sample were selected for MRI measurements.

2. Experimental

Gd_2O_3 nanoparticles were prepared by mixing of aqueous solution of $GdCl_3 \cdot 6H_2O$ and NaOH [2][3]. Reagents were used to set the Na:Cl mole ratio to 1:1, and were dissolved individually in distilled water. These aqueous solutions were mixed to obtain precipitates. The precipitates were washed to remove the Na^+ and Cl^- ions, and dried at about 350 K in a thermostat-controlled oven. Each precursor was annealed in an electric furnace. The annealing treatment was carried out in air or in an Ar atmosphere at temperatures of 773–1273 K for 10 hours.

X-ray powder diffraction measurements (CuKα, λ = 0.154 nm) were performed for each sample. The average sample diameters were analyzed by fitting the theoretical 222 diffraction peak to the measured diffraction peak of each sample. Theoretical diffraction patterns were calculated using the Stokes & Wilson method [8] with the software CSDA (Ver. 1.3, Rigaku Corporation, Japan). The morphology and diameter distribution of the nanoparticles were investigated using a transmission electron microscope (TEM; H-7100; Hitachi, Japan). Magnetization measurements were determined at 300 K under a magnetic field of ±50 kOe using a SQUID magnetometer.

MRI signal measurements were performed using a 0.3-T MRI system (Hitachi Medical Corporation, Japan). The particles annealed at 1173 K in air were dispersed in 0.8 wt% agarose solution, and the concentration of Gd^{3+} ions was set to 1.0 mM. The MRI signal of this sample was measured by the T_1 spin echo sequence (TE = 12 ms) and T_2 spin echo sequence (TR = 2000 ms). For comparison, the same measurements were performed for 0.8 wt% agarose solutions without Gd_2O_3 nanoparticles, and T_1 measurements were performed for solution and a 0.5 mM aqueous solution of gadolinium-diethylenetriaminepentaacetic acid (Gd-DTPA), which is commonly used as a T_1 contrast agent. MR images of Gd_2O_3 solution were took with setting echo time of TE = 12 ms and repetition time of TR = 50, 150, 300, 500, 700, 4000, 6000, 8000, 10000 ms.

3. Results and Discussion

3.1. X-ray diffraction

The X-ray diffraction patterns of each sample are shown in figure 1. It was found that all the samples were single-phase Gd_2O_3. The average diameters of the Gd_2O_3 nanoparticles annealed in air were estimated to be 26, 30, 36, 46, 59, and 66 nm for the samples annealed at 773, 873, 973, 1073, 1173, and 1273 K, respectively. For the samples annealed in an Ar atmosphere, the average diameters were estimated to be 18 and 22 nm for annealing at 823 and 873 K, respectively. These particles were rather smaller than those of the samples annealed in air. The representative TEM data is shown in Fig. 2. Figure 3 shows the relationship between the particle diameter and the annealing temperature. From these results, it was found that the average diameter of Gd_2O_3 nanoparticles could be controlled by adjusting the temperature and atmosphere of the annealing process. The diameter of the Gd_2O_3 nanoparticles (823 K in Ar) was deduced to be 23 ±1.0 nm (statistics: 288 measurements; ±1.0 nm means the standard error of the mean (S.E.M.)) which was good agreement with the result of XRD.

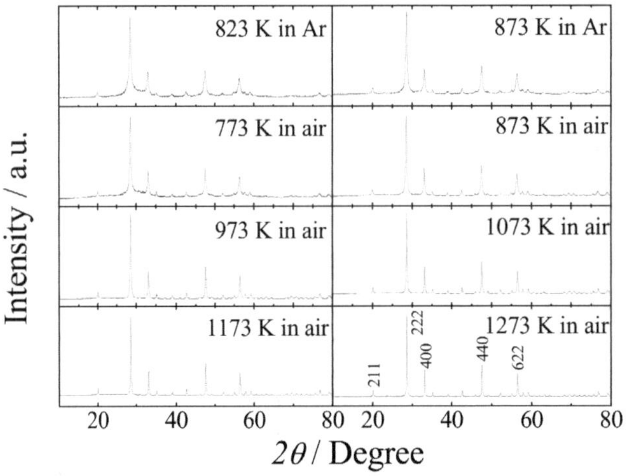

Figure 1. X-ray powder diffraction patterns of the samples annealed at 823 and 873 K in an Ar atmosphere and at 773, 873, 973, 1073, 1173, and 1273 K in air.

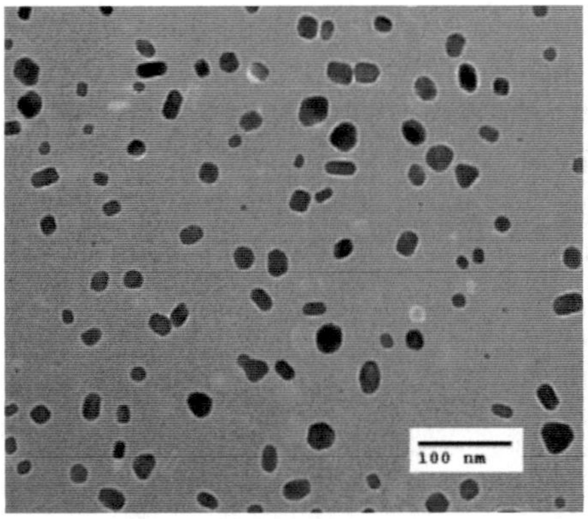

Figure 2. TEM image of Gd_2O_3 nanoparticles. Scale bar is 100 nm.

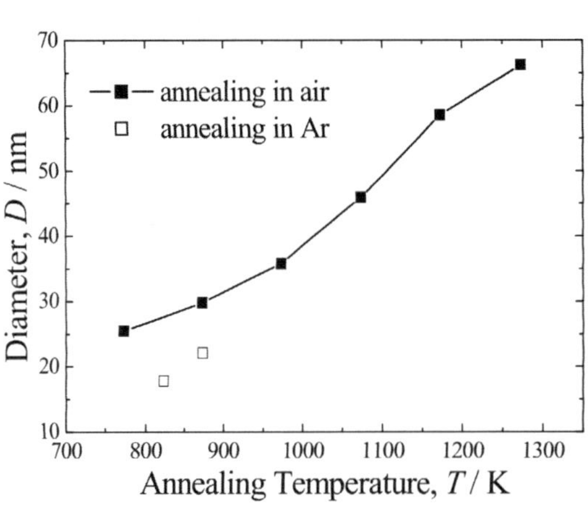

Figure 3. Average diameters estimated from the 222 X-ray diffraction peak of the samples treated under different annealing conditions.

3.2. Magnetization measurements

The results of the magnetization measurements are shown in figure 4. It was found that the Gd_2O_3 nanoparticles of all sizes were paramagnetic at 300 K. The effective magnetic moments μ_{eff} of the Gd^{3+} ions calculated from the magnetization values at 300 K are summarized in table 1. These μ_{eff} values were in good agreement, within experimental error, with the theoretical value of 7.94 μ_B. Temperature dependence of magnetization followed Curie-Weiss law. Paramagnetic Curie temperature of θ_p was determined to be -10–5 K. These results indicated that large magnetic moments of Gd^{3+} ions fluctuated because of thermal energy at 300 K.

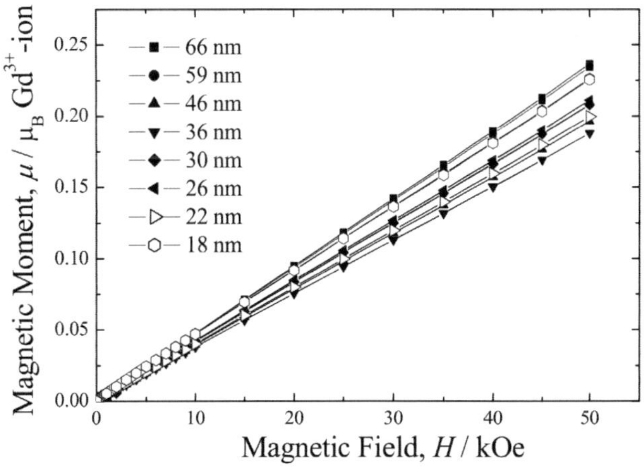

Figure 4. Magnetization of each sample at 300 K, for particle diameters of 18, 22, 26, 30, 36, 46, 59, and 66 nm.

Table 1. Effective magnetic moment (μ_{eff}) of each sample calculated from magnetic measurement at 300 K.

Diameter (nm)	18	22	26	30	36	46	59	66
μ_{eff} (μ_B)	8.05	7.36	7.54	7.31	7.15	7.28	7.49	7.95

3.3. MRI measurements

The T_1 relaxation curves of Gd_2O_3 with diameter of 59 nm and non-Gd_2O_3 solutions are shown in figure 5(a). The T_1 relaxation time was estimated to be 0.196 s in Gd_2O_3 solution and 2.16 s in non-Gd_2O_3 solution. The T_1 relaxation rate (R_1) was calculated to be 5.1 s^{-1} mM^{-1} in Gd_2O_3 solution and 0.46 s^{-1} mM^{-1} in non-Gd_2O_3 solution. The R_1 value of the Gd_2O_3 solution was comparable to the value of 6.1 s^{-1} mM^{-1} for Gd-DTPA in aqueous solution. The phantom images of Gd_2O_3 solution were shown in Fig. 6. The obtained images showed very high contrast in even short repetition time of *TR*, and they became brighter as repetition time of *TR* increased. It was found that the Gd_2O_3 nanoparticles have a strong T_1-shortening capability as MRI contrast agents.

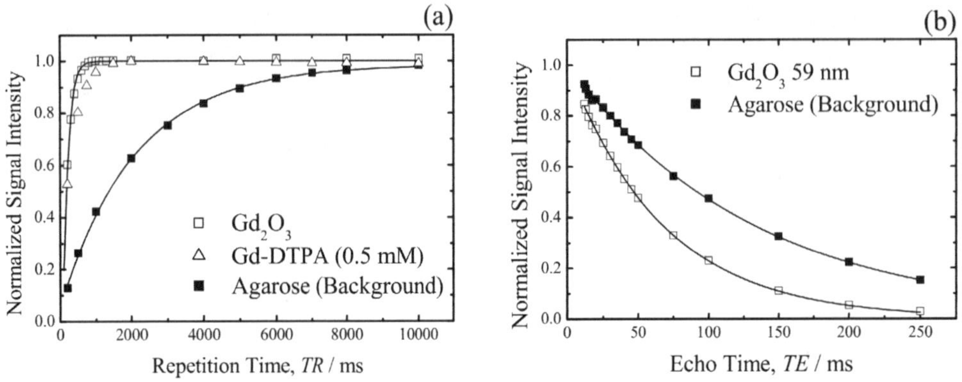

Figure 5. T_1 relaxation curve (a) and T_2 relaxation curve (b) of 0.8 wt% agarose solution with Gd_2O_3 (59 nm) and agarose solution without particles.

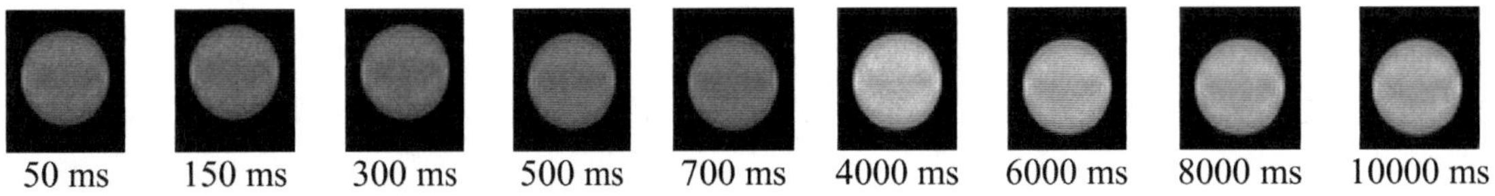

Figure 6. Phantom images of Gd_2O_3 solution with setting echo time of *TE* = 12 ms and repetition time of *TR* = 50, 150, 300, 500, 700, 4000, 6000, 8000, 10000 ms.

The T_2 relaxation curves of Gd_2O_3 and non-Gd_2O_3 solutions are shown in figure 5(b). The T_2 relaxation time was estimated to be 6.67×10^{-2} s in Gd_2O_3 solution and 13.3×10^{-2} s in non-Gd_2O_3 solution. The T_2 relaxation rate (R_2) was calculated to be 15 s^{-1} mM^{-1} in Gd_2O_3 solution and 7.5 s^{-1} mM^{-1} in non-Gd_2O_3 solution. This value of R_2 is enough to emphasize the contrast in T_2-weighted MR images.

Remarkably, these results show that Gd_2O_3 nanoparticles have effects both as T_1- and T_2-shortening MRI contrast agents, especially with their large T_1 relaxation rate. In the case of the Gd^{3+} chelate complex, T_1 relaxation is accelerated by the dynamic local magnetic field, which is derived from the magnetic coupling of the proton and the Gd^{3+} ions [1][9]. It is considered that the Gd^{3+} ions of these nanoparticles exhibit the same relaxation process as that observed in Gd^{3+} chelate complexes. This relaxation process is due to the paramagnetic behavior and large magnetic moment of this Gd_2O_3 nanoparticle system [10].

4. Conclusion

Gd_2O_3 nanoparticles were synthesized by the wet chemical method, and their diameters were controlled between 18 and 66 nm by adjusting the conditions of the annealing process. All the samples showed paramagnetic behavior at 300 K, and their effective magnetic moments were calculated to be in the range 7.15–8.05 μ_B. From the MRI measurements, Gd_2O_3 particles with a diameter of 59 nm were found to have relaxation rate of R_1 = 5.1 s^{-1} mM^{-1} and R_2 = 15 s^{-1} mM^{-1}. From these results, it is concluded that Gd_2O_3 nanoparticles could be used as both T_1- and T_2-shortening MRI contrast agents, and that they have an especially strong ability in the shortening of T_1.

Acknowledgements
This study was partially supported by the Precursory Research for Embryonic Science and Technology of the Japan Science and Technology Agency and the Takahashi Industry Foundation for Economic Research 2010.

References
[1] Westmeyer G G and Jasanoff A 2007 *Mag. Res. Imag.* **25** 1004–10
[2] Ichiyanagi Y, Moritake S, Taira S and Setou M 2007 *J. Magn. Magn. Mater.* **310** 2877–9
[3] Moritake S, Taira S, Ichiyanagi Y, Mrone N, Song S-Y, Hatanaka T, Yuasa S and Setou M 2007 *J. Nanosci. Nanotechnol.* **7** 1–8
[4] Shigeoka D, Katayanagi H, Moro Y, Kimura S, Mashino T and Ichiyanagi Y 2010 *J. Phys.* **200** 122002
[5] Klasson A, Ahren M, Hellqvist E, Soderlind F, Rosen A, Kall P-O, Uvdal K and Engstrom M 2008 *Contrast Media Mol. Imaging* **3** 106–11
[6] Ichiyanagi Y, Uehashi T, Yamada S 2004 *Phys. Status Solidi (c)* **12** 3485-88
[7] Ichiyanagi Y, Yamada S 2005 *Polyhedron* **24** 2813-16
[8] Ida T, Shimazaki S, Hibino H and Toraya H 2004 *J. Appl. Cryst.* **36** 1107–15
[9] Kurki T and Komu M 1995 *Mag. Res. Imag.* **13** 379–85
[10] Lu J, Ma S, Sun J, Xia C, Liu C, Wang Z, Zhao X, Gao F, Gong Q, Song B, Shuai X, Ai H and Gu Z 2009 *Biomaterials* **30** 2919–28

High Temperature Hall sensors using AlGaN/GaN HEMT Structures

S Koide [1*], H Takahashi [1], A Abderrahmane [1], I Shibasaki [#], A Sandhu [1,2,3]

[1]Dept. of Electrical and Electronic Information Engineering, [2]Electronics-Inspired Interdisciplinary Research Institute, Toyohashi University of Technology, 1-1 Hibarigaoka, Tempaku-cho, Toyohashi, Aichi 441-8580, Japan

[3]Dept. of Electrical and Electronic Engineering, Tokyo Institute of Technology, 2-12-1 Ookayama, Meguro-ku, Tokyo 152-8550, Japan

[#]Specially appointed professor at Toyohashi University of Technology

[*]E-mail: koide-s@eiiris.tut.ac.jp

Abstract. Hall effect sensors are the most widely used magnetic sensors, and are commonly fabricated using narrow band-gap semiconductors such as InSb and GaAs. However, the operation of InSb and GaAs-based Hall sensors is unstable at elevated temperatures. Here, we exploited the fact that gallium nitride- (GaN) based wide band-gap (3.4eV) semiconductors have high breakdown electric field strength and are robust at elevated temperatures and tested the high temperature electrical characteristics of AlGaN/GaN heterostructure micro-Hall effect sensors with a two-dimensional electron gas (2DEG). The AlGaN/GaN heterostructures had an electron mobility of 1360 cm^2/Vs and a 2DEG density of 1.0 x 10^{13} cm^{-2}. The supply-current-related sensitivity (SCRS) was $77VA^{-1}T^{-1}$ at room temperature. Notably, the temperature coefficient of the Hall voltage was +0.05 %/°C near room temperature, and SCRS was 67 $VA^{-1}T^{-1}$ at 400°C. The results for AlGaN/GaN were compared with those for AlGaAs/GaAs and AlInSb/InAsSb/AlInSb QW Hall effect sensors.

1. Introduction

Hall effect sensors are ubiquitous, being used in the electronics industry for monitoring rotation in optical memory disks, banknote authentication in vending machines, and for sensing currents in cables [1]. Other notable applications of micro Hall sensors include scanning Hall probe microscopy (SHPM) [2-3] and medical diagnostics based on the detection of magnetic tags attached to target molecules [4-5]. Now, while Hall sensors are widely used at cryogenic and ambient temperatures, there are still niche demands for Hall sensors that are stable at elevated temperatures of about 400°C for applications such as the determination of the Curie temperatures of ferromagnetic materials and electronics systems for space exploration vehicles. More recent demands include the sensing of currents in electric cars, where the ambient is well above room temperature.

Conventional Hall sensors fabricated using Si, InAs or InSb are stable up to temperatures of about 125°C, and above this temperature the sensors are inoperable due to the onset of intrinsic conduction and physical degradation of the semiconducting materials. Silicon carbide could potentially be used for fabricating Hall sensors operating at above 125°C because of its large band gap, high electron saturation velocity, and excellent thermal stability at high temperatures. However, the need for precise control of dopants and the large thickness of the conducting layers of SiC severely limits its sensitivity and stability at high temperatures.

Here, in order to resolve these issues, we describe the fabrication and magnetoresistive characteristics of AlGaN/GaN heterostructure micro-Hall sensors with a two-dimensional electron gas (2DEG). The AlGaN/GaN heterostructures offer the advantages of wide-band-gap materials that are stable to a least 400°C, as well as an extremely thin 2DEG conduction layer for ultrahigh magnetic sensitivity. We compared the temperature dependence of the magnetoresistance of AlGaN/GaN Hall sensors with AlGaAs/GaAs heterostructure and AlInSb/InAsSb/AlInSb quantum well structure-based Hall devices.

2. Experimental

2.1. Fabrication

The AlGaN/GaN heterostructures were grown by metal organic chemical vapor deposition (MOCVD) on sapphire substrates. The epitaxial layer consisted of a 2 μm undoped GaN buffer layer and a 25 nm $Al_{0.25}Ga_{0.75}N$ barrier layer doped with Si to 4 x 10^{18} cm^{-3}. The inset in Figure 1 shows a schematic diagram of a typical structure. The 2DEG layer is located 25 nm below the surface, at the AlGaN to GaN interface. The AlGaN/GaN micro-Hall sensors were fabricated using a photolithography process consisting of three main steps: (a) formation of the mesa and the active "cross" patterns by reactive ion etching (RIE); (b) thermal and electron beam evaporation of Ti/Al/Ni/Au Ohmic contacts; and (c) rapid thermal annealing (RTA) in a nitrogen atmosphere at 850°C for 90 s. The mesa depth etched by RIE was 300nm.

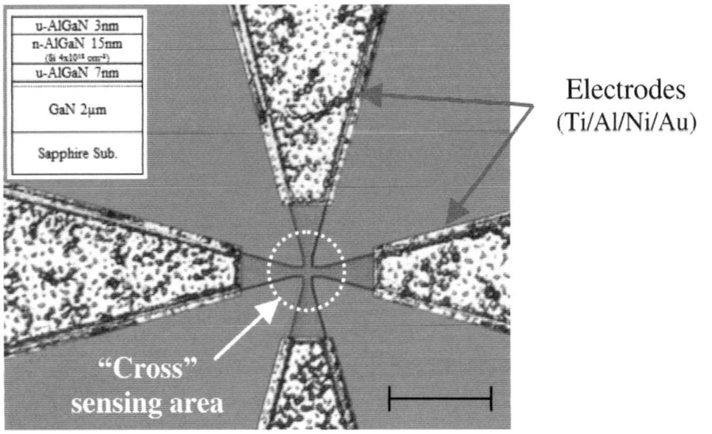

Figure 1. Microscopy image of 5 x 5 μm^2 AlGaN/GaN Hall effect sensor. Inset: AlGaN/GaN Heterostructure grown using MOCVD.

2.2. Measurement evaluation

The electrical characteristics (Hall voltage, sheet carrier concentration, electron mobility and sensitivity) of micro-Hall sensors with active "cross" regions ranging from 1μm x 1μm to 5μm x 5μm were measured from 25°C to 400°C. The electron mobility and sheet carrier concentrations were determined by the van der Pauw technique. The temperature of each of the Hall sensors was controlled by a ceramic heater with a thermocouple installed directly under the Hall sensors. The absolute sensitivity of a Hall effect sensor is defined as its output Hall voltage (V_H) divided by the normal component of the magnetic induction ($B_\perp$): [2][6]

$$S_A = \left|\frac{V_H}{B_\perp}\right| \left(VT^{-1}\right) \tag{1}$$

The ratio of the absolute sensitivity of a transducer to an applied device bias yields a 'relative sensitivity' value. The supply-current-related sensitivity (SCRS) of a cross-shaped Hall sensor can be expressed as follows: [2][6]

$$S_I = \frac{S_A}{I_D} = \left|\frac{1}{I_D}\frac{V_H}{B_\perp}\right| \approx \frac{1}{en_S}\left(VA^{-1}T^{-1}\right) \tag{2}$$

where I_D is the driving current, and n_S is the sheet carrier concentration. From Eq. (2) it can be seen that S_I depends on the sheet carrier concentration.

3. Results and Discussion

Figure 2 (a) shows the magnetic field dependence of the Hall voltage (V_H) for a 5μm x 5μm Hall effect sensor measured at room temperature for several drive currents. The values of V_H show a clear linear dependence over several drive current values and applied magnetic fields.

Table 1 shows the electrical characteristics of the micro-Hall effect sensors at room temperature as determined by the van der Pauw technique. The AlGaN/GaN heterostructures had a 2DEG mobility of 1360 cm^2/Vs and an electron density of 1.0 x 10^{13} cm^{-2}. The Hall coefficient (R_H) was a proportionality factor between the Hall voltage (V_H), the magnetic field (B), and the driving current (I_D). R_H was calculated from the slope of the fitting line.

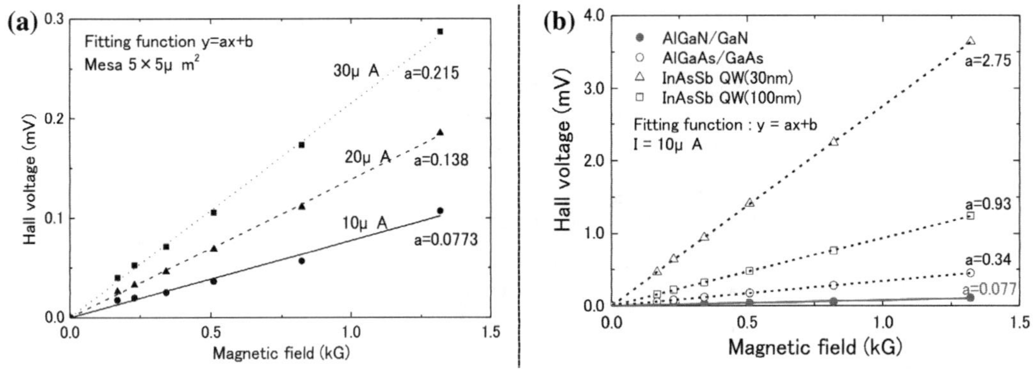

Figure 2. Hall voltage of an AlGaN/GaN Hall effect sensor for several driving currents at room temperature (a) and a comparison of Hall voltage between AlGaN/GaN, AlGaAs/GaAs and InAsSb QW Hall effect sensors.

Table 1. Electrical characteristics of the micro-Hall sensors at room temperature.

Sample name	Mesa size W x L [μm]	Hall coeff. R_H [Ω/G]	Carrier conc. Ns [cm^{-2}]	Sheet resist. Rs [Ω/sq.]	Mobility μ_e [cm^2/Vs]
AlGaN/GaN	5 x 5	0.0077	1.04 x 10^{13}	570	1,360
AlGaAs/GaAs	5 x 5	0.034	1.83 x 10^{12}	660	5,080
InAsSb QW(30 nm)	50 x 50	0.275	2.45 x 10^{11}	970	26,100
InAsSb QW(100 nm)	50 x 50	0.093	7.18 x 10^{11}	220	39,100

Figure 3 shows a comparison of the sensitivity of AlGaN/GaN, AlGaAs/GaAs and InAsSb QW micro-Hall sensors with respect to temperature. The AlGaN/GaN micro Hall sensors operate up to 400°C with an SCRS of 67 $VA^{-1}T^{-1}$, whereas sensors fabricated using the narrower band-gap sensors degrade from ~150°C. The AlGaAs/GaAs Hall sensors showed a relatively high SCRS and were stable up to ~200°C, after which the devices failed. The sensitivity of AlInSb/InAsSb/ AlInSb QW Hall sensors at room temperature was high, at 2540 $VA^{-1}T^{-1}$ and the temperature coefficient was -1.38 %/°C. We will describe the physical interpretation of these results and give examples of potential high temperature applications of 2DEG-AlGaN/GaN Hall sensors.

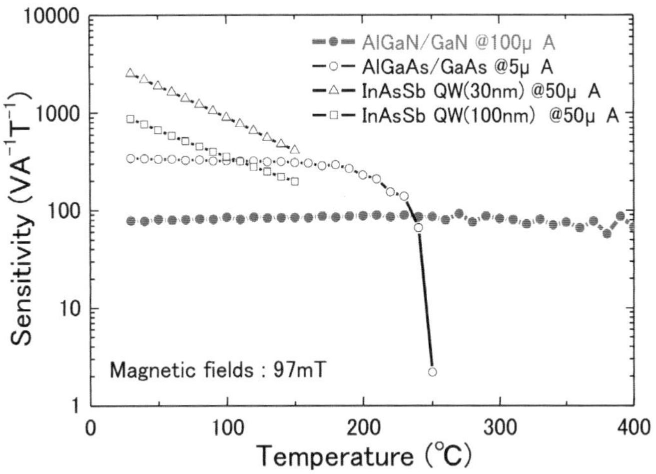

Figure 3. Temperature dependence of current-related magnetic sensitivity. The applied magnetic field was 97 mT

4. Summary

We measured the electrical and magnetic characteristics of AlGaN/GaN micro-Hall effect sensors at temperatures up to 400°C for the development of magnetic field sensors for high temperature applications. The AlGaN/GaN 2DEG Hall sensors were stable between room temperature and 400°C with a current-related magnetic sensitivity of 77 $VA^{-1}T^{-1}$. The AlGaN/GaN devices show promise as components for magnetic measurements systems for high temperature applications. Notably, we could exploit the high-temperature AlGaN/GaN micro-Hall sensors for applications including scanning Hall probe microscopy (SHPM) of ferromagnetic domains at elevated temperatures.

References

[1] Ripka P 2001, *Magnetic Sensors and Magnetometers*, Artech House, New York
[2] Bando M, Dede M, Akram R, Oral A, Shibasaki I, Handa H and Sandhu A 2009, *J. Appl. Phys.* **105**, 07E909
[3] Primadani Z, Osawa H and Sandhu A 2007, *J. Appl. Phys.* **101**, 09K105
[4] Kumagai Y, Abe M, Sakamoto A, Handa H and Sandhu A 2008, *J. Appl. Phys.* **103**, 07A309
[5] Lee W, Joo S, Kim S U, Rhie K, Hong J, Shin K, and Kim K H 2009, *Appl. Phys. Lett.* **94**, 153903
[6] Lu H, Sandvik P, Vertiatchikh A, Tucker J and Elasser A 2006, *J. Appl. Phys.* **99**, 114510

Effects of Proton Irradiation on the Magnetoelectric Properties of 2DEG AlGaN/GaN Micro-Hall Sensors

H Okada[1,2], A Abderrahmane[2] , S Koide[2], H Takahashi, S Sato[3], T Ohshima[3], and A Sandhu[1,2,4]

[1] Electronics-Inspired Interdisciplinary Research Institute (EIIRIS), Toyohashi University of Technology, 1-1 Hibarigaoka, Tempaku-cho, Toyohashi, Aichi 441-8580, Japan
[2] Dept. of Electrical and Electronic Information Engineering, Toyohashi University of Technology, 1-1 Hibarigaoka, Tempaku-cho, Toyohashi, Aichi 441-8580, Japan
[3] Quantum Beam Science Directorate, Japan Atomic Energy Agency (JAEA), 1233 Watanuki-cho, Takasaki, Gunma 370-1292, Japan
[4] Tokyo Institute of Technology, 2-12-1 Ookayama, Meguro-ku, 152-8550, Japan
E-mail: okada@ee.tut.ac.jp

Abstract. The effects of irradiating AlGaN/GaN micro-Hall sensors with 380 keV protons on were investigated by magnetoresistance measurements. The sheet resistance increased after irradiation with a proton dose of 1×10^{13} cm^{-2} due to the decrease of carrier mobility rather than the decrease of sheet carrier density. Our experiments showed that AlGaN/GaN two-dimensional electron gas (2DEG) structures are good strong candidates for Hall sensors operable in harsh radiation environments.

1. Introduction

Hall effect magnetic field sensors fabricated using compound semiconductors such as InSb and GaAs are widely used in the electronics industry for monitoring rotation in equipment such as optical memory disks and for banknote authentication in vending machines. More recently, the availability of gallium nitride-based heterostructures has enabled the fabrication of high performance AlGaN/GaN-based micro-Hall sensors operable at , with reports of applications including medical diagnostics [1]. Recent industrial trends indicate increasing demand for Hall sensors for monitoring magnetic fields in outer space and nuclear power stations, which is extremely challenging because sensors for such applications must withstand extreme fluctuations in temperature as well as harmful radiation in these environments.

The irradiation of semiconductors such as silicon and GaAs with high energy electrons and protons, causes an increase of their resistivity due to the generation of defects in the semiconducting crystals. This behavior has also reported to occur in GaN [2,3]. The effects of proton irradiation on packaged AlGaN/GaN heterojunction-field effect transistors (HFET) have also been reported, with emphasis on irradiation effects on the controllability of Schottky gates [4,5]. To our knowledge, the study of proton irradiation effects on AlGaN/GaN heterostructure for Hall sensors without protective packaging has not been previously reported. Here, we report on the effect of proton irradiation on the magnetoelectric properties of micro-Hall sensors fabricated using 2DEG-AlGaN/GaN heterostructures. The resistivity, sheet carrier density, and electron mobility of the structures were investigated by Hall effect measurements after proton irradiation, and potential application of AlGaN/GaN for Hall sensors operable in harsh radiation environment is discussed.

2. Experimental

Fig.1 shows the structure of the AlGaN/GaN samples used in this study. The high electron mobility transistor (HEMT) structures were grown by metal organic chemical vapor deposition (MOCVD) and consist of an unintentionally doped AlGaN layer and a two-dimensional electron gas (2DEG) at the AlGaN/GaN heterointerface due to highly polar nature of AlGaN and GaN. The sheet carrier density and electron mobility at room temperature were 8×10^{12} cm^{-2} and 1600 cm^2/Vs, respectively. We also studied devices with a Si doped top AlGaN layer.

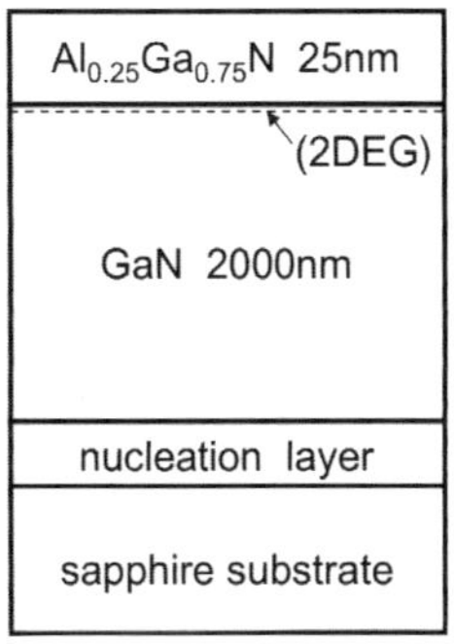

Fig.1. AlGaN/GaN hetero-structure used in this study. Here, unintentionally doped AlGaN layer sample (UID sample), and Si-doped AlGaN layer with $4\times10^{18}cm^{-3}$ (Si-doped sample).

These AlGaN/GaN heterostructures were used to fabrication four-terminal micro-Hall devices with active regions ranging from 1μm x 1μm to5μm x 5μm by photolithography. Details of the fabrication process are given in Ref.1. Notably, the active areas of the resulting Hall sensors were not protected by any form of packaging.

The micro-Hall sensors were irradiated with 380 keV protons using an ion-implantation facility at Takasaki Ion Accelerators for Advanced Radiation Application (TIARA), JAEA, in Japan. The samples were placed in a vacuum chamber and irradiated with a range of proton doses.. According to Stopping and Range of Ions in Matter (SRIM) simulations for calculation of the proton projection range, 98% of the implanted protons at 380 keV penetrated to a distance of 2 μm through the epitaxial nitride layers [6]. Thus the implantation of protons is expected to have a limited effect on the electrical characteristics of the 2DEG, which is 25 nm from the surface..

We prepared samples from same AlGaN/GaN wafer for each proton irradiation conditions in range 1×10^{11} cm^{-2} to 1×10^{16} cm^{-2}. The samples were not annealed before during or after proton irradiation. The sheet resistance, 2DEG electron mobility and sheet carrier density were measured by the Hall effect using a constant current drive at room temperature.

3. Results and Discussion

Fig.2 shows the variation of the sheet resistance with dose of proton irradiation. The sheet resistance increased with increasing proton dose, which suggests the generation of crystalline damage into the AlGaN/GaN structure. A marked increase of sheet resistivity was found above doses of 1×10^{14} cm^{-2}. It was not possible to measure the electrical characteristics of samples irradiated with 1×10^{15} and 1×10^{16} cm^{-2} because they exhibited extremely high resistance.

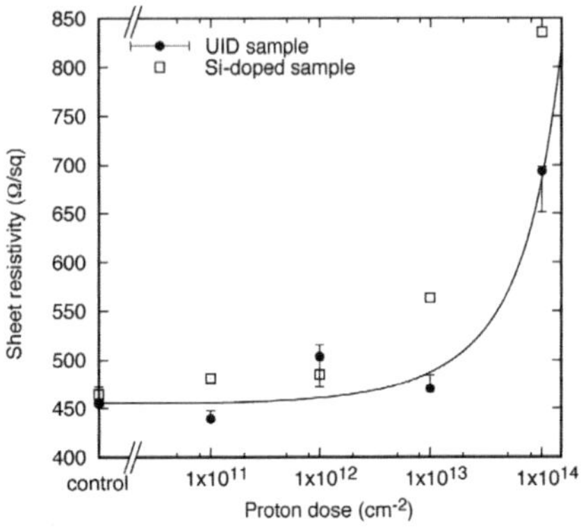

Fig.2. Variation of the sheet resistanceof AlGaN/GaN heterostructures with 380 keV proton dose.

Fig. 3(a) shows the variation of the 2DEG electron mobility and sheet carrier density change with proton dose. Although there was a slight scatter the data point, the electron mobility gradually decreased with increasing proton dose. This tendency was also found in samples with intentionally doped AlGaAN layers. In contrast to the electron mobility, the sheet carrier density shown in Fig.3(b) remained constant with a value similar to the non-irradiated reference sample upto a proton dose of $1x10^{14}$ cm^{-2}.

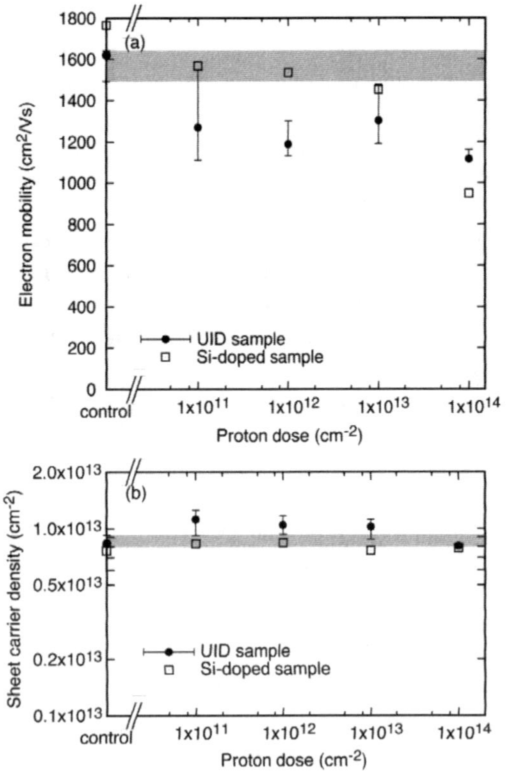

Fig.3. Dependence of (a) electron mobility and (b) sheet carrier density of AlGaN/GaN heterostructure on 380 keV proton dose. Horizontal gray bar indicates values for the non-irradiated reference sample.

The effects of proton irradiation on the crystallinity of AlGaN/GaN heterostructures were examined by X-ray diffraction (XRD) measurements as shown in the in 2θ-ω scan in Fig.4. The results show the XRD peaks of the reference and the $1x10^{14}$ cm^{-2} dose proton samples, which are, assigned as XRD peaks from the GaN layer based on the diffraction angles. These results showed that the XRD peaks remained after irradiation with a dose of $1x10^{14}$ cm^{-2}. It should be noted that an additional shoulder in lower angle side of irradiated sample was observed. A similar shoulder was reported after 2 MeV irradiation by Sonia et al [4], and attributed to the generation of interstitial defects in the heterostructure. Although our XRD results suggest the generation of proton irradiation effects in our AlGaN/GaN structures, it is difficult to directly connect the XRD results with the electrical characteristics because the electrical characteristics are sensitive to the 2DEG region, whereas the XRD signal isfrom whole of the sample. We are planning further studies using depth sensitive characterization techniques such as energy dependent cathodoluminescence.

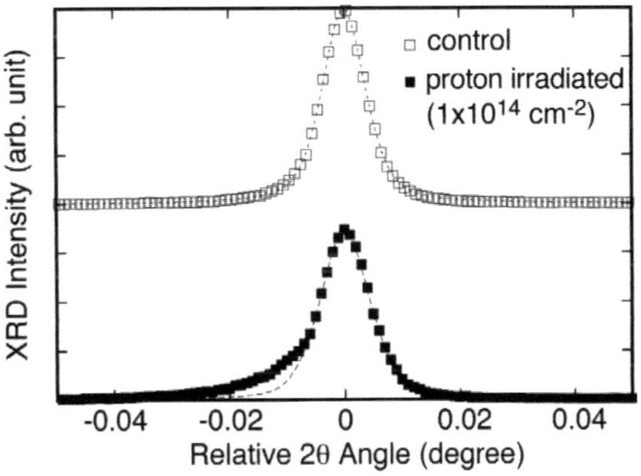

Fig.4. X-ray diffraction spectrum in 2θ-ω scan of reference and proton irradiated samples. Dashed lines are based-on pseudo Voigt functions.

Our experimental results showed that proton irradiation leads to increases in the sheet resistivity of 2DEG, particularly for proton dose of $1x10^{14}cm^{-2}$ due to a decrease in the electron mobility of AlGaN/GaN heterostructure. Several groups have reported on the effects of proton irradiation on n-GaN. For example, Auret *et al* reported the formation of electron traps in n-GaN after 2 MeV proton irradiation based on deep level transient spectroscopy (DLTS) measurements. [2]. They also reported dramatic reduction of the carrier density in n-GaN irradiated by 1 MeV protons [3]. In Ref.[3] agradual decrease of the carrier density was reportedafter relatively low proton dose of $2x10^{12}$ cm^{-2}, which is two-orders of smaller than our results. Similar gradual but clear reductions in carrier densityof n-GaN were reported by Polyakov *et al* for350 keV proton irradiation at a dose of$8x10^{13}cm^{-2}$[7].

Our results in Fig.3(b) showa constant carrier density after irradiation with a proton dose of $1x10^{14}cm^{-2}$. Sonia *et al* also reported the conservation of sheet carrier density in AlGaN/GaN heterostructures by 2 MeV proton irradiation up to the dose of $1x10^{13}cm^{-2}$ [4]. A possible reason for the high stability of carrier density in AlGaN/GaN under high doses of proton irradiation may be the higher carrier concentration of the 2DEG. The initial carrier density reported for n-GaN was in range of 5–12 x 10^{17} cm^{-3}.But in contrast,, the corresponding carrier density of the2DEG in AlGaN/GaN heterostructures is$10^{19}cm^{-3}$ if one assumes the thickness of 2DEG to be 10 nm. Thus, active carriers still remain in the higher carrier density sample remains even after irradiation with high dose of protons. Our results suggest that the high sheet carrier density of AlGaN/GaN structures is an important factor for applications of such Hall sensors in radiation environments., As shown in Figs. 3(a) and (b) the variation of electron mobility and sheet carrier density were not sensitive to Si doping of the top AlGaN layer. These results indicatethat the sheet carrier density and related electrical transport properties of AlGaN/GaN 2DEG Hall sensors after the proton irradiation is dominated by polarization field effects at the heterointerface rather than Si doping of the AlGaN layer.

The increase of sheet resistance of the samples shown in Fig.2 above proton doses of $1x10^{13}$ cm^{-2} is thought to be duc to a decrease of electron mobility rather than a decrease of the sheet carrier density. From SRIM simulations, 380 keV proton irradiation creates defects such as vacancies by displacement of atoms throughout the samples used in this study. The defect density for a proton dose of 10^{14} cm^{-3}

was roughly estimated to be 10^{18} cm^{-3}. This is similar to the nominal carrier density in 2DEG, thus potential fluctuations of defects may affect the electron transport but not the sheet carrier density, which is governed by macroscopic polarization field effects.

Finally, we would like to comment on an advantage of AlGaN/GaN Hall sensors in radiation environment. For Hall sensors operated in constant current mode the output Hall voltage, V_H, is given by $V_H=IB/qNd$, where I is the driving current, B is the magnetic field, q is the electron charge, N is the carrier concentration, and d is the thickness. That is, the magnitude of V_H is independent from the electron mobility. Thus AlGaN/GaN Hall sensors can be used in hard radiation environments under constant current mode operation because the density of the 2DEG is relatively unaffected even at high proton fluences. On the other hand, in case of operation under constant voltage mode, because V_H is a function of electron mobility, the sensitivity of the Hall sensor is affected by irradiation induced damage, as shown in Fig. 3(a). Another advantage of AlGaN/GaN Hall sensors is their higher tolerance against the proton irradiation compared with other heterostructures. Wang *et al* reported 0.5 MeV proton irradiation effects on AlGaAs/GaAs heterostructures [8]. Clear decreases of sheet carrier density were seen after a proton dose of $10^{11}cm^{-2}$, which is much smaller than the doses used in this study on AlGaN/GaN. Our results show that AlGaN/GaN Hall sensors are promising for applications in harsh radiation environments.

4. Conclusion

We investigated the effects of 380 keV proton irradiation on the magnetoelectric properties of AlGaN/GaN micro-Hall sensors. We observed an increase of the sheet resistance after the proton dose of $1x10^{13}$ cm^{-2}, which we explained as being caused by a decrease of the carrier mobility rather than sheet carrier density. Our results show that the AlGaN/GaN 2DEG structures are good candidates for the fabrication of Hall sensor operated in harsh radiation environments.

References

[1] Yamamura T, Nakamura D, Higashiwaki M, Matsui T, Sandhu A 2006 *J. Appl. Phys.* **99**, 08B302

[2] Auret F D, Goodman S A, Koschnick F K, Spaeth J-M, Beaumont B and Gibart P 1999 *Appl. Phys. Lett.* **74** 407

[3] Auret F D, Goodman S A, Hayes M, Legodi M J, Hullavarad S S, Friedland E, Beaumont B and Gibart P 2001 *Nucl. Instr. and Meth. in Phys. Res.* **B 175-155** 292

[4] Sonia G, Richter E, Brunner F, Denker A, Lossy R, Mai M, Lenk F, Bundesmann J, Pensl G, Schmidt J, Zeimer U, Wang L, Baskar K, Weyers M, Wuerfl J, Traenkle G 2008 *Solid-State Electronics* **52** 1011

[5] Lo C F, Chang C Y, Chu B H, Kim H Y, Kim J, Cullen D A, Zhou L, Smith D J, Pearton S J, Dabiran A, Cui B, Chow P P, Jang S, Ren F 2010 *J. Vac. Sci. Technol.* **B 28** L47

[6] Ziegler J *PARTICLE INTERACTIONS WITH MATTER* http://www.srim.org/

[7] Polyakov A Y, Usikov A S, Theys B, Smirnov N B, Govorkov A V, Jomard F, Shmidt N M and Lundin W V 2000 *Solid-State Electronics* **44** 1971

[8] Wang Q, Xu H Q, Omling P, Yang C, Malmqvist K G 2000 *Nucl. Instr. and Meth. in Phys. Res.* **B 160** 33

Microorganism mediated synthesis of reduced graphene oxide films

Y Tanizawa[1,*], Y Okamoto[2], K Tsuzuki[2], Y Nagao[2], N Yoshida[3,4], R Tero[3,4], S Iwasa[2,4], A Hiraishi[2,4], Y Suda[1,4], H Takikawa[1,4], R Numano[3,4], H Okada[1,3,4], R Ishikawa[5] and A Sandhu[1,3,4,5]

[1]Dept. of Electrical and Electronic Information Engineering, Toyohashi University of Technology, 1-1 Hibarigaoka, Tempaku-cho, Toyohashi, Aichi 441-8580, Japan
[2]Dept. of Environmental and Life Sciences, Toyohashi University of Technology, 1-1 Hibarigaoka, Tempaku-cho, Toyohashi, Aichi 441-8580, Japan
[3]Electronics-Inspired Interdisciplinary Research Institute (EIIRIS) , Toyohashi University of Technology, 1-1 Hibarigaoka, Tempaku-cho, Toyohashi, Aichi 441-8580, Japan
[4]Toyohashi Tech Graphene Research Group, Toyohashi University of Technology, 1-1 Hibarigaoka, Tempaku-cho, Toyohashi, Aichi 441-8580, Japan
[5]Electrical and Electronic Engineering, Tokyo Institute of Technology, 2-12-1 Ookayama, Meguro-ku, 152-8550

E-mail: tanizawa-y@eiiris.tut.ac.jp

Abstract. The wide-ranging industrial application of graphene and related compounds has led researchers to devise methods for the synthesis of high quality graphene. We recently reported on the chemical synthesis, patterning, and doping of graphene films by the chemical exfoliation of graphite into graphene oxide (GO) with subsequent chemical reduction into graphene films [1, 2]. Here, we describe a hybrid approach for the synthesis of reduced graphene sheets, where chemically derived GO was reduced by microorganisms extracted from a riverside near the University. Our procedure enabled the production of ~100 μm sized reduced graphene sheets, which showed excellent Raman spectra associated with high quality reduced graphene. We give a detailed account of the relationship between the type of microorganisms and the properties of the resulting reduced graphene.

1. Introduction

Chemically-derived graphene, composed of a single layer of carbon, now has an available route for bulk production. GO seems to be a much more efficient, low-cost, and bulk production pathway to incorporate graphene sheets into hybrids [3].

Currently, the most intensely researched approach for large-scale graphene production is through the chemical reduction of chemically derived GO, which can be obtained from natural graphite powder. GO may be reduced to graphene, either by exposure to pure hydrazine or by means of a thermal treatment. However, these methods have some drawbacks. First, hydrazine vapors are highly toxic. Second, thermal reduction is a very complex phenomenon because of the thermal-energy-induced multi-step removal processes of intercalated H_2O molecules and oxide groups of –COOH (carboxyl group), -OH(hydroxyl group), and >O(epoxy group) [3]. These approaches may cause environmental or economic concerns when large-scale production is considered.

Here, inspired by a recent report describing how graphene oxide behaves as a terminal electron acceptor for bacteria [4], GO is reduced by microbial action (e.g. bacteria) in the process of breathing or electron transport. In this study, we describe a hybrid approach for the synthesis of reduced graphene sheets, where chemically derived graphene oxide was reduced by microorganisms extracted from a river bank near the Tempaku Campus of Toyohashi University of Technology, in Aichi Prefecture.

2. Experimental

2.1. Synthesis of graphene oxide

GO was synthesized by the modified Hummer's method [5, 6] using natural graphite powder (Ito Kokuen Co., Ltd, Japan) cleaned using $K_2S_2O_8$, P_2O_5 and H_2SO_4, and oxidized in $KMnO_4$ and H_2SO_4. After overnight sedimentation and washing with 3wt% HCl aq. and MilliQ water, the resulting graphite oxide was exfoliated into GO without ultra-sonication. The resulting graphite oxide was then exfoliated into GO using centrifugation (10000rpm, 45min) instead of ultra-sonication, because if we use ultra-sonication, the size of GO sheets become smaller and smaller. Finally, a GO aqueous dispersion was purified by dialysis for one week to remove residual metal ions, acids and to adjust pH to neutral.

2.2. Microorganism-mediated reduction of graphene oxide

A microbial culture was extracted for the reduction of GO, which was obtained by repeating several iterations of transfer-cultivation of river sediments with a medium containing various mineral salts and organic substrates under anaerobic conditions. Silicon substrates with GO sheets attached to their surfaces were put into the microbial cultures in serum bottles. The silicon-GOs were incubated in the microbial culture at 28°C for 3 days in the dark. After incubation, they were fully washed with MilliQ and dried.

The GO was reduced when electrons from a sodium acetate solution were passed to the GO by anaerobic breathing of the environmental microorganisms. These microbes have the ability to use a large array of organic and inorganic compounds as terminal electron acceptors in their respiratory pathway [4, 7].

3. Results and Discussion

The GO sheets and the reduced GO sheets were characterized by drop-casting a solution containing GO sheets onto a thermally oxidized SiO_2/Si substrate. Then the SiO_2/Si substrate was slightly tipped at a shallow angle to prevent agglomeration of GO sheets in the liquid suspension those attached to the SiO_2/Si substrate during drying. The GO sheets dropped onto the substrates were examined under an optical microscope, Raman spectroscopy (excited by 532 nm ×10 objective lens, NRS-1000 JASCO CO., Ltd, Japan) and atomic force microscope (in tapping mode using Si tips). The thickness of SiO_2 was about 90 nm in order to obtain high contrast images of the GO sheets under visible light [1, 8, 9].

Figure 1 shows optical images of chemically derived GO sheets. There are a few layers of GO sheets, which were several hundreds of micrometers in size, with signs of overlapped layers.

Figure 2 shows optical images of the surface of GO sheets on a SiO_2/Si substrate after bacterial respiration when a reduction medium was added to the solution. The bright dots are believed to be bacteria or metallic precipitates from the reduction medium. The reduction of GO was visibly apparent in the solutions due to the formation of black precipitates, and in the case of GO fixed to silicon substrates, as a sharp increase in contrast under optical microscopic observations.

Figure 3 shows an AFM image of reduced GO via bacterial respiration. If we can make perfectly reduced GO, the thickness of the reduced GO sheet would be the same as that of graphene, but we did not observe this effect. We observed some bacteria, which have flagella and which appear as white dots of about 2μm in size.

Figure 4 shows the Raman spectra of reduced GO produced without the use of a reduction agent. The GO exhibited two peaks centered at 1347 and 1610 cm^{-1}, corresponding to the D and G bands, the latter corresponding to the in-phase vibration of the graphite lattice [10]. The prominent D peak is the disorder band from the structural imperfections created by the attachment of hydroxyl and epoxide groups on the carbon basal plane or edge [11]. The reduced GO, where the D band peak shifts from 1347 to 1320 cm^{-1}, and the G band became a broad peak. The intensity ratio of D/G was greater than that observed before reduction, due to reduction of the GO. These results indicate that the number of sp^2 domains increased due to the reduction process [1]. In the former paper, C.Tung V et al observed the same phenomenon. However, fully reduced samples would exhibit small D/G ratios if large-scale aromaticity were re-established [12].

We give a detailed account of the relationship between the type of bacteria and the properties of the resulting reduced graphene.

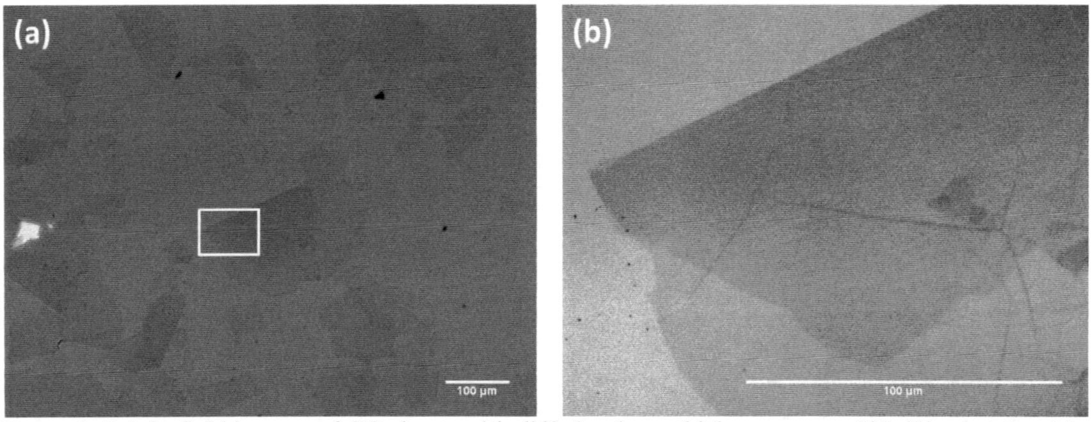

Figure 1. Bright field images of GO sheets with differing layer thicknesses on a SiO_2/Si substrate. (a) Optical microscope image and (b) close up of the white box area.

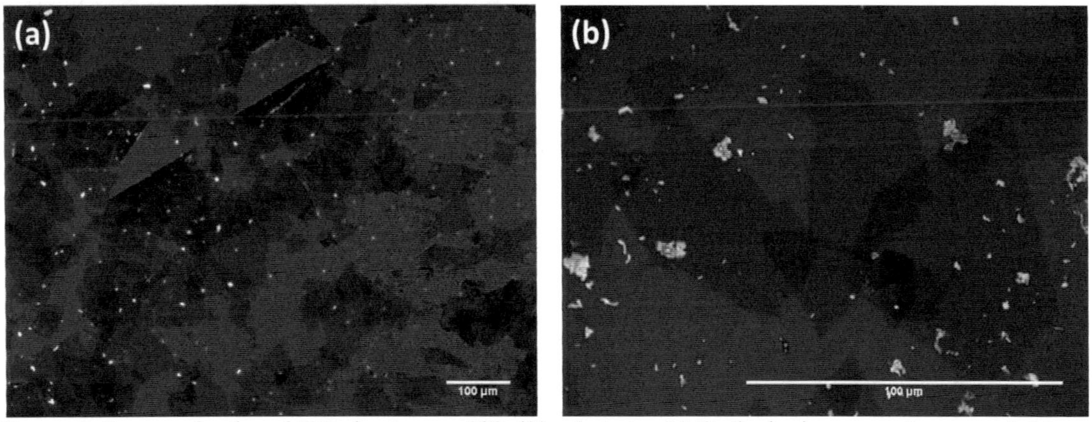

Figure 2. Image of reduced GO sheets on a SiO_2/Si substrate. (a) Optical microscope image and (b) high magnification.

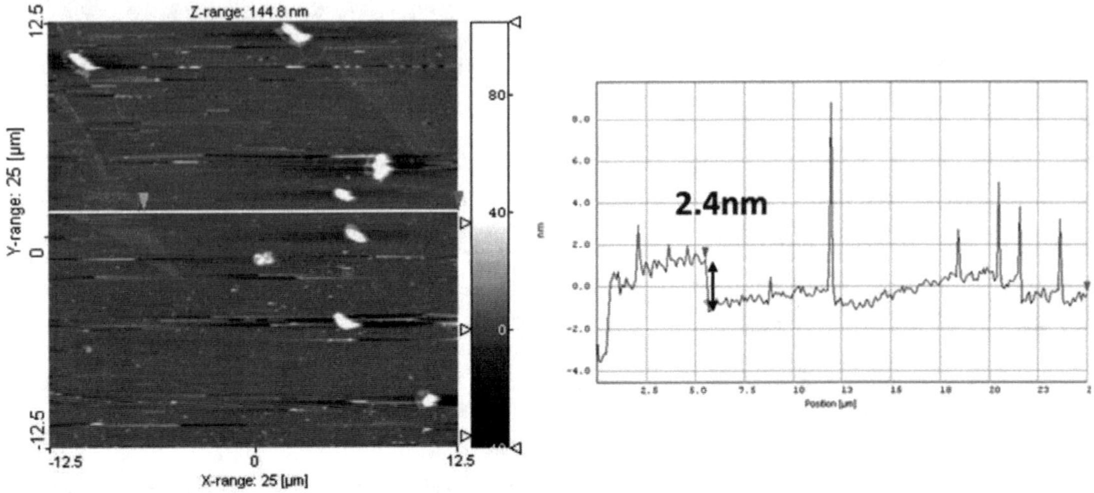

Figure 3. AFM image of reduced GO sheets on a SiO_2/Si substrate. The white dots that are than 2μm are bacteria.

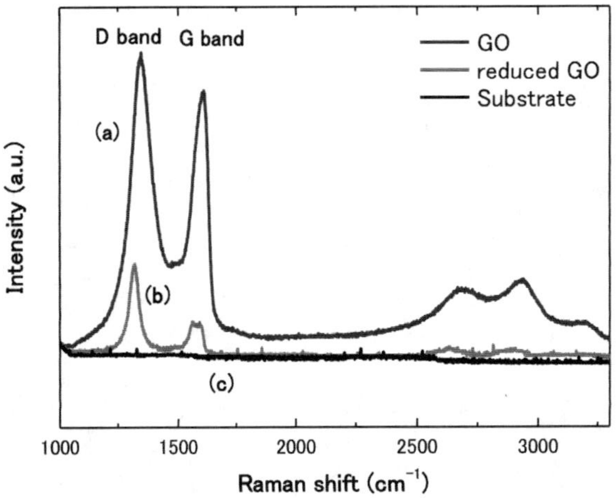

Figure 4. Raman spectra of (a) chemically synthesized GO sheets, (b) reduced GO sheets after bacterial reduction without the addition of reducing agent and (c) the surface of SiO_2.

4. Conclusions

Bacteria extracted from a riverside in Toyohashi city were used to reduce graphene oxide sheets without the addition of a reducing agent, opening up the possibility of a facile means of reducing graphene oxide dispersed in a bacterial suspension or immobilized on solid substrates.

Acknowledgments

The Toyohashi Tech Graphene Research Group (TT-GRG) would like to thank Professor Makoto Ishida for his support during this research. This work was partly funded by EIIRIS. The graphite powder was provided from Ito Kokuen Co., Ltd.

References

[1] Ishikawa R, Bando M, Morimoto Y, Park S Y and Sandhu A 2010 *Japanese Journal of Applied Physics* **49** 06GC02

Journal of Physics: Conference Series **352** (2012) 012011 doi:10.1088/1742-6596/352/1/012011

[2] Ishikawa R, Bando M, Morimoto Y and Sandhu A 2011 *Nanoscale Research Letters* **6** 111

[3] Hur S H 2011 Physics and Application of Graphene Experiments ed Mikhailov S (Rijeka) Chapter 5, pp73-136

[4] Xu L Q, Liu Y L, Neoh K-G, Kang E-T and Fu G D 2011 *Macromolecular rapid communications* **32** 684–8

[5] Tung V C, Allen M J, Yang Y and Kaner R B 2008 *Nature nanotechnology* **4**

[6] Gao Y, Chen X, Xu H, Zou Y, Gu R, Xu M, Jen A K-Y and Chen H 2010 *Carbon* **48** 4475–82

[7] Wang G, Qian F, Saltikov C W, Jiao Y and Li Y 2011 *Nano Research* **4** 563–70

[8] Blake P, Hill E W, Castro Neto a H, Novoselov K S, Jiang D, Yang R, Booth T J and Geim a K 2007 *Applied Physics Letters* **91** 063124

[9] Nagashio K, Nishimura T, Kita K and Toriumi A 2009 *Applied Physics Express* **2** 025003

[10] Kudin K N, Ozbas B, Schniepp H C, Prud'homme R K, Aksay I a and Car R 2008 *Nano letters* **8** 36–41

[11] Yang D, Velamakanni A, Bozoklu G, Park S, Stoller M, Piner R, Stankovich S, Jung I, Field D and Ventricejr C 2009 *Carbon* **47** 145–52

[12] Tung V C, Allen M J, Yang Y and Kaner R B 2009 *Nature nanotechnology* **4** 25–9

Electronic structure and magnetic properties of $^{31}P@C_{60}$-SWCNT as peapods

A Suzuki and T Oku

Department of Materials Science, The University of Shiga Prefecture

2500 Hassaka, Hikone, Shiga 522-8533 Japan

E-mail: suzuki@mat.usp.ac.jp

Abstract. Design of the NMR quantum computer of 1D spin chains based on $^{31}P@C_{60}$-SWCNT as peapods was studied. Electronic structure and magnetic properties of $^{31}P@C_{60}$ encapsulating fullerenes within SWCNT as peapods was investigated by quantum chemical calculation. Characterization of chemical shifts of ^{13}C, principle *g-tensor* and *A-tensor* of hyperfine structure for phosphorus atom with electronic spin S=3/2 and nuclear spin I=1/2 were studied. The excited energy and wavelength at 1512 nm and 728 nm confirmed assignment of the first and second states as van Hove transitions. Molecular design of carbon peapods is important to control quantum spin qubits, splitting by spin-local interaction and dipole-dipole interaction based on *p*-orbital spin density distribution under hybrization of molecular orbital at excited state and grand state.

1. Introduction

Electronic structure of single-walled-carbon nanotube (SWCNT) encapsulating fullerenes have considerably been studied after the discovery of C_{60}@SWCNT as peapods [1]. A scalable spin quantum computer that combines aspects of magnetic properties with the advantageous features of endohedral fullerenes within SWCNT has been proposed. Design of spin labels inside for possible molecular spintronics, which contains of 1D spin chains of magnetic endohedral fullerenes of $^{14}N@C_{60}$ and $^{31}P@C_{60}$ within SWCNT have been proposed [2]. The magnetic parameters of principle *g-tensor, A-tensor* for hyperfine coupling constant (*hfc*) of nitrogen, phosphorus atoms, chemical shift of ^{13}C in $^{14}N@C_{60}$ within SWCNT as peapods for chiral index had been investigated by ESR, ENDOR, NMR and DFT [3,4]. The magnetic endohedral fullerenes of $^{14}N@C_{60}$ and $^{31}P@C_{60}$ within SWCNT have a high advantage of the NMR quantum computer to control spin gate with decoupling pulses in relaxation time. For instant, Morton reported the NMR quantum-computer controlled un-perturbation Rabi oscillation of spin polarization under influence of decoupling process [5, 6]. In this present work, electronic structure and magnetic properties of ^{31}P atom encapsulating fullerenes within SWCNT as peapods was investigated by quantum chemical calculation. Characterization of chemical shifts of ^{13}C, principle *g-tensor* and *A-tensor* for hyperfine structure of phosphorus atom with electronic spin S=3/2 and nuclear spin I=1/2 were studied. The magnetic properties were discussed on the basis of spin local interaction related on *p*-orbital spin density distribution under hybrization of molecular orbital at grand state and excite state. Especially, magnetic properties of ^{31}P and ^{14}N atoms encapsulating fullerenes within SWCNT as peapods based on the electronic structure at grand state and excited state was focused on controlling quantum spin qubits for NMR quantum computer.

2. Quantum chemical calculation

The molecular structures were assembled by CS ChemDraw, CS Chem3D (Cambridge Soft) and Nanotube Modeler (JCrystal Soft). Molecular orbital calculations were carried out by molecular mechanics calculations (MM2) and semi-empirical molecular orbital calculations (Hamiltonian: Parameterized Model Revision 3: PM3). In addition, the isolated molecular structures were optimized by *ab-initio* quantum calculation using unrestricted Hartree–Fock (UHF) and DFT using UB3LYP with hybrid function LANL2DZ and STO-3G*, 3-31G* and 6-31G* as basis set because of computational limitations (Gaussian 03 Inc.). HOMO, LUMO, LUMO+1 and the HOMO-LUMO band gap (E_g) were calculated. Phosphorus atomic charges of $^{31}P@C_{60}$ within SWCNT were estimated by Mulliken population analysis. Continuously, chemical shift (δ) of ^{13}C, principle *g-tensor* (g_{xx}, g_{yy}, g_{zz}) and *A-tensor* (A_{xx}, A_{yy}, A_{zz}) in hyperfine coupling constant (*hfc*) of nitrogen atom were calculated by DFT using NMR / GIAO with hybrid function UB3LYP and 3-31G* as basis set. Excited energy and wavelength were calculated by TD-DFT.

3. Results and Discussion

Electronic structure of molecular orbital at HOMO an LUMO and energy levels of $^{31}P@C_{60}$-SWCNT (9, 9) and $^{14}N@C_{60}$-SWCNT (9, 9) was investigated by DFT with hybrid function UB3LYP using STO-3G* as basis set. Figure 1 shows electron structure and energy levels of $^{31}P@C_{60}$-SWCNT (9, 9). Electron spin density distribution of *p*-orbital in phosphorus atom was localized at HOMO. The electron spin density distribution at LUMO was delocalized on surface of carbon tube. As listed in Table 1, energy levels at HOMO and LUMO and energy gap of $^{31}P@C_{60}$-SWCNT were -1.58 eV, -1.09 eV and 0.49 eV, which narrowed as compared with the levels of $^{14}N@C_{60}$-SWCNT and $^{31}P@C_{60}$ and $^{14}N@C_{60}$. Mulliken atomic charge for phosphorus atom in $^{31}P@C_{60}$-SWCNT was estimated to be -1.9×10^{-2} e, which indicated negative charge transfer from C_{60} to phosphorus atom. In reference case, the Mulliken atomic charge for the phosphorus atom in $^{31}P@C_{60}$ was 5.7×10^2 e, which suggested charge transfer from phosphorus to C_{60}. Gross orbital populations of *3s* orbital and *3p* orbital in the phosphorus atom inside C_{60} within SWCNT were 2.03, 1.03, 1.03 and 1.03, respectively. These results indicate localization of spin orbital distribution at *3s* orbital and principle *3p* orbital: p_{xx}, p_{yy}, p_{zz} in the phosphorus atom inside C_{60} within SWCNT. In contrast case using $^{14}N@C_{60}$-SWCNT, gross orbital populations of *2s* orbital and *2p* orbital in nitrogen atom inside C_{60} within SWCNT were 0.94, 0.64, 0.64 and 0.64, respectively. There was considerable delocalization of *2s* orbital and *2p* orbital in nitrogen atom. As reference cases using $^{31}P@C_{60}$ and $^{14}N@C_{60}$, the gross orbital populations of nitrogen and phosphorus atom inside C_{60} suggested delocalization of *s* orbital and *p* orbital. Comparison between $^{31}P@C_{60}$–SWCNT and $^{31}P@C_{60}$ supports the assertion that spin distribution at *3p* orbital on phosphorus atom is well screened within the C_{60} cage around SWCNT. On the basis of comparison between $^{31}P@C_{60}$-SWCNT and $^{31}P@C_{60}$, coverage of SWCNT with $^{31}P@C_{60}$ on the electronic structure was influenced by π orbital interaction of carbon cage between C_{60} and SWCNT with hybrization of *p*-orbital spin density distribution in ^{31}P atom covered with C_{60} and SWCNT. As reported in Ref. 4, we will expect considerable influence on geometrical structure of $^{31}P@C_{60}$ within SWCNT varied with chirality and diameter on the electronic structure.

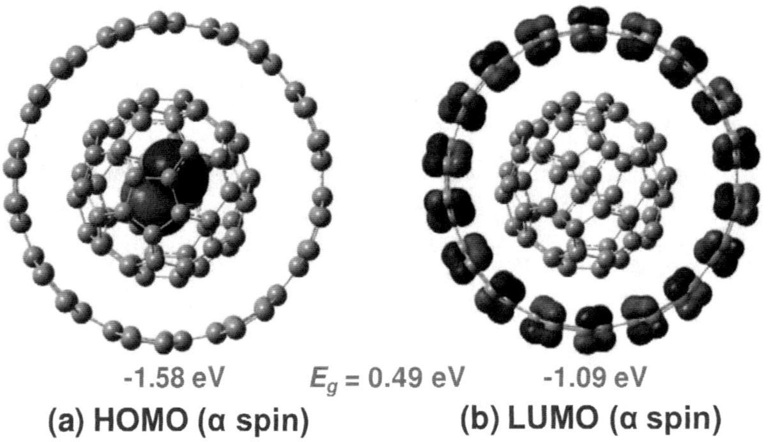

Figure 1. Electronic structure of $^{31}P@C_{60}$-SWCNT at HOMO (a) and LUMO (b).

Table 2 lists EPR parameters of principal *g-tensor*, *A-tensor* in hyperfine coupling constant (*hfc*) for carbon peapods, calculated by DFT/UB3LYP/STO-3G*.The magnetic parameters of principal *g-tensor* (g_{xx}, g_{yy}, g_{zz}), *A-tensor* (A_{xx}, A_{yy}, A_{zz}) of *hfc* for $^{31}P@C_{60}$-SWCNT (9, 9) were estimated to be 1.96509, 1.99151, 2.00298, 5.7, 5.5 and 3.5, respectively. The calculated results would be mainly attributed to hyperfine coupling interaction based on isotropic and anisotropic spin density distribution.

Journal of Physics: Conference Series **352** (2012) 012012 doi:10.1088/1742-6596/352/1/012012

Table 1. Electronic structure of carbon peapods, fullerenes, and SWCNT, as calculated by DFT using UB3LYP/ STO-3G* as basis function.

Molecule	HOMO/LUMO (eV)	E_g (eV)	Mulliken atomic charge / e	Gross orbital populations *N: $2s^2 2p^3$, P: $3s^2 3p^3$*
$^{31}P@C_{60}$-SWCNT	-1.58/-1.09	0.49	-1.9×10^{-2}	2.00, 1.03, 1.03, 1.03
$^{14}N@C_{60}$-SWCNT	-4.05/-2.89	1.16	3.8×10^{-5}	0.94, 0.64, 0.64, 0.64
$^{31}P@C_{60}$	-6.41/-3.68	2.73	5.7×10^{-2}	1.49, 0.70, 0.70, 0.70
$^{14}N@C_{60}$	-6.60/-3.57	3.03	-4.5×10^{-3}	0.93, 0.64, 0.64, 0.64
SWCNT (9, 9)	-4.08/-2.84	1.24	-	-

The hyperfine coupling interaction was the sum of isotropic (A_{iso}) and anisotropic interaction (A_{aniso}). The isotropic interaction of A_{iso} corresponded to Fermi contact interaction based on *s* spin localization on the phosphorus atom covered with the C_{60} cage within SWCNT. The anisotropic hyperfine interaction of A_{aniso} in *hfc* for $^{31}P@C_{60}$-SWCNT was approximately 10 times as large as those of $^{14}N@C_{60}$-SWCNT. This result was strongly related to magnetic interaction between the electron and nuclear spin on the nucleus under a control of angle-dependence on *p*-orbital spin density distribution. Nuclear quadropole interaction is based on multiplication of nuclear quadropole moment *Q* and the electro field gradient (*EFG*) tensor generated by charge distribution around nucleus. There was a slight different value of *Q* for $^{31}P@C_{60}$-SWCNT and $^{14}N@C_{60}$-SWCNT as listed in Table 2. The chemical shift of $^{14}N@C_{60}$-SWCNT compared with $^{31}P@C_{60}$-SWCNT will be shifted by nuclear quadropole interaction between *Q* and *EFG* tensor, as shown in Fig. 2.

Table 2. EPR parameters of principal *g-tensor* and *A-tensor* in *hfc* (DFT/UB3LYP/STO-3G*).

$^{31}P@C_{60}$-SWCNT(9, 9)					$^{14}N@C_{60}$-SWCNT (9, 9)				
g-tensor	*hfc/G*	A_{iso}/G	A_{aniso}/G	*Q/DA**	*g-tensor*	*hfc/G*	A_{iso}/G	A_{aniso}/G	*Q/DA**
1.96509	5.7	-9.13	-0.050	-759	2.00098	6.0	10.43	-0.002	-813
1.99151	5.5		-0.034	-759	2.00098	6.0		0.000	-819
2.00298	3.5		0.084	-729	2.00093	3.8		0.002	-771

* *Q/DA: Quadropole moment / Debye-Ang.*

Magnetic mechanism will be discussed on the basis of the quantum calculation. Especially, geometrical effect of magnetic fullerenes within SWCNT varied with chirality and diameter on chemical shift of ^{13}C, principal *g-tensor*, *A-tensor* for *hfc* of central atom and excited state transition will be considered. The magnetic properties would be originated in the spin-local interaction and dipole-dipole interaction based on *p*-orbital spin density distribution of central atom covered with C_{60} and inner surface on SWCNT varied with chirality and diameter. Spin Hamiltonian is formulated with different terms as noted by Eq. (1). These terms represent the following interaction: electron and nuclear Zeeman interaction, hyperfine structure of the spin-nuclear interaction, the fine structure of the spin-spin interaction and the nuclear quadropole interaction. The magnetic parameters of principle *g-tensor* and *A-tensor* for *hfc* are based on electron spin density distribution and dipole-dipole interaction. Magnetic parameter of *g-tensor* is sum of isotropic and anisotropic parts as noted by Eq. (2)-(3). Anisotropic *g-tensor* is related with extent of spin-local interaction under hybrization of molecular orbital at grand state and excited states. The *A-tensor* of *hfc* is sum of isotropic and anisotropic terms as noted by Eq. (4)-(5). The isotropic *A-tensor* (A_{iso}) as Fermi contact interaction is in proportion as the *s* spin density distribution. The anisotropic *A-tensor* (A_{aniso}) as dipole-dipole interaction is under a control of angle dependence on *p*-orbital electron spin distribution. The 3*p* spin density distribution for phosphorus atom rather than nitrogen atom in $^{14}N@C_{60}$ within SWCNT was considerably localized as listed in Table 1.

$$H = g_s \beta_s S H - g_N \beta_N I H + S A I + S D S + I Q I \tag{1}$$

$$g\text{-}tensor = g_e + \Delta g = g_e \delta_{ij} - 2 \lambda \Lambda_{ij} \tag{2}$$

$$g_e = 2.002319, \quad g_{ii} = g_e \left\{1 + \sum_m \frac{|\langle \varphi_m | L_i | \varphi_p \rangle|}{\varepsilon_p - \varepsilon_m}\right\} \tag{3}$$

$$A\text{-}tensor = A_{iso} + A_{aniso} \tag{4}$$

$$A_{iso} = \frac{8\pi}{3} g_n \mu_N \rho(o), A_{aniso} = g_n \mu_N \int \frac{(3\cos^2\theta - 1)}{r^3} \rho(r) dr \tag{5}$$

The chemical shifts of ^{13}C with multicity in optimized molecular structure were calculated. The chemical shift is constructed with a total sum of four terms as noted by Eq. (6).

$$\sigma = \sigma_{dia} + \sigma_{para} + \sigma_{FC} + \sigma_{SD} \tag{6}$$

where σ_{dia} is diamagnetic term, σ_{para} is paramagnetic term, σ_{FC} is Fermi contact interaction, σ_{SD} is spin dipole-interaction. The chemical shift will be mainly in dependence on paramagnetic effect based on the spin-dipole interaction.

The chemical shifts of ^{13}C with multicity for $^{31}P@C_{60}$-SWCNT and $^{14}N@_{60}$-SWCNT had been quantitatively analyzed by *ab-nitio* DFT calculation. Figure 2 show the chemical shift of ^{13}C with multicity for $^{31}P@C_{60}$-SWCT and $^{14}N@C_{60}$-SWCNT. The chemical shift of ^{13}C for $^{31}P@C_{60}$-SWCNT was shifted with multicity and multi-splitted in the range of 80-100 ppm. The chemical shifts of ^{13}C were confirmed as part of SWCNT and C_{60} at 80 ppm and 100 ppm. In contrast, the chemical shift of ^{13}C for $^{14}N@C_{60}$-SWCNT was shifted with multicity and multi-splitted in a wide spin-spin coupling as part of SWCNT and C_{60} at 110 ppm and 130 ppm. The chemical shifts for both cases using carbon peapods mainly had influence on the paramagnetic effect under the spin-dipole interaction.

Journal of Physics: Conference Series **352** (2012) 012012 doi:10.1088/1742-6596/352/1/012012

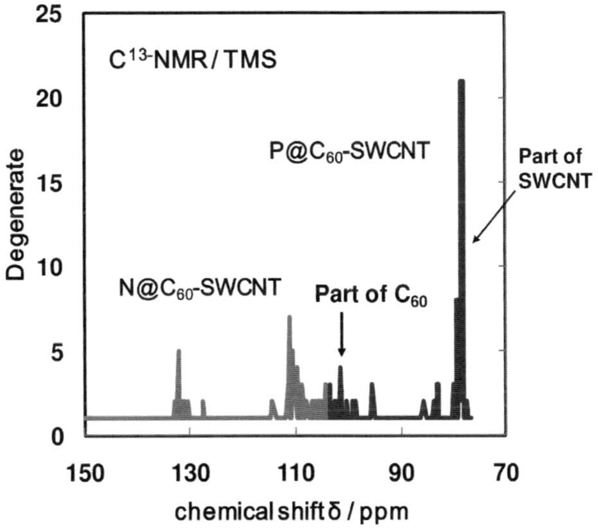

Figure 2. Chemical shift of ^{13}C with multicity for $^{31}P@C_{60}$-SWCNT and $^{14}N@C_{60}$-SWCNT.

In general case, the chemical shift of ^{13}C with multicity for optimized structure with paramagnetic moment can be explained by Karplus-Pople equation [7] as noted by Eq. (7). The chemical shift mainly consists with diamagnetism and paramagnetic parts in first and other terms as noted by Eq. (7). The chemical shift strongly range for *p*-orbital nuclei dominated by paramagnetic parts. The paramagnetic term has an influence on spin-local interaction under hybrization of molecular orbital at excited state and grand state. The spin-local interaction is based on the *p*-orbital spin density distribution. The chemical shift of ^{13}C with multicity will be influenced by hybrization of molecular orbital at energy levels between excited state and grand state in second term as noted by Eq. (7).

$$\sigma = \frac{e^2}{3mc^2}\langle 0\left|\sum_k \frac{1}{r_k}\right|0\rangle$$

$$-\frac{2}{3}\left(\frac{eh}{2mc}\right)^2 \sum_n \frac{1}{E_n-E_0}\left\{\langle 0|\sum_k L_k|n\rangle \langle n\left|\sum_k \frac{L_k}{r_k{}^3}\right|0\rangle + \langle 0\left|\sum_k \frac{L_k}{r_k{}^3}\right|n\rangle \langle n|\sum_k L_k|0\rangle\right\} \quad (7)$$

Table 3 list excited energy level and optical wavelength of $^{31}P@C_{60}$-SWCNT and $^{14}N@C_{60}$-SWCNT, calculated by TD-DFT. The excited energy for $^{31}P@C_{60}$-SWCNT was in the range of 0.8199 eV-1.7042 eV, which converted to the wavelength in the range of 1512-728 nm. There were multi-state energy levels from HOMO to LUMO, LUMO+1 and band gaps, which would be assigned as first and second state as Van Hove transition at 0.8199 eV (1512 nm) and 1.6920 eV (733 nm). Similarly, the excited energy with wavelength for $^{14}N@C_{60}$-SWCNT confirmed assignment of the first and second state as van Hove transition on Raman spectroscopy with density of state using tight binding model as Van Hove transition at 0.728 eV (1703 nm) and 1.169 eV (1061 nm) as listed in Table 3. The calculated results made agreement with experimental results using UV-vis-NIR spectroscopy [8]. Consequently, the magnetic properties were summarized with spin local interaction, dipole-dipole interaction, nuclear quadropole interaction and hybrization of molecular orbital at excited state and grand state. The spin density distribution and the hybrization of molecular orbital at excited state are important to control the spin gate for NMR quantum computer. Molecular design of $^{14}N@C_{60}$-SWCNT rather than $^{31}P@C_{60}$-SWCNT as the NMR quantum computer is advantage to quantum-calculate at multiple-state.

Table 3. Excited energy levels and wavelength of $^{31}P@C_{60}$-SWCNT (9, 9) and $^{14}N@C_{60}$-SWCNT (9, 9), calculated by TD-DFT.

$^{31}P@C_{60}$-SWCNT (9, 9)			$^{14}N@C_{60}$-SWCNT (9, 9)		
Energy (eV)	Wavelength (nm)	Oscillator Strength	Energy (eV)	Wavelength (nm)	Oscillator Strength
1.7042	728	0.0006			
1.6920	733	0.0012			
1.5606	794	0.0007	1.169	1061	0.0007
0.8949	1386	0.0002			
0.8679	1429	0.0002			
0.8199	1512	0.0002	0.728	1703	0.0001

4. Conclusion

Design of the NMR quantum computer of 1D spin chains based on $^{31}P@C_{60}$-SWCNT as peapods was studied. Comparison between $^{31}P@C_{60}$ and $^{14}N@C_{60}$ within SWCNT as peapods on electronic structure and magnetic properties was investigated. Electronic structure and magnetic properties of $^{31}P@C_{60}$ encapsulating fullerenes within SWCNT as peapods was investigated by quantum chemical calculation. Characterization of chemical shifts of ^{13}C, principle *g-tensor* and *A-tensor* of *hfc* in phosphorus atom with electronic spin S=3/2 coupled with nuclear spin I=1/2 were employed. The magnetic properties were originated in spin-local interaction, dipole-dipole interaction based on *p*-orbital spin density distribution, nuclear quadropole interaction and hybridization of molecular orbital at excited state and grand state. The calculated chemical shifts were mainly dominated by paramagnetic term with spin local interaction and hybrization of molecular orbital at excited and grand state. The excited energy levels and wavelength confirmed assignment of the first and second van Hove transition, which made agreement with the experimental results of UV-vis-NIR, Raman using tight binding model. Molecular design of $^{14}N@C_{60}$-SWCNT rather than $^{31}P@C_{60}$-SWCNT as the NMR quantum computer is advantage to quantum-calculate with controlling quantum spin-qubits at multiple state.

References

[1] Bailey S W D and Lambert C J 2007 *Physica E* **40** 99

[2] Benjamin S C, Ardavan A, Andrew G, Briggs D, Britz D A, Gunlycke D, Jefferson J, Jones M A G, Leigh D F and Lovett B W *et al.* 2006 *J. Phys. Cond. Matt.* **18** S867

[3] Morton J J L, Tyryshkin A M, Ardavan A, Benjamin S C, Porfyrakis K, Lyon S A and Briggs G A D, 2006 *Phys. Stat. Sol.* (b) **243** 3028

[4] Suzuki A and Oku T, 2011 *Physica B* **406** 3274

[5] Morton J J L, Tyryshkin A M, Ardavan A, Benjamin S C, Porfyrakis K, Lyon S A and Briggs G A D, 2006 *Nature Physics* **2** 40

[6] Brown R M, Tyryshkin A M, Porfyrakis K, Gauger E M, Lovett B W, Ardavan A, Lyon S A, Briggs G A D and Morton J J L, 2011 *Phys. Rev. Lett.* **106** 110504

[7] Pople J A, 1957 *Proc. Roy. Soc.* **A239** 549

[8] Iizumi Y, Okazaki T, Liu Z, Suenaga K, Nakanishi T, Iijima S, Rotas G and Tagmatarchis N, 2010 *Chem. Commun.* **46** 1293

A Solution of Time Dependent Schrödinger Equation by Quantum Walk

Hideo Sekino, Masayuki Kawahata and Shinji Hamada

[1]Toyohashi University of Technology

Toyohashi Aichi 441-8580, Japan

sekino@tut.jp

Abstract. Time Dependent Schrödinger Equation (TDSE) with an initial Gaussian distribution, is solved by a discrete time/space Quantum Walk (QW) representing consecutive operations corresponding to a dot product of Pauli matrix and momentum operators. We call it as Schrödinge Walk (SW). Though an Hadamard Walk (HW) provides same dynamics of the probability distribution for delta-function-like initial distributions as that of the SW with a delta-function-like initial distribution, the former with a Gaussian initial distribution leads to a solution for advection of the probability distribution; the initial distribution splits into two distinctive distributions moving in opposite directions. Both mechanisms are analysed by investigating the evolution of the both amplitude components. Decoherence of the oscillating amplitudes in central region is found to be responsible for the splitting of the probability distribution in the HW.

1. Introduction

Quantum information is propagated by unitary transformations of the quantum states which is called q-bit. Expectation value of a particle to be in position x at time t is evaluated by the spatial evolution of the probability amplitude for the particle given at initial time t_0. Conventional technology to provide such evolution of the probability amplitude is to solve Time Dependent Schrödinger Equation (TDSE) for the quantum particle in a Hilbert space composed by real-space and internal degree of freedom $H_p \otimes H_c$. TDSE is a partial differential equation which is equivalent to a finite difference equation with the difference of infinitesimal length. As is understood, however, infinitesimal length is not allowed as any meaningful size of quantum particles which carry information. That is, the measure of propagation of the quantum information by the solution of the differential equation with finite difference approximation is bounded by the finite length used. The any simulation with finite momentum (the ratio of physical space interval to time interval times mass) is meaningful for evaluating the transportation of information contained in a space of finite size. With this in mind, we introduce a quantum walk (QW) which leads to a solution of the TDSE with an initial Gaussian distribution and further analyze the solution by comparison with Hadamard Walk (HW) which leads to a solution for advection of the probability distribution.

2. Theoretical method

QW is typically used in the context of quantum computation where time-progression of quantum state is simulated using consecutive application of unitary operations to the initial state. Natural

Published under licence by IOP Publishing Ltd

Journal of Physics: Conference Series **352** (2012) 012013 doi:10.1088/1742-6596/352/1/012013

interpretation of wavefunction represented in physical space is the probability amplitude distribution of a quantum particle on physical space. A dynamics of quantum information within a space smaller than the minimum length allowed for quantum mechanics cannot be simulated using techniques based on the finite difference approximation where the unit length is smaller than the quantum minimum. More precisely, we cannot simulate the dynamics of the probability amplitude within an area smaller than the area of $\Delta x \cdot \Delta p$. We focus the dynamics of quantum information and simulate it using QW techniques. A typical QW is performed by consecutive application of unitary operation to a state from which we evaluate an observable quantum property at later time. We introduce a QW to generate a wavefunction at time t by consecutively applying a unitary operation in the composite Hilbert space $H_p \otimes H_c$ to the initial probability distribution of a quantum particle at t_0. Here H_p is physical space and the probability amplitude is represented on a set of discrete grid and H_c is a space spanned by internal degree of freedom. The latter can be spin of electron or chirality of photon for example and often called as coin space. Since we focus on the TDSE, it is probably more appropriate to interpret the internal degree of freedom considered here as spin of electron. Since the physical space is represented on grid points, an initial distribution at a single grid point mimics delta function. However it is impossible to describe a quantum dynamics within the area smaller than $\Delta x \cdot \Delta p$ by the simulation using such initial distribution. In order to provide appropriate solutions of TDSE by such discrete time/space QW, we take a QW of regular strategy[1] but with an initial probability amplitude distribution of Gaussian instead.

2.1 Lévy-Leblond Equation

Schrödinger Equation (SE) usually does not contain spin coordinate and it is commonly understood that the notion of spin arises from consideration of relativity beyond SE. However, we could formulate an equation which has a coordinate for extra degree of freedom than physical space for a non-relativistic particle which obeys TDSE. The equation was introduced[2] and is called Lévy-Leblond Equation (LLE).

$$H_{LL} \equiv \begin{pmatrix} V & \sigma \cdot p \\ \sigma \cdot p & -2m \end{pmatrix} \begin{pmatrix} \phi \\ \chi \end{pmatrix} = E \begin{pmatrix} \phi \\ 0 \end{pmatrix} \tag{1}$$

Where $\boldsymbol{\phi}$ and χ are upper and lower components of wavefunction spinors and each of them has two components. Here m is a mass of the particle and p and V are momentum and potential energy operators and $\boldsymbol{\sigma}$ stands for Pauli matrix vector. Eliminating the lower component from the time-dependent version of the LLE, we have a solution for the upper component.

$$\phi(r,t) = \exp[-\{ \frac{i}{2m}(\sigma \cdot p)^2 + V \}t]\varphi(r) = \exp[-\{ \frac{i}{2m} p^2 + V \}t]\varphi(r) \tag{2}$$

Typically numerical solution of the TDSE is achieved by consecutively applying the operation $-ip^2/2m + V$ as in the last expression of Eq. (2), to the wavefunction using certain approximation scheme for the exponential ansatz with sufficiently small time step for guaranteeing the precision of the approximation. We have identical solutions for the upper and lower components of the spinor $\boldsymbol{\phi}$ and therefore the equation reduces to a solution of scalar wavefunction. We here employ the first expression in Eq. (2) for solving the TDSE by QW. For a particle which has a definite spin along a direction, the corresponding component of the σ operator is diagonalized and the operation $\sigma \cdot p$ provides momenta -p and +p for the lower and upper components.

$$(\sigma \cdot p)\varphi(r) = -i\sigma_z \frac{\partial}{\partial z}\phi(r) = \begin{pmatrix} -i\frac{\partial}{\partial z} & 0 \\ 0 & +i\frac{\partial}{\partial z} \end{pmatrix}\phi(r) \tag{3}$$

However, the spin is not always aligned along certain axis without imposing an external perturbation to enforce it and thus σ is not diagonalized all the time even started with a definite spin state. Indeed, any spin component is smaller than the absolute value of spin itself, meaning that no spin is aligned completely parallel to any axis. A spin with a definite value of a component to an axis can be interpreted to be rotating around the axis with certain angle kept. When we simulate the dynamics of a quantum particle by QW in a combined Hilbert space using a finite time step, the spin state should be altered during the boost corresponding to a momentum operation in Eq. (3). For the second operation of $\sigma \cdot p$, the spin is not aligned along the axis even if it was before the first operation of $\sigma \cdot p$. Therefore we employ two different components of the Pauli matrix together with the diagonal component σ_z in order to investigate the effect of the rotation during the walk.

$$\exp(i\sigma_x \cdot p) = \begin{pmatrix} \cos p & i\sin p \\ i\sin p & \cos p \end{pmatrix} \tag{4a}$$

$$\exp(i\sigma_y \cdot p) = \begin{pmatrix} \cos p & \sin p \\ -\sin p & \cos p \end{pmatrix} \tag{4b}$$

We refer (4a) to TDSE walk (SW) and (4b) to Hadamard walk.(HW). The operation corresponding the $\sigma_z p_z$ is performed by a QW as in the typical QW where shift operators corresponding to –p and +p momenta are applied to the wavefunction represented on a grid in the discrete physical space, together with the spin rotating operations *P* and *Q* which correspond to the upper and lower components of (4a) or (4b) respectively.

2.2 Numerical method

For the preliminary study of the QW for solving TDSE by the above walks, we set $p = \pi/4$ for (4a) and (4b) operations. For investigating the basic behaviour of the QW defined above, we execute QWs with an initial distribution where only one grid has a value with all the others zero. We call the initial condition as delta-function (DF) distribution. Next, we execute QWs with a Gaussian distribution as an initial condition.

3. Results and discussion

It is known that HWs result in non-symmetric evolution of the probability for certain initial condition. We study here a symmetric HW with an initial condition $(1/\sqrt{2},\ i/\sqrt{2})^T$ and SW with an initial condition$(1/\sqrt{2},\ 1/\sqrt{2})^T$. We investigate the evolution of each component of the amplitude together with the evolution of the probability distribution. Those are shown in the pictures of second- and third-rows in Fig.1. The ones indicated by alpha and beta in the figures mean the upper and lower components of the amplitudes. The corresponding results of SW are summarized in Fig 2. Both walks provide identical probability evolution and the peaks of consisting amplitudes moves similarly. Namely the peaks of upper and lower components of the amplitudes move in an opposite direction with same speed. However careful observation provides significant information on the difference between those walks. While each component of the HW has gradients in the magnitude (opposite signs to each other) of their amplitudes, there seems no spatial inhomogeneity in their oscillatory behaviour. On the other hand, while the magnitudes of the SW amplitudes evolve non-symmetrically as those of the HW amplitudes, spatial inhomogeneity of the oscillatory behaviour is observed for the case. There is finer structure of the oscillation in the central region than in the peripheral front area of the SW amplitudes distribution where the peaks are.

Fig. 1 Symmetric Hadamard Walk with delta-function initial distribution at 99th step

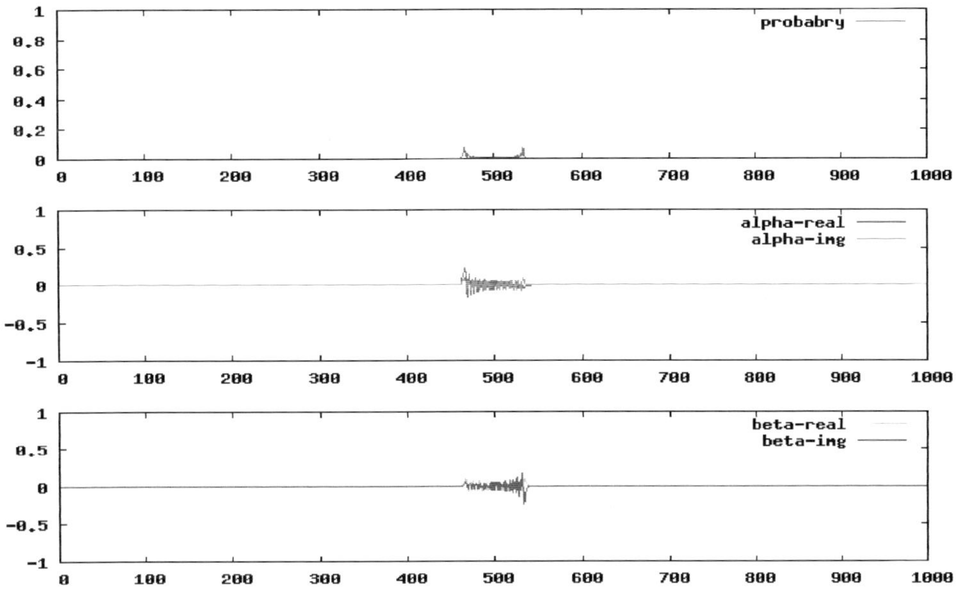

Fig 2. Symmetric Schrödinger Walk with delta-function initial distribution at 99th step

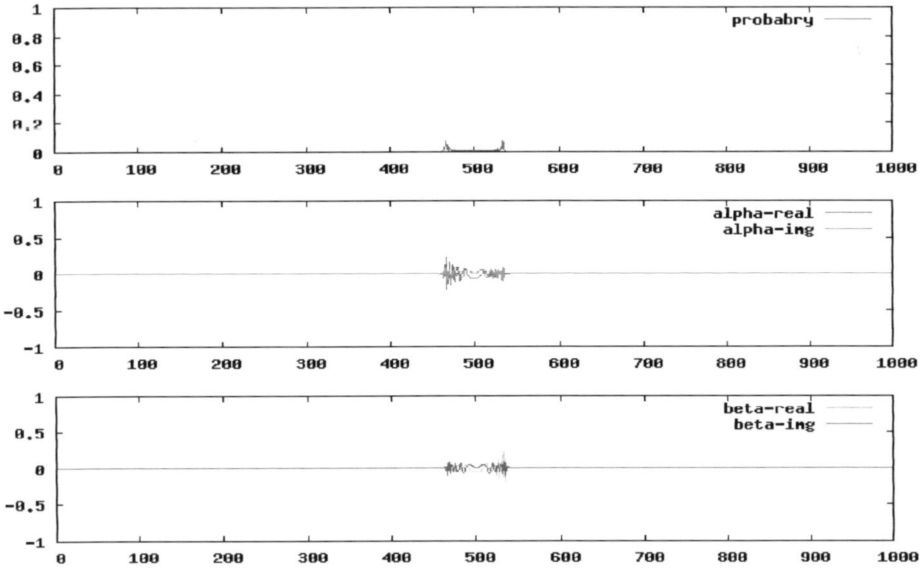

We show the results of the two QWs with an initial Gaussian distribution in Fig. 3 and Fig. 4. In the HW, the initial Gaussian peak of the probability distribution splits into two and each peak moves in opposite directions with same speed (Fig. 3). On the contrary, the SW with the initial Gaussian distribution provides an evolution of the original single Gaussian shape kept. The widening of the probability distribution is quite slow and the evolution of the probability seems stable along the walk (Fig. 4). The remarkable difference of the two walks is explained by the difference in the oscillatory structures of the amplitudes. The spatial averaging of the amplitudes results in a cancelation in magnitude for oscillating amplitudes. Two oscillating components of the HW spread in opposite directions and cancelation of the magnitude by the Gaussian superposition is enhanced in the central region. Consequently the original Gaussian peak splits to two distinct distributions which move in opposite directions. The spatial decoherence of amplitudes in the central region is responsible for the distinctive dynamics of the probability distribution in the HW.

Fig. 3 Symmetric Hadamard Walk with Gaussian initial distribution at 400th step

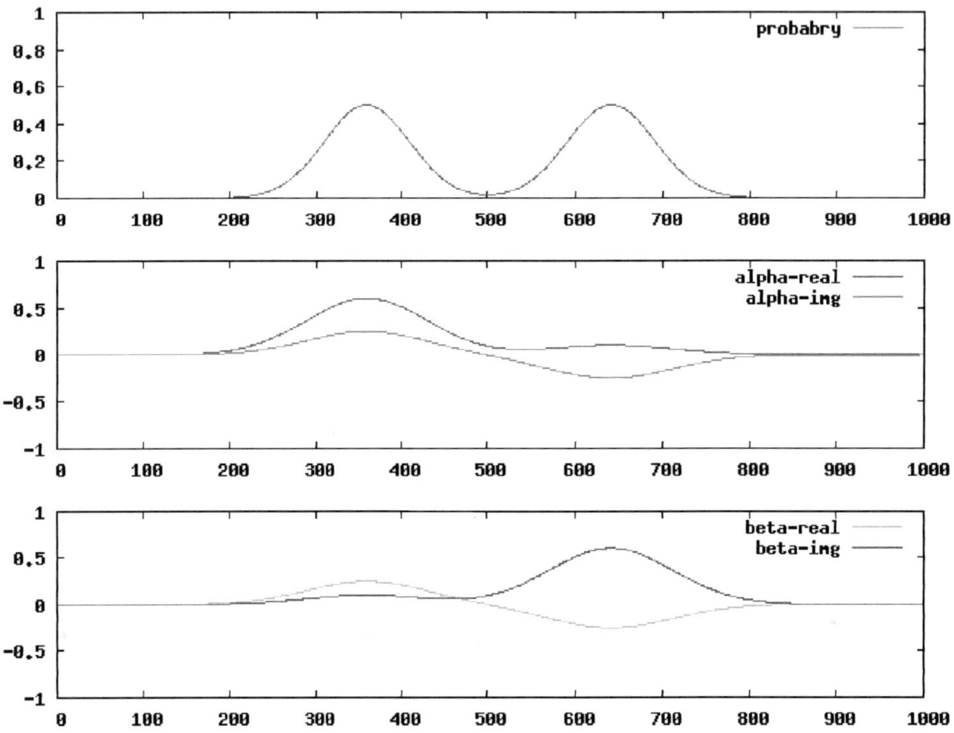

As is seen in Fig 2, the two amplitudes of SW evolve also non-symmetrically with the peaks moving in opposite directions, but the oscillation wavelength of the amplitudes is longer in central region than in the peripheral region where the peaks of components are. The Gaussian superposition of the amplitudes results in much lesser cancelation of the magnitudes in the central region. Consequently the initial Gaussian probability distribution evolves without splitting the original single peak shape. The evolution of the initial Gaussian distribution obtained by the SW is exact and identical to the solution by the method based on finite difference approximation for the TDSE partial differential equation.

Fig. 4 Symmetric Schrödinger Walk with Gaussian initial distribution at 30000th step

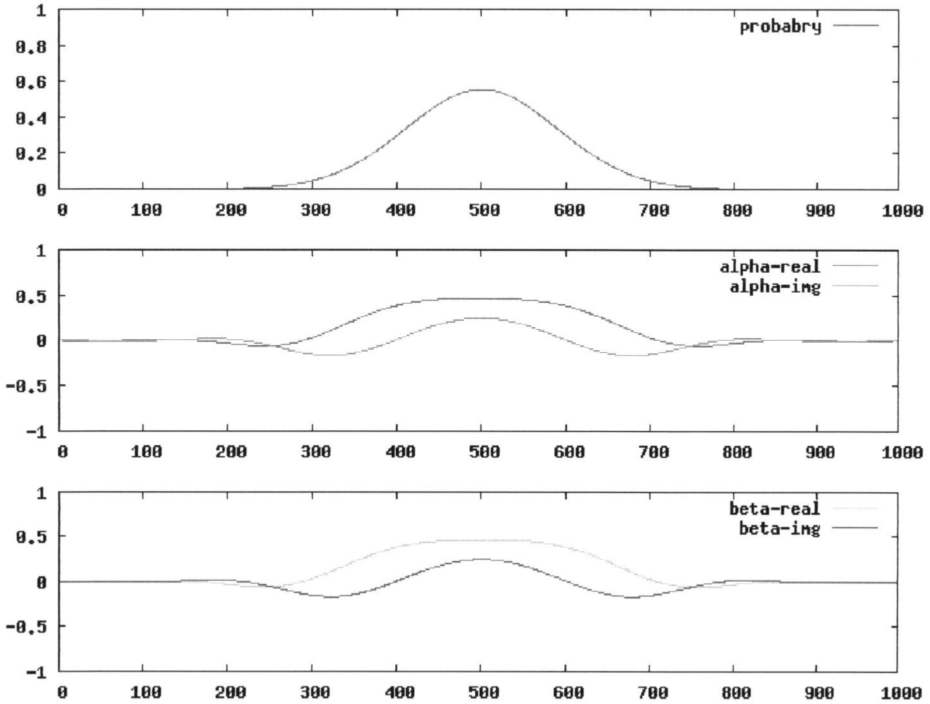

4. Conclusion

We perform a QW which leads to a solution for the TDSE with an initial Gaussian distribution by introducing a Pauli matrix component which mixes the upper and lower components of the coin space (rotation of spin) together with a boost of the quantum information in the real discrete space. The QW reproduces the exact results and the one by a conventional finite difference method. Another Pauli matrix component with certain delta-function like initial distribution leads to symmetric HW. Although two walks provide identical evolutions of probability distribution, the spatial oscillatory structures of amplitudes in the two walks are different. Consequently the probability distributions of the two walks evolve in markedly different ways: the SW evolves with the single peak structure of the initial Gaussian distribution shape kept while the HW results in splitting of the initial Gaussian distribution into two distinctive distributions moving in opposite directions. The latter evolution is considered as advection process.

We introduce Lévy-Leblond Equation which is identical to Schrödinger Equation with multi-components in order to interpret the QW in the combined Hilbert space. While LLE can be derived without any consideration of relativity, it can be considered as zeroth order approximation of Dirac equation[3]. Preliminary study[4] was done for a one-dimensional Dirac equation where the width of the wave packet is referred in connection with the velocity of quantum information. Wide wave packets evolved as a single peaked distribution while narrow wave packets provided an evolution of two peak structure for the probability distribution. Similar observation is done in our SW based upon the LLE formulation.

Acknowledgment
This research was supported in part by the Ministry of Education, Cluture, Sports, Science and Technology (MEXT), Japan Development & Application of Advanced High-Performance Supercomputer.

References
[1] N.Konno, *Quantum Walks* (Sangyo Tosho 2008):Quantum Potential Theory: Structure and Applications to Physics at the Alfried Krupp Kolleg Greifdwald, Feb. 26- March 10, 2007 http://www.math-inf.uni-greifswald.de/algebra/qpt/
[2] J.M.Lévy-Leblond, Commun.Math.Phys., **6** (1967) 286
[3] W.K.Kutzelnigg, Z.Phys.D, **11** (1989) 15
[4] F.W.Strauch, Phys.Rev. A **73** (2006) 054302, Erratum, Phys. Rev. A **73** (2006) 069908

Journal of Physics: Conference Series **352** (2012) 012014 doi:10.1088/1742-6596/352/1/012014

A new implementation of dynamic polarizability evaluation using a multi-resolution multi-wavelet basis set

Hideo Sekino[1], Yukina Yokoi[1] and Robert J. Harrison[2]

[1]Toyohashi University of Technology

Toyohashi Aichi 441-8580, Japan

sekino@tut.jp

[2]Oak Ridge National Laboratory and University of Tennessee

Knoxville Tennessee 37996, USA

Abstract. An algorithm to evaluate the response function on the space spanned by Multi-resolution Multi-wavelet (MRMW) basis functions is implemented in a new efficient quantum chemical program system. Dynamic electric dipole polarizability of a molecule is evaluated for a range of applied optical frequencies as a preliminary application of the implementation. The results obtained by the MRMW provide the one in near completeness limit. We also compare a few examples of the dynamic polarizabilities evaluated using conventional Gaussian basis set functions with the one evaluated by the MMMW basis functions. For small frequencies, there is no significant difference among them and large Gaussian results are close to the MRMW value, but they diverge from the MRMW value when the applied frequency approaches to the resonance region.

1. Introduction

Electric polarization of matter caused by optical field provide a variety of interesting properties in materials and the response of the electrons against the electro-optical perturbation in the consisting molecules is responsible for the electric polarization of the matter. Quantitative evaluation of the polarization needs quantum mechanics on the electrons in molecules. While interactions among electrons and electron with nuclei are important in the quantitative evaluation of the polarizability, precision of the model space to which infinite dimensional Hilbert space is truncated is an another very important issue. We have developed an algorithm for computing response functions[1] such as electronic polarizability as well as electronic wavefunction[2,3] using Multi-resolution Multi-wavelet MRMW) basis function[4,5]. The results by the preliminary program written in Python provided the evaluation of static polarizabilities for a series of small molecules in near basis set limit[1]. While the results computed indicated the algorithm to be superior over conventional Gaussian basis set technology, the computational resources such as storage space required for the MRMW algorithm is more demanding. New implementation in the program based on new massively parallel environments has been awaited. We implement an algorithm for evaluating dynamic polarizability in the quantum chemical electronic structure program MADNESS[6]. While static polarizabilities and hyperpolarizabilities are very important index for some materials designing where fast off-resonance

Journal of Physics: Conference Series **352** (2012) 012014 doi:10.1088/1742-6596/352/1/012014

response upon optical exposure to the molecules determines the functionality of the material qualitatively, dynamic polarizabilities are the one which should be compared with experiments. The latter is much more important especially for observations at near resonance region since the magnitude of the response is much more enhanced there. However the evaluation of dynamic polarizability involves more elaboration of the quantum chemical technology than that for static polarizability (frequency zero limit). We briefly discuss the theoretical framework of the evaluation of the dynamic polarizability on the near complete space spanned by the MRMW basis functions. We provide MRMW dynamic polarizbility of hydrogen molecule computed for a range of applied frequencies and compare the one computed by the different Gaussian basis functions against the MRMW results.

2. Theoretical method

Polarizablity is evaluated as a response coefficient of the dipole polarization which is defined by

$$\mu_\xi(E_\zeta(\omega)) = \mu_\xi^0 + \alpha_{\xi\zeta}(\omega) \bullet E_\zeta(\omega) + ... \tag{1}$$

and is determined from the first-order density operators $\rho(+\omega)$, $\rho(-\omega)$

$$\alpha_{\xi\zeta}(\omega) = -Tr\langle \xi \bullet (\rho_\zeta(+\omega)) \rangle \tag{2}$$

$$\rho(\omega) = \rho^{(0)} + \rho(+\omega)\lambda e^{+\omega} + \rho(-\omega)\lambda e^{-\omega} + \bullet\bullet\bullet \tag{3}$$

Here $\rho^{(0)}$ is a zer-th order density operator.

The pth response function is defined as a set of parameter functions

$$u_p^{(+)} = (1-\rho^0)\rho(+\omega)\rho_p^0 \tag{4a}$$

$$u_p^{(-)} = (1-\rho^0)\rho(-\omega)\rho_p^0 \tag{4b}$$

to represent the first-order density operators. Here ρ_p^0 is the pth orbital contribution to the zero-th order density operator ρ^0 generated from the zero-th order orbitals. The equation of motion for the perturbed density operators can then conveniently expressed as the coupled equations with respect to $u_p^{(+)}, u_p^{(-)}$.

$$(F^0 u_p^{(+)} - u_p^{(+)}\varepsilon_p^0) + \{\frac{d}{2} + \frac{\delta g}{\delta \rho}[\rho^0] * (\sum_i^{N_{occ}} u_i^{(+)}\phi_i^0 + \sum_i^{N_{occ}} \phi_i^0 u_i^{(-)})\} = \omega u_p^{(+)} \tag{5a}$$

$$(F^0 u_p^{(-)} - u_p^{(-)}\varepsilon_p^0) + \{\frac{d}{2} + \frac{\delta g}{\delta \rho}[\rho^0] * (\sum_i^{N_{occ}} u_i^{(-)}\phi_i^0 + \sum_i^{N_{occ}} \phi_i^0 u_i^{(+)})\} = -\omega u_p^{(-)} \tag{5b}$$

Where F^0, $\phi_p^0, \varepsilon_p^0$, d and $\frac{\delta g}{\delta \rho}[\rho]$ are zero-th order Hamiltonian, orbital, orbital energy, dipole operator and coulomb and exchange interactions for Hartree Fock (HF) and exchange-correlation functional for Density Functional Theory (DFT). A*B indicates the convolution of A and B. Note that no operator such as density operator is stored as an array and all the unknowns such as response functions are stored in one-dimensional arrays during the solution of the above equations.

Equations (5a) and (5b) are solved using Green's function technique for the kinetic energy term with the free space boundary condition.

$$u_p^{(+)} = G(\omega) * [V^0 u_p^{(+)} + \{\frac{d}{2} + \frac{\delta g}{\delta \rho}[\rho^0] * (\sum_i^{N_{occ}} u_i^{(+)}\phi_i^0 + \sum_i^{N_{occ}} \phi_i^0 u_i^{(-)})\}] \tag{6a}$$

$$u_p^{(-)} = G(-\omega) * [V^0 u_p^{(-)} + \{\frac{d}{2} + \frac{\delta g}{\delta \rho}[\rho^0] * (\sum_i^{N_{occ}} u_i^{(-)}\phi_i^0 + \sum_i^{N_{occ}} \phi_i^0 u_i^{(+)})\}] \tag{6b}$$

Here G(ω) is the Green's function

$$G(\omega) = \left[\varepsilon_p^0 + \omega - (-\frac{1}{2}\nabla^2) \right]^{-1} \tag{7}$$

and depends on the applied frequency.

The straightforward iterative solution of (6a) and (6b) converges slowly and not efficient especially for frequencies of near resonance. We employ Krylov subspace algorithm[7] in order to improve the efficiency.

3. Results and discussion

We compute the dynamic dipole polarizability of the hydrogen molecule for many different frequencies (from 0.365 to 0.46 a.u.). The longitudinal component of Hartree Fock polarizability is plotted along the different frequencies (Fig. 1). The residual $\varepsilon = 1 \times 10^{-4}$ and the order of polynomial k=6 are set for all the calculations. The values computed using conventional Gaussian function are plotted together. We also plot the deviation of the Gaussian results from the one obtained by the MRMW in Fig. 2. For the off-resonance region, all the results obtained by Gaussian basis sets are almost identical and coincides with the one by the MRMW except for relatively small Gaussian basis sets such as aug-cc-pVDZ and d-aug-cc-pVDZ. However, we observe the divergence of those values from the MRMW value in near resonance region even for larger basis sets. It seems that the addition of d-function to the large basis sets overshoots the MRMW value whereas enlargement of the basis by splitting valence area improves the Gaussian results towards the MRMW one. However, there is no definite criterion on the systematic improvement of the predicted polarizability as in the MRMW case. Intuitive construction by addition of extra functions to large Gaussian basis sets sometimes causes serious problems such as incapability of convergence because of over-completeness and the resulting overwhelming computational time. The MRMW basis functions are completely local and no issue relating to over-completeness arises. Systematic and guaranteed improvement of the space is expected by only setting a threshold for the residuals of the target property.

Fig. 1

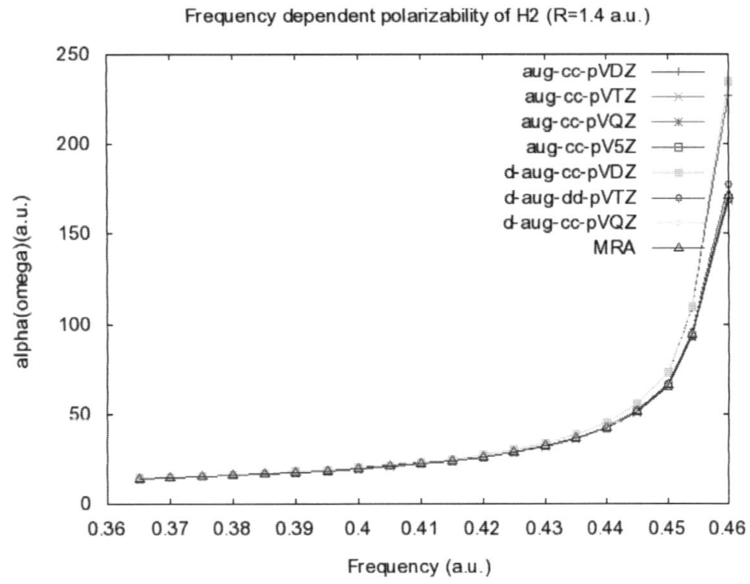

Fig. 2

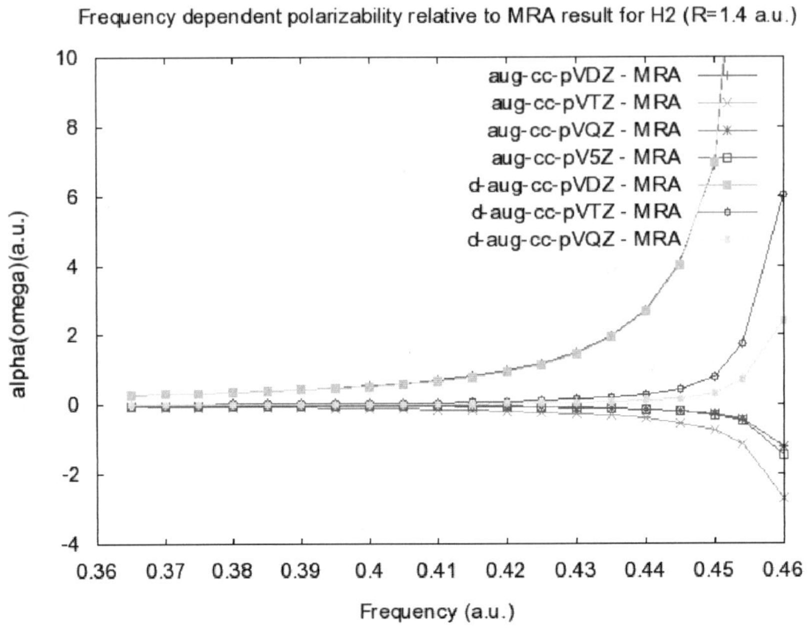

4. Conculsion

A code based on the response function algorithm is implemented in MADNESS which is a framework for scientific simulation in many dimensions using adaptive multi-resolution methods in multi-wavelet bases. The preliminary application for the evaluation of dynamic polarizability shows potential problems of Gaussian as incomplete basis set in the near resonance region.

While we have already shown the strength of the MRMW algorithm in quantum chemical computation a while ago, the memory handling issue arising from the huge number of polynomials representing the quantum properties such as wavefunction and response functions with sufficient precision was problematic for older generation of scientific computational environments. The situation is rapidly being altered because of the availability of inexpensive processors and memories with high quality. MADNESS is an object-oriented designed program and is created from the beginning as a parallel processing program for computers with up to millions of cores running on the Cray XT5 at Oak Ridge National Laboratory and the IBM Laboratory and the IBM Blue Gene at Argonne National Laboratory. We expect the new implementation will enable us to evaluate interesting properties such as dynamic polarizability and hyperpolarizabilities of larger molecules at near resonance with high precision.

Acknowledgment

This research was supported in part by the Ministry of Education, Culture, Sports, Science and Technology (MEXT), Japan Development & Application of Advanced High-Performance Supercomputer. Y.Y. thanks to Japan Student Service Organization (JASSO) for providing her an opportunity of visiting to the University of Tennessee.

References

[1] H. Sekino, Y. Maeda, T. Yanai and R. J. Harrison, J.Chem.Phys., **129** (2008) 034111
[2] R. J.Harrison, G.I.Fann, T.Yanai,Z.Gan and G. Beylkin, J. Chem.Phys. **121** (2004) 11587
[3] T.Yanai, G.I.Fann, Z.Can, R.J.Harrison and G.Beylkin, J.Chem.Phys. **121** (2004)2866
[4] B.Alpert, G.Beylkin, D.Gines and L.Vozovoj, J.Comput. Phys., **182** (2002) 149

[5] B.Alpert, SIAM J.Math.Anal. **24** (1993) 246
[6] MADNESS (Multiresolution Adaptive Numerical Environment for Scientific Simulation)
A quantum chemical program system using Multiresolution Multiwavelet basis function
http://www.csm.ornl.gov/ccsg/html/projects/madness.html
[7] R.J.Harrison, J.Comput.Chem. **25** (2002) 328

Detecting Ferrite Nanobeads for Sentinel Lymph Node Mapping with a Highly Sensitive Hall Differential Magnetic Field Sensor

M. Abe[1,2*], T. Ueda[2], T. Masaki[3], Y. Kitamoto[3], N. Matsushita[1], and H. Handa[2,4]

[1]Materials and Structures Laboratories, [2]Solutions Research Laboratory, [3]Graduate School of Science and Engineering, and [4]Department of Biological Information, Tokyo Institute of Technology, 4259 Nagatsuta-cho, Midori-ku, Yokohama 226-8503, Japan

*amasanor@pe.titech.ac.jp

Abstract. We fabricated a novel type of Hall differential magnetic field sensor for anti-cancer sentinel lymph node (SLN) mapping using ferrofluid as a marker. A pair of Hall devices are mounted on both end surfaces of a ferrite core (10 mm ϕ ×32 mm) of an electromagnetic coil which generates an AC exciting magnetic field at 2.5 kHz. The signals are retrieved by a digital phase sensitive detection circuit. Mapping a ferrofluid (Resovist[R]) sample of 100μg in Fe atomic amount (comparable to that accumulated in human SLNs) was attained when the sample was placed within 6 mm distance from the sensor head. The detectable distance is limited primarily due to the magnetic induction effect of the metal XYZ stage which held the sample.

1. Introduction

Distant metastasis of cancer is caused by cancer cells which are isolated from primary tumors and drifted along with blood flow and lymph flow. Cancer cells drifting along with lymph flow are trapped inside lymph nodes, where lymph cells attack to kill them. If the cancer cells survive, metastasis occurs at the lymph nodes.

As such the concept of "sentinel lymph node" (SLN) was established: SLN is the first lymph node that filters the fluid draining away from the primary tumor. If the SLN is not metastasized, cancer is not spread to down stream lymph nodes. This phenomenon has been exploited in "SLN mapping and biopsy", or "SLN navigation surgery". Tracers are injected as a marker from near the primary tumor and the SLN is identified by detecting the marker that has migrated along the lymph channel. The SLN is then resected for pathological examination of the staging of the metastasis. If the SLN is not metastasized, resection of down stream lymph nodes can be avoided, and the patient's quality of life is much improved.

SLN navigation surgery has become a standard treatment for breast cancer and melanoma at present, and it is gaining popularity for other kinds of cancers [1].

In conventional SLN mapping a radioisotope and a pigment are used at the same time as the marker. The isotope is detected by a γ-ray probe and pigment is used for secondary eye observation. However, they have drawbacks: the pigment can be allergy provoking, and the radio-isotopes are strictly regulated by law and small-scale medical agencies can not afford to use them. To solve the problems researchers tried to replace the marker with ferrite nanobeads and detect them with magnetic field sensors, as shown in Fig.1. They first used magnetoresistance (MR) devices for the sensing and clinical tests were performed [2, 3]. Then, sensors utilizing SQUIDs were developed to improve the sensitivity [4, 5]. However, SQUIDs are expensive and complicated to use.

To address these issues we invented sentinel lymph node mapping by exploiting the magneto-acoustic effect [6,7]. The magnetic beads trapped and accumulated in the SLN were excited by an AC magnetic field to generate acoustic waves, which were detected by a microphone. This gave sensitivity high enough for SLN mapping.

However, we found problems with the acoustic detection. In actual SLN surgery, conclusive identification of SLN is made by directly touching the γ-ray probe head to the lymph nodes in open wound or to the resected lymph nodes put on a back table. This is impossible by using the microphone probe, because the direct contact causes acoustic impedance mismatch; impedance matching is obtained when the lymph nodes are embedded in continuous media such as living body.

Therefore we developed a novel type of Hall differential magnetic field sensor [8]. In this paper we will describe the principle of our novel sensor, report their performance evaluated from our preliminary experiments, and discuss the outlook for the new sensor.

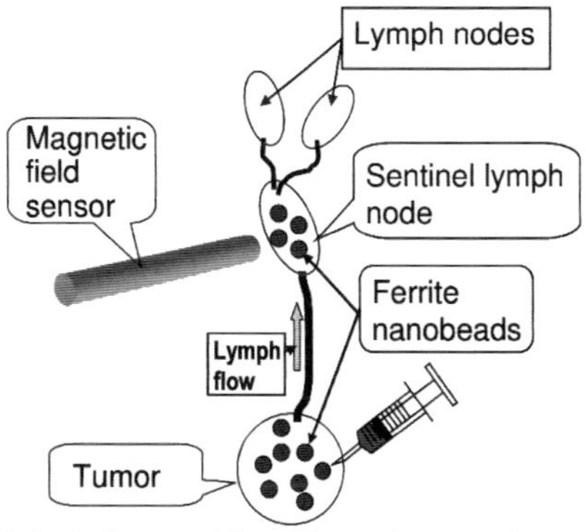

Fig.1 Sentinel lymph node mapping by detecting ferrite nanobeads with a magnetic field sensor.

2. Principle and experimental

Figure 2 shows our new sensor, which is molded in a gun type case. The sensor head has a pair of Hall devices (Honeywell SS495A) mounted on both end surfaces of a ferrite cylindrical core (10 mm ϕ ×32 mm) of an electromagnet coil. This enables differential detection of near field signals from the magnetic beads, and magnetic (far) fields of environmental noise and earth magnetism are cancelled. The ferrite core concentrates magnetic fluxes generated by the exciting coil and by the magnetic beads. These features enhance the sensitivity of the sensor.

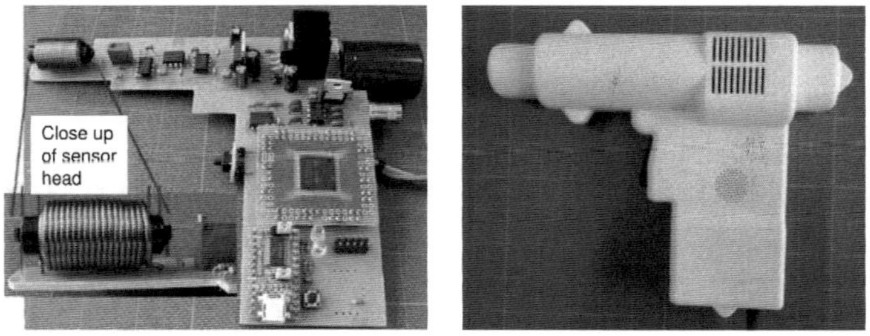

Fig. 2 New type of Hall differential field sensor, circuit board and case.

Applying a continuous AC current at 2.5 kHz to the exciting coil (which is resonated with capacitors connected in parallel) a 250 Oe rms field is generated on the sensor head as measured by the Hall device on the head. The signals from the magnetic beads are retrieved by a digital phase sensitive detection circuit with a 24 bit A/D conversion resolution. Data acquired over 50 msec were averaged over 400 cycles (sampling time: 20 sec), which improved the S/N ratio by about 26 dB (=20 sec/50 msec).

We evaluated the sensitivity of the sensor by detecting Resovist^{R}, an FDA approved intravenous injection MRI contrast agent of ferrite nanobeads ferrofluid. The ferrofluid was adsorbed on the tip of a cotton swab to make a dot-like sample which was less than 1 mm in size. The sample was roughly 100 μ g in Fe atomic amount, a value comparable to that accumulated in human axillary SLNs in SLN surgery.

The cotton swab containing the sample was mounted on an XYZ stage via a plastic bar (ca. 20 cm in length), as shown in Fig.3, in order to minimize the magnetic induction effect of bulk metal of the XYZ stage. This length was chosen to limit signal deterioration due to mechanical vibration. Also the background signal (obtained when the sample was removed), which was mainly caused by the XYZ stage, was subtracted from the raw signal. The background signal was nearly independent of the distance, *x*, of the sample from sensor head.

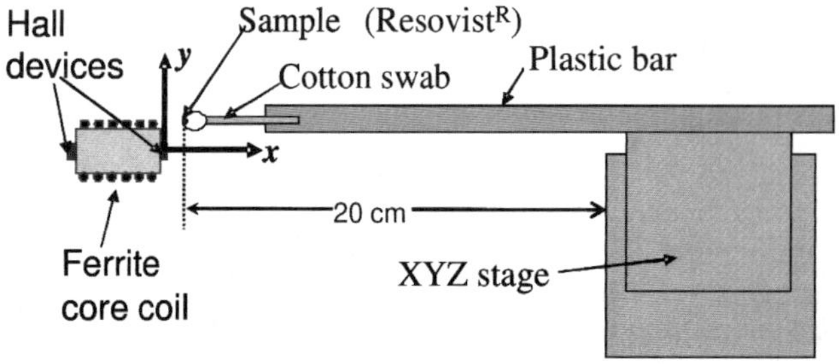

Fig. 3 Experimental setup for mapping a dot-like sample of ferrofluid (Resovist^R) on the tip of a cotton swab. The sample is moved by XYZ stage with respect to the sensor head (Hall device).

3. Results and discussion

Figure 4(a) shows the output signal plotted in logarithmic scale as a function of *x*. Here we fixed the lateral position (*y* and *z*) of the sample to the place where the signal reached maximum, roughly along the coil axis. The maximum detectable distance *x* (corresponding to depth of SLN from body surface) with S/N ratio=1 was determined to be 13 mm. The exciting magnetic field decreased to 12 Oe rms at x=13 mm.

Figure 4(b) shows the output signal plotted as a function of *y* (a lateral distance), obtained when *x* was fixed at values between 4 mm and 10 mm; when *x* was smaller than 4 mm we could not retrieve signals which were too strong to fall in the dynamic range of the circuit. When x=4 mm and 6 mm the signal exhibited a nearly symmetrical curve making a peak around the coil axis. Thus our sensor is capable of mapping the Resovist^R sample at 6mm depth. For x=8 mm and 10 mm the signal curves did not exhibit a peak and mapping was impossible. This is because the signal is dominated by the magnetic induction effect of the XYZ stage

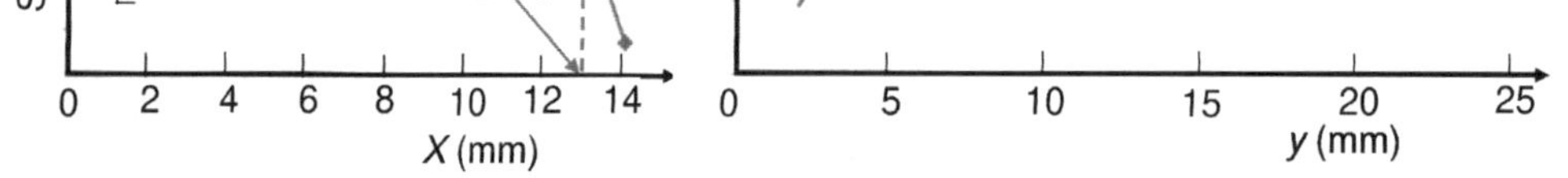

Fig. 4 Signal obtained for ferrofluid (Resovist^R) sample: (a) plotted as a function of distance (*x*) of sensor head from sample, obtained when lateral position (*y* and *z*) was fixed around coil axis, and (b) plotted as a function of a lateral distance (*y* in relative value), obtained when *x* was fixed to various values.

4. Concluding remarks

For the Resovist[R] sample of an amount comparable to that accumulated in human SLNs, we obtained maximum detectable distance (x) of 13 mm when the sample was placed roughly along the coil axis (Fig. 4(a)). Since in actual SLN mapping maximum detectable depth of 10 mm is required, our sensor has a potential for clinical application. However, for mapping the sample in the lateral directions (y and z) maximum detectable distance decreased to ca. 6 mm (Fig. 4(b)) due to the magnetic induction from the metal XYZ stage, albeit it being separated from the sensor head by about 20 cm. Therefore in clinical SLN mapping metal materials must be separated from the sensor much longer than 20 cm.

The sensitivity of our new sensor will be improved by introducing the spread-spectrum noise reduction technique [9] to the digital phase sensitive detection circuit. If the signal spectrum is spread from present 20 Hz (corresponding to 50 msec sampling time) to e.g. 200 kHz, the information capacity increases by factor 10,000 (=200 kHz/20 Hz) and thus we can improve the S/N ratio by nominally 40 dB ($=(10{,}000)^{1/2}$) at maximum. We are currently investing our efforts to take advantage of this.

On a final note, our sensor might be easily integrated into an endoscope by separating the sensor head from main circuit and mounting it on the endoscope edge.

References

[1] Chen S L, Iddings D M, Scheri R P and Bilchik A J 2006 *CA Cancer J. Clin.* **56** 292

[2] Nakagawa T, Minamiya Y, Katayose Y, Saito H, Taguchi K, Imano H, Watanabe H, Enomoto K, Sageshima M, Ueda T and Ogawa J 2003 *J. Thorac. Cardiovasc. Surg.* **126** 563

[3] Minamiya Y, Ito M, Katayose Y, Saito H, Imai K, Sato Y and Ogawa J 2006 *Ann. Thoracic Surgery* **81**, 327

[4] Tanaka S, Hirata, A, Saito Y, Mizoguchi T, Tamaki Y, Sakita I and Monden M 2001 *IEEE Trans. Appl. Superconductivity* **11** 665

[5] Joshi T, Pankhurst QA, Hattersley S, Brazdeikis A, Hall-Craggs M, De Vita E, Bainbridge A, Sainsbury R, Sharma A and Douek M 2007 *Breast Cancer Res. Treatment* **106 (suppl 1)** S129

[6] Kakegawa K, Ueda T, Tada M, Nakagawa T, Handa H and Abe M 2008 *Abstracts of 53rd Magnetism and Magnetic Materials Conf.* (Austin) CG-07

[7] Abe M, Kakegawa K, Ueda T, Nakagawa T, Tada M and Handa H 2009 *Digests of IEEE International Magnetics Conf.* (Sacramento) CB-03

[8] Abe M, Ueda T, Masaki T, Kitamoto Y, Matsushita N and Handa H 2011 *Digests of IEEE International Magnetics Conf.* (Taipei) EG-03

[9] Dixon C D 1994 *Spread Spectrum Systems: With Commercial Applications* (Wiley, John & Sons)

Reduced Graphene Oxide as the Support for Lipid Bilayer Membrane

K Tsuzuki[1], Y Okamoto[1], S Iwasa [1], R Ishikawa[2], A Sandhu[2,3] and R Tero[3]

[1] Department of Environmental and Life Sciences, Toyohashi University of Technology, 1-1 Hibarigaoka, Tempaku-cho, Toyohashi, Aichi, 441-8580, JAPAN
[2] Department of Electrical and Electronic Engineering, Tokyo Institute of Technology, 2-12-1 O-okayama, Meguro, Tokyo, 152-8552, JAPAN
[3] Electronics-Inspired Interdisciplinary Research Institute, Toyohashi University of Technology, 1-1 Hibarigaoka, Tempaku-cho, Toyohashi, Aichi, 441-8580, JAPAN

E-mail: k061503@edu.tut.ac.jp (K.T), and tero@eiiris.tut.ac.jp (R.T)

Abstract. Supported lipid membranes of dioleoylphosphatidylcholine (DOPC) were formed on reduced graphene oxide (r-GO) by the vesicle fusion method. Graphene oxide prepared by chemical exfoliation was deposited on a thermally oxidized SiO_2/Si substrate, and was reduced by hydrazine vapour. Flat lipid membranes were formed on the r-GO surface after the r-GO/SiO_2/Si substrate was incubated in the DOPC vesicle suspension. We propose a structural model in which a lipid bilayer and monolayer stacked on r-GO based on the results of fluorescence microscope and atomic force microscope observation.

1. Introduction

Graphene is single atomic-layered carbon material, and the applications in various fields are anticipated because of its intriguing optical [1,2], magnetical [3], and mechanical properties [4], as well as the application for electronic materials with high electron mobility [5]. Recently several applications of graphene and its derivatives including graphene oxide (GO) to biosensing were reported: the detection of DNA hybridization and protein-peptide binding using fluorescence quenching effect of graphene [6,7], detection of biomolecules by using graphene-based field emission transistor [8], and the application of GO as a fluorescence probe in cells [9]. Plasma membranes are the reaction fields of the variety of biological reactions involved in diseases and the transportation of information and materials in and out of cells. Lipid bilayers are the fundamental structure of cell membranes, and examining the molecular behavior of lipids and proteins, and their ensemble using artificial lipid bilayers is important to ascertain physical and chemical processes involved in cell membrane reactions. Molecular orientation and two dimensional assembly of biomolecules such as lipids, peptides and proteins in and on lipid bilayers were investigated with various imaging and spectroscopic methods [10-21].

Our research interest is in the development of a new method to obtain molecular information in lipid bilayers using graphene. In this paper, we describe the formation of lipid membranes on reduced GO (r-GO) prepared by chemical exfoliation and chemical reduction. Graphene oxide prepared by the chemical exfoliation of graphite has amphipathic property because of the co-existence of hydrophilic areas containing hydroxyl, carboxyl, and epoxide groups, and hydrophobic areas similar to perfect graphene [2]. The elimination of these hydrophilic functional groups by the chemical (e.g. using hydrazine) and thermal reduction increases the hydrophobicity of the GO surface. A supported lipid bilayer (SLB) is one of artificial lipid bilayer systems [22 - 24] formed at solid-liquid interfaces. Hydrophilicity of the substrate surface is one of critical factors for the SLB formation [25 - 27], because the lipid bilayer is a self-assembled structure constructed by hydrophilic/hydrophobic interaction. Therefore the efficiency of the SLB formation will be influenced by the reduction of GO. We examined the formation, structure and properties of the lipid membranes on r-GO, which we prepared by exposing GO to hydrazine vapor, in detail with atomic force microscope (AFM) and fluorescence microscope.

2. Experiment

The GO flakes was prepared from graphite powder (Z+80, Ito Kokuen Co., Ltd., Japan) by chemical exfoliation according to the modified Hummer's method [28,29]. The graphite powder was stirred in the mixture of sulfonic acid, peroxydisulfuric acid and phosphorus pentoxide at 80°C for 4.5 h, then filtrated and washed with water. The residue was dried overnight, and added to the mixture of sulfonic acid and potassium permanganate below 10°C. After stirring the mixture at 35°C for 2 h, we added water to the mixture, and the reaction was quenched by hydrogen peroxide. After the overnight sedimentation, the sediment was washed with 3% HCl aq. and water, and exfoliated by using a centrifuge and an ultrasonic bath sonicator, to obtain a GO solution. The GO solution was dropcast on a SiO_2/Si (100) substrate and dried. As-delivered Si wafer was thermally oxidized under oxygen atmosphere at 1000°C for 135 min, and the SiO_2/Si substrate was cleaned by boiling in piranha solution for 10 min, and sonication in 0.02 M KOH aq. for 10 min. The thickness of the SiO_2 layer was 92 nm, which is suitable to the observation of graphene flakes with high contrast using a bright field microscope [30]. The GO/SiO_2/Si substrate was exposed to hydrazine vapor in a glass dish with a cover on a hot plate at 100°C [31], to reduce GO.

The solutions of dioleoylphosphatidylcholine (DOPC) (Avanti Polar Lipid) in chloroform, and a fluorescence-labelled lipid (BOPIPY-H-PC: Ex/Em=534/552 nm) (Invitrogen Corp.) in ethanol were mixed at 100:1 molar ratio. The solution of the lipid mixture in a glass vial was dried by N_2 flow and overnight evacuation. We put a buffer solution (100 mM KCl, 25 mM HEPES/ pH 7.4 NaOH, chemicals were purchased from Wako Pure Chemical Industries, Ltd.) into the glass vial with the vacuum-dried lipid film, and agitated the vial to prepare the suspension of multilamellar vesicles (MLV) [14]. The MLV suspension was frozen and thawed in liquid nitrogen and a 45°C water bath repeatedly, extruded through a 100 nm polycarbonate filter, and sonicated to obtain unilamellar vesicles. The r-GO/SiO_2/Si substrate was incubated in the sonicated DOPC+BODIPY-H-PC vesicle suspension with $CaCl_2$ (10 mM) at 45°C for 60 min. The excess vesicles in the liquid phase were washed out by exchanging the suspension with the buffer solution without vesicles.

The GO flakes were observed by scanning electron microscope (SEM) (S-3000N, Hitachi High-Tech Corp.) under high vacuum ($<10^{-3}$ Pa), at 25 kV and in the secondary electron mode. The r-GO/SiO_2 surface after the incubation in vesicle suspension was observed in the buffer solution using an AFM and a fluorescence microscope. An epifluorescence microscope (Olympus IX51) was used for the bright-field and fluorescence observation. The GO/SiO_2/Si and r-GO/SiO_2/Si surfaces were observed in air. The r-GO/SiO_2 surface after the incubation in the vesicle suspension was observed in the buffer solution with a 60× water-immersion objective lens. Fluorescence recovery after photobleaching (FRAP) was performed by the irradiation of excitation light which intensity was 280 times stronger than that at the observation. The AFM observation (Agilent PicoScan2500) of the GO and r-GO flakes on SiO_2 surfaces were performed in acoustic AC mode in air by using a cantilever with spring constant (C) of 28 N/m (SI-DF40, Seiko Instruments Inc.). The r-GO/SiO_2 surface after the incubation in the vesicle suspension was observed in magnetic AC mode in the buffer solution by using a magnetically coated cantilever (TYPE I MAC Lever, Agilent, C = 0.6 N/m).

3. Results and discussion

The SEM image of the GO flakes prepared by modified Hummer's method is shown in figure 1A. The maximum size of the GO flakes was 323.38 μm. The size of the original graphite particle was 258.2 μm on average (data not shown), and some GO flakes exfoliated remaining the original sheet size in the graphite particle. In the previous reports, in general the size of GO flakes is in the order of μm to 10 μm [28,29,31,32] and the reports about the GO flakes over 100 μm is still very few [33,34]. The size of the GO flake was decreased with the time of sonication for rapid exfoliation. The size of the GO flakes was partly controlled by the sonication time, to obtain appropriate flake size for the AFM observation on a scale of 10-15 μm. We found that the fragments of the GO flakes had characteristic stable size (figure 1B). As-prepared GO flakes without sonication distributed at the areas around 25000 μm^2 (corresponds to ~160 μm × 160 μm) and 2500 μm^2 (corresponds to ~50 μm × 50 μm) (figure 1B, red histogram). The flake size mainly distributed at 2500 μm^2 and 375 μm^2 (corresponds to ~19 μm × 19 μm) after the sonication for 0.5-1 h (figure 1B, yellow and green histograms). After the sonication for 2 h, the flake size had monodisperse distribution around 250 μm^2 (corresponds to ~16 μm × 16 μm) (figure 1B, blue histogram).

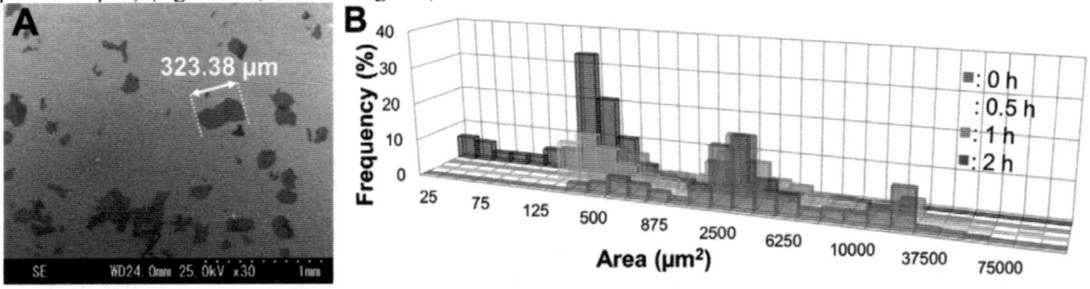

Figure 1. (A) SEM image of the GO flakes on SiO_2/Si. (B) Histogram of the area of the GO flakes evaluated from SEM images, and its dependence on the time of ultrasonic fragmentation: (red) without sonication, (yellow) 0.5 h, (green) 1 h, and (blue) 2 h.

The chemical exfoliation by the modified Hummer's method has an advantage that a quantity of GO can be obtained as an aqueous solution at low cost. The size of the graphene flakes can be roughly controlled once a large GO flake is prepared (figure 1). The GO flakes contain sp^3-carbon regions and holes, which interfere electron mobility, thus the chemical exfoliation is regarded to be inferior to other methods [35,36] such as chemical vapor deposition(CVD), mechanical exfoliation of graphite, and epitaxial growth, for the application to electronic devices. For the uses in biological applications, however, the defects in GO and r-GO may not be a serious matter [6,7,9]. Thus we prepared GO using the modified Hummer' method to construct graphene-supported lipid bilayer system.

Figure 2A shows the AFM topography of GO observed in air. Thickness of the adsorbed GO on the SiO_2 surface was 1.45±0.23 nm (n=121) (figure 2C). Thickness of the second GO layer on first the GO flakes was 1.34±0.26 nm (n=30), at the region two GO flakes overlapped. Previous reports showed that the thickness of single GO layer on solid substrates is 1-1.5 nm, although the inter-layer distance in graphite is 0.3 nm. Graphene oxide measures thicker than perfect graphene because the functional groups, such as hydroxyl, epoxy, and carboxyl groups, protrude from its two-dimensional carbon framework, and also because linearly aligned sp^3 carbons makes a crinkle [37]. Thickness of GO on solid substrates is also sensitive to the roughness and adsorbed water layer on the substrate surface [38,39]. The sonication for 2 h, which fragmented GO flakes from 25000 μm^2 to 500 μm^2 by breaking interlayer C-C covalent bonds (figure 1B), would be sufficient to exfoliate the intra-layer π-π stacking almost completely. Therefore we conclude that the flakes observed in figure 2A were single-layered GO.

Figure 2B shows the AFM topography of r-GO after it was exposed to hydrazine vapor for 30 min. Thickness of the r-GO flakes adsorbed on the SiO_2/Si surface was 2.86±0.36 nm (n=89) (figure 2D), and the thickness of the second r-GO layers on the first r-GO flakes was 1.18±0.18 nm (n=15). The first layer of r-GO (figure 2D) was 1.4 nm thicker than GO before reduction (figure 2C). On the other hand, second r-GO layer was 0.3 nm thinner than the second GO layer before reduction. We think at present that the first layer of r-GO was thickly observed because of the effect of substrate surface and/or the adsorbed water between SiO_2 and r-GO on the reduction process [38,39]. We dropcast GO solution on the SiO_2 surface pre-treated with hydrazine, the thickness of the first and second GO layers was 2.05±0.27 nm (n=125) and 1.56±0.18 nm (n=17), respectively. Thus the apparent height of the first GO layer strongly depended on the surface condition of the substrate surface. We suppose that the second layer of r-GO was thinner than the GO without reduction because the recovered double bonds made increase the π-π interaction between the r-GO layers.

Figure 2. AFM topographies of (A) the GO flakes before reduction, and (B) the r-GO flakes after exposure to hydrazine vapor at 100°C for 30 min. (C, D) The cross section profiles at the white lines in (A) and (B), respectively.

Figure 3A and 3B show the bright-field and fluorescence images of as-prepared GO on SiO_2/Si, respectively. Fluorescence signal (figure 3B) was observed at the corresponding position of the GO flakes, which was observed in the bright-field image (figure 3A). Figure 3C-F show the bright-field and fluorescence images of GO after the reduction with hydrazine vapor. The fluorescence intensity from the r-GO flakes after exposure to hydrazine vapor for 10 min (figure 3D) and 30 min (figure 3F) was less than 1% of that from the GO flakes before reduction (figure 3B).

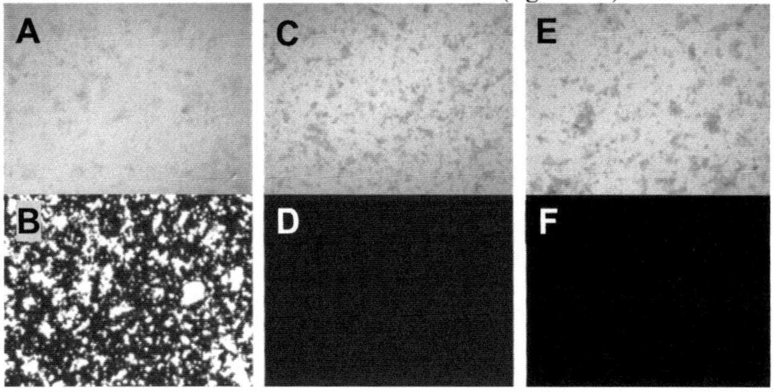

Figure 3. Bright-field images and fluorescence images of GO and r-GO obtained at the same position of each surface. (A) Bright-field image and (B) fluorescence image of GO before reduction. (C) Bright-field image and (D) fluorescence image of r-GO after exposure to hydrazine vapor for 10 min. (E) Bright-field image and (F) fluorescence image of r-GO after exposure to hydrazine vapor for 30 min. Fluorescence observation was performed with the irradiation of excitation light 69 times brighter, and the exposure time was 10 times longer than those for the SLB observation in figure 4A.

Figure 4 shows the fluorescence image of the r-GO/SiO_2/Si surface incubated in the sonicated vesicle suspension of DOPC containing 10 mM of Ca^{2+}. The bare SiO_2 regions were bright, while the regions of the r-GO flakes were dark. The homogeneous fluorescence intensity from the SiO_2 regions indicates the formation of SLB on SiO_2, in accordance with previous studies [25-27]. The fluorescence from the r-GO regions was considerably weaker than that from the SiO_2 region, probably because of the fluorescence quenching effect of r-GO [1,2,40,41]. We could not distinguish the state of the DOPC bilayer on r-GO, whether planar membrane or adhesive vesicular layer [14], only from the static fluorescence image in figure 4Aa. Therefore we performed FRAP at the SiO_2 region surrounded by r-GO (dotted circle in figure 4Aa). The FRAP process shown in figure 4b-d indicates that lipid molecules diffused freely between the SiO_2 regions and the surrounding r-GO regions, and that fluid and continuous SLB were formed on both r-GO and SiO_2. Figure 4B shows the AFM topography of the r-GO/SiO_2 surface after the incubation in the DOPC vesicle suspension, observed under the same condition as figure 4A. The surfaces on the r-GO and SiO_2 regions had flat morphology, and adsorbed vesicles were rarely observed. The region on the first r-GO layer was 3.70±0.48 nm (n=60) higher than the SiO_2 region, and the region on the second r-GO layer was 1.51±0.42 nm (n=40) higher than that on the first r-GO layer.

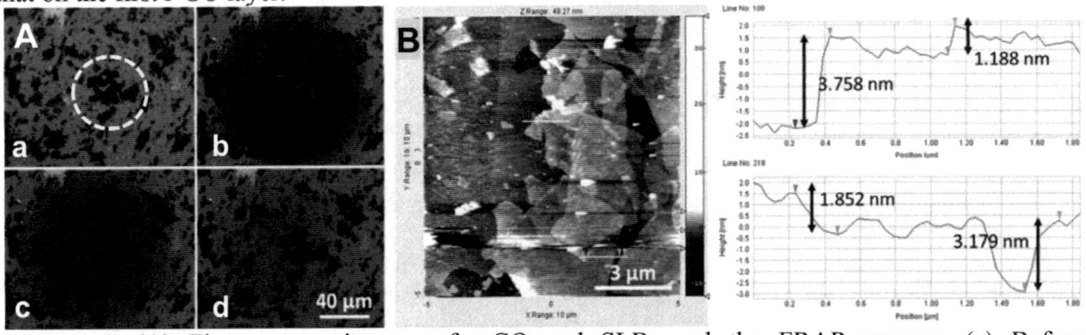

Figure 4. (A) Fluorescence images of r-GO and SLB, and the FRAP process. (a) Before fluorescence bleaching. (b) 0 s, (c) 1 min, and (d) 5 min after photobleaching. (B) AFM topography of the same sample as (A), and the cross section profiles at the white lines.

We propose that stacking layers of lipid bilayer/monolayer were formed on r-GO as shown in figure 5, according to the results of FRAP and AFM image in figure 4. It is well known that single SLB forms on SiO_2 surfaces by the vesicle fusion method [42-45]. The thickness of the PC-SLB in liquid crystal phase observed by AFM is 4-5 nm [14,27]. The thickness of the first r-GO layer on SiO_2 was 2.9 nm in this study (figure 3), and the height difference between the r-GO and SiO_2 regions was 3.7 nm. Therefore the lipid membrane formed on r-GO was 0.8 nm thicker than the single SLB on SiO_2. We attributed this additional 0.8 nm thickness to a lipid monolayer on r-GO, and suggest that a lipid bilayer stacked on the lipid monolayer/r-GO. Previous reports showed that a lipid monolayer forms from vesicles on hydrophobic surfaces covered with alkyl chains (e.g. octadecyltrichlorosilane (OTS)-SAM/SiO_2 and alkanethiol/Au). A lipid monolayer will be formed on r-GO in a similar way, because the hydrophilic GO turns to hydrophobic by chemical or thermal reduction due to the removal of hydrophilic functional groups (epoxy, hydroxyl, and carboxyl groups) [46,47]. The model in which lipid molecules with their hydrophobic tails on r-GO is also consistent with the experimental result of lipid-coated carbon nanotube [48], and the molecular dynamics simulation of graphene incorporated in a lipid bilayer [49]. The thickness of the lipid monolayer on r-GO (0.8 nm) was smaller than the half of the bilayer thickness (2 nm). This was because the lipid molecules in the monolayer were not closely-arranged. Tilted molecular orientation is also suggested in the lipid monolayer on alkyl-SAM [26,50]. Lipid molecules in the monolayer on r-GO will be inclined because of unevenness and heterogeneity of the r-GO surface. The result of FRAP in figure 4A suggests that the lipid bilayer on the lipid monolayer/r-GO had fluidity and was continuous to the SLB on SiO_2. The structural model of the planar lipid membranes on the r-GO/SiO_2 surface and height difference among each region was summarized in figure 5.

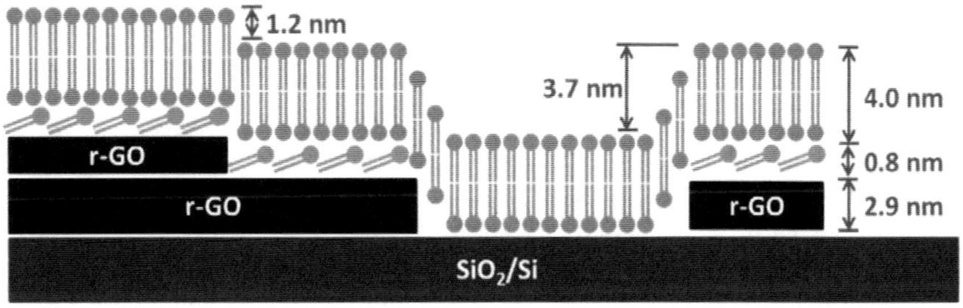

Figure 5. Schematic of the structure of lipid membranes on r-GO/SiO_2

4. Summary

Graphene oxide with maximum size of 323 μm was prepared by modified Hummer's method, and the size of the GO flakes was controlled at the range of 16 μm to 160 μm by ultrasonic fragmentation. We observed the change of the morphology and fluorescence property of GO by hydrazine reduction by using AFM and fluorescence microscope. We formed planar DOPC membranes on r-GO by vesicle fusion method. From the AFM observation and FRAP results, we revealed that continuous and fluid planar lipid membranes were formed on r-GO as well as on SiO_2. We proposed a structural model of lipid bilayer/lipid monolayer/r-GO based on the AFM topography.

Acknowledgment

Graphite was kindly provided by Ito Kokuen Co., LTD. We appreciate the kind support on SiO_2/Si preparation from Mr. H. Nakazawa and Prof. K. Sawada in TUT. This work was partly supported by the Grant-in-Aid for Scientific Research (KAKENHI) (#23685003) from MEXT, Japan.

References

[1] Kim J, Kim F and Huang J 2010 *Mater. Today* **13** 28–38
[2] Loh K P, Bao Q, Eda G and Chhowalla M 2010 *Nat. Chem.* **2** 1015–24
[3] Lu J, Zhang H, Shi W, Wang Z, Zheng Y, Zhang T, Wang N, Tang Z and Sheng P 2011 *Nano Lett.* **11** 2973–7
[4] Lee C, Wei X, Kysar J W and Hone J 2008 *Science* **321** 385–8
[5] Geim A K and Novoselov K S 2007 *Nat Mater* **6** 183–91
[6] Lu C-H, Yang H-H, Zhu C-L, Chen X and Chen G-N 2009 *Angew. Chem. Int. Ed.* **48** 4785–7
[7] He S, Song B, Li D, Zhu C, Qi W, Wen Y, Wang L, Song S, Fang H and Fan C 2010 *Adv. Funct. Mater.* **20** 453–9
[8] Ang P K, Jaiswal M, Lim C H Y X, Wang Y, Sankaran J, Li A, Lim C T, Wohland T, Barbaros O and Loh K P 2010 *ACS Nano* **4** 7387–94
[9] Sun X, Liu Z, Welsher K, Robinson J T, Goodwin A, Zaric S and Dai H 2008 *Nano Res.* **1** 203–12
[10] Tero R, Sazaki G, Ujihara T and Urisu T 2011 *Langmuir* **27** 9662–5
[11] Sumino A et al 2011 *Biomacromolecules* **12** 2850–8
[12] Mao Y, Shang Z, Imai Y, Hoshino T, Tero R, Tanaka M, Yamamoto N, Yanagisawa K and Urisu T 2010 *Biochim. Biophys. Acta* **1798** 1090–9
[13] Shang Z, Mao Y, Tero R, Liu X, Hoshino T, Tanaka M and Urisu T 2010 *Chem. Phys. Lett.* **497** 108–14
[14] Tero R, Ujihara T and Urisu T 2008 *Langmuir* **24** 11567–76
[15] Ujihara T, Suzuki S, Yamauchi Y, Tero R and Takeda Y 2008 *Langmuir* **24** 10974–80
[16] Lei S, Tero R, Misawa N, Yamamura S, Wan L and Urisu T 2006 *Chem. Phys. Lett.* **429** 244–9

[17] Shinozaki Y, Sumitomo K, Furukawa K, Miyashita H, Tamba Y, Kasai N, Nakashima H and Torimitsu K 2010 *Appl. Phys. Express* **3** 027002
[18] Kunze A, Sjövall P, Kasemo B and Svedhem S 2009 *J. Am. Chem. Soc.* **131** 2450–1
[19] Yoshii N and Okamura E 2009 *Chem. Phys. Lett.* **474** 357–61
[20] Meier R J, Csiszár A and Klumpp E 2006 *J Phys Chem B* **110** 20727–8
[21] Hull M C, Cambrea L R and Hovis J S 2005 *Anal. Chem.* **77** 6096–9
[22] Hirano-Iwata A, Niwano M and Sugawara M 2008 *Trends Anal. Chem.* **27** 512–20
[23] Richter R P, Bérat R and Brisson A R 2006 *Langmuir* **22** 3497–505
[24] Castellana E T and Cremer P S 2006 *Surface Science Reports* **61** 429–44
[25] Isono T, Tanaka H and Ogino T 2007 *e-J. Surf. Sci. Nanotech.* **5** 99–102
[26] Tero R, Takizawa M, Li Y-J, Yamazaki M and Urisu T 2004 *Langmuir* **20** 7526–31
[27] Tero R, Watanabe H and Urisu T 2006 *Phys. Chem. Chem. Phys.* **8** 3885–94
[28] Ishikawa R, Bando M, Morimoto Y, Park S Y and Sandhu A 2010 *Jpn. J. Appl. Phys.* **49** 06GC02
[29] Tung V C, Allen M J, Yang Y and Kaner R B 2009 *Nat. Nanotechnol.* **4** 25–9
[30] Suganuma K, Watanabe S, Gotou T and Ueno K 2011 *Appl. Phys. Express* **4** 021603
[31] Zhu Y, Murali S, Cai W, Li X, Suk J W, Potts J R and Ruoff R S 2010 *Adv. Mater.* **22** 3906–24
[32] Chen T, Zeng B, Liu J L, Dong J H, Liu X Q, Wu Z, Yang X Z and Li Z M 2009 *J. Phys.: Conf. Ser.* **188** 012051
[33] Su C-Y, Xu Y, Zhang W, Zhao J, Tang X, Tsai C-H and Li L-J 2009 *Chem. Mater.* **21** 5674–80
[34] Zhou X and Liu Z 2010 *Chem. Commun.* **46** 2611–3
[35] Subrahmanyam K S, Panchakarla L S, Govindaraj A and Rao C N R 2009 *J. Phys. Chem. C* **113** 4257–9
[36] Kim K S, Zhao Y, Jang H, Lee S Y, Kim J M, Kim K S, Ahn J-H, Kim P, Choi J-Y and Hong B H 2009 *Nature* **457** 706–10
[37] Schniepp H C, Li J-L, McAllister M J, Sai H, Herrera-Alonso M, Adamson D H, Prud'homme R K, Car R, Saville D a and Aksay I a 2006 *J Phys Chem B* **110** 8535–9
[38] Tsukamoto T and Ogino T 2009 *Appl. Phys. Express* **2** 075502
[39] Xu K, Cao P and Heath J R 2010 *Science* **329** 1188–91
[40] Swathi R S and Sebastian K L 2009 *J. Chem. Phys.* **130** 086101
[41] Swathi R S and Sebastian K L 2008 *J. Chem. Phys.* **129** 054703
[42] Keller C A and Kasemo B 1998 *Biophys. J.* **75** 1397–402
[43] Jenkins A T A, Bushby R J, Evans S D, Knoll W, Offenhäusser A and Ogier S D 2002 *Langmuir* **18** 3176–80
[44] Lingler S, Rubinstein I, Knoll W and Offenhäusser A 1997 *Langmuir* **13** 7085–91
[45] Tero R, Takizawa M, Li Y-J, Yamazaki M and Urisu T 2004 *Applied Surface Science* **238** 218–22
[46] Gao W, Alemany L B, Ci L and Ajayan P M 2009 *Nat. Chem.* **1** 403–8
[47] Gao X, Jang J and Nagase S 2010 *J. Phys. Chem. C* **114** 832–42
[48] Wu Y, Hudson J A S, Lu Q, Moore J M, Mount A S, Rao A M, Alexov E and Ke P C 2006 *J. Phys. Chem. B* **110** 2475–8
[49] Titov A V, Král P and Pearson R 2010 *ACS Nano* **4** 229–34
[50] Parikh A N, Beers J D, Shreve A P and Swanson B I 1999 *Langmuir* **15** 5369–81

Fabrication of Supported Lipid Bilayer on Graphene Oxide

Y Okamoto[1], K Tsuzuki[1], S Iwasa [1], R Ishikawa[2], A Sandhu[2,3] and R Tero[3]

[1] Department of Environmental and Life Sciences, Toyohashi University of Technology, 1-1 Hibarigaoka, Tempaku-cho, Toyohashi, Aichi, 441-8580, JAPAN
[2] Department of Electrical and Electronic Engineering, Tokyo Institute of Technology, 2-12-1 O-okayama, Meguro, Tokyo, 152-8552, JAPAN
[3] Electronics-Inspired Interdisciplinary Research Institute, Toyohashi University of Technology, 1-1 Hibarigaoka, Tempaku-cho, Toyohashi, Aichi, 441-8580, JAPAN

E-mail: y093506@edu.tut.ac.jp (Y.O), and tero@eiiris.tut.ac.jp (R.T)

Abstract. Planar lipid bilayers of dioleoylphosphatidylcholine was prepared on graphene oxide (GO), as a fundamental platform for biosensing in plasma membrane model using graphene. The GO flakes were prepared according to modified Hummer's method, and deposited on thermally oxidized SiO_2/Si surfaces. We found that planar lipid bilayers were reproducibility formed on the GO/SiO_2/Si surface in the presence of Ca^{2+} ions, while unruptured vesicles remained on the GO surface without Ca^{2+} ion. The results of atomic force microscope observation and fluorescence recovery after photobleaching experiment revealed that double lipid bilayers were spontaneously formed on the GO surfaces.

1. Introduction

A lipid bilayer is the fundamental structure of plasma membranes. The lipid bilayer is not only a static wall between the inside and outside of cell, but also has important functions, such as the transportation of small molecules, ions, and signals into and out of cells through specific channels and endo- and exocytosis, through the two-dimensional organization and molecular diffusion [1,2].The behavior of lipids and proteins in plasma membranes and their relation with the functions during reactions have been studied extensively [2-4]. Supported lipid bilayers (SLBs), which are artificial bilayer membrane at solid-liquid interfaces, have been investigated as cell membrane models to study the physicochemical properties of lipid bilayers, and as the platform for membrane proteins [5,6]. Recently, some new methods were reported to obtain detailed information of biochemical molecules using graphene and graphene oxide (GO) [7-11]. Our purpose is the development of a new method to obtain detailed information of biomolecules on and in plasma membranes, using graphene-supported lipid bilayers (G-SLBs). We prepared the SLB of phosphatidylcholine on GO by the vesicle fusion method [12-16].

At the formation of SLB by the vesicle fusion method, the efficiency of the formation of planar membranes depends on the material, shape and chemical states of substrate surfaces [15-17]. It is necessary to establish a reproducible condition for the formation of SLB on GO. GO has amphiphilicity, because GO consists of the hydrophilic sp^3-carbon regions including –OH, –O–, and –COOH groups, and of the hydrophobic sp^2-carbon regions, which are close to perfect graphene [18-20]. A lipid bilayer is a self-assembled structure of amphiphilic lipid molecules. Therefore, the behavior of lipid bilayers on the amphiphilic GO surface is an interesting subject from the view point of interfacial chemistry. In this manuscript, we describe the formation of SLB by vesicle fusion method, and the effect of vesicle size and Ca^{2+} ion.

2. Experiments

2.1. Preparation of graphene oxide

GO was prepared from graphite powder (Z+80, Ito Kokuen Co., Ltd., Japan) by chemical exfoliation according to the modified Hummer's method [21-23], then the GO suspension was sonicated for 60 min to decrease the size of GO flakes. The GO suspension was dropcast on a thermally oxidized SiO_2/Si(100) substrate (87 nm thick SiO_2) [24] cleaned by boiling in piranha solution for 10 min followed by sonication in 0.02 M KOH(aq.) for 10 min. The GO/SiO_2/Si substrate was observed with an optical microscope and an atomic force microscope (AFM). The width of the GO flakes before sonication was 213 μm at maximum (figure 1a). The size of the GO flakes became 1-10 μm after the sonication. The height of the GO flakes on the SiO_2/Si surface obtained from AFM topography was 1.74 ± 0.20 nm (n=40) (figure 1b, and 1c).

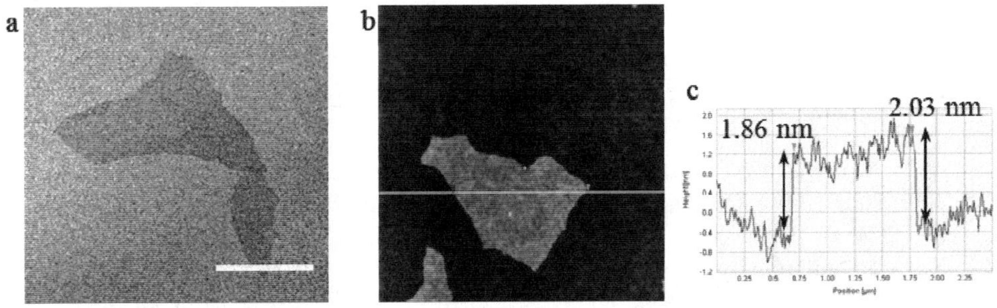

Figure 1. (a) Bright field image of GO on SiO_2/Si. Scale bar correspond to 100 μm. (b) AFM topography of GO after sonication dropcast on SiO_2/Si. Image size is 2.5×2.5 μm^2. (c) Cross section profile at the white line in (b).

2.2. Preparation of vesicle suspension and formation of SLB on GO/SiO_2/Si substrate

The chloroform solution of dioleoylphosphatidylcholine (DOPC) (Avanti Polar Lipid) and the ethanol solution of fluorescence-labelled lipid (BOPIPY-H-PC: Ex/Em = 534/552 nm) (Invitrogen) were mixed at 100:1 molar ratios in a glass vial. The lipid mixture was dried with N_2 flow, followed by overnight evacuation. A multilamellar vesicle (MLV) suspension were prepared by suspending the vacuum-dried lipid film into a buffer solution (100 mM KCl, 25 mM HEPES/pH 7.4 NaOH, chemicals were purchased from Wako Pure Chemical Industries, Ltd., Japan) [13]. The MLV suspension was repeatedly frozen and thawed in liquid nitrogen and a water bath at 45 °C, respectively, and was extruded through a 100 nm polycarbonate filter to obtain unilamellar vesicles. The GO/SiO_2/Si substrate was incubated in DOPC + BODIPY-H-PC vesicle suspension at 45 °C for 60 min. The excess vesicles in the liquid phase were washed out by exchanging the suspension with the fresh buffer solution.

2.3. Apparatus

An epifluorescence microscope (Olympus IX51) was used in bright-field and fluorescence observation. The GO flakes on the SiO_2/Si surface was observed in air. The GO/SiO_2/Si surface after the incubation in vesicle suspension was observed in buffer solution with a 60× water-immersion objective lens. We performed fluorescence recovery after photobleaching (FRAP) by irradiating excitation light 225 times brighter than for observation.

In AFM observation (Agilent PicoScan2500), the GO/SiO_2/Si surface was observed by using a cantilever with spring constant (C) of 28 N/m (SI-DF40, Seiko Instruments Inc.) in acoustic AC mode in the air. The GO/SiO_2/Si surface after the incubation in vesicle suspension was observed by using magnetically coated cantilever (TYPE I MAC Lever, Agilent, C = 0.6 N/m) in magnetic AC mode in the buffer solution.

3. Results and discussion

3.1. GO/SiO$_2$ surface incubated in DOPC vesicle suspension

Figure 2a shows the AFM topography of the GO/SiO$_2$ surface after the incubation in the 100 nm-extruded vesicle suspension. Generally, a full-coverage SLB can be obtained on SiO$_2$ using 100 nm-extruded vesicle suspensions [12,13]. The GO regions in figure 2a, however, were covered with 13 nm-high protrusions, which were assigned to unruptured vesicles [14]. The transformation from vesicles to SLB did not proceed on GO. From the previous results, it is known that the formation of SLB are assisted if vesicle sizes were decreased by sonication, and/or Ca^{2+} ion was added into buffer solutions [12,13]. Therefore, we studied the effect of the pre-sonication of the vesicle suspension and of the addition of Ca^{2+} ion in the buffer solution on the formation of the DOPC-SLB on the GO/SiO$_2$ surface (figure 2).

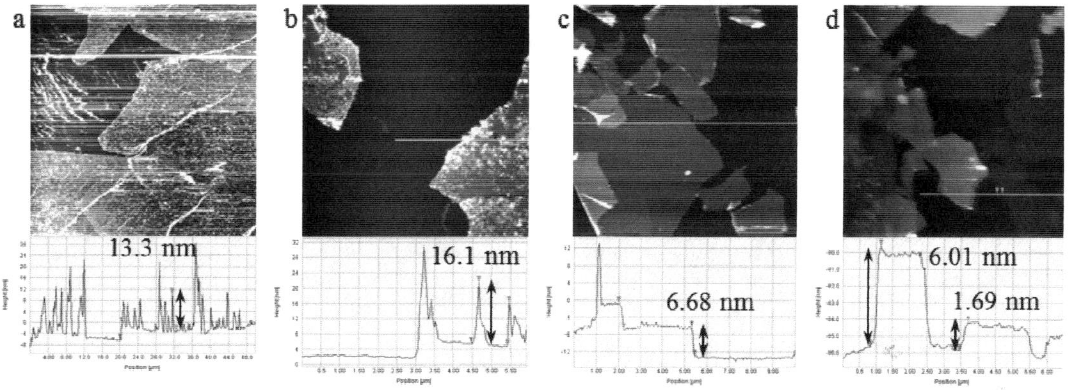

Figure 2. AFM topographies of the GO/SiO$_2$ surfaces after incubated at typical condition, and the cross section profiles at the white line in each AFM topography. (a) Vesicle suspension without sonication in the absence of CaCl$_2$. (b) Sonicated vesicle suspension in the absence of CaCl$_2$. (c) Vesicle suspension without sonication in the presence of 5 mM CaCl$_2$. (d) Sonicated vesicle suspension in the presence of 5 mM CaCl$_2$. Image sizes are (a) 50×50 μm^2, and (b-d) 10×10 μm^2.

Figure 2b shows the AFM topography of the GO/SiO$_2$/Si surface after the incubation in sonicated vesicle suspension. Although a few flat regions were observed on GO, the majority of the GO surface was covered with 16 nm-high protrusion, which were also assigned to the unruptured vesicles. Figure 2c and 2d show the AFM topographies of the GO/SiO$_2$/Si surface after the incubation in vesicle suspension containing 5 mM of CaCl$_2$.We observed flat morphology on both GO and SiO$_2$ regions with both the 100 nm-extruded vesicle suspension (figure 2c) and the sonicated vesicle suspension (figure 2d). We found few large protrusions observed in figure 2a and 2b. The surface of each GO flake was flat, the height within each GO flake was constant, and the GO regions were always higher than the SiO$_2$ regions. These results suggest that SLB was formed on GO. In figure 2c, the height of the GO regions from the SiO$_2$ regions was 6.86 ± 0.73 nm (n = 40). In figure 2d, the height of the GO regions from the SiO$_2$ regions were 1.55 ± 0.28 nm (n = 60), or 5.61 ± 0.58 (n = 60).

Figure 3 shows the fluorescence images of the SLB/GO/SiO$_2$ surface incubated under the same condition as figure 2d. The SLB formed on GO was remarkably darker than that on SiO$_2$. This result confirmed that the fluorescence from the BODIPY-H-PC in the DOPC-SLBs was quenched by GO [25-27]. We performed FRAP to study the fluidity of the SLB on GO. After the SiO$_2$ regions surrounding by GO (figure 3a, dotted circle) was photobleached, the fluorescence at the SiO$_2$ regions recovered with time (figure 3b-3d). This result indicates that lipid molecules diffused freely between the SiO_2 regions and the surrounding GO regions, therefore that fluid and continuous SLB were formed on GO, as well as on SiO_2.

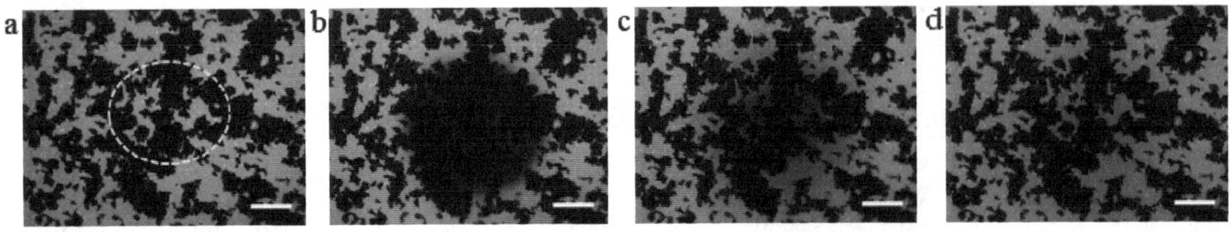

Figure 3. Fluorescence images of the GO-supported lipid bilayers and FRAP process. (a) Before fluorescence bleaching. (b) 0 s, (c) 180 s, and (d) 600 s after photobleaching. Scale bars correspond to 20 μm.

3.2. Structural model of the SLB/GO system

Figure 4 shows the structural model of GO-SLB system, which we propose based on the height observed in the AFM topography (figure 2d) and the result of FRAP (figure 3). The SLB on the SiO_2 regions had fluorescence, and FRAP proceeded. This result indicates that single lipid bilayer was formed on SiO_2 similarly to previous studies [13,15]. Single lipid bilayer was also formed on GO, because the height of the GO regions, which was 1.6 nm from SLB/SiO_2, was close to the height of GO in figure 1. On the other hand, the GO regions 5.6 nm higher than the SLB/SiO_2 was 4.0 nm higher than the single SLB/GO regions. This value corresponded to the thickness of single lipid bilayer observed by AFM. Previous AFM studies show that the thickness of DOPC-SLB observed with tapping mode is 4-5 nm [28-31]. Hence we conclude that two layers of lipid bilayer were formed on GO. Furthermore, the results of FRAP measurement (figure 3) suggests that the lipid bilayers on GO and SiO_2 had fluidity and continuity.

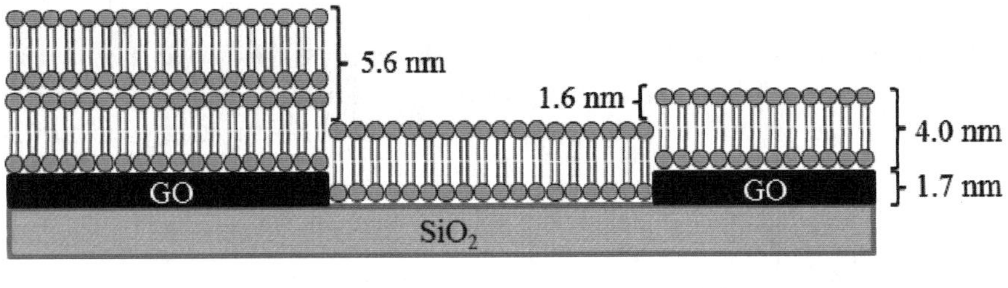

Figure 4. Structural model of the SLB/GO/SiO_2 system.

The spontaneous formation of double SLBs on GO by vesicle fusion is a unique phenomenon to GO. Generally, only single SLB is formed on homogeneously hydrophilic oxide surfaces (e.g. SiO_2, mica, TiO_2, etc.) by vesicle fusion method. The reason why only single SLB is formed on the oxide surfaces is not clear, even though multilamellar lipid bilayers are thermally stable [31]. The process of the transformation from adsorbed vesicle to planar membrane at the vesicle fusion method has highly kinetic aspects [12,13]. The layer number of the formed SLB may be related to the membrane formation mechanism of the vesicle fusion method. The GO surface is heterogeneous, because it consists of the hydrophilic regions including hydroxyl, epoxy and carboxyl groups, and of the hydrophobic regions close to perfect graphene [20]. The transformation from vesicles to SLB may proceed through a different mechanism from the homogeneously hydrophilic oxide surfaces. If we consider the application of SLBs as a platform for membrane protein research, direct interaction between the solid surface and the proteins incorporated in the SLB, especially their extramembrane regions, should be avoided. One of strategies for this problem is "tethered lipid bilayer", lipid bilayers linked with the substrate surface to make a space between the lipid bilayer and substrate [32-35]. The double SLB system on GO may provide another methodology: the first SLB on GO may work as a buffer between the substrate and the extramembrane region of the proteins incorporated in the second lipid membrane, because PC-SLB prevents non-specific adsorption of proteins.

4. Summary

We prepared DOPC-SLB on the GO/SiO_2/Si surface by the vesicle fusion method. Fluid and planar lipid bilayers were formed on GO after the incubation of GO on the thermally oxidized SiO_2/Si substrate in $CaCl_2$-containing DOPC vesicle suspension. From the AFM observation, we found that not only single lipid bilayer, similar to inorganic substrates, but also double lipid bilayers were formed on GO. We proposed a structural model of the GO-SLB system. This GO-SLB system is the fundamental platform for the measurement of biomolecules in the plasma membrane model using graphene.

Acknowledgment

Graphite was kindly provided by Ito Kokuen Co., LTD. We appreciate the kind support on SiO_2/Si preparation from Mr. H. Nakazawa and Prof. K. Sawada in TUT. This work was partly supported by the Grant-in-Aid for Scientific Research (KAKENHI) (#23685003) from MEXT, Japan.

References

[1] Hirayama H 1998 *J. Memb. Sci.* **139** 109–24
[2] Anderson R G W and Jacobson K 2002 *Science* **296** 1821–5
[3] Jacobson K, Mouritsen O G and Anderson R G W 2007 *Nat. Cell Biol.* **9** 7–14
[4] Lingwood D and Simons K 2010 *Science* **327** 46–50
[5] Hirano-Iwata A, Niwano M and Sugawara M 2008 *Trends Anal. Chem.* **27** 512–20
[6] Cha T, Guo A and Zhu X-Y 2006 *Biophys. J.* **90** 1270–4
[7] Lu C-H, Yang H-H, Zhu C-L, Chen X and Chen G-N 2009 *Angew. Chem. Int. Ed.* **48** 4785–7
[8] He S, Song B, Li D, Zhu C, Qi W, Wen Y, Wang L, Song S, Fang H and Fan C 2010 *Adv. Funct. Mater.* **20** 453–9
[9] Ang P K, Jaiswal M, Lim C H Y X, Wang Y, Sankaran J, Li A, Lim C T, Wohland T, Barbaros O and Loh K P 2010 *ACS Nano* **4** 7387–94
[10] Wu M, Kempaiah R, Huang P-J J, Maheshwari V and Liu J 2011 *Langmuir*
[11] Wang X, Wang C, Qu K, Song Y, Ren J, Miyoshi D, Sugimoto N and Qu X 2010 *Adv. Funct. Mater.* **20** 3967–71
[12] Reviakine I and Brisson A 2000 *Langmuir* **16** 1806–15
[13] Tero R, Ujihara T and Urisu T 2008 *Langmuir* **24** 11567–76
[14] Tero R, Watanabe H and Urisu T 2006 *Phys. Chem. Chem. Phys.* **8** 3885
[15] Isono T, Tanaka H and Ogino T 2007 *e-J. Surf. Sci. Nanotech.* **5** 99–102
[16] Tero R, Takizawa M, Li Y-J, Yamazaki M and Urisu T 2004 *Langmuir* **20** 7526–31
[17] Richter R P, Bérat R and Brisson A R 2006 *Langmuir* **22** 3497–505
[18] Gómez-Navarro C, Meyer J C, Sundaram R S, Chuvilin A, Kurasch S, Burghard M, Kern K and Kaiser U 2010 *Nano Lett.* **10** 1144–8
[19] Erickson K, Erni R, Lee Z, Alem N, Gannett W and Zettl A 2010 *Adv. Mater.* **22** 4467–72
[20] Loh K P, Bao Q, Eda G and Chhowalla M 2010 *Nat. Chem.* **2** 1015–24
[21] Ishikawa R, Bando M, Morimoto Y, Park S Y and Sandhu A 2010 *Jpn. J. Appl. Phys.* **49** 06GC02
[22] Gao Y, Chen X, Xu H, Zou Y, Gu R, Xu M, Jen A K-Y and Chen H 2010 *Carbon* **48** 4475–82
[23] Tung V C, Allen M J, Yang Y and Kaner R B 2009 *Nat. Nanotechnol.* **4** 25–9
[24] Nagashio K, Nishimura T, Kita K and Toriumi A 2009 *Appl. Phys. Express* **2** 025003
[25] Kim J, Kim F and Huang J 2010 *Mater. Today* **13** 28–38
[26] Swathi R S and Sebastian K L 2009 *J. Chem. Phys.* **130** 086101

[27] Swathi R S and Sebastian K L 2008 *J. Chem. Phys.* **129** 054703
[28] Goksu E I, Vanegas J M, Blanchette C D, Lin W-chen and Longo M L 2009 *Biochim. Biophys. Acta* **1788** 254–66
[29] Marquês J T, Viana A S and De Almeida R F M 2011 *Biochim. Biophys. Acta* **1808** 405–14
[30] Milhiet P-E, Gubellini F, Berquand A, Dosset P, Rigaud J-L, Le Grimellec C and Lévy D 2006 *Biophys. J.* **91** 3268–75
[31] Domenici F, Panichelli D and Castellano a C 2009 *Colloids Surf. B* **69** 216–20
[32] Tero R, Misawa N, Watanabe H, Yamamura S, Nambu S, Nonogaki Y and Urisu T 2005 *e-J. Surf. Sci. Nanotech.* **3** 237–43
[33] Lei S, Tero R, Misawa N, Yamamura S, Wan L and Urisu T 2006 *Chem. Phys. Lett.* **429** 244–9
[34] Knoll W, Park H, Sinner E, Yao D and Yu F 2004 *Surf. Sci.* **570** 30–42
[35] Terrettaz S, Mayer M and Vogel H 2003 *Langmuir* **19** 5567–9

Fabrication and characterization of poly[diphenylsilane]-based solar cells

M. Iwase[1], T. Oku[1], A. Suzuki[1], T. Akiyama[1], K. Tokumitsu[1], M. Yamada[2] and M. Nakamura[2]

[1]Department of Materials Science, The University of Shiga Prefecture, Shiga 522-8533, Japan
[2]Osaka Gas Chemicals Co., Ltd, Osaka 550-0023, Japan
E-mail: ze21miwase@ec.usp.ac.jp, oku@mat.usp.ac.jp

Abstract. Poly[diphenylsilane] (PDPS)-based photovoltaic cells were fabricated by using mixture solution of PDPS, phosphorus and boron. An influence of phosphorus and boron doping into PDPS on the performance of the photovoltaic device was investigated. The solar cell using fluorine doped tin oxide glass plates provided short-circuit current density of 0.12 mA/cm^2 and open-circuit voltage of 0.28 V under simulated sunlight. Energy levels, formation mechanism and microstructure of the solar cells were discussed.

1. Introduction

Most used energy on the earth is fossil fuels in the present time, and burning fossil fuels affects on the natural environment by production of greenhouse gases. Problems of energy supply and use are related not only to global warming but also to such environmental concerns as air pollution and forest destruction. In addition, fossil fuel is limited, and other energy resources should be developed.

A mainstream of solar cell is silicon solar cells which have a high energy conversion efficiency and long lifetime. However, they are expensive and heavy-weight. On the other hand, researches on organic thin film solar cells are advanced since they have easy fabrication process, low cost, light-weight and flexible substrates. However, they provide low energy conversion efficiency and short lifetime.

Polysilanes have high hole mobility as organic semiconductors because of its σ-conjugated polymers, and it is possible to use the materials for organic thin film solar cells [1-3]. Although the polysilanes could be applied to p-type semiconductors on organic thin film solar cells, few studies on polysilane solar cells have been reported [4-6]. The purpose of the present study is to fabricate photovoltaic cells by using mixture solution of poly[diphenylsilane] (PDPS), phosphorus and boron. PDPS have high hole mobility as organic semiconductors because of its σ-conjugated polymers, and it is possible to use the materials for organic thin film solar cells

2. Experimental procedures

A thin layer of polyethylenedioxythiophen doped with polystyrene-sulfonic acid (PEDOT:PSS) (Sigma Aldrich) was spin-coated on pre-cleaned indium tin oxide (ITO) glass plates (Geomatec Co., Ltd, ~10Ω/□) or fluorine doped tin oxide (FTO) glass plates (Luminescence Technology Co. ~14Ω/□). ITO substrates have the characteristic that the resistance increase when annealed at a high temperature of 300 °C. Therefore, the FTO substrate with stability at 300 °C was used for Cell 2. Then, semiconductor layers were prepared on the PEDOT:PSS layer by spin coating using the mixture solution of PDPS (Osaka Gas Chemical Co., Ltd) and diethylmethoxyborane (Sigma Aldrich) in 1 ml o-dichlorobenzene. After annealing at 250 °C for 10 min or 300 °C for 10 min in N_2 atmosphere, a next layer was prepared by spin-coating using the mixture solution of PDPS and phosphoric acid (Sigma Aldrich) in 1 ml o-dichlorobenzene. After annealing at 250 °C for 10 min or 300 °C for 10 min in N_2 atmosphere, aluminum (Al) metal contacts were evaporated as the top electrode. A schematic diagram of the present solar cells is shown in Figure 1.

The current density-voltage (J-V) characteristics (Hokuto Denko, HSV-100) of the solar cells were measured under illumination at 100 mW/cm^2 by using an AM 1.5 solar simulator (San-ei Electric, XES-301S). The solar cells were illuminated through the side of the ITO or FTO substrates, and the illuminated area was 0.16 cm^2. Optical absorption of the solar cells was investigated by means of UV visible spectroscopy (Hitachi Ltd., U-4100). The microstructures of the PDPS thin films were investigated by X-ray diffractometry (XRD, Philips X'Pert-MPD System) with CuKa radiation at 40 kV operating voltage and 40mA operating current.

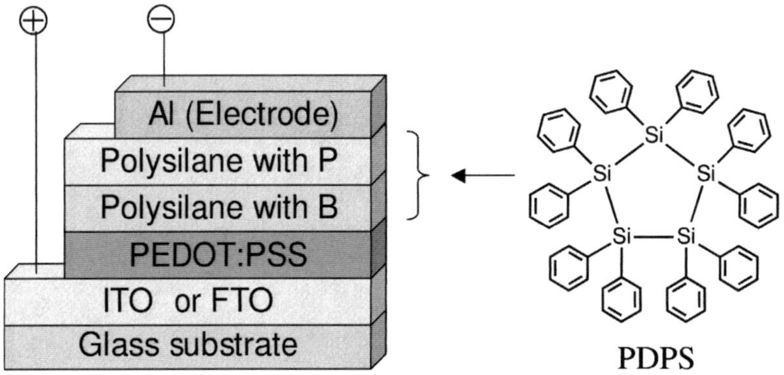

Figure 1. Cross-sectional structure of heterojunction solar cells and atomic structure of PDPS.

3. Results and discussion

Measured current density-voltage (J-V) characteristic of solar cells under illumination is shown in Figure 2. Photocurrents were observed under the illumination, and each structure showed characteristic curves with short-circuit current and open-circuit voltage.

Measured parameters of the solar cells are summarized in Table 1. Cell 2 provided the higher power conversion efficiency (η) of 0.010 %, fill factor (FF) of 0.31, short-circuit current density (J_{sc}) of 0.12 mA/cm^2 and open-circuit voltage (V_{oc}) of 0.28 V.

Figure 3 shows UV-visible-IR absorption spectra of solar cells. The PDPS:phosphorus/PDPS:boron structure shows broad absorption in range of 300 and 1800 nm.

X-ray diffraction pattern of the annealed PDPS thin film is shown in Figure 4. A broad peak at 2θ of ~28.5 ° would be a reflection of Si 111, which would indicate an amorphous Si structure. Other several broad peaks are due to PDPD or decomposed PDPS, and this thin film would be a nanocomposite film with amorphous Si and PDPS.

Table1. Measured parameters of solar cells

Cell	Glass substrate	Annealing temperature (°C)	V_{oc} (V)	J_{sc} (mA/cm^2)	FF	η (%)
Cell 1	ITO (10 Ω/□)	250	0.23	0.019	0.25	1.1×10^{-3}
Cell 2	FTO (14 Ω/□)	300	0.28	0.12	0.31	0.010

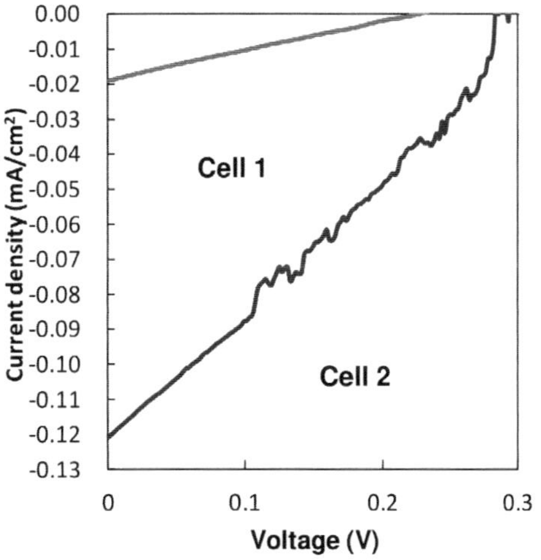

Figure 2. Measured J-V characteristic of solar cells under illumination.

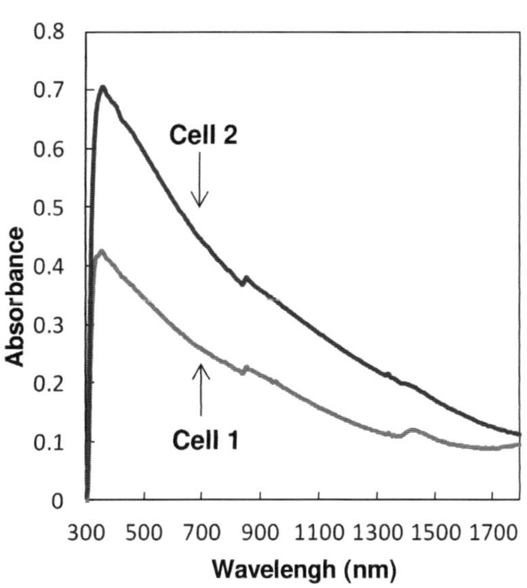

Figure 3. UV-visible-IR absorption of PDPS thin films with phosphorus and boron.

Energy level diagram of the solar cells is summarized as shown in Figure 5(a). Previously reported values were used for the energy levels [4,5]. The incident of light is from ITO or FTO side. Energy barrier would exist near the semiconductor/metal interface. Electronic charge separation was caused by light irradiation from the ITO or FTO substrate side. Separated holes could be transferred from the valence band of Si doped with B to the ITO or FTO, and separated electrons could be transferred from the conduction band of Si doped with P to the Al electrode. To improve the efficiency, control of microstructure by annealing would be important. In addition, there would be energy barrier at the semiconductor/metal interfaces as indicated by band bending in Fig. 5(a), and the heavy doping of B or P would reduce the contact resistance.

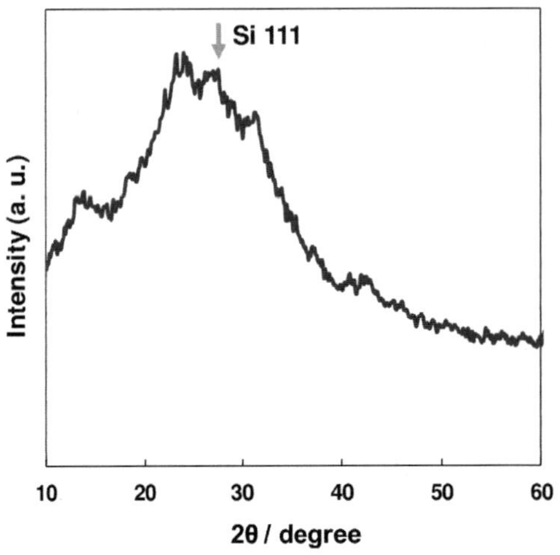

Figure 4. XRD pattern of PDPS thin film after annealing at 300 °C for 10 min.

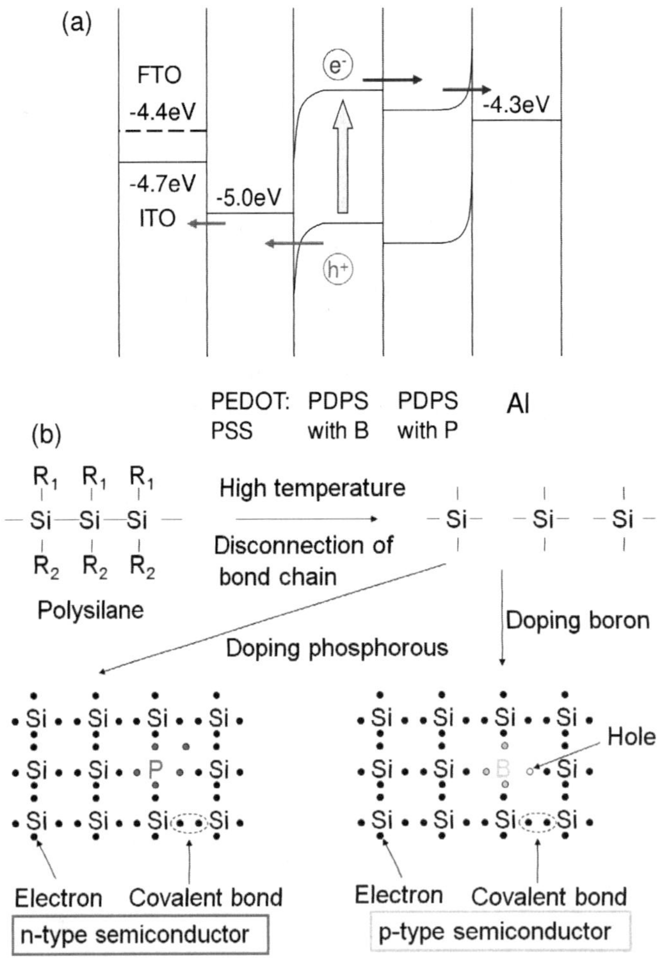

Figure 5. (a) Energy level diagram and (b) doping process of the present solar cells.

Doping process is summarized as shown in Figure 5(b). Polysilanes decompose to form amorphous silicon when heated to high temperature. Amorphous silicon doped with boron would function as p-type semiconductor. On the other hand, amorphous silicon doped with phosphorus would function as n-type semiconductor. Energy levels would be different between PDPS with B and PDPS with P. Several broad peaks were observed in the X-ray diffraction pattern of the annealed PDPS films, and the present solar cells would have a nanocomposite structure of amorphous Si and PDPS. Further annealing would be effective for the decomposition of PDPS and crystallization of amorphous Si. To improve the efficiency, control of microstructure by annealing would be important. Optimization of annealing temperatures and electrical conductivity due to doping or microstructural change would increase the efficiencies of solar cells.

4. Conclusions

Polysilane-based solar cells were fabricated by using mixture solution of PDPS, phosphorus and boron, and characterized by electrical measurements. A device using FTO substrate provided η of 0.010 %, FF of 0.31, J_{sc} of 0.12 mA/cm^2 and V_{oc} of 0.28 V. The solar cell showed broad absorption in range of 300 and 1800 nm. Microstructure analysis showed formation of a nanocomposite structure amorphous Si and PDPS. Energy levels of the molecules and carrier transport mechanism were discussed.

References

[1] Lee J, Seoul C, Park J and Youk H J, 2004 Synth. Met. **145** 11
[2] Silence M S, Scott C J, Hache F, Ginsburg J E, Jenkner K P, Miller D R, Twieg J R and Moerner E W, 2003 J. Opt. Soc. Am. B **10** 2306
[3] Kakimoto M, Kashihara H, Kashiwagi T and Takiguchi T. 1997 Macromolecules. **7816** 30
[4] Rybak A, Jung J, Ciesielski W and Ulanski J, 2006 Mater. Sci. Poland **24** 527
[5] Shimoda T, Matsuki Y, Furusawa M, Aoki T, Yudasaka I, Tanaka H, Iwasawa H, Wang D, Miyasaka M, Takeuchi Y, 2006 Nature **440** 783
[6] Haga Y, Harada Y, 2001 Jpn. J. Appl. Phys. **40** 855
[7] Oku T, Motoyoshi R, Fujimoto K, Akiyama T, Jeyadevan B, Cuya J, 2011 J. Phys. Chem. Solids **72** 1206
[8] Oku T, Noma T, Suzuki A, Kikuchi K, Kikuchi S, 2010 J. Phys. Chem. Solids **71** 551

Fabrication and characterization of polysilane/C_{60} thin film solar cells

Junya Nakagawa[1], Takeo Oku[1], Atsushi Suzuki[1], Tsuyoshi Akiyama[1], Katsuhisa Tokumitsu[1], Masahiro Yamada[2], Mika Nakamura[2]

[1] Department of Materials Science, The University of Shiga Prefecture, 2500 Hassaka, Hikone, Shiga 522-8533 Japan
[2] Osaka Gas Chemical Co. Ltd., 3 Tiyozaki Nishiku, Osaka 550-0023 Japan

E-mail: zv22jnakagawa@ec.usp.ac.jp, oku@mat.usp.ac.jp

Abstract. Polysilane/fullerene bulk heterojunction solar cells were fabricated on indium tin oxide electrode by a spin-coating method, and performance and microstructures of the solar cell were investigated. Poly-methyl-phenylsilane, poly-phenylsilane, deca-phenyl-cyclopentasilane and [6,6]-phenyl C_{61}-butyric acid methyl ester (PCBM), poly[3-hexylthiophene] (P3HT) were used for the present cells. A device based on the P3HT and PCBM provided better efficiency, fill factor, and short-circuit current compared to those of a device based on P3HT and fullerene.

1. Introduction

Solar cell technology using organic semiconductors has been developed as next generation systems. Organic thin film solar cells have advantages of low cost, flexible and light weight product, and photovoltaic and optical properties of the organic solar cells have been studied [1-4]. Polysilane has high hole mobility as p-type semiconductor, and has been applied as electrical conductive materials and photovoltaic systems [5-7].

The purpose of the present work is to fabricate and characterize bulk heterojunction solar cells with polysilane and fullerene (C_{60}). C_{60} is a good electronic acceptor. Light induced carrier separation with charge transfer was investigated by experimental measurements. Electronic structures and energy levels of the molecules were calculated by molecular orbital calculation. Optical absorption of the solar cells was measured by UV-vis spectroscopy. Microstructure analysis was investigated by transmission electron microscopy and atomic force microscopy. Photovoltaic mechanism was discussed on the basis of the experimental results. The purpose of the present work is to fabricate and characterize polysilane/C_{60}-based solar cells.

2. Experimental

Three-types polysilanes were used in the present work as follows: poly-methyl-phenyl-silane (PMPS), poly-phenyl-silyne (PPSi), poly-diphenyl-silane (PDPS) [8]. Figure 1 shows molecular structures of PMPS, PPSi and PDPS to fabricate solar cells and structure of bulk-heterojunction solar cells. Indium tin oxide (ITO) grass plates (Geomatec, ~10Ω/□) were cleaned by an ultrasonic bath with acetone and methanol, and were dried by nitrogen gas. A thin layer of polyethylenedioxythiophen doped with polystyrene-sulfonic acid (PEDOT:PSS, Sigma Aldrich) was spin-coated on the ITO substrates. Then, semiconductor layers were prepared on a PEDOT:PSS layer by spin coating using a mixed solution of (1) C_{60} (Material Technologies Research, 99.98%), (2) C_{60} and P3HT (Sigma Aldrich) [9], (3) PCBM (Frontier Carbon) and P3HT [1, 10-13], PMPS (Osaka Gas Chemical Co, Ltd), PPSi (Osaka Gas Chemical Co, Ltd) or PDPS (Osaka Gas Chemical Co, Ltd) in 1 mL o-dichlorobenzene. Total weight of C_{60}:PMPS, C_{60}:PPSi or C_{60}:PDPS were 10 mg, and weight ratio of C_{60}:PMPS, C_{60}:PPSi or C_{60}:PDPS was 8:2. Aluminium (Al) metal contacts were evaporated as a top electrode. Finally, the devices were annealed at 140°C for 30 min in N_2 atmosphere.

Current density-voltage (J-V) characteristics (Hokuto Denko Corp., HSV-100) of the solar cells were measured both in the dark and under illumination at 100 mW/cm^2 by using an AM 1.5 solar simulator (San-ei Electric, XES-301S) in N_2 atmosphere. The solar cells were illuminated through the side of the ITO substrates, and the illuminated area is 0.16 cm^2. Optical absorption of the solar cells was investigated by means of UV-visible spectroscopy (JASCO, V-670ST).

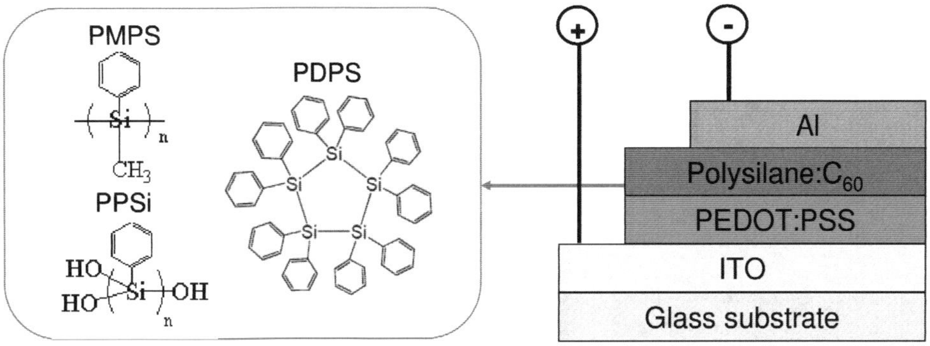

Figure 1. Polysilanes to fabricate solar cells and structure of bulk-heterojunction solar cells.

3. Results and Discussion

Measured parameters of PMPS:C_{60}, PPSi:C_{60} and PDPS:C_{60}, PMPS:P3HT:C_{60}, PPSi:P3HT:C_{60} and PDPS:P3HT:C_{60} bulk heterojunction solar cells are summarized in Table 1. A solar cell with PMPS-B:P3HT:C_{60} structure provided the highest power conversion efficiency (η) of 2.7×10^{-3} %, fill factor (FF) of 0.20, short-circuit current density (J_{SC}) of 0.06 mA/cm^2 and open-circuit voltage (V_{OC}) of 0.22 V, which is better than those of other devices. It is considered that the efficient charge separation would be increased by interface area of PMPS-B:P3HT:C_{60} sample. The molecular weights of PMPS-A and PMPS-B are 22100 and 1800, respectively.

Table 1. Experimental parameters of polysilane:C_{60} solar cells.

Sample	V_{OC} (V)	J_{SC} (mA/cm^2)	FF	η (%)
PMPS-A:C_{60}	0.060	2.0×10^{-4}	0.26	3.1×10^{-6}
PMPS-A:C_{60}:P3HT	0.10	9.0×10^{-3}	0.29	2.6×10^{-4}
PMPS-B:C_{60}	0.10	4.5×10^{-2}	0.25	1.2×10^{-4}
PMPS-B:C_{60}:P3HT	0.22	6.0×10^{-2}	0.20	2.7×10^{-3}
PPSi:C_{60}	2.5×10^{-3}	7.0×10^{-5}	0.26	4.6×10^{-8}
PPSi:C_{60}:P3HT	0.17	1.2×10^{-2}	0.20	4.1×10^{-4}
PDPS:C_{60}	0.050	5.0×10^{-3}	0.24	6.0×10^{-5}
PDPS:C_{60}:P3HT	0.13	1.5×10^{-3}	0.20	3.5×10^{-5}

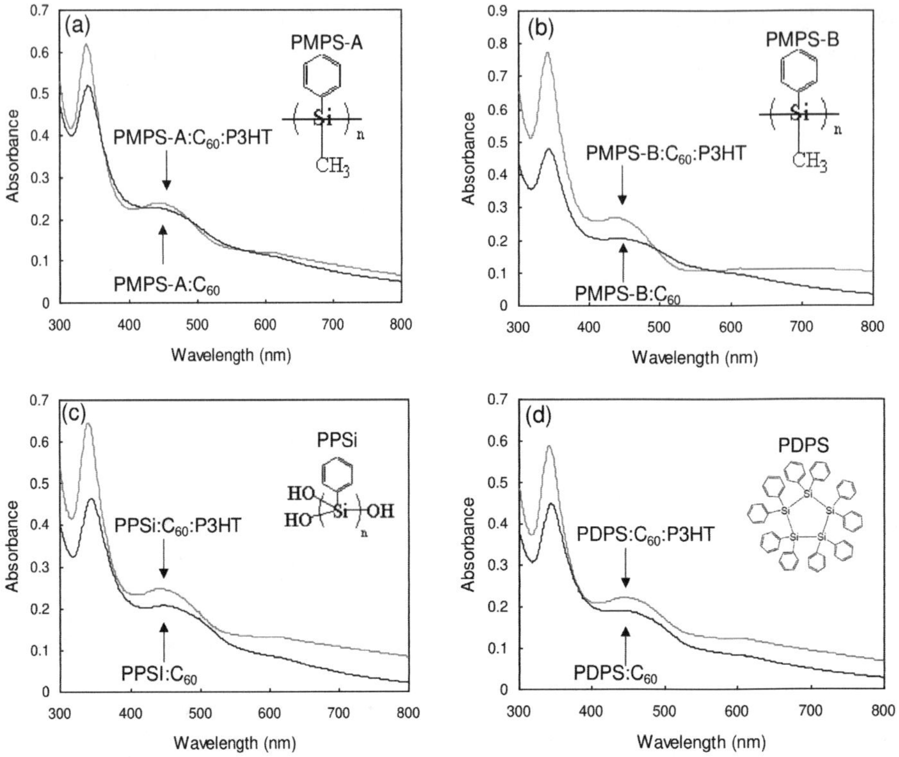

Figure 2. UV-vis absorption spectra of (a,b) PMPS:C_{60}, (c) PPSi:C_{60} and (d) PDPS:C_{60} bulk heterojunction solar cells.

Figure 2 shows UV-visible absorption spectra of the present solar cells. Polysilane:C_{60} structures provided higher photo-absorption in the range of 300-400 nm and 400-500 nm. The absorption peaks at ~350 nm and at ~450 nm for the polysilane:C_{60} structures is due to C_{60}. Moreover, polysilane:P3HT:C_{60} structures provided higher photo-absorption than that of polysilane:C_{60} structures. The increased absorption peaks is due to P3HT.

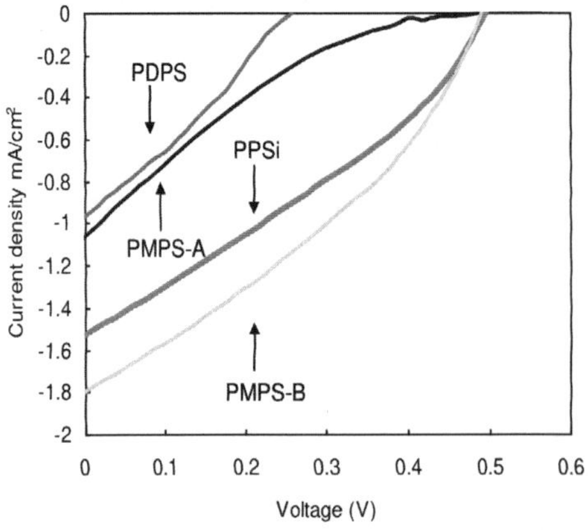

Figure 3. J-V-characteristics of PMPS:PCBM:P3HT, PPSi:PCBM:P3HT and PDPS:PCBM:P3HT bulk heterojunction solar cells under illumination.

Table 2. Experimental parameters of polysilane:PCBM solar cells.

Sample	V_{OC} (V)	J_{SC} (mA/cm^2)	FF	η (%)
PMPS-A:PCBM:P3HT	0.54	1.1	0.14	0.080
PMPS-B:PCBM:P3HT	0.53	1.8	0.33	0.31
PPSi:PCBM:P3HT	0.50	1.5	0.32	0.24
PDPS:PCBM:P3HT	0.27	0.9	0.27	0.070

The J-V characteristics of the PMPS:P3HT:PCBM, PPSi:P3HT:PCBM and PDPS:P3HT:PCBM structure under illumination is shown in Figure 3. Measured parameters of these bulk heterojunction solar cells are summarized in Table 2. A solar cell with PMPS-B:P3HT:C_{60} structure provided the highest η of 0.31 %, FF of 0.33, J_{SC} of 1.8 mA/cm^2 and V_{OC} of 0.53 V, which is better than those of other devices. It is considered that the efficient charge separation would be increased by interface area of PMPS-B:P3HT:PCBM sample. In the present study, the PMPS:C_{60} bulk heterojunction solar cell provided η of 0.32 %, FF of 0.31, J_{SC} of 17 mA/cm^2 and V_{OC} of 0.37 V. PMPS has a characteristic of cross-linked and mixed with organic layer with polymer chain [5], which results in the increase of carrier transport. PPSi has a characteristic of compatibility, and uniform spin-coating would be possible. Although PDPS has a high refractive index, spin-coating was not so uniform, which results in the low carrier transport.

An energy level diagram of polysilane:C_{60} and polysilane:PCBM:P3HT solar cells is summarized as shown in Figure 4. Previously reported values were used for the energy levels [14]. For the polysilane:PCBM:P3HT solar cells, the combination of P3HT and polysilane would increase the carrier separation, which would results in the increase of efficiency. The incident direction of light is from ITO side. Energy barrier would exist near the semiconductor/metal interface. Electronic charge is transferred by light irradiation from the ITO substrate side. Holes are transported to an ITO substrate and electrons are transported to an Al electrode. A relation between V_{OC} and polymer oxidation potential was reported as $V_{OC}=(1/e)(|E^{Polysilane}HOMO|-|E^{C60}LUMO|)-0.3$ (V), where e is the elementary charge. The value of 0.3 V is an empirical factor, and this is enough for efficient charge separation.

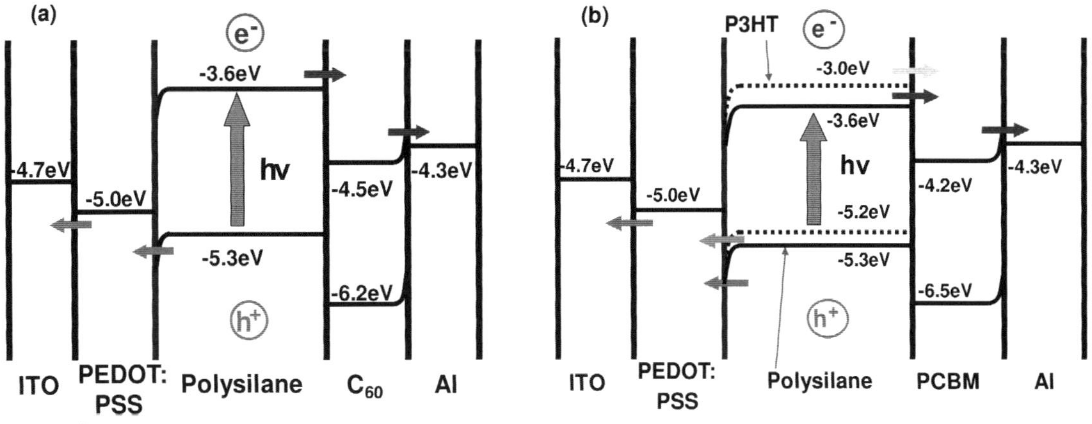

Figure 4. Energy level diagrams of (a) polysilane:C_{60} and (b) polysilane:PCBM:P3HT bulk heterojunction solar cells.

Since PMPS has the excellent cross-linked feature, cross-links are formed with other molecules in a solvent. This promotes the charge transfer within a solar cell, which would results in the improvement of conversion efficiency as listed in Table1 and Table 2. In addition, PMPS-B with a lower molecular weight provided higher conversion efficiency compared to the PMPS-A, which would be due to the easy cross-linked reaction.

PPSi has a network type polysilane that has a silanol group (Si-OH) at the end, and it has excellent compatibility and cross-links. When heated, this polysilane is cured, making cross-links with other reactive compounds, and it is possible to suppress changes in the film thickness. Although combination of compatibility and cross-links would be effective for film formation, the carrier transport would be disturbed by oxygen atoms at the cross-links, and the conversion efficiency was low compared with the PMPS-B.

Since PDPS has a cyclic structure, its refractivity is so high to confine light compared to the other polysilanes in the present work. When it is used, higher conversion efficiency would be expected. However, in the present work, since the charge transfer was not so smooth compared to the other polysilanes, which would result in the low efficiency.

4. Conclusion

Polysilane:C_{60}, polysilane:P3HT:C_{60} and polysilane:P3HT:PCBM solar cells were fabricated and characterized. Four types of polysilane were used in the present work. A device based on PMPS:P3HT:C_{60} structure provided the highest η of 0.31 %, FF of 0.33, J_{SC} of 1.8 mA/cm^2 and V_{OC} of 0.53V, which was better than those of other devices in the present work. Energy levels of the molecules were investigated and discussed.

References

[1] Bagienski W, Gupta M, 2011 Sol. Energy Mater. Sol. Cells **95** 933
[2] Tang C, 1986 Appl. Phys. Lett. **48** 183
[3] Mi S, Hyuk J, Jin J, 2010 Sol. Energy Mater. Sol. Cells **94** 152
[4] Zhao D, Ke L, Li Y, Tan S, 2011 Sol. Energy Mater. Sol. Cells **95** 921-926.
[5] Lee J, Seoul C, Park J, York J, 2004 Synth. Metals **145** 11
[6] Silence S, Scott J, Hache F, Ginsbrug E, Jenkner P, Miller P, Twieg R, Moermer W, 2003 J. Opt. Soc. Am. B **10** 2306
[7] Jung J, Ciesielski W, Ulanski J, 2006 Mater. Sci. -Poland **24** 527
[8] Cherry B.R., Alam T.M., 2004 Polymer, **45**, 16, 5611-5618
[9] Liu C, Holman Z, Kortshagen R, 2009 Nano Lett. **9** 449
[10] Baek W, Yang H, Yoon T, Kang C, Lee H, Kim Y, 2009 Sol. Energy Mater. Sol. Cells **93** 1263
[11] Jin S, Naidu S, Jeon H, 2007 Sol. Energy Mater. Sol. Cells **91** 1187
[12] Lange A, Wegener M, Boeffel C, Fischer B, Wedel A, Neher D, 2010 Sol. Energy Mater. Sol. Cells **94** 1816
[13] Hauch J, Schilinsky P, Choulis S, Childers R, Biele M, Brabec C, 2008 Sol. Energy Mater. Sol. Cells **92** 727
[14] Oku T, Motoyoshi R, Fujimoto K, Akiyama T, Balachandran J, Cuya J, 2011 J. Phys. Chem. Solids **72** 1206

Journal of Physics: Conference Series **352** (2012) 012020 doi:10.1088/1742-6596/352/1/012020

Low temperature synthesis of FePt alloy nanoparticles by polyol process

K Fukuda[1], S Fujieda[1], K Shinoda[1], S Suzuki[1] and B Jeyadevan[2]

[1]IMRAM, Tohoku University, Katahira, 2-1-1, Japan
[2]Department of Engineering, The University of Shiga Prefecture, Hikone City, Japan

E-mail: fujieda@tagen.tohoku.ac.jp

Abstract. Low temperature synthesis of equiatomic FePt alloy nanoparticles with the FCC structure was carried out by the polyol process in the presence of NaOH at above 383 K. The Pt content in the FePt alloy particles decreased with decreasing reaction temperature and equiatomic FCC-FePt alloy nanoparticles were successfully obtained at 383 K. When the FCC-FePt alloy nanoparticles synthesized at 383 K were annealed at 573 K, a partially ordered phase with the $L1_0$ structure was observed. For comparison, FCC-FePt alloy nanoparticles were synthesized at 468 K without adding NaOH. Results showed that the degree of atomic ordering of the sample synthesized at 383 K with the addition of NaOH was higher than those obtained at 468 K after annealing at low temperatures. Thus, it is concluded that the temperature employed during the synthesis of FCC-FePt alloy nanoparticles affects the ordering transformation temperature.

1. Introduction

Equiatomic $L1_0$-FePt alloy shows large magnetic anisotropy that depends on the degree of atomic ordering [1]. The control of such degree of equiatomic $L1_0$-FePt alloy nanoparticles is important for advanced magnetic recording media applications [2,3].

Liquid phase synthesis of FePt alloys has been actively investigated because well-dispersed equiatomic FePt alloy nanoparticles could be obtained by adjusting the synthesis conditions [2]. For example, thermal decomposition of $Fe(CO)_5$ and reduction of $Pt(C_5H_7O_2)_2$ in octyl ether and 1,2-alkanediol are a well-known method to produce FePt alloy nanoparticles [2,3]. Other liquid phase methods that use reducing agents such as polyols [4], hydrazine [5] or borohydride [6] have also been reported. Since disordered FCC-FePt alloy nanoparticles are prepared with these methods, heat treatment is necessary to obtain large magnetic anisotropy [2,3]. However, the heat treatment is carried out at high temperatures, resulting in the agglomeration and particle growth of nanoparticles [7]. Therefore, it is necessary to lower the ordering transformation temperature from the FCC to the $L1_0$-type structure to keep the size of the as-synthesized FePt alloy nanoparticles invariable.

There are several reports on the results of lowering the ordering transformation temperature [8-10]. These results show that the ordering transformation temperature of the FCC-FePt alloy nanoparticles can be decreased by adding Ag salts in the synthesis step [8] and also by annealing the FePt particles in high pressure H_2 gas [9]. On the other hand, the ordering transformation temperature of FePt nanoparticles has been found to decrease when a low molar ratio between ethylene glycol (EG) to Pt salts, EG/Pt, is used in their synthesis [4]. It has been proposed that the ordering transformation temperature could be lowered by decreasing the synthesis temperature because the reduction rate and the consequent formation of FePt alloy nanoparticles are retarded [4].

In general, the reduction rate of metallic ions decreases with decreasing reaction temperature. In the case of the FePt system, the Pt ions are easily reduced compared to Fe ions, so it is expected that Pt metal can be obtained at lower temperatures than that needed to obtain Fe metal. Therefore, it is necessary to enhance the reduction rate of Fe ions to synthesize equiatomic FePt alloy nanoparticles at low temperatures [12-14]. Recently, metal Fe and FeCo alloy nanoparticles have been synthesized at low temperatures by the polyol process in the presence of hydroxyl ions [11,14]. By the addition of hydroxyl ions, the continuous dehydration of ethylene glycol and formation of acetaldehyde is enhanced resulting in the reduction of metallic ions [14]. Hence, it has been considered that the concentration of hydroxyl ions plays a major role in enhancing the reduction of the metallic ions such as Fe ions [11,14]. In this study, we extended these results in the synthesis of FCC-FePt alloy nanoparticles at relatively low temperatures and furthermore investigated the effect of reaction temperature on the ordering transformation temperature of FePt particles.

2. Experimental

The synthesis of FePt particles was carried out using $FeCl_2{\cdot}4H_2O$ and $Pt(C_5H_7O_2)_2$ as precursors. Ethylene glycol was used as solvent and reducing agent, and NaOH was added in order to introduce hydroxyl ions to the reaction solution. Assuming that the reduction rate of Pt ions is higher than that of Fe ions, the initial molar ratio of Fe salts to Pt salts introduced to the polyol solution was set at 7 to 3. Reaction time was 1 h and reaction temperature was set at 373, 383, 403, 423 and 443 K. The crystal phases of precipitated particles were analyzed by X-ray diffractometry (XRD). The morphology and the size of precipitated particles were observed by transmission electron microscopy (TEM). To determine the Pt contents, the precipitated particles were dissolved in aqua regalis and then diluted. The obtained solution was then analyzed by inductively coupled plasma atomic emission spectrometry (ICP-AES). The precipitated equiatomic FCC-FePt alloy nanoparticles were annealed in Ar-5%H_2 at 473-1173 K for 1 hr. After heat treatment, XRD measurements were conducted to evaluate the degree of atomic ordering.

3. Results and discussion

The reduction of metal salts was not possible at temperatures below 373 K; successful synthesis of precipitates was possible at reaction temperatures above 383 K. Figure 1(a) shows a TEM image of the precipitated particles synthesized at 383 K. Well-dispersed spherical particles with an average diameter of 2 nm were obtained. It was observed that the particle size of the precipitate increased to

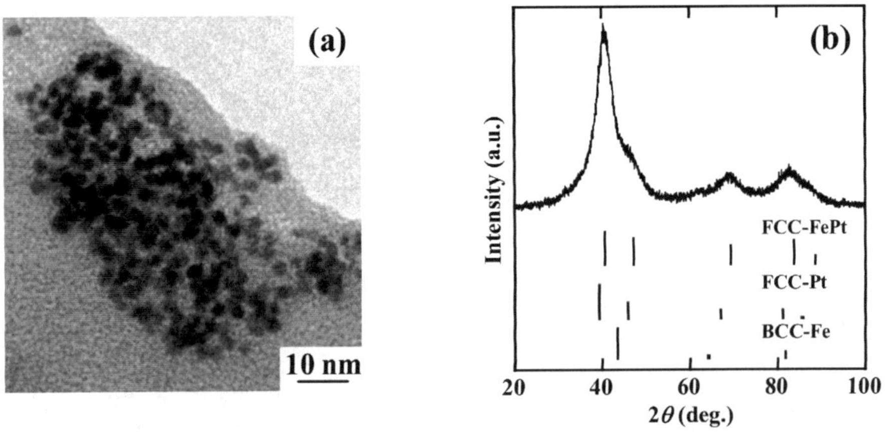

Figure 1 (a) Transmission electron microscopy image and (b)X-ray diffraction pattern of the particles synthesized at 383 K.

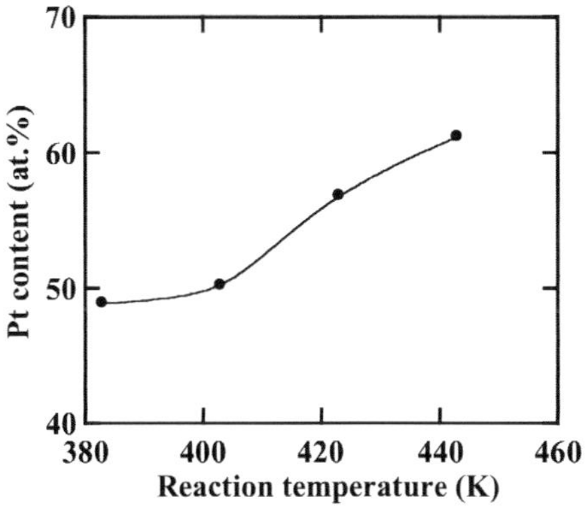

Figure 2 Reaction temperature dependence of the Pt content in the FePt alloy particles.

about 5 nm when the reaction temperature was increased to 443 K. It is noted that cubic-shaped Fe particles of several tens of nanometers in size have been synthesized by the NaOH-polyol process at 403 K [11]. However, the shape and size of the precipitated particles obtained in this work, as shown in Fig. 1(a), differed from those of the Fe metal fine particles. In addition, when the synthesis was carried out without NaOH, no particles were obtained at below 403 K. This suggests that not only the reduction of Fe ions but also the reduction of Pt ions is accelerated by hydroxyl ions in polyol solution. Therefore, the precipitated nanoparticles shown in Fig. 1(a) are expected to be FePt alloy.

Figure 1(b) shows the XRD pattern of the precipitated nanoparticles synthesized at 383 K together with reference data of Fe metal with the BCC structure (JCPDS#06-0696), Pt metal with the FCC structure (JCPDS#04-0802) and equiatomic FePt alloy with the FCC structure (JCPDS#29-0718) from the International Centre for Diffraction Data (ICCD) database. The diffraction pattern is assigned to the FCC single phase, and diffraction peaks assigned to the BCC phase are not observed. In addition, since the diffraction pattern of the precipitated nanoparticles is observed at higher angles in comparison with that of Pt metal, the lattice constant of the precipitated nanoparticles is smaller than that of Pt. Accordingly, the precipitated nanoparticles are considered to be FePt alloy with the FCC structure.

The magnetic anisotropy of $L1_0$-FePt alloy depends on the alloy composition [15]. The reaction temperature dependence of the Pt content in FePt alloy nanoparticles is shown in Fig. 2. Although the initial molar ratio of Fe salts to Pt salts was 7 to 3, the Pt content in the FePt alloy nanoparticles synthesized at 443 K was 62 at%. The reduction rate of Pt ions is regarded to be much higher than that of Fe ions at 443 K. The Pt content in FePt alloy nanoparticles decreases with decreasing reaction temperature. As a result, the equiatomic FePt alloy particles are obtained at reaction temperatures below 403 K, suggesting that the reduction rate of Pt ions becomes almost equivalent to that of Fe ions at this temperature. This suggests that the reduction rate of Pt ions has a strong temperature dependence in comparison with that of Fe ions.

$Fe_{51}Pt_{49}$ alloy particles with the FCC structure synthesized at 383 K were annealed at different temperatures, T_a. The XRD results of these particles are presented in Fig. 3(a) together with reference data of equiatomic FePt alloy with the FCC structure (JCPDS#29-0718) and FePt alloy with the $L1_0$ type structure (JCPDS#43-1359) from the ICCD database. It has been reported that equiatomic FePt alloy nanoparticles are synthesized at relatively high temperatures such as 468 K by the reduction of $Fe(C_5H_7O_2)_3$ and $Pt(C_5H_7O_2)_2$ in ethylene glycol without NaOH [4]. According to this report, the FePt alloy nanoparticles with the FCC structure were synthesized for comparison with the $Fe_{51}Pt_{49}$ alloy particles synthesized at 383 K with NaOH. The Pt content of the particles synthesized at 468 K was about 53 %, which was close to that of the particles synthesized at 383 K. Also, the average size of the obtained particles was about 2.5 ± 0.5 nm, which was quite similar to that of the particles synthesized at 383 K. The XRD patterns after annealing the equiatomic FePt alloy nanoparticles synthesized at 468 K without NaOH are shown in Fig. 3(b). The diffraction peak position of (111) for the $L1_0$-FePt alloy is higher than that for the FCC-FePt alloy. This is believed to be due to the

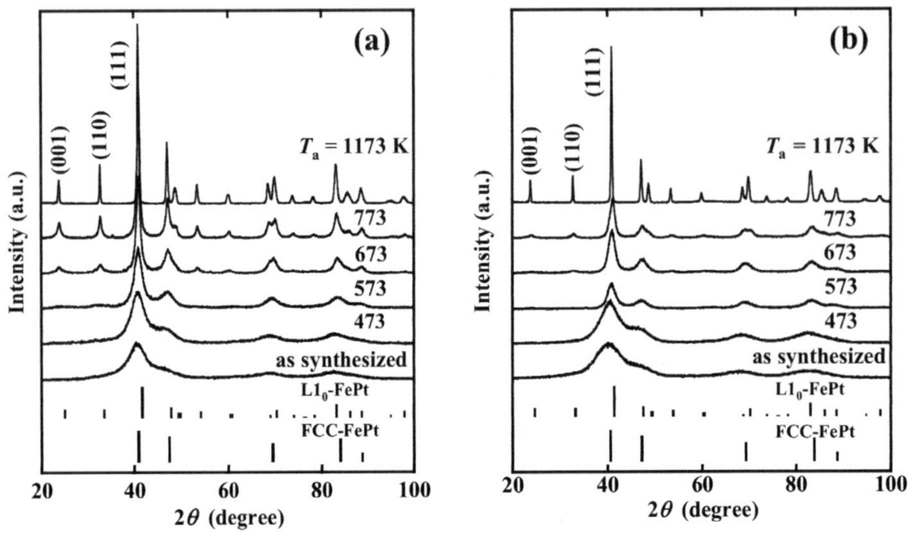

Figure 3 X-ray diffraction patterns of FCC-FePt alloy nanoparticles synthesized at (a) 383 K with NaOH and (b) 468 K without NaOH before and after heat treatment at T_a = 473, 573, 673, 773 and 1173 K.

decrease in the c/a ratio caused by the change in crystal structure from cubic to tetragonal. In Fig. 3(a), the (111) peak position of the FCC structure is slightly shifted to a higher angle in the alloy particles annealed at T_a = 473 K. In addition, in the alloy particles annealed at T_a = 573 K, peaks assigned as (001) and (110) of $L1_0$-FePt were observed. These peaks are superlattice lines, suggesting that atomic ordering occurred at T_a = 573 K. On the other hand, in the case of FePt alloy nanoparticles synthesized at 468 K without NaOH, diffraction peaks related to atomic ordering were not detected in the alloy particles annealed at T_a = 573 K, although the (111) peak was shifted to a higher angle. In these particles, atomic ordering was observed after annealing at T_a = 673 K.

To compare the behavior of atomic ordering due to heat treatment in detail, ordering parameter S is evaluated. The value of S^2 is calculated by dividing the ratio of integrated peak intensity of (110) to (111), i.e., $I_{(110)}/I_{(111)\text{measured}}$ by $I_{(110)}/I_{(111)\text{ideal}}$ for an ideally ordered phase, as shown in Eq. (1).

$$S^2 = \frac{I_{(110)}/I_{(111)\text{measured}}}{I_{(110)}/I_{(111)\text{ideal}}}, \tag{1}$$

where the value of $I_{(110)}/I_{(111)\text{ideal}}$ is estimated theoretically. Figure 4 shows T_a dependence of S for the

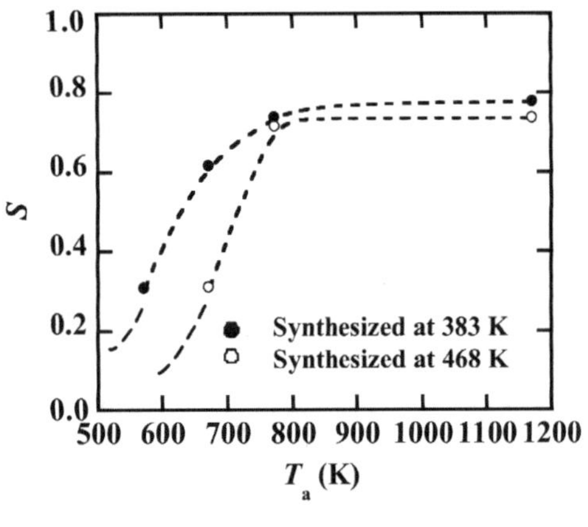

Figure 4 Annealing temperature (T_a) dependence of the ordering parameter S of FCC-FePt alloy nanoparticles synthesized at 383 K with NaOH and at 468 K without NaOH.

equiatomic FePt alloy nanoparticles synthesized at 383 K with NaOH and at 468 K without NaOH. Compared with the values of S at a relatively low T_a such as 673 K, the value of S for the FePt alloy nanoparticles synthesized at 383 K is much larger than that for the particles synthesized at 468 K without NaOH. At T_a = 773 K or higher, there is a little difference in the value of S for each annealed particle. Consequently, the lower reaction temperature in the synthesis of equiatomic FePt alloy nanoparticles led to the lowering of ordering transformation temperature from the disordered FCC structure to the ordered $L1_0$ type structure.

The ordered phase with the $L1_0$-type structure is thermodynamically stable for equiatomic FePt alloy below about 1573 K. It is believed that the disordered phase with the FCC structure of the particles synthesized at 383 and 468 K in polyol solution is a consequence of fast rate of reduction in synthesis, implying that the slight atomic order is caused by the low reduction rate in synthesis. In other words, small difference in atomic ordering between the particles synthesized at 383 and 468 K are expected, although superlattice line due to the atomic ordering cannot be detected by x-ray measurements as shown in Fig. 3. In addition, the amounts of defects such as vacancy and dislocation for the former are expected to be larger than that for the latter because of low reaction temperature. Note that such defects promote the diffusion of Fe and Pt atoms in particles during heat treatment [10]. The effects mentioned above are speculated to cause the lowering of ordering transformation temperature.

4. Conclusion

Equiatomic FCC-FePt alloy nanoparticles were successfully synthesized at 383 K by the polyol process by adding NaOH. When FCC-$Fe_{51}Pt_{49}$ alloy particles were annealed at 573 K, a partially ordered phase with the $L1_0$ structure was observed. Compared to the FePt alloy nanoparticles synthesized at 468 K without the addition of NaOH, the degree of atomic ordering of nanoparticles synthesized at 383 K increased after annealing at low temperatures. In conclusion, the ordering transformation temperature of FCC-FePt alloy nanoparticles is affected by the reaction temperature in synthesis.

5. Acknowledgement

This work was partly supported by a Grant-in-Aid for Young Scientists (B), No. 23760620, from the Japan Society for the Promotion of Science (JSPS).

References

[1] Brück E, Xiao QF, Thang PD, Toonen MJ, de Boer FR and Buschow KHJ 2001 *Physica B* **300** 215
[2] Sun S 2006 Adv. Mater. **18** 393
[3] Sun S, Murray CB, Weller D, Folks L and Moser A 2000 *Science* **287** 1989
[4] Jeyadevan B, Hobo A, Urakawa K, Chinnasamy CN, Shinoda K and Tohji K 2003 *J. Appl. Phys.* **93** 7574
[5] Gibot P, Tronc E, Chanéac C, Jolivet JP, Fiorani D and Testa AM 2005 *J. Magn. Magn. Mater.* **290-291** 555
[6] Sun S, Anders S, Thomson T, Baglin JEE, Toney MF, Hamann HF, Murray CB and Terris BD 2003 *J. Phys. Chem. B* **107** 5419
[7] Thomson T, Lee SL, Toney MF, Dewhurst CD, Ogrin FY, Oates CJ and Sun S 2005 *Phys. Rev. B* **72** 064441
[8] Kang S, Harrell JW and Nikles DE 2002 *Nano. Lett.* **2** 1033
[9] Nakaya M, Kanehara M, Yamauchi M, Kitagawa H and Teranishi T 2007 *J. Phys. Chem. C* **111** 7231
[10] Sung YN, Lee MK, Kim KE and Kim TG 2007 *Chem. Pyis. Lett.* **443** 319
[11] Joseyphus RJ, Shinoda K, Kodama D and Jeyadevan B 2010 *Mater. Chem. Phys.* **123** 487
[12] Aizawa S, Tohji K and Jeyadevan B 2008 *5th International Workshop on Water Dynamics 22*

[13] Chinnasamy CN, Jeyadevan B, Shinoda K and Tohji K 2003 *J. Appl. Phys.* **93** 7583
[14] Joseyphus RJ, Matsumoto T, Takahashi H, Kodama D, Tohji K and Jeyadevan B 2007 *J. Solid State Chem.* **180** 3008
[15] Sakuma A 1994 *J. Phys. Soc. Jpn.* **63** 305

Synthesis of multiferroic $DyFeO_3$ nanoparticles and study of their magnetic properties

L. Zhu, N. Sakai, T. Yanoh, S. Yano, N. Wada, H. Takeuchi, A. Kurokawa, and Y. Ichiyanagi*
Graduate School of Engineering, Yokohama National University,
79-5 Tokiwadai, Hodogaya-ku, Yokohama 240-8501, JAPAN

E-mail: yuko@ynu.ac.jp

Abstract. We synthesized $DyFeO_3$ nanoparticles having sizes of 53, 65, 87, and 93 nm by mixing aqueous solutions of metal chlorides and sodium hydroxide. We found that the particle size increased with the annealing temperature. The magnetization of each particle was measured by using a SQUID magnetometer under a H = ±50 kOe magnetic field at temperatures ranging from 5 to 300 K. Samples having sizes of 65, 87, and 93 nm were observed to have a spin arrangement transition temperature T_N of around 55 K. The magnetic properties of the particles changed drastically depending on their size. For example, the sample having a size of 53 nm did not have a T_N value.

1. Introduction

Multiferroic materials are defined as materials that simultaneously exhibit both ferromagnetic and ferroelectric ordering.[1]-[3] Multiferroic materials have attracted considerable attention because of the coupling between the magnetic and the dielectric properties in these systems as well as their controllability under applied magnetic or electric fields. Thus far, many multiferroic materials have been discovered, for example, metal oxides such as $BiFeO_3$, $BiMnO_3$, and $TbMnO_3$.[4] Because these materials exhibit a strong interplay between the ferromagnetic and the ferroelectric ordering, they find applications in information storage, spintronics, and sensors. In particular, $BiFeO_3$, which simultaneously exhibits ferroelectric and antiferromagnetic ordering at room temperature, has attracted increasing attention. However, $BiFeO_3$ magnetization value is weak, and therefore, it presently finds few applications in practical devices. We noted that $DyFeO_3$ [5] and $BiFeO_3$ have the same crystal structure [6]-[10], and therefore, we synthesized $DyFeO_3$ magnetic nanoparticles and measured their magnetic properties and dielectric constant in order to develop novel multiferroic materials.

2. Experiment

$DyFeO_3$ nanoparticles were produced by mixing aqueous solutions of $DyCl_3{\cdot}6H_2O$, $FeCl_2{\cdot}4H_2O$, and NaOH. The mole ratio of the prepared reagent was Dy:Fe:OH = 1:1:7. The obtained precipitates were washed three times with distilled water and dried at around 350 K for 48 h. The as-prepared samples were heat-treated in a furnace in air at annealing temperatures of 973–1273 K. Each sample was examined by Cu-Kα X-ray (λ = 0.154 nm) powder diffraction (XRD), following which magnetization measurements were carried out using a SQUID magnetometer (Quantum Design, MPMS) under external magnetic fields of -50 to +50 kOe at temperatures ranging from 5 to 300 K. The dielectric constants were measured using an RF impedance/material analyzer (Agilent E4991A).

3. Result and Discussion

Figure 1 shows the X-ray diffraction patterns. All the peaks of the samples annealed at 973–1273 K can be readily indexed to a pure orthorhombic phase (space group: PBNM(62)) of $DyFeO_3$ with calculated lattice constants of a = 0.5296 nm, b = 0. 5603 nm, and c = 0.763 nm. The average particle sizes were estimated to range from 53 to 93 nm using the Scherer formula and the half-width of the diffraction peaks. The particle size increased with the annealing temperature. Table 1 lists the particle sizes observed at various annealing temperatures.

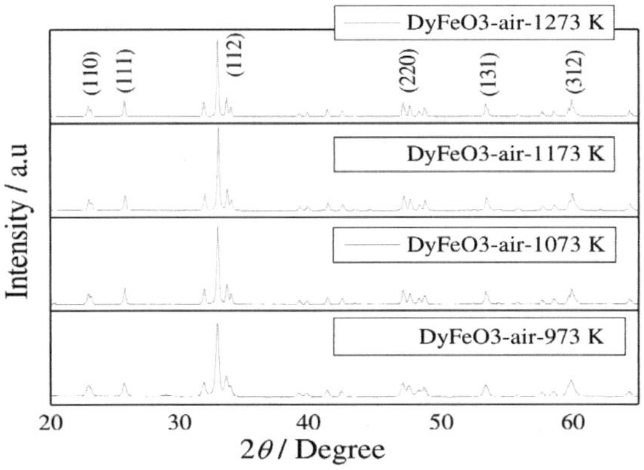

Figure 1. Powder X-ray diffraction patterns of samples annealed at 973–1273 K.

Annealing Temperature/ K	Particle size/ nm
973	53
1073	65
1173	87
1273	93

Table 1. Particle sizes of $DyFeO_3$ samples at various annealing temperatures, as estimated by powder X-ray diffraction.

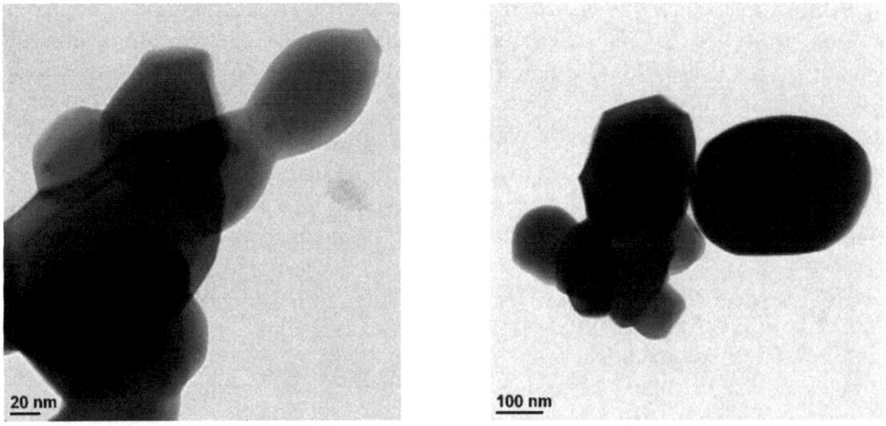

Figure 2. TEM images of $DyFeO_3$-53 nm (left-hand side) and $DyFeO_3$-93 nm (right-hand side).

Although the particles were aggregated, from the TEM images (Figure 2), their individual size can be estimated to be approximately 53 nm (left-hand side) and 93 nm (right-hand side).

The magnetization measurements were performed using an MPMS SQUID magnetometer under an external magnetic field of -50 to +50 kOe at temperatures ranging from 5 to 300 K.
Figure 3 shows the *M–H* curves for each particle size measured under an external magnetic field of -50 to +50 kOe at a temperature of 5 K; the inset shows an enlarged view of the region around the origin of the figure. The coercive force can be observed at 35~115 Oe for 4 samples (Table 2); therefore, each particle exhibits ferromagnetic behavior, and the coercive force increases with the particle size at a temperature of 5 K.

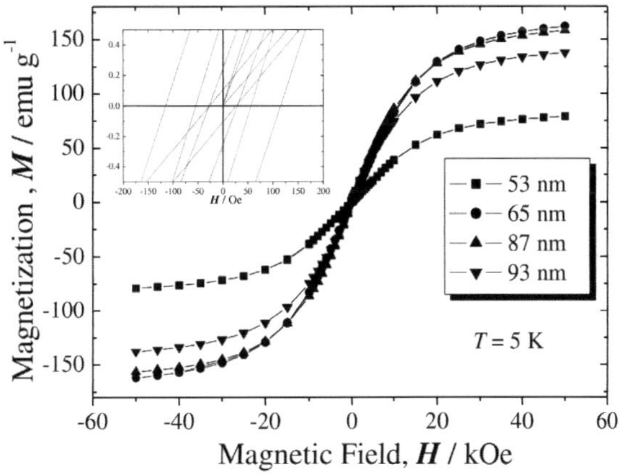

Figure 3. Magnetization curves for particles having sizes of 53, 65, 87, and 93 nm.

Particle size/ nm	Coercive force/ Oe
53	35
65	43
87	60
93	115

Table 2. Relationship between particle size and coercive force.

Figure 4 shows the temperature dependence of both the field-cooled (FC) (closed marks) and the zero-field-cooled (ZFC) (open marks) magnetization in each sample under an external magnetic field of 100 Oe. As seen in this figure, the peaks of samples having sizes of 65, 87, and 93 nm appear at around 53 K; therefore, we consider that these samples have antiferromagnetic ordering. The peaks of these samples broaden with a decrease in the particle size. In the sample having a size of 53 nm, the peak completely disappears; in other words, this sample exhibits paramagnetic behavior. We have speculated about the mechanism of this change in magnetic ordering. In particles having smaller size, the spins cannot be completely aligned along the interior of the particle. Surface spins are randomly oriented perpendicular to the surface of the particles. On account of the greater number of random spins than aligned spins, the particles exhibit paramagnetic behavior. However, $DyFeO_3$ particles do not have different Neel temperatures depending on the size; samples having a size of 53 nm also exhibit antiferromagnetic behavior.

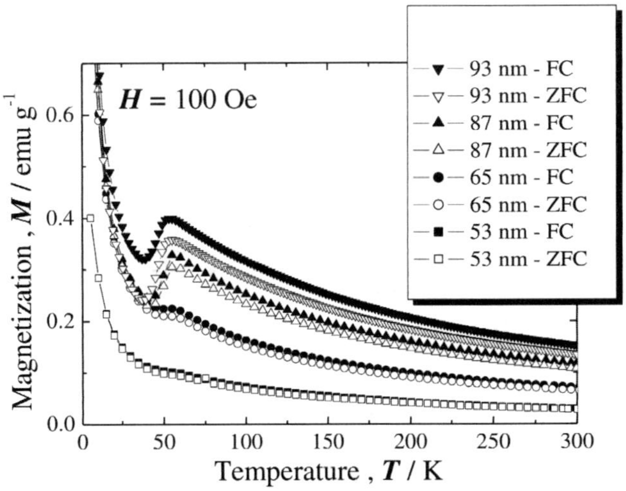

Figure 4. Temperature dependence of FC (closed marks) and (open marks) ZFC magnetization in samples having sizes of 53, 65, 87, and 93 nm.

The Curie-Weiss law is given by the following equation:

$$\frac{1}{\chi} = \frac{1}{C}(T - \theta) \qquad (1)$$

Figure 5 shows the temperature dependence of the inverse magnetic susceptibility. The figure indicates that particles with sizes of 65, 87, and 93 nm will exhibit antiferromagnetic ordering because the Weiss temperatures are -51, -73, and -84 K, respectively. However, the $DyFeO_3$ nanoparticle having a size of 53 nm exhibits paramagnetic ordering because the Weiss temperature is 0 K.

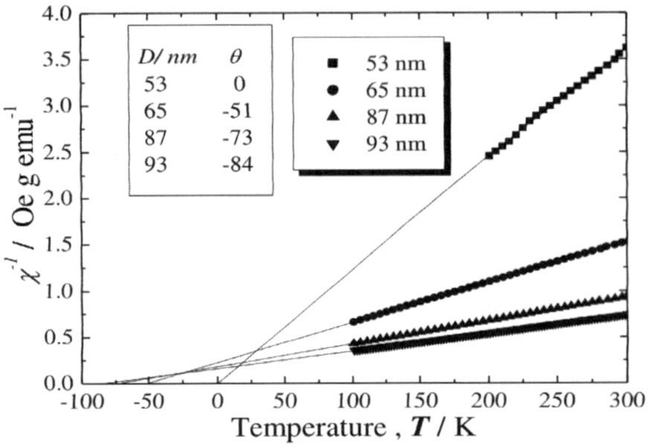

Figure 5. Temperature dependence of inverse magnetic susceptibility.

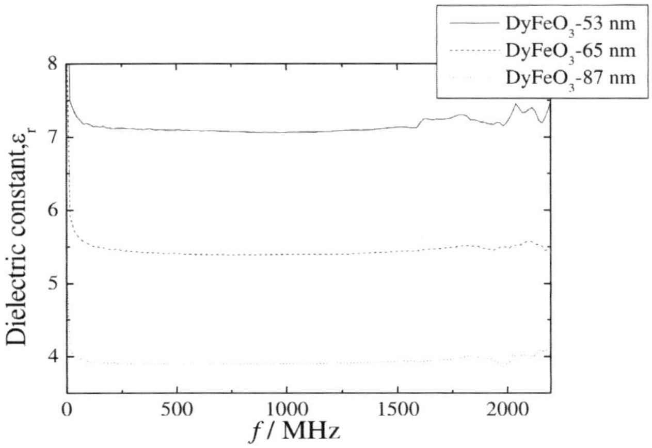

Figure 6. Frequency dependence of dielectric constant

The frequency dependence of the dielectric constant of $DyFeO_3$ particles having sizes of 53, 65, and 87 nm is shown in Figure 6. The dielectric constant is observed to be between 3.4 and 8, which is smaller than the more practical values in the case of $BiFeO_3$, of between 18 and 22. In addition, we found that the dielectric constant increases as the particle size decreases. In particles having smaller size, there was an increase in the number of conduction electrons at the particle surface. The interactions among these conduction electrons led to an increase in the dielectric constant.

4. Conclusions and Future Work

We synthesized $DyFeO_3$ nanoparticles by a simple method, and we observed the dependence between various magnetic properties and the particles size. Samples having sizes of 65, 87, and 93 nm exhibit antiferromagnetic ordering, and the sample having a size of 53 nm exhibits paramagnetic behavior. We also analyzed the dielectric constant. The dielectric constant increases as the particle size decreases. However, the value of the dielectric constant was smaller than that of $BiFeO_3$, which has the same crystal structure. In the future, we will synthesize Bi-doped $DyFeO_3$ in order to realize a multiferroic material with high magnetization and high dielectric constant.

Acknowledgements

This study was partially supported by the Precursory Research for Embryonic Science and Technology program of the Japan Science and Technology Agency.

References

[1] Treves D *J. Appl. Phys.* **36** 1033
[2] Ito N, Takahashi, Takamatsu, Tanaka D and Saitoh 2000 ***J. Appl. Phys.*** 39 1371–77
[3] White R L *J. Appl. Phys.* **40** 1061
[4] Zhou H, Yiu Y, Aronson M C and Wong S S 2008 ***J. Solid State Chem.*** **181** 1539–45
[5] Tokunaga Y, Iguchi S, Arima T and Tokura Y *Phys. Rev. Lett.* **101** 097205
[6] Ichiyanagi Y and Uehashi T 2004 *Phys. Stat. Sol. C* **1** 3385
[7] Ichiyanagi Y, Kondoh H and Ohta T 2003 *Chem. Phys. Lett.* **379** 345
[8] Kubota M, Kanazawa Y, Nasu K, Moritake S, Kawaji H, Atake T and Ichiyanagi Y 2008 *J. Therm. Anal. Cal.* **92** 461
[9] Ichiyanagi Y, Uehashi T, Yamada S, Kanazawa Y and Yamada T 2005 *J. Therm. Anal. Cal.* **81** 541
[10] Ichiyanagi Y and Uehashi T 2004 *Phys. Stat. Sol. C* **1**(12) 3485

Fabrication and characterization of CuO/ZnO solar cells

H. Kidowaki, T. Oku, T. Akiyama
Department of Materials Science, The University of Shiga Prefecture,
2500 Hassaka, Hikone, Shiga 522-8533, Japan

E-mail: oku@mat.usp.ac.jp

Abstract. Cuprous oxide (CuO) and zinc oxide (ZnO) heterojunction solar cells fabricated on indium tin oxide-coated glass were studied. CuO and ZnO films were deposited using a galvanostatic method. Structural, morphological and optoelectronic properties of the CuO/ZnO heterojunction were studied by using X-ray diffraction, atomic force microscopy and light current-voltage characteristics.

1. Introduction

Solar cell technology for future energy resources has been progressed recently. Silicon is used as the semiconductor material for conventional solar cells, but silicon is expensive and the cost reduction of the solar cells is one of the most important issues. Oxide semiconductors are one of the alternatives to silicon solar cells, and copper oxides such as CuO and Cu_2O are one of the candidate materials. The features of copper oxide semiconductors are relatively higher optical absorption, low cost of raw materials and non-toxic. CuO and Cu_2O are p-type semiconductors with band gap energies of 1.5 eV and 2.0 eV, respectively, which are close to the ideal energy gap of 1.4 eV for solar cells and allows for good solar spectral absorption. The highest efficiency of ~2 % for Cu_2O solar cells has been obtained by using the high temperature annealing method [1]. Cu_2O/ZnO thin film solar cells have also been studied and reported [2,3]. However, the solar cells with a CuO/ZnO structure have not been reported.

The purpose of the present work is to fabricate and characterize solar cells with CuO/ZnO structures. The band gap energy of CuO is ~1.5 eV, which is closer to the ideal band gap of 1.4 eV. Zinc oxide (ZnO) is an n-type semiconductor with a wide band gap of ~3.37 eV, which can be applied to solar cells [4].

2. Experimental procedures

CuO layers were prepared on pre-cleaned indium tin oxide (ITO) glass plate by electro deposition using platinum as counter electrode. Copper(II) sulfate ($CuSO_4$, 0.4 mol/L, Wako 97.5%) and l-lactic acid (3 mol/L, Wako) were dissolved into distilled water. pH of the electrolyte solution was adjusted to 12.5 by adding NaOH. The temperature of electrolyte solution was kept at 65 °C during electro deposition. Preparation of CuO layers were carried out at voltages of +0.70 V and quantity of electric charge of 2.2 C cm^{-2}. After the deposition, the sample was rinsed with water and transferred into the ZnO electro deposition bath. ZnO layers were galvanostatically electro-doposited from 0.025 M aqueous solution of $Zn(NO_3)$ on the ITO/CuO substrate. The sample thickness ranged between 1 and 2 μm depending on the deposition time and current. Finally, the substrates were rinsed with water, dried with air and quickly transferred into a thermal evaporator for the vacuum deposition of the aluminum (Al) back contact. Structure of heterojunction solar cells were denoted as ITO/CuO/ZnO/Al, with a schematic illustration as shown in Fig. 1.

Current density-voltage (J-V) characteristics (Hokuto Denko Corp., HSV-100) of the solar cells were measured both in the dark and under illumination at 100 mW/cm^2 by using an AM 1.5 solar simulator (San-ei Electric, Electric, XES-301S). The solar cells were illuminated through the side of the ITO substrate, and the illuminated area was 0.16 cm^2. Optical absorption of the solar cells was investigated by means of UV visible spectroscopy (Hitachi, Ltd., U-4100). Microstructures of the copper oxides were investigated by X-ray diffractometer (XRD, PHILIPS X'Pert-MPD System) with CuKα radiation operating at 40 kV and 40 mA.

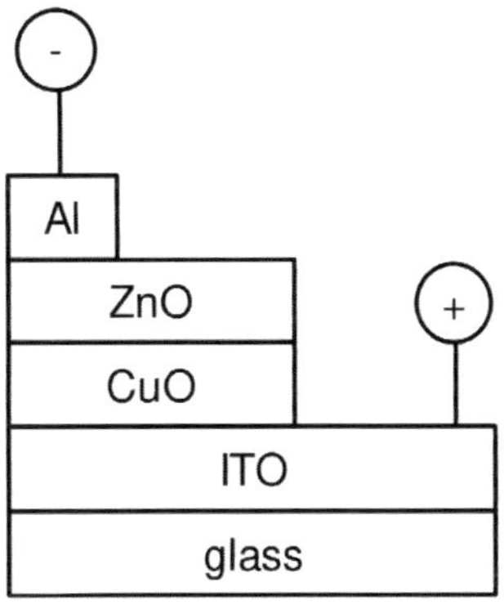

Figure 1. Structure of ITO/CuO/ZnO/Al heterojunction solar cells

3. Results and discussion

The measured parameters of the solar cells are summarized in Table 1. Thicknesses of CuO were ~1 μm and ~2 μm for 5 min and 10 min, respectively. A solar cell with a CuO/ZnO structure provided a power conversion efficiency (η) of 1.1×10^{-4} %, fill factor (FF) of 0.25, short-circuit current density (J_{sc}) of 1.6 mAcm^{-2} and open-circuit voltage (V_{oc}) of 2.8 × 10^{-4} V. The photocurrent was observed under illumination, and the CuO/ZnO structure showed characteristic curves with short-circuit current and open-circuit voltage. Figure 2 shows measured optical absorption of thin films. The CuO and ZnO thin fims shows high optical absorption in the range of 400 nm and 800 nm.

Transmittance spectrum of 100 nm thick CuO film, deposited on ITO, is presented in Fig. 3(a). From this spectrum, the optical absorption coefficients (α) of this film was determined from the spectral transmittance using the next equation, $\alpha = 1/d \cdot \ln(1/T)$. Where d is the film thickness and T is the transmittance [5]. For determination of the optical band gap energy (E_g), the method based on the relation of $\alpha h\nu = A(h\nu - E_g)^{n/2}$ was used, where n is a number that depends on the nature of the transition. In this case, its value was found to be 1, which corresponds to direct band to band transition.

Figure 3(b) is a Tauc plot, which shows $(\alpha h\nu)^2$ versus $h\nu$ for the CuO film. The intersection of the straight line with the hv-axis determines the optical band gap energy E_g [6]. It was found to be ~2.1 eV which is lower than the ideal band gap of the CuO crystal. Because of the small band gap energy, the open-circuit voltage would be low.

Table 1. Measured parameters of CuO/ZnO solar cells.

Deposition time of CuO (min)	V_{oc} (mV)	J_{sc} (mA/cm^2)	FF	η (%)
5	0.28	1.9	0.25	1.1×10^{-4}
10	0.18	0.017	0.25	1.1×10^{-7}

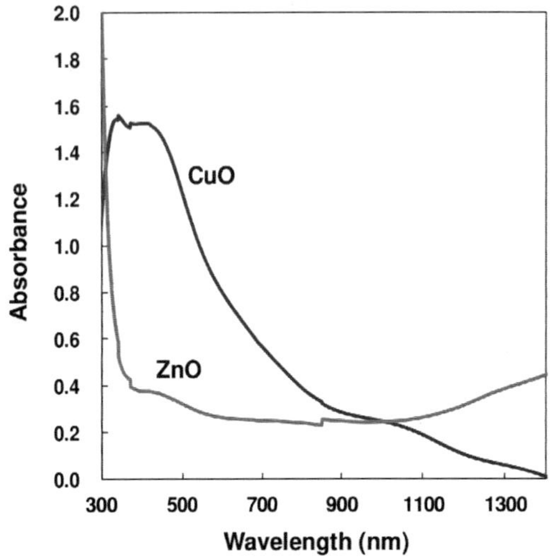

Figure 2. Optical absorption of thin films prepared by electrodeposition.

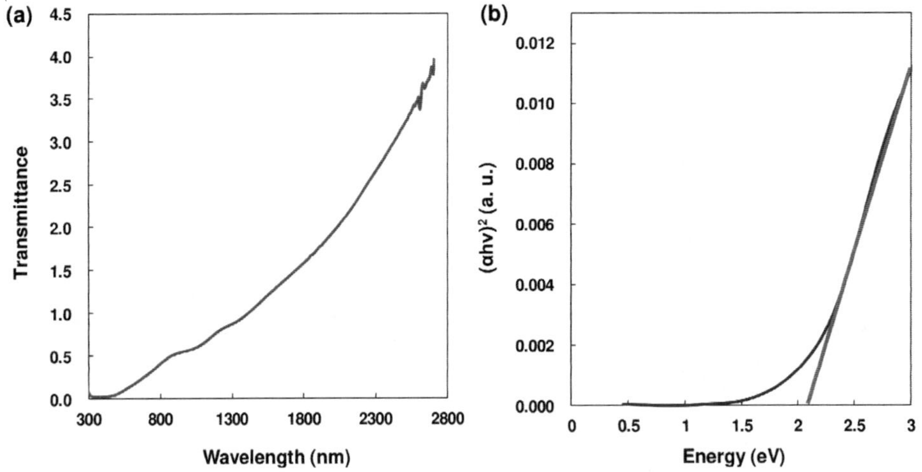

Figure 3. (a) Optical transmission spectrum of CuO film and (b) Tauc plot for CuO film.

All the crystalline components in the CuO and ZnO thin films were investigated by XRD, as shown in Fig. 4. Diffraction peaks corresponding to CuO and ZnO are observed in thin films, which consisted of cupric phase with monoclinic system (space group of C2/c and lattice parameter of a=0.4653 nm, b=0.3410 nm, c=0.5018 nm, β=99.481°). The particle size was estimated using Scherrer's equation: $D = 0.9\lambda/B\cos\theta$, where λ, B, and θ represent the wavelength of the X-ray source, the full width at half maximum (FWHM), and the Bragg angle, respectively [7]. The crystallite sizes of CuO and ZnO were determined to be 49.0 nm and 82.0 nm, respectively. Lattice constants of CuO and ZnO were summarized as listed in Table 1 and Table 2, comparing with reported values [8,9]. From the lattice constants, the crystal structures of CuO and ZnO have some crystal distortions. To increase the efficiency of the CuO/ZnO solar cells, small grain size of ZnO and higher crystallinity of CuO would be necessary.

Energy level diagram of the CuO/ZnO solar cell is summarized as shown in Fig. 5. Previously reported values were used for the energy levels [7,8]. It has been reported that V_{oc} is nearly proportional to the band gap of the semiconductors, and control of the energy level is important to increase efficiency [9]. Compared to silicon with an indirect transition band structure, CuO with a direct transition band structure is more suitable for the optical absorption property. In addition, the ultrathin film of the CuO layers could provide efficient charge injection because of the high optical absorption.

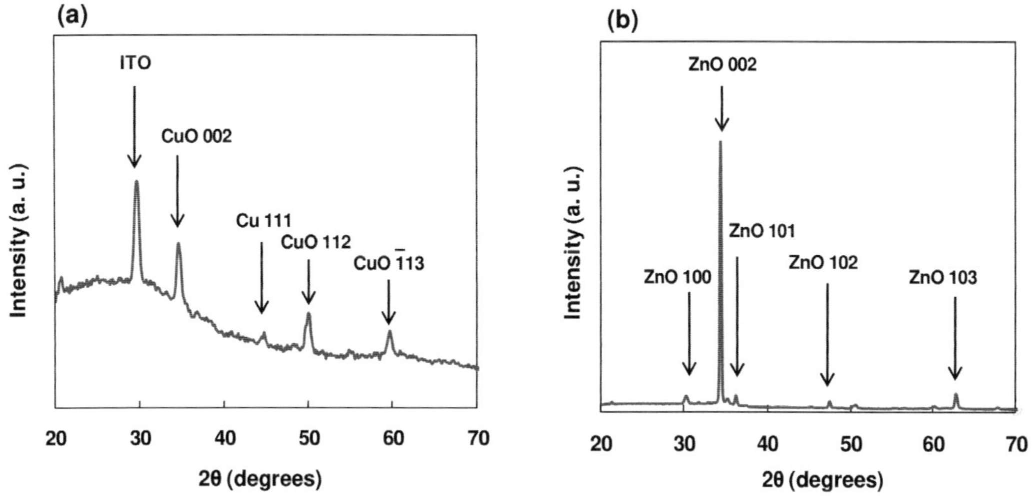

Figure 4. XRD patterns of (a) CuO and (b) ZnO thin films.

Table 2. Lattice constants of CuO.

	a (nm)	b (nm)	c (nm)
Present data	0.5122	0.4732	0.5182
Reference [8]	0.4679	0.3426	0.5127

Table 3. Lattice constants of ZnO.

	a (nm)	c (nm)
Present data	0.2816	0.5213
Reference [9]	0.3257	0.5213

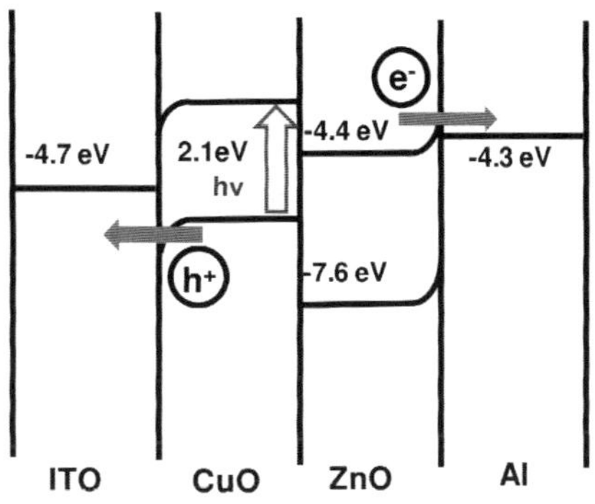

Figure 5. Energy diagram of ITO/CuO/ZnO/Al heterojunction solar cells..

In the present work, the V_{oc} is not so high compared to Fig. 5. Microstructures of CuO and ZnO thin films in the present work were found to have some crystal distortion, which would result in the reduction of electrical transport. If the crystal qualities of the CuO and ZnO thin films are increased, V_{oc} would be improved.

4. Conclusion

ITO/CuO/ZnO/Al solar cells were produced and characterized, which provided η of 1.1×10^{-4} %, FF of 0.25, J_{sc} of 1.6 mAcm^{-2} and V_{oc} of 2.8×10^{-4} V. The CuO/ZnO structure showed high optical absorption in the range of 400 nm and 800 nm, and the E_g of CuO was found to be ~1.2 eV from the Tauc plot, which is smaller than that of the ideal band gap of the CuO crystal and the open-circuit voltage would be decreased. A crystallite size of CuO was determined to be 49.0 nm, and higher crystallinity of CuO would increase the efficiency of the CuO/ZnO solar cells. The energy level of the present solar cell was proposed, and separated holes could transfer from the valence band of the CuO to the ITO, and separated electrons could transfer from the conduction band of the CuO to the Al electrode, respectively. Formation of the CuO/ZnO active layer with homogeneous distributed CuO nanopaticles would improve the efficiencies of the solar cells.

References

[1] Mittiga A, Salza E, Sarto F, Tucci M, Vasanthi R, 2006 *Appl. Phys. Lett.* **88** 163502
[2] Jeong S S, Mittiga A, Salza E, Masci A, Passerini, S 2008 *Electerochem. Acta* **53** 2226
[3] Cui J, Gibson U J, 2010 *J. Phys. Chem.* **114** 6408
[4] Vinodkumar R, Letthy K J, Beena D, Detty A P, Nava I, Nayar U V, Mahadevan V P, Ganesan V, Rdddy V R, 2010 *Sol. Energy. Mater. Sol. Cells* **94** 68
[5] Chen J W C, Perng D C, Fang J F, 2011 *Sol. Energy. Mater. Sol. Cells* **95** 2471
[6] Georgieva V, Ristov M, 2002 *Sol. Energy. Mater. Sol. Cells* **73** 67
[7] Oku T, Noma T, Suzuki A, Kikuchi K, Kikuchi S, 2010 *J. Phys. Chem. Solids* **71** 551
[8] Domnina M I, Filatov S K, Zyuzyukia I I, and Vergasova L P, 1986 *Neorgan. Materialy.* **22** 1743
[9] Albertsson, Abrahams S C, and Kvick A 1989 *Acta Cryst.* **B45** 34
[10] Oku T, Motoyoshi R, Fujimoto K, Akiyama T, Jeyadevan B, Cuya J, 2011 *J. Phys. Chem. Solids* **72** 1206
[11] Park N G, Kang M G, Ryu K S, Kim K M, Chang S H, 2010 *J. Photochem. Photobiol. A: Chem.* **161** 105

Asia-Pacific Interdisciplinary Research Conference 2011 (AP-IRC 2011) IOP Publishing
Journal of Physics: Conference Series **352** (2012) 012023 doi:10.1088/1742-6596/352/1/012023

Microstructure analysis of spherical silicon solar cells coated with anti-reflection films

Y. Ono[1], T. Oku[1], T. Akiyama[1], Y. Kanamori[2], Y. Ohnishi[2], Y. Ohtani[2], M. Murozono[2]

[1] Department of Materials Science, The University of Shiga Prefecture, Shiga 522-8533, Japan
[2] Clean Venture 21 Corporation, Kyoto 601-8121, Japan

E-mail: ze21yono@ec.usp.ac.jp, oku@mat.usp.ac.jp

Key words: Solar cells, Spherical Silicon, SnO_2

Abstract. Microstructures of spherical silicon solar cells were investigated by transmission electron microscopy and X-ray diffraction. Optical absorption and photoluminescence of the solar cells were also measured by UV-visible and fluorescence spectroscopy. F-doped SnO_2 anti-reflection coating layers were investigated by XRD, and the nanostructures and grain sizes were determined. The present work indicated a guideline for spherical silicon solar cells with higher efficiencies.

1. Introduction

Solar cells are expected to be clean energy devices instead of fossil fuels in the view point of no emission of greenhouse gas. However, production cost of silicon solar cells is high, and the cost reduction of the silicon solar cells is one of the most important issues. From the view point of the cost reduction, spherical silicon solar cells are a promising technology [1-3]. Silicon spheres with a diameter of ~1.0 mm are used as p-n junction solar cells, and consumption of silicon is lower and more flexible compared to the conventional silicon solar cells. However, the conversion efficiency of the spherical silicon solar cells should be improved further compared to the crystalline silicon solar cells.

To improve the conversion efficiency, various properties are mandatory as follows; high quality silicon balls, optical confinement structures, inactivation of defects by passivation, low resistances of conductive anti-reflection coating, and effective reflecting mirror structure. The purpose of the present work is to investigate microstructures of spherical silicon with F-doped SnO_2 for anti-reflection coating layers. Combination of microstructure analysis and optical property characterization of the spherical silicon with anti-reflection films will give us a guide line for improvement of the efficiency.

2. Experimental

The spherical silicon used in this experiment was supplied by Clean Venture 21 Co., Ltd [4,5], and the surface of the spherical silicon was coated by SnO_2 as anti-reflective coating (ARC) with a film thickness of ~100 nm [6]. The microstructure of SnO_2:F film was investigated by X-ray diffraction (XRD, Philips X'Pert-MPD System). Optical absorption of the solar cells was also measured by UV-visible spectroscopy (Hitachi U-4100). Photoluminescence measurements were performed using (Jasco FP-6600) at room temperature employing a Xenon lump.

3. Result and discussion

X-ray diffraction pattern of spherical silicon coated by SnO_2:F is shown in Fig. 1(a). Diffraction peaks corresponding to SnO_2 crystal are observed [7-10]. The SnO_2:F were found to be a cassiterite type with a tetragonal rutile structure as shown in Fig. 1(b), and lattice parameters were a = 0.4681 nm and c = 0.3156 nm in the present work, whereas the reported values of SnO_2 lattice parameters are a = 0.4737 nm and c = 0.3185 nm [11], as summarized in Table 1. The differences of lattice parameters of SnO_2:F are 1.2 % and 0.9 % for a and c-axes, respectively. The difference between the present values and reported values would be due to the fluorine ions (F^-) and crystal distortion. In the SnO_2:F, oxygen ions (O^{2-}) were estimated to be substituted by the fluorine ions (F^-) as following reasons [4]: similar ionic size (F^-: 0.136 nm, O^{2-}: 0.140 nm), the energy of the Sn-F bond (~26.75 D_0/kJmol^{-1}) is comparable to that of the Sn-O bond (~31.05 D_0/kJmol^{-1}), and Coulomb forces that bind the lattice together are reduced, since the charge on the F^- is only half of the charge on the O^{2-}. Here, D_0 is the unit of bond-dissociation energy. The particle size of the SnO_2:F was estimated using Scherrer's formula: $D=0.9/B\cos\theta$, where λ, B, and θ represent the wavelength of X-ray source, the full width at half maximum, and the Bragg angle, respectively. The crystallite size of SnO_2:F was determined to be 57.5 nm. Since the grain boundaries would disturb the light incidence and carrier transport by scattering the light and electronic carrier, single crystal or crystals with larger grain size would be desirable. Although it seems to be difficult to deposit thin films with a single crystal on the spherical Si surface, further improvement would be expected by optimizing deposition and annealing conditions.

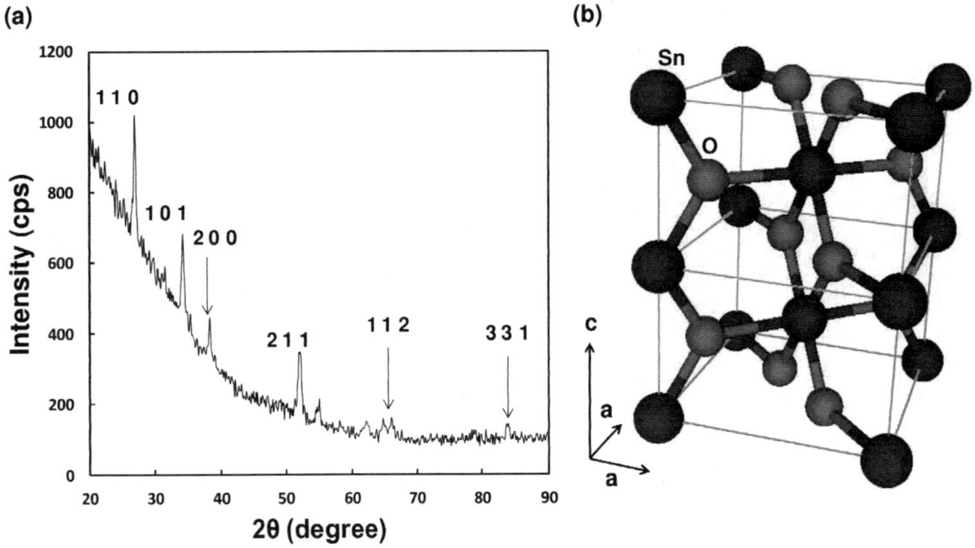

Figure 1. (a) XRD pattern of SnO_2 doped with F. (b) Atomic structure model of SnO_2.

Table 1 Lattice parameters of SnO_2:F and SnO_2.

	SnO_2	SnO_2:F	Difference (%)
a (nm)	0.4737	0.4681	-1.182
c (nm)	0.3185	0.3156	-0.911

Figure 2 shows measured optical absorption of the spherical silicon with and without ARC. Spherical silicon with ARC shows higher absorption than that of without ARC in the range from 400 nm to 1100 nm, especially around ~600 nm, which indicates the effectiveness of the ARC.

Figure 3 shows photoluminescence (PL) spectrum of spherical silicon with ARC and without ARC. The PL emission of the spherical silicon without ARC is higher than that of spherical silicon with ARC in all range from 400 nm to 900 nm. PL emission observed around 3.3 eV would be due to the impurities contained in silicon, which might contribute formation of exitons that increase efficiency. The spherical silicon coated with SnO_2 shows the decrease of the PL emission, which indicated the confinement of light by the ARC. On the other hand, intensity of the PL emission from the spherical silicon without ARC is higher compared to that of spherical silicon with ARC, which also indicated the effectiveness of ARC.

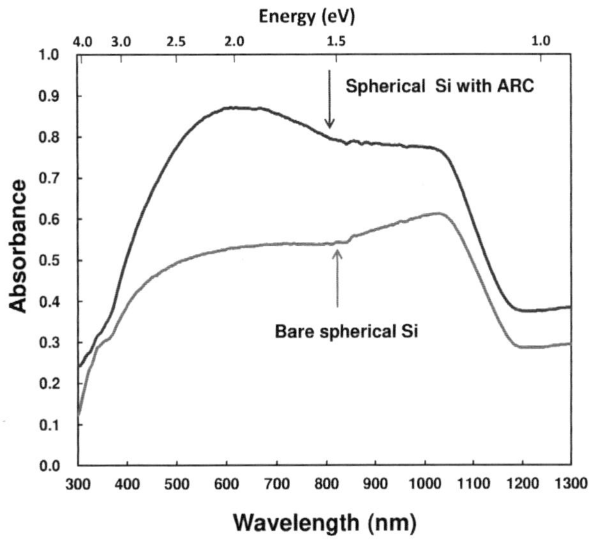

Figure 2. Optical absorption of spherical silicon with ARC and without ARC.

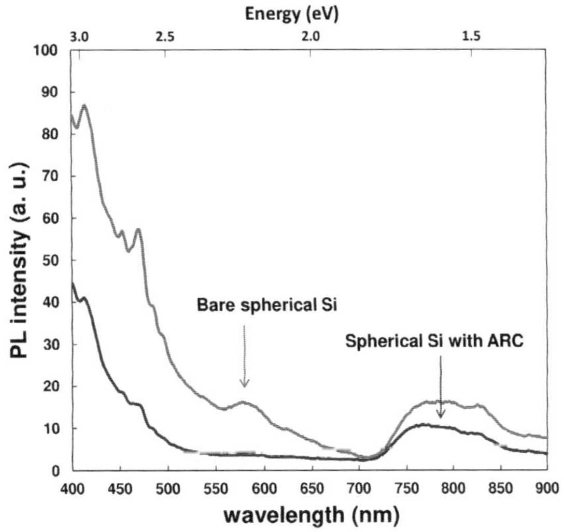

Figure 3. Photoluminescence spectra of spherical silicon with ARC and without ARC.

The solar cell with ARC provided power convergent efficiency (η) of 11.5 %, a fill factor (FF) of 0.697, J_{sc} of 28.1 mA/cm^2, V_{oc} of 0.584 V. About 20 % increase of efficiency was observed by introducing the ARC. Schematic structure and energy level diagram of spherical silicon solar cell is shown in Figure 4(a) and 4(b), respectively. 4. The spherical Si solar cell with a reflector cup is completed by mounting of the spherical Si on a reflector cup consisting of an Ag-electroplated Al substrate. Electronic carriers of electrons and holes are separated at the p-n junction by light irradiation. Separated holes could be transferred from the valence band of p-Si to FTO of antireflective coating, and separated electrons could be transferred from the conduction band of n-Si to the Al electrode. As shown in Fig. 4(a), charge separation is strongly dependent of incident angle of light. To improve the efficiency, microstructure control of antireflective coating by annealing and deposition conditions would be important. In addition, there would be energy barrier at the semiconductor/metal interfaces as indicated by band bending in Fig. 4(b) [12], which results in contact resistance. Heavy doping or control of interfacial microstructure by annealing would reduce the contact resistance.

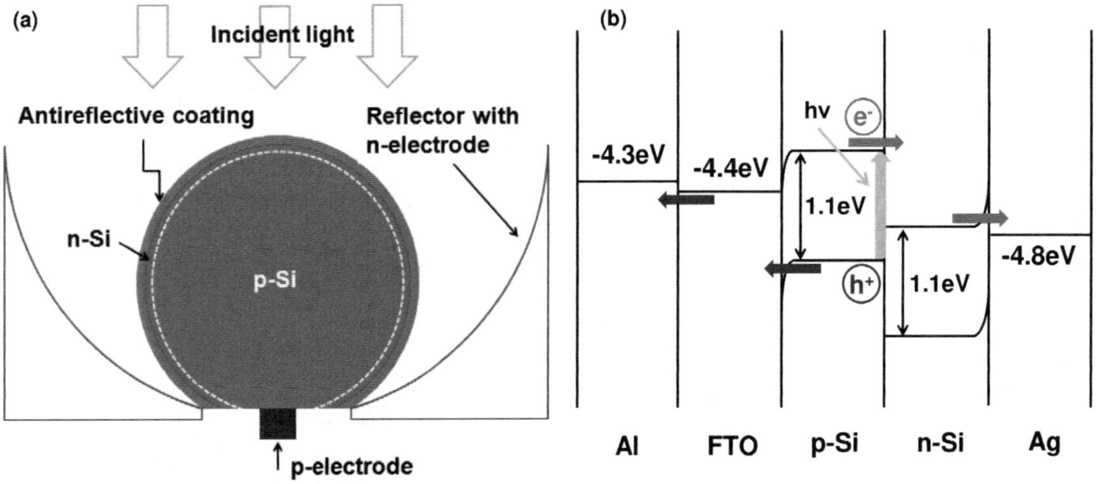

Figure 4. (a) Schematic structure and (b) energy level diagram of spherical silicon solar cell.

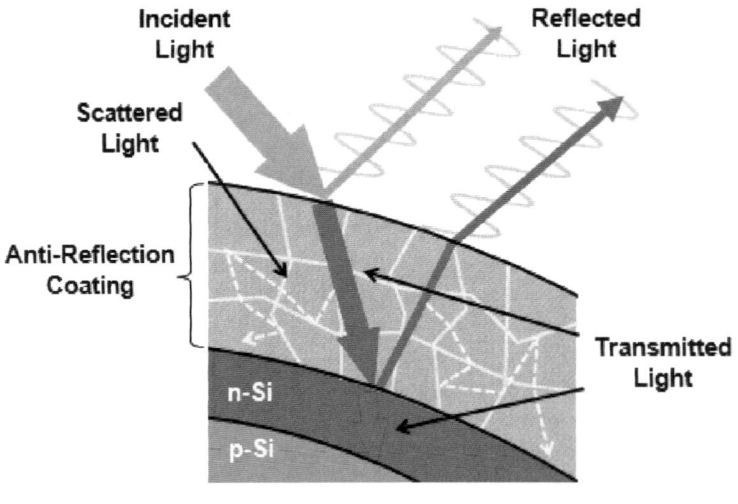

Figure 5. Mechanism of anti-reflection in the present spherical silicon with ARC.

Figure 5 shows a mechanism of ARC. The principle of the ARC is based on the phase changes. Destructive interference between the two reflected lights occurs, cancelling both lights before they exit the surface. The transmission properties of the ARC are dependent on the wavelength of light, the substrate's index of refraction, the index of refraction of the coating, the thickness of the coating, and the angle of the incident light. Therefore, the best ARC condition should be investigated further to reduce the reflected light. According to the previous report [13], the best condition of the thickness of the coating and the index of refraction of the coating are 83 nm and 2.0 respectively.

It should be pointed out that the optimization of the thickness of ARC, and the analysis of the impurities contained in spherical silicon are considered to be important factors for higher efficiencies of the spherical silicon solar cell, which should be investigated further.

4. Conclusions

Microstructures of the spherical silicon solar cells with anti-reflection films were investigated by using XRD, which indicated lattice parameters of a = 0.4681 nm and c = 0.3156 nm, and crystalline sizes of SnO_2:F was determined to be 57.5 nm. Spherical silicon with ARC showed higher optical absorption compared to that of spherical silicon without ARC. Fluorescence measurement indicated that the impurities would be contained at the center of spherical silicon, and the incident and luminescent light is confined by the ARC. The optimization of the thickness of ARC, and the analysis of the impurities contained in spherical silicon are required to improve the efficiency of the spherical silicon solar cells.

References

[1] Minemoto T, Takakura H and Hamakawa Y 2006 Sol. Energy Mater. Sol. Cells. **90** 3576
[2] Liu Z, Masuda A and Nagai T 2007 Sol. Energy Mater. Sol. Cells. **91** 1805
[3] Gharghi M 2006 IEEE Trans. Elect. Dev. **53** 1355
[4] Liu Z, Asai K, Masuda A, Nagai T, Akashi Y, Murozono M and Kondo M 2007 Jpn. J. Appl. Phys. **46** 5695
[5] Minemoto T, Okamoto C, Murozono M, Takakura H and Hamakawa Y 2006 Rare Metals Spec. Iss. **25** 121
[6] Hayashi S, Minemoto T, Takakura H and Hamakawa Y 2006 Rare Metals Spec. Iss. **25** 115
[7] Kumar V, Govind A and Nagarajan R 2011 Inorg. Chem. **50** 5637
[8] Zhang B, Tian Y, Zhang J X, Cai W 2011 J. Opt. Adv. Mat . **13** 89
[9] Elangovan E, Ramamurthi K 2005 App. Surf. Sci. **249** 183
[10] Liu Z, Masuda A and Kondo M 2009 J. Cryst. Growth. **311** 4116
[11] Baur W H 1956 Acta. Crystallogra. **9** 515
[12] Oku T, Wakimoto H, Otsuki A and Murakami M 1994 J. Appl. Phys. **75** 2522
[13] Mizuta T, Ikuta T, Minemoto T, Takakura H, Hamakawa Y and Numai T 2006 Sol. Energy Mater. Sol. Cells. **90** 46

Microstructures and photovoltaic properties of fullerene-based organic-inorganic hybrid solar cells

T. Oku, A. Takeda, A. Nagata, K. Fujimoto, T. Akiyama, A. Suzuki

Department of Materials Science, The University of Shiga Prefecture,
Shiga 522-8533, Japan
E-mail: oku@mat.usp.ac.jp

Abstract. C_{60}-based bulk heterojunction solar cells were fabricated, and the electronic and optical properties were investigated. C_{60} were used as n-type semiconductors, and copper oxides, $CuInS_2$ and diamond were used as p-type semiconductors. Electronic structures of the molecules were investigated by molecular orbital calculation, and energy levels of the solar cells were discussed. Nanostructures of the solar cells were investigated by transmission electron microscopy, electron diffraction and X-ray diffraction, which indicated formation of mixed nanocrystals.

1. Introduction

Fullerene-based solar cells have been investigated and reported recently. These organic solar cells have a potential for use in lightweight, flexible, inexpensive, and large-scale solar cells. However, significant improvements of photovoltaic efficiencies are mandatory for use in future solar power plants. One of the improvements is donor-acceptor proximity in the devices by using blends of donor-like and acceptor-like molecules or polymers, which are called donor-acceptor bulk-heterojunction solar cells [1-3]. Organic-inorganic hybrid structures would be other improvements for the organic solar cells [4].

The purpose of the present work is to fabricate and characterize fullerene-based organic-inorganic hybrid solar cells by using different organic polymers. In the present work, $CuInS_2$, diamond, and copper oxides were used for p-type semiconductors, and C_{60} was used for n-type ones. Device structures were produced, and efficiencies and optical absorption were investigated. Electronic structures were also investigated by molecular orbital calculations.

2. Experimental procedures

Cu_2O and CuO layers were prepared on pre-cleaned indium tin oxide (ITO) glass plates by an electrodeposition method using platinum counter electrode. Copper (II) sulfate ($CuSO_4$, 0.4 mol/L, Wako 97.5%) and l-lactic acid (3 mol/L, Wako) were dissolved into distilled water. The electrolyte pH was adjusted to 12.5 by adding NaOH. The electrolyte temperature was kept at 65 °C during electrodeposition. The electrodeposition of Cu_2O and CuO layers were carried out at voltages of -0.35 and +0.70, respectively, and quantity of electric charges of 2.2 Ccm^{-2} [4].

C_{60} layers with thickness of ~100 nm were prepared on the Cu_2O and CuO layers by vacuum evaporation from C_{60} powder (Material Technologies Research, 99.98 %). For all the present copper oxides semiconductor solar cells, aluminum (Al) metal contacts of thickness ~100 nm were deposited as top electrodes, and annealed at 140 °C for 20 min in N_2. The present solar cells were denoted as ITO/Cu_2O/C_{60}/Al and ITO/CuO/C_{60}/Al as shown in Fig. 1.

CIS solution for p-type semiconductors was also produced by dissolving CuI (Sigma Aldrich Corp., 99.99%) and $InCl_3$ (Sigma Aldrich Corp., 99.99%) in a mixture of triphenylphosphite (1 mL) (Sigma Aldrich Corp., 97%) and acetonitrile (2 mL) (Nacalai Tesque, Inc., 99.5%), dropping bis(trimethylsilyl)sulfide (Tokyo Chemical Industry Co., Ltd., >95%) [5,6]. The solution for n-type semiconductors was prepared by dissolving C_{60} in o-dichlorobenzene. A thin layer of polyethylenedioxythiophene doped with polystyrene-sulfuric acid (PEDOT:PSS) (Sigma Aldrich) was spin-coated on a pre-cleaned fluorine dope tin oxide (FTO) glass plates (Asahi Glass, ~9.3 Ω/□). Then, semiconductor layers were prepared on a PEDOT:PSS layer by spin coating, and annealed at 120 °C for 10 min in N_2 atmosphere. The FTO was used because of the high temperature annealing process. The thickness of the blended device was ~150 nm. Al metal contacts were evaporated as a top electrode, and annealed at 140 °C for 20 min in N_2 atmosphere.

A diamond:C_{60} BHJ solar cell was also prepared by spin coating using a mixed solution of C_{60} (Material Technologies Research, 99.98 %) and diamond powder (New metals & Chemicals Co. Ltd., >95 %) in 1,2-dichlorobenzene. Although the diamond powder did not have high purity, the impurities would not be activated, and the carrier concentration would be low. Total weight of diamond:C_{60} was 18 mg, and weight ratio of diamond:C_{60} was 1:8.

Current density-voltage (J-V) characteristics were measured (Hokuto Denko Corp., HSV-100), both in the dark and under illumination at 100 $mWcm^{-2}$ using an air mass (AM) 1.5 solar simulator (San-Ei Electric, XES-301S). The solar cells were illuminated through the substrate side, with an illuminated area of 0.16 cm^2. Optical absorption of the solar cells was investigated by means of UV-visible absorption spectroscopy (Hitachi Ltd., U-4100). The microstructures of the copper oxides thin films were investigated by X-ray diffractometry (XRD, Philips X'Pert-MPD System) with CuKα radiation at 40 kV operating voltage and 40 mA operating current. Transmission electron microscopy (TEM, Hitachi H-8100, 200 kV operating voltage) were also carried out for nanostructure analysis. Thermodynamical calculations for the copper oxide reactions were carried out by HSC Chemistry (Outokumpu Research Oy. Poli, Finland).

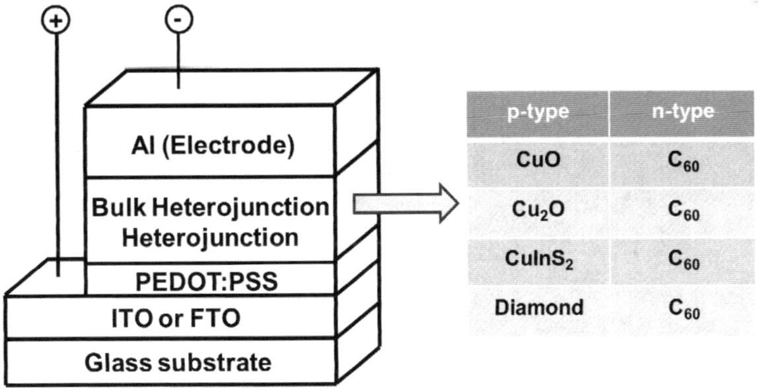

Figure 1. Device structure of solar cells.

3. Results and discussion

J-V characteristics of the present solar cells under illumination showed photocurrents with short-circuit currents and open-circuit voltages. Measured parameters of the present solar cells are summarized in Table 1. Cu_2O/C_{60} structures fabricated by the electrodeposition method provided better power conversion efficiency (η) of ~4×10^{-3} % compared to CuO/C_{60} solar cells. The Cu_2O/C_{60} solar cell prepared by the electrodeposition gave an η of 4.2×10^{-3} %, a fill factor (FF) of 0.25, short-circuit current density (J_{SC}) of 67 μAcm^{-2} and open-circuit voltage (V_{OC}) of 0.20 V.

Journal of Physics: Conference Series **352** (2012) 012024 doi:10.1088/1742-6596/352/1/012024

Table 1. Measured parameters of the present solar cells.

Sample	V_{OC} (V)	J_{SC} (μAcm^{-2})	FF	η (%)
CuO/C_{60}	0.24	1.8	0.25	9.0×10^{-5}
Cu_2O/C_{60}	0.20	67	0.25	4.2×10^{-3}
$CuInS_2:C_{60}$	0.18	16	0.28	8.0×10^{-4}
Diamond:C_{60}	0.023	5.3	0.35	4.3×10^{-5}

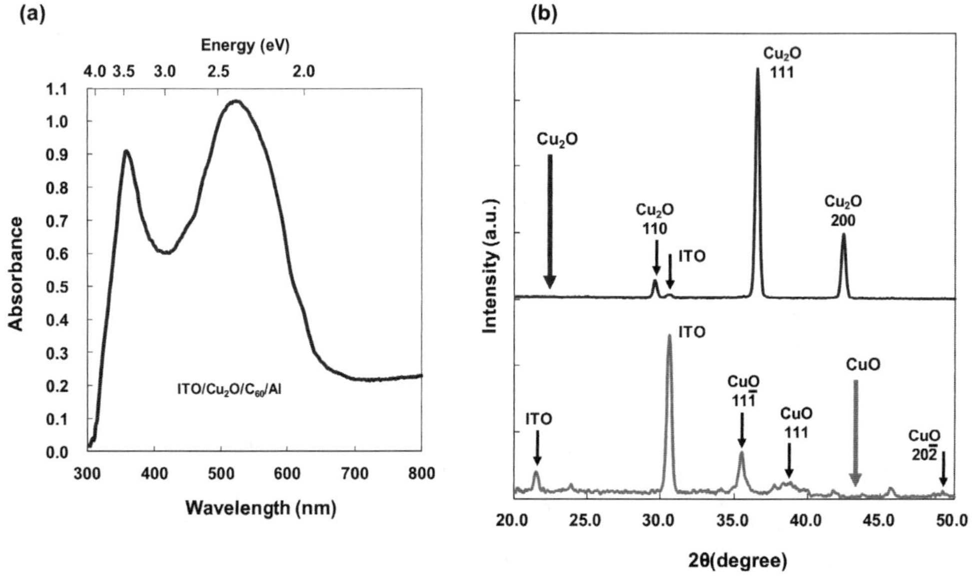

Figure 2. (a) Optical absorption of Cu_2O/C_{60} solar cells. (b) XRD patterns of Cu_2O and CuO thin films prepared by electrodeposition.

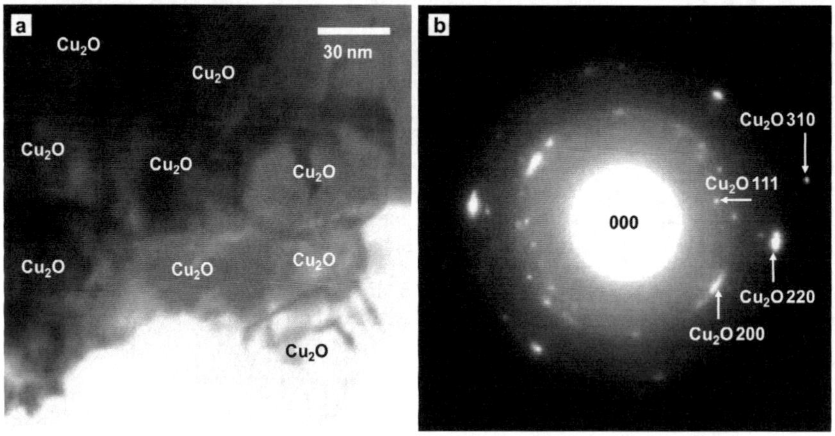

Figure 3. (a) TEM image of Cu_2O and (b) electron diffraction pattern of C_{60} heterojunction layer.

Figures 2(a) show optical absorption of the Cu_2O/C_{60} and CuO/C_{60} solar cells. The solar cells prepared by the electrodeposition indicated higher absorption in the range of 300-700 nm compared to those by spin-coating, which would be due to the film thickness. Absorption peak at ~360 nm and ~500 nm were due to copper oxides, and peaks at ~340, ~440 and ~610 nm corresponded to C_{60}.

Crystalline components in the Cu_2O and CuO thin films were investigated by XRD, as shown in Fig. 2(b). Diffraction peaks corresponding to Cu_2O and CuO were observed for the Cu_2O and CuO thin films, and they consisted of cuprite phase with cubic system (space group of Pn3m and lattice parameters of a = 0.4250 nm) and cupric phase with monoclinic system (space group of C2/c and lattice parameter of a = 0.4653 nm, b = 0.3410 nm, c = 0.5018 nm, β= 99.48°) [7]. The particle sizes were estimated using Scherrer's equation: $D = 0.9\lambda/B\cos\theta$, where λ, B, and θ represent the X-ray wavelength, full width at half maximum, and Bragg angle, as listed in Table 1. The crystallite sizes of Cu_2O and CuO produced by the electrodeposition are ~40 nm, which are larger compared to those prepared by spin-coating. The increase of crystallinity which was due to the thicker film thickness led to an increase in open-circuit voltage. Crystallite sizes of Cu_2O and C_{60} of the Cu_2O:C_{60} bulk heterojunction thin film were determined to be 7.2 nm and 25.7 nm, respectively.

Figure 3(a) and 3(b) are a TEM image and a selected area electron diffraction pattern of a Cu_2O layer prepared by the electrodeposition, respectively. The TEM image indicated Cu_2O nanocrystal structures with sizes of 40~50 nm, which agreed well with the calculated XRD results. The comparatively larger crystallite sizes of the Cu_2O resulted in unclear Debye-Scherrer rings in Fig. 3(b), which indicated the higher crystallinity of the Cu_2O nanoparticles.

Ellingham diagram of copper oxides for an O_2 gas molecule (per mol) was investigated by thermodynamical calculation. The copper element would be oxidized to CuO and Cu_2O completely by annealing at elevated temperatures, and Cu_2O would be more stable compared to CuO from the Ellingham diagram [4].

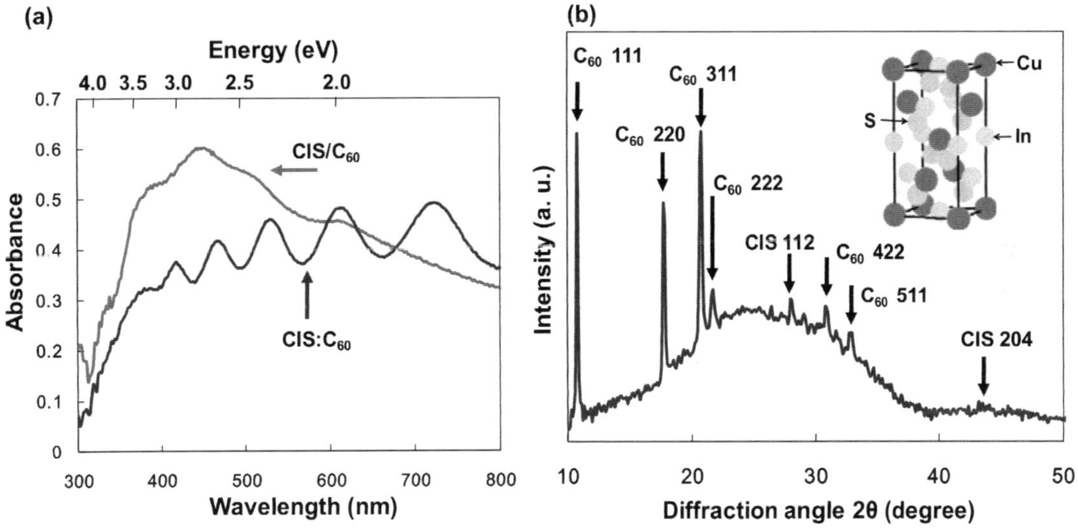

Figure 4. (a) Optical absorption spectra of heterojunction and bulk heterojunction solar cells. (b) X-ray diffraction pattern of CIS:C_{60} thin film.

Figure 4(a) shows a measured optical absorption of the solar cells based on CIS. These solar cells show a wide optical absorption range from 400 to 800 nm, and the heterojunction solar cell show as higher optical absorption range from 350 nm to 550 nm than that of the bulk heterojunction. Since the FTO substrate was set as an incident side, the optical absorption of the CIS layer was high for the heterojunction structure. On the other hand, optical absorption of the bulk heterojunction would be lower compared to that of the heterojunction structure because C_{60} were mixed with the CIS layer [6].

An X-ray diffraction pattern of CIS:C_{60} bulk heterojunction is shown in Fig. 4(b). Several diffraction peaks are observed, which correspond to 112, 204 of CIS and 111, 220, 311, 222, 422, 511 of C_{60}. The average particle sizes of $CuInS_2$ and C_{60} were calculated from Scherrer's formula to be 5 nm and 13 nm, respectively. The 204 peak of CIS is too small to be used for the calculation of the CIS grain size, and only one peak of 112 was used for the calculation.

Figure 5(a) is a TEM image of CIS, and many CIS particles are observed. Figure 5(b) is an electron diffraction pattern of CIS. Debye-Scherrer rings are observed in the diffraction pattern, which shows crystallite structures of CIS particles. An interfacial structure of CIS and C_{60} was observed by TEM. Filtered Fourier transform of the HREM image of CIS:C_{60} bulk heterojunction layer is shown in Fig. 5(c). Figure 5(d) is an inverse Fourier transform of (c), and arrows show the interface of CIS and C_{60}. Lattice fringes of {101} of CIS and {111} of C_{60} were observed. The enlarged image of a part of C_{60} in (b) is shown in Fig 5(e). Arrangements of C_{60} molecules are observed in the image. CIS and C_{60} have size distribution, and the crystal sizes of them observed in the TEM image are larger compared to the averaged sizes.

Optimization of the nanocomposite structure with CIS and C_{60} would increase the efficiencies of the bulk heterojunction solar cell. From the present TEM observation, CIS and C_{60} were not mixed in a molecular scale. If the mixture structure of CIS and C_{60} is improved to a nanoscale, it is believed that the area of the p-n junction interfaces is increased, and the efficiency would be improved. In addition, it is important to search the most suitable mixture ratio of the p-type and n-type semiconductors for bulk heterojunction solar cells.

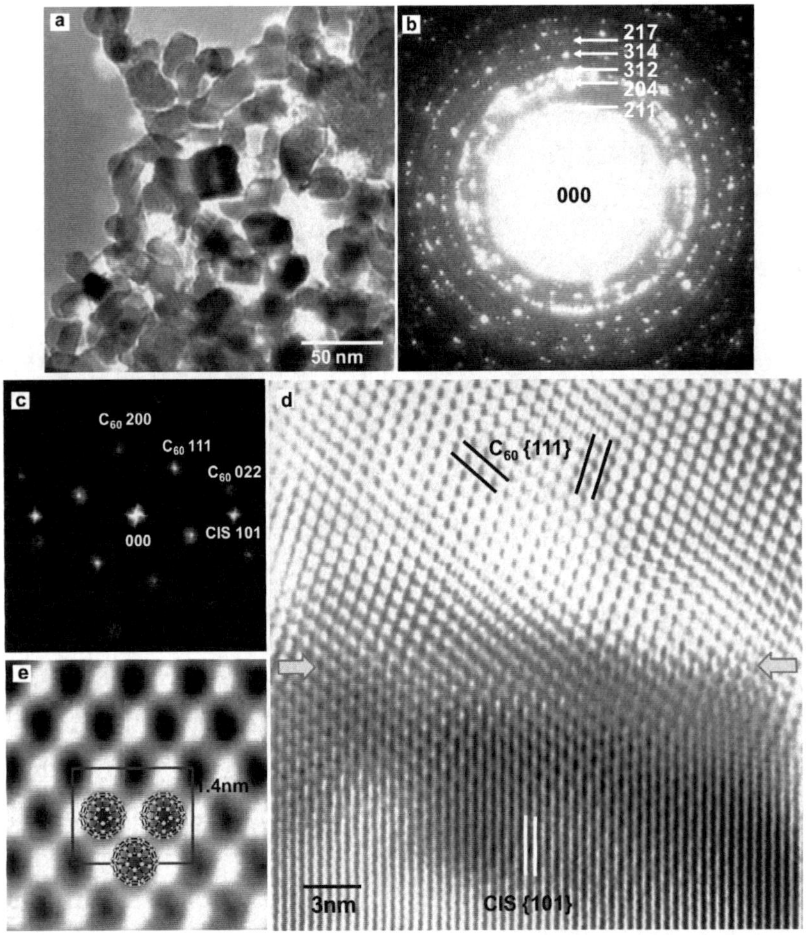

Figure 5. (a) TEM image of CIS nanoparticles and (b) electron diffraction pattern of CIS. (c) Filtered Fourier transform of HREM image of CIS:C_{60} bulk heterojunction layer. (d) Inverse Fourier transform of (c). (e) Enlarged image of a part of C_{60} in (d).

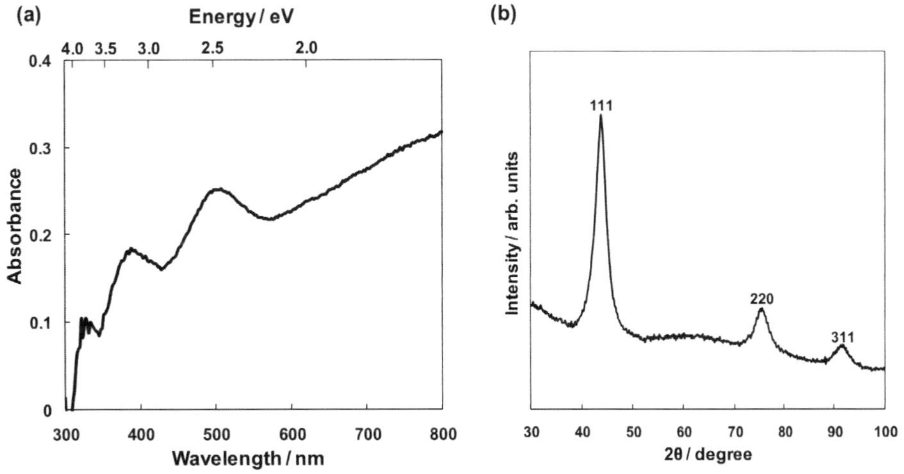

Figure 6. (a) Optical absorption spectra of diamond:C_{60} layer. (b) X-ray diffraction patterns of diamond powder,

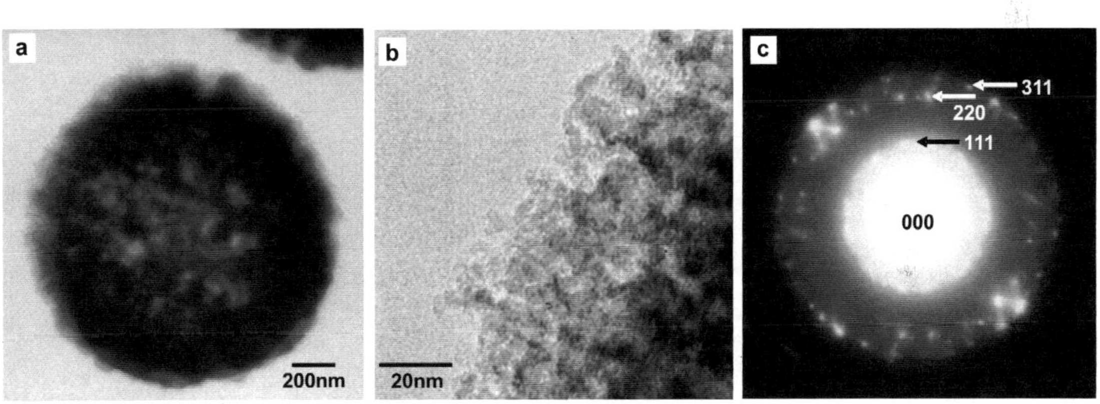

Figure 7. (a) TEM image, (b) enlarged image of a part of (a) and (c) electron diffraction pattern of diamond:C_{60} layer.

Figure 6(a) shows optical absorption spectra of the nanodiamond-based thin films. In Figure 6(a), the diamond:C_{60} nanocomposite structure provided photo-absorption in the range of 350 to 500 nm, and shows high absorption at 339, 402 and 506 nm, which correspond to 3.7, 3.1 and 2.5 eV, respectively. Absorption peaks of the C_{60} were confirmed within the range from 300 to 400 nm, and an absorption peak of 506 nm corresponds to the diamond.

Figure 6(b) shows X-ray diffraction patterns of diamond powder. In Fig. 6(b), diffraction peaks of the diamond powder were confirmed as 111, 220 and 311 of the diamond structure. A grain size of diamond powder was determined to be 12 nm, which was calculated by Scherrer's equation. An increase of photo-absorption above ~600 nm would be due to the nanostructure of diamond particles, which will be discussed later.

A TEM image, an enlarged image and an electron diffraction pattern of the diamond:C_{60} composite layer are shown in Fig. 7(a), 7(b) and 7(c), respectively. Diamond powder has an fcc structure with a lattice parameter of a=0.357 nm. C_{60} has also an fcc structure with a lattice parameter of a=1.42 nm. In the electron diffraction pattern of Fig. 7(c), expansion of C_{60} reflections was observed, which indicates a disordered structure of the composite layer. In the present TEM observation, diamond and C_{60} were not mixed well in nanoscopic scale, and the fabricated thin film would show low conversion efficiency.

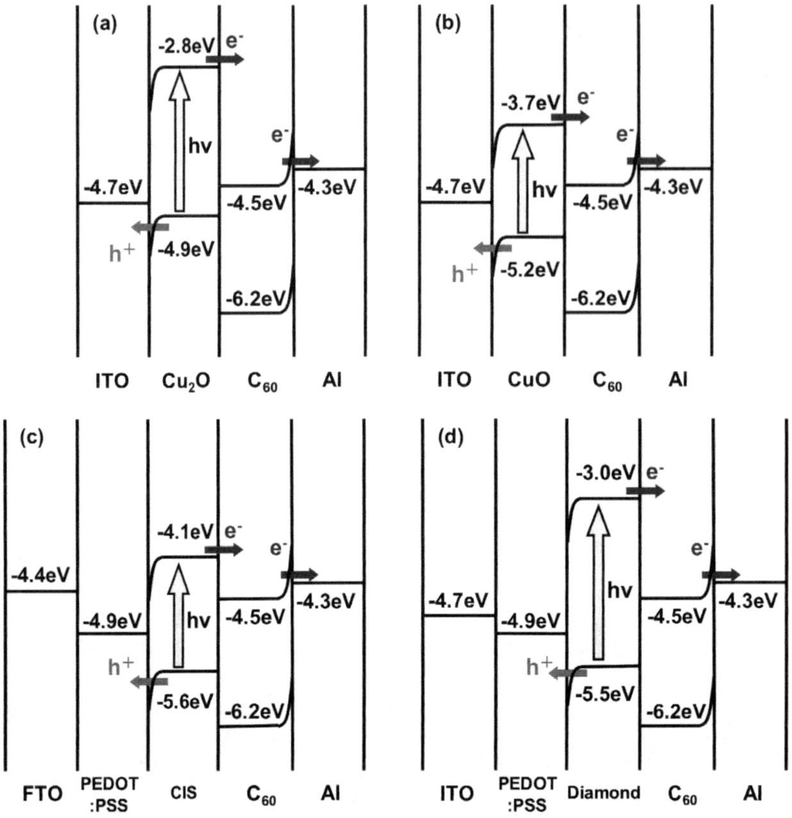

Figure 8. Energy level diagram of (a) Cu_2O/C_{60}, (b) CuO/C_{60}, (c) CIS/C_{60} and diamond/C_{60} solar cells.

Energy level diagram of ITO/Cu_2O/C_{60}/Al and ITO/CuO/C_{60}/Al solar cells is shown in Fig. 8(a) and (b), respectively, in which previously reported values of energy levels were used [8,9]. Separated holes could transfer from the valence band of the Cu_2O to the ITO, and separated electrons could transfer from the conduction band of the Cu_2O to the Al electrode, respectively. It has been reported that V_{OC} is nearly proportional to the semiconductor bandgap [10], control of the energy levels is important for increasing cell efficiency. Compared with a Si semiconductor with an indirect transition band structure, Cu_2O and CuO have direct transition bandgaps and greater optical absorption. Thus, ultrathin films of copper oxide layers could potentially provide efficient charge injection.

In the present study, Cu_2O and CuO were selected as a p-type semiconductor oxide for solar cells. These copper oxides are advantageous as a low-cost reagent with a simple fabrication process. Their lower conversion efficiencies compared to the Si may have been due to presence of heterogeneous grain size copper oxides in the active layer. Defects produced by inadequate crystallinity may have caused carrier recombination, and the formation of higher quality copper oxide thin films may overcome this problem.

An energy level diagram of CIS/C_{60} solar cells is shown in Fig. 8(c). Previously reported values were used for the energy levels of the figures by adjusting them to the present work [8]. When light is incident from the FTO side, excitation by the light absorption happens in the p-n interface, and electrons and holes are produced by charge separation. Carriers would transport from -4.5 eV to -4.3 eV by hopping conduction. Improvement of the present bulk heterojunction solar cells would be possible by the introduction of a buffer layer, change of annealing conditions, and the improvement of the microstructure is also necessary to obtain high efficiency.

The evaporation method provided high quality thin films, but a high vacuum and high temperature process are necessary. Although CISCuT method is a productive process, it requires a high temperature process [11]. On the other hand, the present spin coating method is simpler compared to the other formation methods. In addition, we can apply the spin coating method to plastic substrates without high vacuum and high temperature processes.

An energy level diagram of nanodiamond-based solar cells is shown in Fig. 8(d). Previously reported values were also used for the energy levels [12]. An energy gap of diamond estimated from Fig. 6(a), which corresponds to absorbance of 506 nm, is used for the model. From a theoretical calculation [13], nanodiamonds are composed of three layers; a diamond core (sp^3), a middle core (sp^{2+x}) and a graphitized core (sp^2). Therefore, a band gap of the nanodiamond is decreased by the existence of the sp^{2+x} bonding [14].

4. Conclusions

C_{60}-based solar cells with copper oxides, $CuInS_2$ and diamond were fabricated and characterized. Devices based on the Cu_2O/C_{60} structure were fabricated by an electrodeposition method, which provided better η of ~4×10^{-3} % compared to CuO/C_{60} solar cells. XRD and TEM results indicated the presence of Cu_2O and CuO nanoparticles. Formation of higher quality copper oxide thin films, introduction of a new electrode and optimization of deposition condition would improve the efficiencies of the solar cells. Energy level diagram was proposed, and separated holes could transfer from the valence band of the Cu_2O to the ITO, and separated electrons could transfer from the conduction band of the Cu_2O to the Al electrode, respectively.

Chalcopyrite/fullerene solar cells were also fabricated and characterized. A device of bulk heterojunction structure based on $CuInS_2$ and C_{60} provided η of 8.0×10^{-4} %, FF of 0.28, Voc of 0.18 V, which is better than those of the heterojunction structure. The present solar cells showed a high optical absorption in the range of 300 nm to 800 nm. Microstructures of the solar cells were observed by TEM and XRD, which indicated average particle sizes of $CuInS_2$ and C_{60} of 5 nm and 13 nm, respectively. A carrier transfer mechanism was discussed by the energy level diagram.

Nanodiamond-based solar cells were fabricated and characterized. Diamond:C_{60} nanocomposite structure provided photo-absorption in the range of 350 to 500 nm, and provided η of 4.3×10^{-5}%, FF of 0.35, J_{sc} of 5.3 μA/cm^2 and V_{oc} of 0.023V. Nanostructures of the thin films were investigated by TEM and X-ray diffraction, and the grain size of diamond were determined to be 12 nm. Optimization of blended structures with diamond would increase the efficiencies of the thin films. Photovoltaic behavior including charge transfer and mobility can be described on the basis of the energy diagram of the bulk heterojunction solar cells from the present J-V measurements, optical absorption and microstructure analysis. Optimization of blended structures with C_{60} would increase the efficiencies of solar cells.

References

[1] Yu G, Heeger A J, 1995 J. Appl. Phys. **78** 4510
[2] Padinger F, Rittberger R S, Sariciftci N S, 2003 Adv. Funct. Mater. **13** 85
[3] Oku T, Noma T, Suzuki A, Kikuchi K, Kikuchi S, 2010 J. Phys. Chem. Solids **71** 551
[4] Oku T, Motoyoshi R, Fujimoto K, Akiyama T, Jeyadevan B, Cuya J, 2011 J. Phys. Chem. Solids **72** 1206
[5] Arici E, Sariciftci N S, Meissner D, 2003 Adv. Funct. Mater. **13** 165
[6] Takeda A, Oku T, Suzuki A, Kikuchi K, Kikuchi S, 2009 J. Ceram. Soc. Jpn. **117** 967.
[7] Park N G, Kang M G, Ryu K S, Kim K M, Chang S H, 2004 J. Photochem. Photobiol. A **161** 105.
[8] Oku T, Nagaoka S, Suzuki A, Kikuchi K, Hayashi Y, Inukai H, Sakuragi H, Soga T, 2008 J. Phys. Chem. Solids **69** 1276
[9] Oku T, Takeda A, Nagata A, Noma T, Suzuki A, Kikuchi K, 2010 Energies **3** 671.
[10] Green, M A, Emery K, King D L, Hishikawa Y, Warta W, 2006 Photovolt. Prog. Res. Appl. **2**

14, 455

[11] Winkler M, Griesche J, Konovalov I, Penndorf J, Wienke J. Tober O. 2004 Solar Energy **77** 705

[12] Nagata A, Oku T, Kikuchi K, Suzuki A, Yamasaki Y, Ōsawa E, 2010 Prog. Nat. Sci. Mater. Inter. **20** 38

[13] Korobova M V, Avramenko N V, Bogachev A G, Rozhkova N V, Ōsawa, E, 2007 J. Phys. Chem. C **111** 7330

[14] Ōsawa E, Ho D, Huang H, Korobov M V, Rozhkova N N 2009 Diam. Rel. Mater. **18** 904.

Design of Trench Gate GaN Power MOSFET using Al_2O_3 Gate Oxide

Ey Goo Kang[1], **Yong Tae Kim**[2]
[1] Department of Photovoltaic Engineering, Far East University
Eumseong 369-700, Korea.
[2] Semiconductor Materials and Devices Laboratory, Korea Institute of Science and Technology,Seoul 136-791, Korea

E-mail:ytkim@kist.re.kr

Abstract. A trench gate type 600V GaN power MOSFET structure with Al_2O_3 gate oxide has been designed and simulated for obtaining optimum device and process parameters for fabrication. As a result, when the trench gate depth is 2.4 ㎛, the breakdown voltage is 620V and the on resistance is $0.4\Omega cm^2$, which is relatively very low on resistance comparing with the same trench gate Si power MOSFET structure.

1. Introduction

Recently, as a key component of power conversion and control devices in the power electronic industry of high voltage, multi-functions and large capacity rising as one of the most attracting fields in the electric and electronic industry, the power semiconductor element has faced a chance for rapid growth along with the development of a power bipolar transistor and is popularly used as thyristor, power MOSFET, IGBT (Insulated Gate Bipolar Transistor), MCT(MOS Controlled Thyristor), and SIT(Static Induction Thyristor), etc.[1-2]

Practically, these power semiconductor devices are expected to be useful for the systematization of the express subway and electric car in transportation fields, the FACTS(Flexible AC Transmission System), HVDC(High Voltage DC) transmission and missile related equipment system in power fields, and the design, fabrication, etc. of a micro-airplane, a military information system, as well as the existing home appliances and industrial equipment.[3-5]

Meanwhile, since large capacity, chip size and durability of power chip are seriously concerned for the above mentioned applications, compound power semiconductors such as GaN and SiC are newly demanded. The compound power semiconductor takes advantage of a wide band gap and seems to be the next generation power element due to its far more excellent thermal properties and on resistance properties than the existing Si. However, considering that it is very hard to prepare a p type deep junction (pillar structure) in GaN and SiC it is very important to suggest a method to avoid the formation of p type deep junction.

Therefore, this work has proposed a structure of GaN power MOSFET, which is operated as a normally off FET, and designed Al_2O_3 gate insulator instead of the SiO_2 gate dielectric layer by considering the surface condition of GaN. Particularly, in this work, we have tried to investigate

---

[2] To whom any correspondence should be addressed.

Published under licence by IOP Publishing Ltd

effects of trench type gate on the optimal electrical performance of GaN power device to avoid p type junction doping that is known as a very hard process in the GaN.

Experiment Method

The proposed device structure is shown in Fig. 1. In this structure, N drift and P base layers are consisted on a N type GaN substrate to avoid the formation of P junction as mentioned in the introduction and a vertical channel is suggested by preparing the trench gate MOSFET structure.

Initially, the thickness of a gate oxide film is fixed as 0.1 ㎛, the concentration and thickness of a N+ source are $3\times10^{18}cm^{-3}$ and 0.5 ㎛, respectively. These values are given as the fixed conditions for the GaN power MOSFET operated as a normally off FET. The simulation was carried out by 2D simulation method using Athena and Atlas, Silvaco. From the proposed structure and the general relationship between a depletion layer and impurity concentration, optimal design and process parameters should be simulated for targeting about 600 V breakdown voltage. At first, the electric field applied to the N drift layer is calculated with the concentration of the N drift layer, predicting the critical electric field is 3.3MV/cm as shown in Fig. 2. Based on this calculation, when the concentration of the N drift layer is $5\times10^{16}cm^{-3}$ the electric field applied to the N drift layer is reliable since it is less than the maximum critical electric field of 3.3MV/cm. And, the relation between the thickness and the concentration of N drift layer reveals that the thickness of the N drift layer is 3.0 ㎛

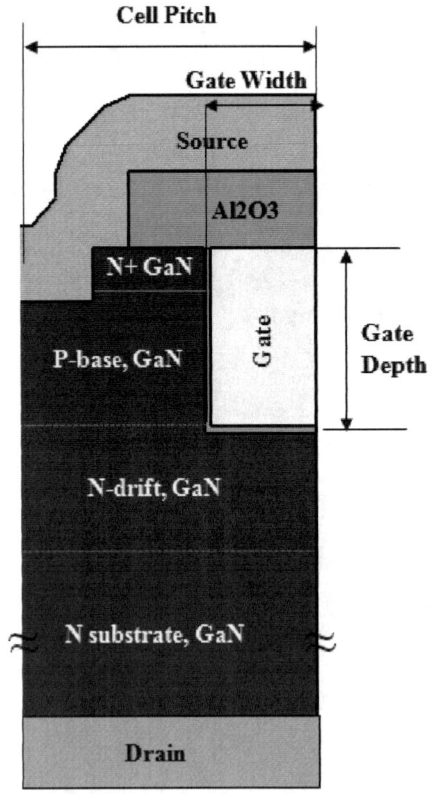

Figure 1. The proposed structure of a trench gate and Al_2O_3 gate oxide GaN power MOSFET

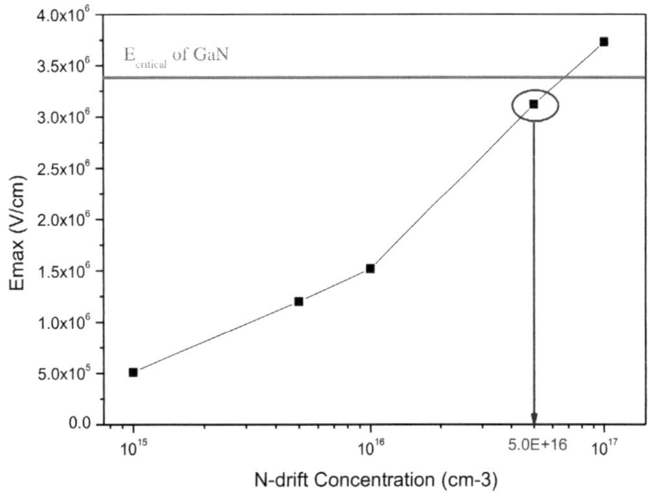

Figure 2. The relation between N- drift concentration and critical electrical field

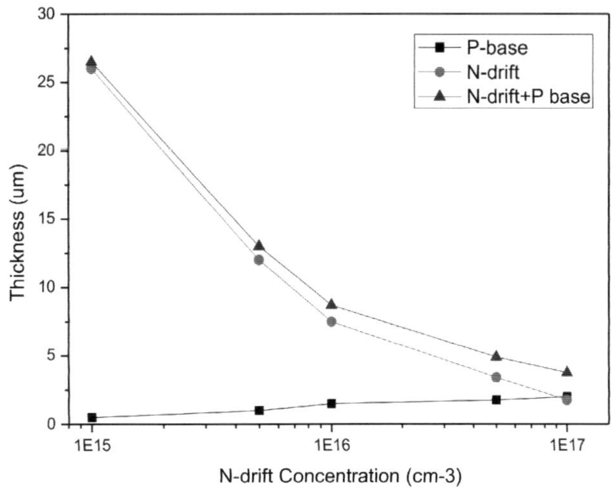

Figure 3. The relation between N- drift concentration and N-drift Thickness

at the optimum concentration. However, to realize a trench gate structure it is necessary to adjust the thickness of the P base layer with the N drift layer.In this work, the thickness of the P base layer is changed from 1.5, 1.75, and 2 ㎛, respectively and then, the thickness of the N drift layer is simulated as shown in Fig. 3. Therefore, the thickness of the N drift layer is fixed at 2 ㎛ as an optimal design parameter by considering the relation between the electric field and the concentration of a trench gate. As a result, the target breakdown voltage is 690V and Fig. 4 shows clearly how the thickness variations of the P base and N drift layers are influenced on the breakdown voltage. Finally, the surrounding design and process parameters are optimized as presented in Table 1.

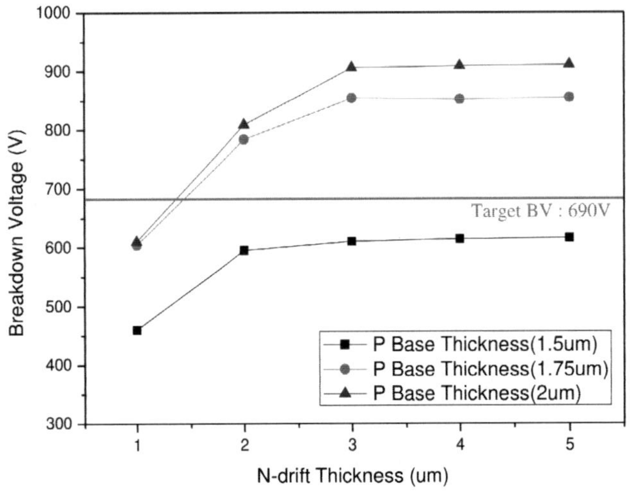

Figure 4. The characteristics of breakdown voltage according to the thickness of N drift and P base layers

Table 1. Design and process parameter of simulation conditions.

Division	Concentration(cm^{-3})	Thickness (㎛)
gate depth	-	1.5
Gate oxide film	-	0.1, 0.2, 0.3
p base	1×10^{17}	1.75
n+ source	3×10^{18}	0.5
n drift	5×10^{16}	2.0, 3.0
n+ substrate	3×10^{18}	150

Journal of Physics: Conference Series **352** (2012) 012025 doi:10.1088/1742-6596/352/1/012025

2. Result and Consideration

Table 2. The electrical characteristics according to gate oxide thickness.

Oxide Thickness	0.1um		0.2um		0.3um	
N-drift Thickness	Vth	BV	Vth	BV	Vth	BV
2um	3.5V	570V	7V	620v	11v	700V
3um	3.5V	640V	7v	680v	11v	790V

As a device parameter, it is important to obtain threshold and breakdown voltage depending on the variations of Al_2O_3 gate thickness and N drift layer. Table 2 summarized the obtained results. As shown in the table, when the thickness of a gate oxide film is 0.2 ㎛ and the thickness of the N drift layer is 2.0 ㎛, the breakdown voltage is 620V, which is increased to 680V that is still less than the target voltage at the N drift layer of 3.0 ㎛. However the threshold voltage is not changed and fixed at 7V constantly. Increasing the gate oxide thickness, the threshold and breakdown voltages rise up to 11 and 700V even at the optimum N drift layer.

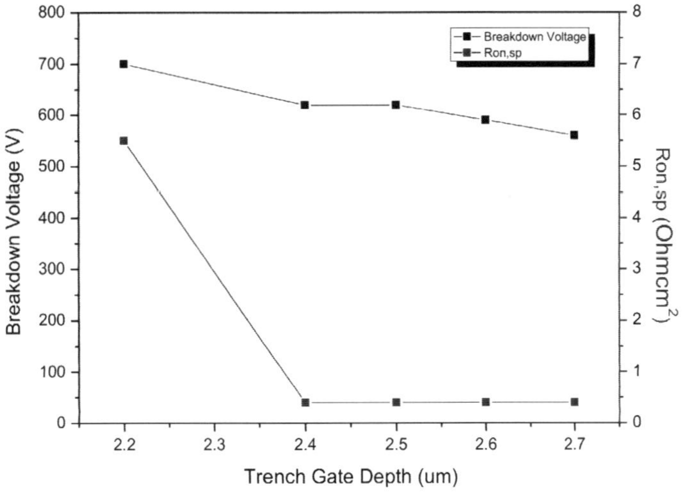

Figure 5. The electrical characteristics according to the trench gate depth

It is important to understand how the breakdown voltage and the on resistance change with the trench gate depth. As shown in Fig.5, when the trench depth is 2.3 ㎛, the breakdown voltage is 660V, and the on resistance is 3 Ωcm^2. However, when the trench depth is 2.4 ㎛, the breakdown voltage is slightly decreased to 620V and the on resistance is abruptly decreased to $0.4\Omega cm^2$, which is a very low on resistance comparing with the same trench gate Si power MOSFET structure. This result suggests that the optimum trench depth in the proposed device structure is 2.4 ㎛. Comparing the effect of gate oxide material on the threshold voltage, Fig. 6 and 7 are the simulated results showing that the threshold voltage depends on the both materials, SiO_2 and Al_2O_3. Using the Al_2O_3 gate oxide, the threshold voltage reduces from 10 V, which is the threshold voltage of SiO_2 gate oxide case, to 7V.

Journal of Physics: Conference Series **352** (2012) 012025 doi:10.1088/1742-6596/352/1/012025

Since both cases are operated with a normally off type and then, the Al_2O_3 gate oxide case indicates that the device is fast turned on. Finally, we can simulated the I-V characteristics of a trench gate and Al_2O_3 gate oxide GaN power MOSFET using the optimized parameters and 150 μm thickness of a GaN substrate. Fig.8 shows that when Vgs is 15V, the on resistance is $0.45m\Omega cm^2$. If the Vgs increases from 5 to 15V, the average on resistance becomes to be slightly decreased to $0.4m\Omega cm^2$.

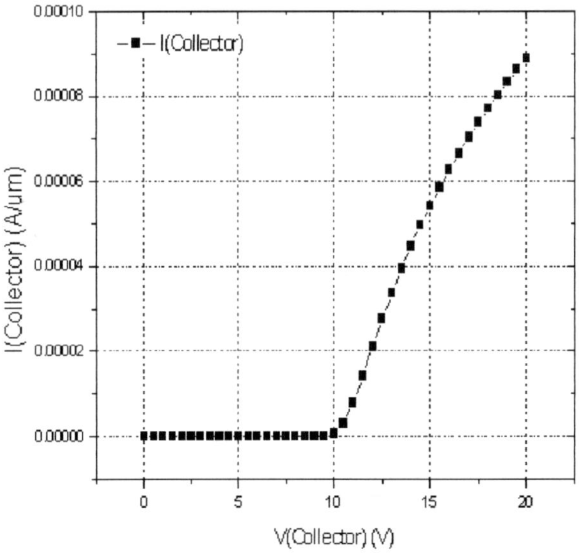

Figure 6. The characteristic of threshold voltage depending on 0.1 μm SiO_2 gate oxide

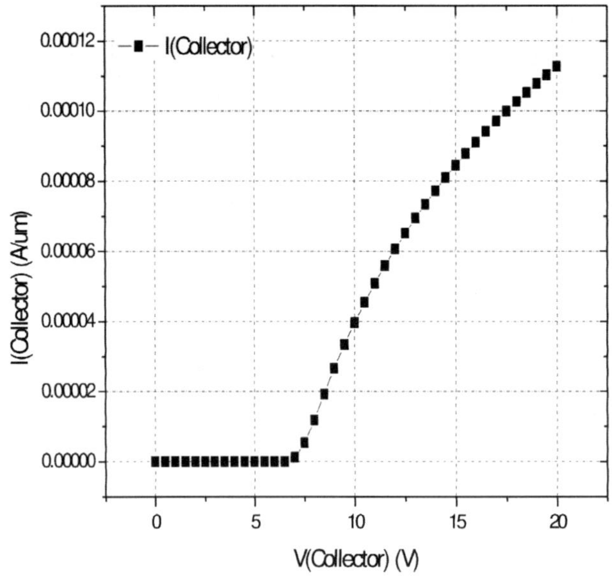

Figure 7. The characteristic of threshold voltage depending on 0.2 μm Al_2O_3 gate oxide

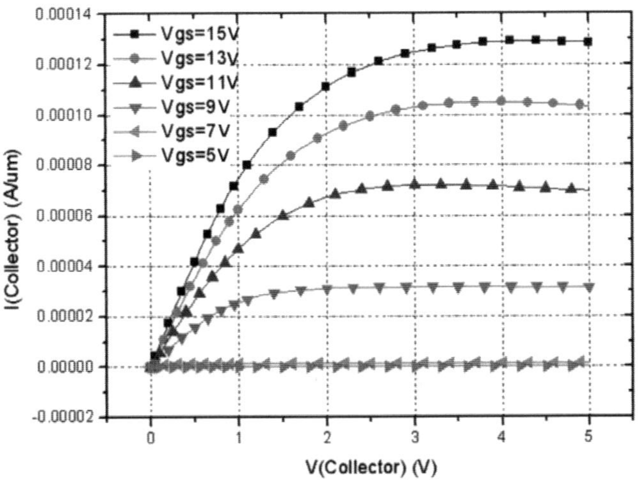

Figure 8. The I-V characteristics of a trench gate and Al_2O_3 gate oxide GaN power MOSFET

Fig.9 shows the characteristic of breakdown voltage, which is one of the most important properties of a power semiconductor. Using the same optimum design parameters, it is found that the breakdown voltage is 620V, which is suitable for the application of 600V breakdown range.

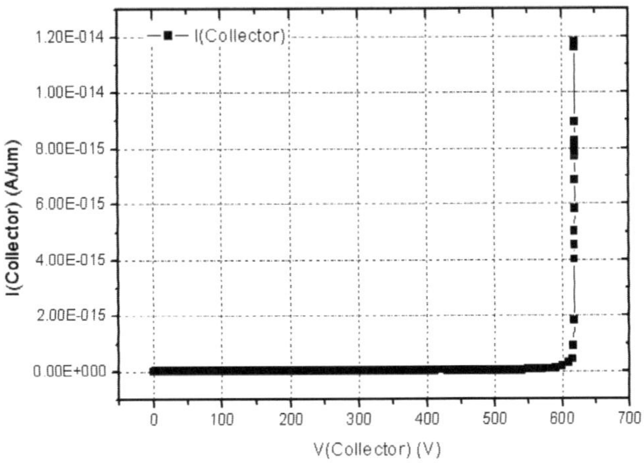

Figure 9. The breakdown characteristic of a trench gate and Al_2O_3 gate oxide GaN Power MOSFET

Journal of Physics: Conference Series **352** (2012) 012025 doi:10.1088/1742-6596/352/1/012025

3. Conclusion

We have proposed a trench gate type GaN power MOSFET that is operated by normally off type and tried to optimize design and process parameters targeting at the application of 600V class inverters. Especially, using Al_2O_3 as a gate oxide layer, the electrical characteristics such as the I-V characteristics, the on resistance, and the breakdown voltage are simulated. As a result, it is found that the breakdown voltage is 620V and the on resistance is $0.4m\Omega cm^2$. This work will suggest a method to avoid the formation of p junction in the GaN power MOSFET structure, and the optimum parameters are also very useful to implement the GaN power switch element of 600V class, which has a great demand for green technology, so called electric cars, system connection, and inverters.

4. References

[1] I. Akasaki and H. Amano: Jpn. J. Appl. Phys. 36 (1997) 5393
[2] Y.-F. Wu, D. Kapolnek, J. P. Ibbetson, P. Parikh, B. P. Keller, and U. K. Mishra: IEEE Trans. Electron Devices 48 (2001) 586
[3] Y.-F. Wu, D. Kapolnek, J. P. Ibbetson, P. Parikh, B. P. Keller, and U. K. Mishra: IEEE Trans. Electron Devices 48 (2001) 586.
[4] K. Joshin, T. Kikkawa, H. Hayashi, T. Maniwa, S. Yokokawa, M.Yokoyama, N. Adachi, and M. Takikawa: IEDM Tech. Dig., 2003,p. 983.
[5] Y. Ohmaki, M. Tanimoto, S. Akamatsu, and T. Mukai: Jpn. J. Appl. Phys. 45 (2006) L1168.

Journal of Physics: Conference Series **352** (2012) 012026 doi:10.1088/1742-6596/352/1/012026

Fabrication and properties of thermoelectric oxide thick films deposited with aerosol deposition method

Y. Nakamura*, Y. Matsufuji, and M. Inoue

Toyohashi University of Technology, 1-1 Hibarigaoka, Tempaku-cho, Toyohashi, Aichi, 441-8580, Japan

E-mail: nakamura@ee.tut.ac.jp

Abstract. Thermoelectric materials can directly convert thermal energy to electric energy via Seebeck effect and have attracted attention for clean energy harvesting material. $Ca_3Co_4O_{9+\delta}$ (Co349) oxide shows a high thermoelectric performance. Since the Co349 is a misfit-layered oxide, the c-axis oriented structure should be required to achieve high thermoelectric performance. In this study, the Co349 thick film was deposited with aerosol deposition (AD) method. The effect of process parameters on the microstructure and thermoelectric properties were investigated. The thermal strain caused by the difference in thermal expansion coefficient should be less than 0.2%. The deposition rate of 2 μm/min could be attained by controlling particle size distribution. The *c*-axis oriented Co349 film was obtained after the annealing at the temperature higher than 700°C. Seebeck coefficient of 170 (μV/°C) and electrical conductivity of 110 (S/cm) were achieved at 700 °C in the film annealed at 900 °C for 1 hour. The thermoelectric performance could be kept up to the thick film of 55μm. These values were as high as those of the usual bulk samples. This suggests that the AD process is effective to fabricate thick film type thermoelectric module.

1. Introduction

Large amount of thermal energy is not effectively used and thrown away as a waste heat. Since this heat is usually dispersed widely and relatively small scale, it is difficult to harvest effectively. Power generation by thermoelectric (TE) energy conversion is promising method to improve efficiency of energy resources because TE power generation system can directly convert waste heat into electricity without driving parts. In particular, TE oxides, which generally show high chemical stability even at high temperatures in air, are promising materials for power generation systems using high-temperature waste heat.

Among the thermoelectric materials, $Ca_3Co_4O_{9+\delta}$ (Co349) oxides were found to show high thermoelectric performance, e.g. $ZT\sim1$ in single crystal [1], while this dimension less figure of merit, ZT of polycrystalline sample decreased to about 1/3. This *p*-type cobaltite, Co349, is a misfit-layered oxide, which consists of alternating layers of a Ca_2CoO_3 rock-salt-type layer and a CdI_2-type CoO_2 layer stacked in the ***c***-direction. Due to this layered structure, electrical properties of Co349 such as Seebeck coefficient (S) and electrical conductivity (σ) could be enhanced by forming *c*-axis aligned textured structure using hot pressing (HP) [2,3] and spark plasma sintering (SPS) [4,5].

For practical usage of TE power generation system, downsized and highly integrated TE modules are required as well as lower production costs. In the case of bulk application, the assembly processes using small pieces are not easy. On the other hand, thin films produced by vapor growth methods has relatively slow growth rates and is not easy to form well grain-aligned structure without using single crystalline substrates. Hence, the development of a novel production method, which can achieve higher deposition rates and high degrees of grain orientation, is expected for production of highly integrated TE modules with low production cost.

The aerosol deposition (AD) is a technique to prepare high-density ceramic thick film by ejecting aerosol consisting of a mixture of fine ceramic particles and gas from the nozzle to the substrate [6-9]. In this method, the ejecting ceramic powder was synthesized by conventional solid-state reaction method. The dense thick film of PZT and α-Al_2O_3 could be obtained with high deposition rate of several μm/min by proper control of particle size and ejecting condition. In addition, the patterned film could be also formed using a mask deposition. Using the AD method, planer thick film type and multi-layered thick film type TE modules [10] could be achieved.

In this work, the Co349 thick films were fabricated with AD method at room temperature. The effect of process parameters on microstructure and properties was investigated.

2. Experimental

Calcined powder was prepared by conventional solid-state reaction method from the raw materials of $CaCO_3$, Bi_2O_3, and Co_3O_4. These powders were weighted in the molar ratio of Ca:Bi:Co = 2.7:0.3:4.0, and were mixed. The mixed powder was calcined at 900°C for totally 60h with intermediate grindings. The calcined powder was finally crashed by ball milling for 24 and 48 h to obtain an appropriate particle size. The Co349 powders were mixed with a N_2 carrier gas to form an aerosol flow, and the gas flow rate of 1 L/min was used. This aerosol flow was transported through the nozzle of 10mm x 0.5 mm into an evacuated deposition chamber and deposited on substrates of Si, quartz, sapphire, SUS410, and $(GdCa)_3(GaMgZr)_5O_{12}$ (SGGG) that were located 10 mm from the nozzle. The film areas were limited in 3mm x 20mm using metal mask. The deposited films were annealed for crystallization at temperatures up to 900 °C for 1 h in air. The crystallinity of the films was analyzed by X-ray diffraction (XRD), and the microstructure was observed with scanning electron microscopy (SEM). Electrical conductivity was evaluated by conventional four-probe method, and Seebeck coefficient was obtained in the temperature range from 400 °C to 700°C in a direction parallel to the surface of the substrate.

3. Results and discussion

3.1. Effect of substrate

The effect of the difference in thermal expansion coefficient between substrate and Co349 was examined by heating the deposited film to 750 °C for 1h. After this heat treatment, the films deposited on quartz and Si substrates showed cracks and were detached from the substrates. The results were summarized in Table 1 with the thermal expansion coefficients of the substrates and the thermal strain caused by cooling from 750°C to room temperature (R.T.) where the thermal expansion coefficient of $\alpha_{349} = 9\times10^{-6}$ 1/K was used for Co349 to calculate the thermal strain during cooling. As shown in Table 1, the films with the thermal strain larger than 0.47% were detached from the substrate, while the films with thermal strain less than 0.14% did not show any cracks and detachment. In addition, the Co349 films deposited on sapphire substrate showed neither detachment nor cracks after heating up to 900°C. This suggests that the deposited Co349 films show neither cracks nor detachment from the substrate in the case of the thermal strain less than 0.2%, which is the destruction strain commonly used. In the following experiments, we used the SGGG substrate whose thermal expansion coefficient is almost equal to Co349.

Table 1 Thermal expansion coefficient of substrate and the existence of detachment of film from substrate after heat treatment at 750°C for 1h.

Substrate	Thermal expansion coefficient, α_{sub} (1/K)	Thermal strain (750°C to R.T.)	Detachment
quartz	0.5×10^{-6}	0.61%	Yes
Si	2.41×10^{-6}	0.47%	Yes
sapphire	7.0×10^{-6}	0.14%	No
SGGG	9.0×10^{-6}	0%	No
SUS	10.4×10^{-6}	-0.10%	No

3.2. Effect of particle size distribution

Figure 1 shows the SEM image of the Co349 powder used for AD deposition. Before ball milling, the particle size was in the range between 3 μm and 15 μm. After ball milling for 24 h, the minimum particle size decreased to about 1 μm however large particles of larger than 10 μm still existed. After ball milling for 48 h, most of the particle had a diameter from 0.1 to several μm and some of them showed the particle size about 7 μm. As shown in this image, the number of small particle with the size of a few microns increased as increasing ball milling time, while the relatively large particles about 5-7 μm were still existed after ball milling for 48 h.

Figure 1 SEM image of the Co349 powder used for AD deposition. (a) before ball mill, (b) after ball milling for 24h, and (c) after ball milling for 48 h.

Using these powders, Co349 films were deposited with AD process, and the deposition rate was evaluated. In the case of using the powder ball milled for 48 h, a pressed powder structure that was weak and easily falling to pieces was obtained, and a dense adhered film was not attained. In the AD processing, since the dense film is formed by the reduction of crystallite particle size by fracture or plastic deformation at room temperature, the particle speed, which depends on the gas flow rate, should be higher than a critical particle speed [7]. The small particles in the powder ball milled for 48 h may not have enough particle speed to form dense film. To overcome this issue, the increase in gas flow rate would be required. However, the large particles with high speed damaged the substrate and the formed film. Hence, the selection of gas flow rate and the simultaneous control of the particle size distribution are required for the formation of dense film.

The relation between deposition time and film thickness is shown in Figure 2. The dense film could be obtained in the case of using the powder without ball milling, while the deposition rate of this film was smaller than that using the smaller powder of ball milled for 24 h. In both cases, the thickness of the films increased linearly as increasing deposition time. The deposition rate of about 2 μm/min was obtained by using smaller particles ball milled for 24 h. These results suggest that the appropriate particle size exists in the range about a few μm, and the larger particles larger than about 10 μm would damage the deposited film. This would be a reason for small deposition rate in the case of non-ball milled particle that included larger particle.

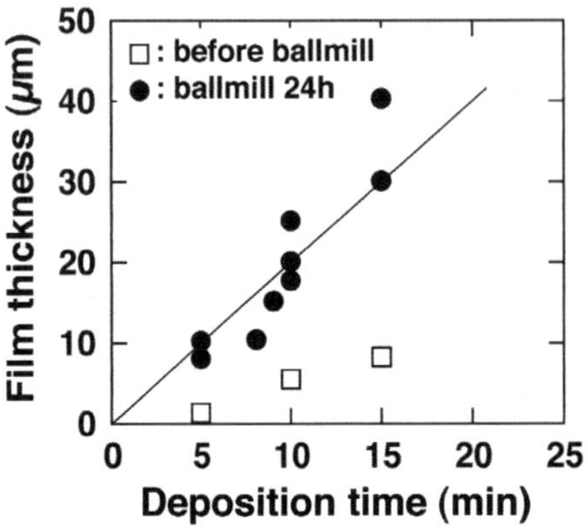

Figure 2 The relation between the deposition time and film thickness.

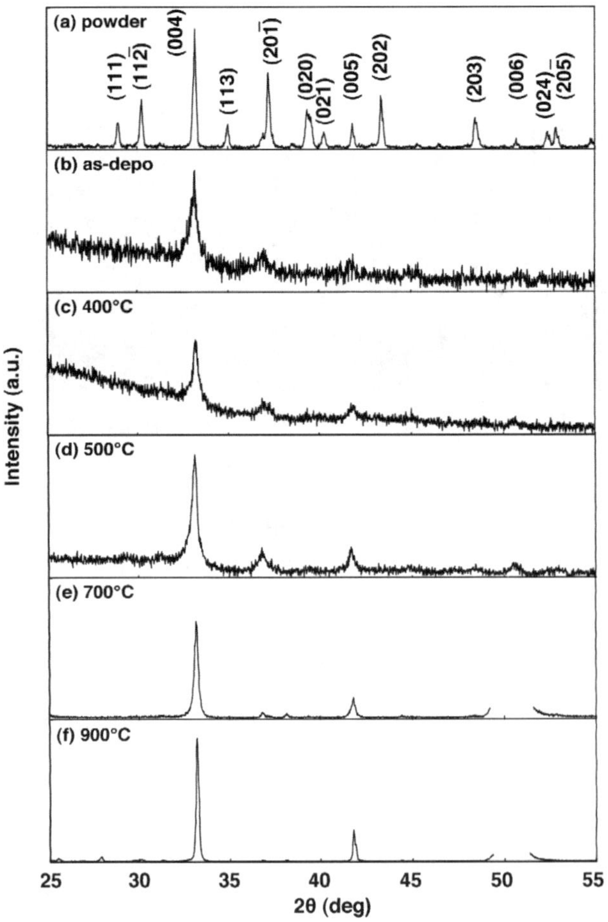

Figure 3 XRD patterns of (a) calcined powder, (b) as-deposited film, and crystallized film sintered at (c) 400 °C, (d) 500 °C, (e) 700 °C, and (f) 900 °C for 1 h, respectively.

3.3. Effect of crystallized temperature

Figure 3 (a) shows the XRD pattern of the calcined powder, and Fig. 3 (b) showed that of as-deposited film, respectively. As shown in Fig. 3 (b), the as-deposited film showed weak and broad peaks. This means the formation of nano-crystalline structure and/or the introduction of large strain into the deposited Co349 crystal during the collision of the powders with the substrate.

To improve the crystallinity of the film, the films were sintered for 1 h at the temperatures from 400 °C to 900 °C. The XRD patterns of the sintered films were shown in Fig. 3 (c)-(f). As shown in Fig. 3 (c), the XRD peaks of the sample sintered at 400 °C for 1 h was almost the same as those of as-deposited film. However, the XRD peaks of the sample sintered at 500 °C became sharp, and the improvement of the intensity was also observed. In the case of the samples sintered at the temperature higher than 500 °C, the peak intensity increased as increasing sintering temperature. This means that the sintering at the temperature higher than 500 °C is required to recover the crystallinity of deposited Co349 film. In addition, it should be mentioned that the almost *c*-axis oriented Co349 films could be obtained by AD method in the samples sintered at the temperature higher than 700 °C since the intensity of the peaks except for (0 0 *l*) faces was very weak in those samples.

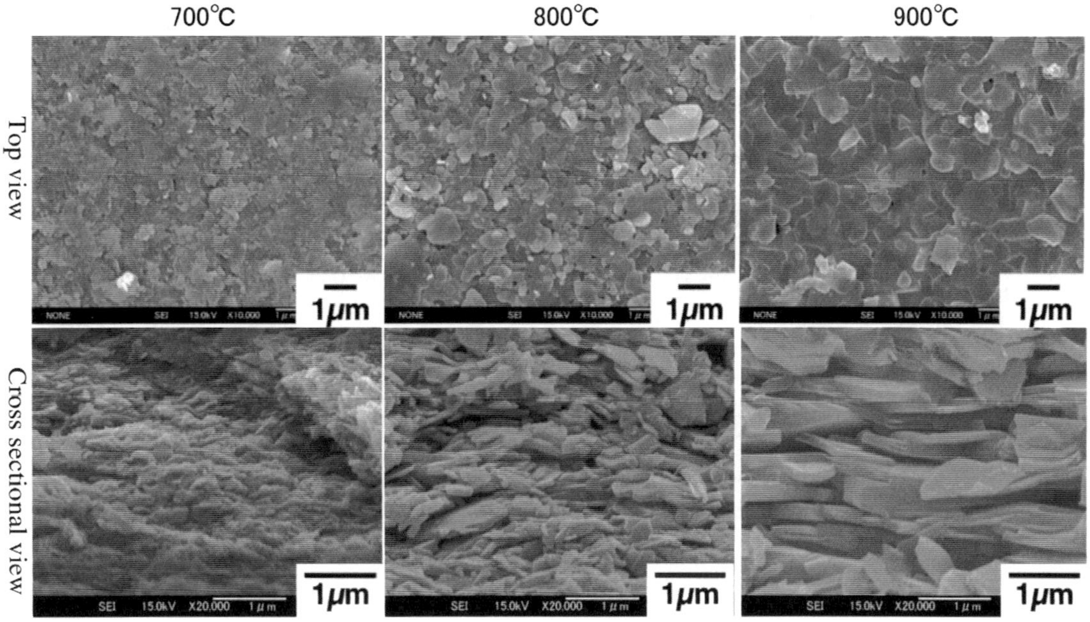

Figure 4 Top view and cross sectional SEM images of the samples sintered at 700-900 °C for 1 h.

Figure 4 shows the top view and cross sectional images of the samples sintered at 700-900 °C for 1 h, where the horizontal direction of the cross sectional view is in parallel to the substrate face. The sample sintered at 700 °C showed a dense structure and was consisted of the particles with submicron scale. As increasing the sintering temperature, the Co349 grains were grown as platelets, and the sample sintered at 900 °C consisted of the grains grown to 1 μm or larger. In addition, the sintered film was consisted of the plate like crystals piled up in layers as clearly seen in the sample sintered at 900 °C. This also supports that those samples had the *c*-axis oriented structure as suggested in XRD analysis.

A reason why the *c*-axis oriented structure could be obtained by AD process is thought to be the platelet shape of the Co349 crystal. The Co349 crystal easily breaks into plate like shape due to the anisotropic structure when the powder was collided with the substrate. The crashed powder with plate like shape would be piled up as the layered structure since the wide *ab*-face easily becomes parallel to the substrate surface. Such the grains would become large during sintering and formed the *c*-axis aligned layered structure. In the end, the high temperature sintering enhanced the grain growth, and the well textured structure was achieved in the sample sintered at 900 °C.

The thermoelectric properties of Seebeck coefficient and electrical conductivity were evaluated in those sintered samples with the thickness of 7 μm. The results were shown in Figure 5 with the properties of bulk samples fabricated with HP [2] and SPS [5] methods. The sample sintered at high temperature showed the large thermopower (Seebeck coefficient), while the difference among the samples was not so large. The sample sintered at 900 °C showed the highest thermopower of about S = 170 μV/°C at 700 °C. On the other hand, the electrical conductivity of the sample sintered at 700 °C was only about σ = 60 S/cm, while that of the sample sintered at 900 °C was higher than 110 S/cm. The thermoelectric properties of the sample sintered at 900 °C were as high as those of bulk samples fabricated with HP and SPS method since the dense and well aligned large grain structure was achieved in the sample. As shown in this figure, the difference in electrical conductivity among the samples sintered at different temperature was larger than those in Seebeck coefficient. This would be explained by the connection between grains. Seebeck voltage is the potential difference generated in Co349 grain and was affected by the crystallinity of the Co349 grains. The crystallinity of the Co349 grains would be recovered during the annealing over 700 °C, and the difference among the samples would be small. On the other hand, the electrical conductivity is the transport property and is affected by the crystallinity of Co349 and the connection between the grains. The results shown in Fig. 5 (b) suggested that the grain connection was not enough in the sample annealed at 700°C and became good as increasing the annealing temperature.

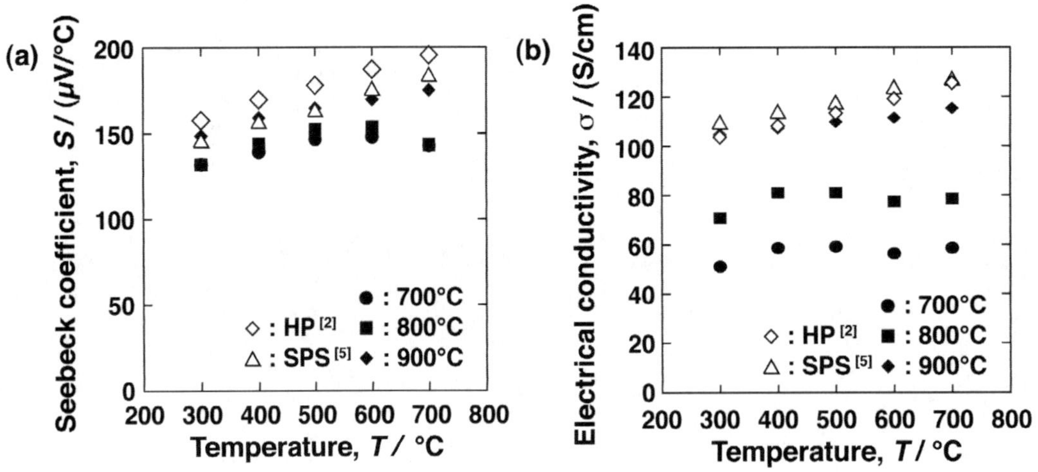

Figure 5 Thermoelectric properties of the samples sintered at 700-900°C for 1h. (a) Seebeck coefficient, (b) electrical conductivity.

3.4. Effect of film thickness

The effect of film thickness on the thermoelectric properties is shown in Figure 6. The Seebeck coefficient of the samples were almost the same value irrespective of film thickness after sintering at 900 °C for 1h as shown in Fig. 6 (a). On the contrary, The samples with the thickness larger than 5 μm showed almost the same electrical conductivity, however the electrical conductivity of the 1 μm thick sample was only about 40 S/cm which was almost less than half of that of the other samples.

To see the reason for this low electrical conductivity, the cross sectional structure of 7 μm thick sample was observed by SEM and is shown in Figure 7. The Co349 grains in the region about 1 μm thick from the interface between substrate and AD film was small and poor textured compared to upper region in the AD film. The rough interface of the substrate would suppress the grain growth along the surface of the substrate. This rough interface may be caused by the damage due to the collision of large particles in the early stage of AD deposition. As a result, the c-axis textured structure did not be developed in the thin 1 μm thick film, and this poor structure resulted the low electrical conductivity. However, the effect of rough interface was limited near the interface. The well developed textured structure was achieved in the region thicker than 2 μm. Since the effect of poor textured initial layer was not so large in the thick film, the electrical conductivity of those samples showed almost the same value. So the increase in thickness of Co349 film would have less effect to decrease the thermoelectric properties. However, the Seebeck coefficients of the thick films were sharply decreased at 700 °C, while that of the 1μm thick film was not show such the decrease. The reason of this sudden reduction is not clear. Since the electrical conductivity of those samples did not show any change at 700 °C, the microstructure and composition would not be different so much in these samples. Further investigation is needed to clarify this change at 700°C.

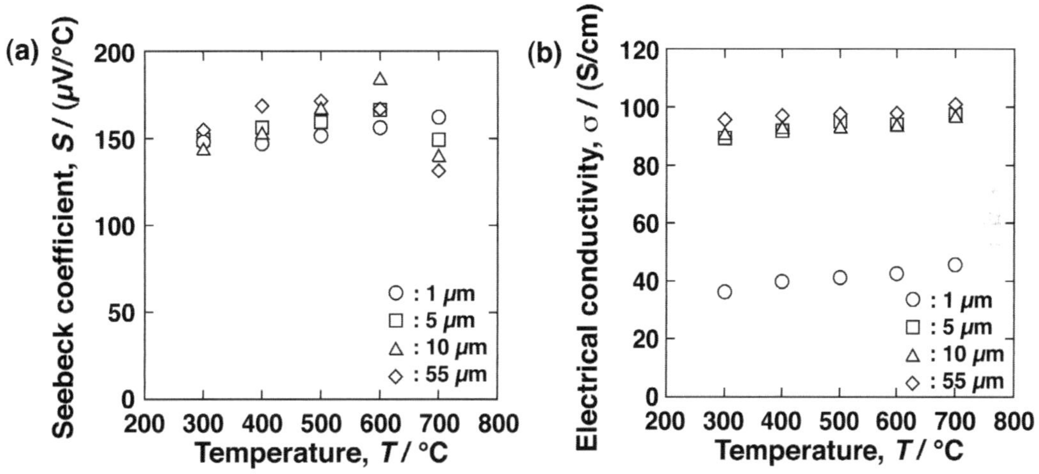

Figure 6 Thermoelectric properties of the samples with different film thickness. (a) Seebeck coefficient, and (b) electrical conductivity.

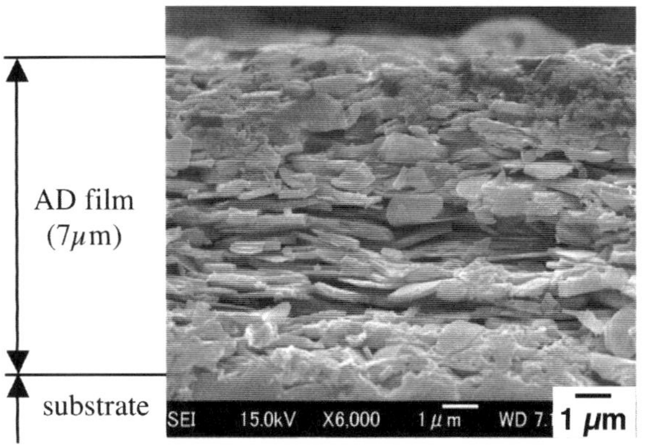

Figure 7 The cross sectional view of the sample sintered at 900°C.

4. Summary

The Co349 thick film was fabricated by AD method, and the effect of process parameters on the microstructure and thermoelectric properties were investigated. The Co349 film deposited with AD method showed the dense *c*-axis oriented structure after sintering at the temperature higher than

700 °C. The film annealed at 900 °C showed the thermoelectric properties as high as those of the usual bulk samples fabricated by HP and SPS method. This high thermoelectric performance could be kept in the thick film of 55μm. This suggests that the AD process is a good candidate to fabricate thick TE films and to realize multi-layered thick film type TE modules.

References

[1] Shikano M and Funahashi R 2003 *Appl. Phys. Lett.*, **82** 1851
[2] Xu G, Funahashi R, Shikano M, Matsubara I, and Zhou Y *Appl. Phys. Lett.* **80** 3760
[3] Prevel M, Lemonnier S, Klein Y, Hebert S, and Chateigner D. 2005 *J. Appl. Phys.* **98** 093706
[4] Liu Y, Lin Y, Shi Z, Nan CW, and Shen Z 2005 *J. Am. Ceram. Soc.* **88** 1337
[5] Liu Y, Lin Y, Jiang L, Nan CW, and Shen Z 2008 *J. Electroceram.* **21** 748
[6] Akedo J and Lebedev M 1999 *Jpn. J. Appl. Phys.* **38** 5397
[7] Akedo J 2006 *J. Am. Ceram. Soc.* **89** 1834
[8] Akedo J 2008 *J. Therm. Spray Technol.* **17** 181
[9] Shin KH, Mizoguchi M, and Inoue M 2007 *J. Magnetics* **12** 129
[10] Okamoto T, Horii S, Uchikoshi T, Suzuki TS, Sakka Y, Funahashi R, Ando N, Sakurai M, Shimoyama J, and Kishio K, *Appl. Phys. Lett.* **89** 081912

Development of Cell Culture Microdevice Actuated by Piezoelectric Thin Films for Delivering Mechanical Vibratory Stimuli to Cells

Y Yamada[1], G Umegaki[1], T Kawashima[1], M Nagai[1], T Shibata[1], T Masuzawa[2], T Kimura[3] and A Kishida[3]

[1] Department of Mechanical Engineering, Toyohashi University of Technology, Toyohashi 441-8580, JAPAN
[2] Department of Mechanical Engineering, Ibaraki University, Hitachi 316-8511, JAPAN
[3] Institute of Biomaterials and Bioengineering, Tokyo Medical and Dental University, Chiyoda-ku 101-0062, JAPAN

E-mail: y-yamada@mems.me.tut.ac.jp

Abstract. In order to realize a cell culture microdevice actuated by piezoelectric thin films for on-chip regulation of cell functions, this paper reported on a feasibility study by using the microdevice with KOH-etched cavities surrounded by four (111) sidewalls as microchambers in order to introduce cells to be cultured. As a result, the vibration characteristic of the PZT actuator was improved by using an electric field -150 kV/cm at 70 °C for 30 min in poling process. A feasibility study on cell culture for delivering mechanical vibratory stimuli to cells revealed the microdevice could be applicable to the culture with actual biological cells. In addition, it was found that O_2-plasma treated parylene-C process could be applicable for obtaining homogeneous surface of cell culture microdevice.

1. Introduction

Recent advances in microelectromechanical systems (MEMS) technology have opened up new possibilities in cell biology, neurobiology, pharmacology, and tissue engineering [1]. Among them, actual evidence suggests that cell functions such as cell adhesion, cell multiplication, and gene transfer can be enhanced by delivering mechanical vibratory stimuli to cells [2, 3]. This approach, which is unnecessary to provide any accelerating agents, might be free of the potential risk of this infection considering biological applications such as gene therapy and tissue engineering. However, the influencing factors of mechanical vibratory stimuli (amplitude, frequency, etc.) to cells are not obvious. Therefore, this paper focused on mechanical vibratory stimulation with nanometer-order displacement and frequency of less than 10 kHz. In order to realize screening and optimization of the factors on a chip-based system, we have been developing a cell culture microdevice actuated by piezoelectric thin films for on-chip regulation of cell functions.

This paper reports on characterization of a cell culture microdevice with KOH-etched cavities surrounded by four (111) sidewalls as microchambers in order to introduce cells to be cultured. In addition, a feasibility study on cell culture for delivering mechanical vibratory stimuli to cells was executed by using the fabricated microdevice. Moreover, deposition of parylene-C film on silicon (Si) substrates was investigated in order to obtain a homogeneous substrate surface for fabrication of a prototype of a cell culture microdevice proposed here.

2. Cell culture microdevice actuated by piezoelectric thin film

Figure 1 shows the cell culture microdevice proposed here has a number of microactuators with a lead zirconate titanate (PZT) thin film. The PZT actuator with 1 μm in thickness is fabricated on a silicon nitride (Si_3N_4) diaphragm ($250 \times 250 \mu m^2$ in size with 1 μm in thickness) formed on a Silicon substrate (p-type, (100) single crystal) with 20×20 mm^2 in size. Biological cells are seeded onto the PZT actuator covered by an insulation layer, giving mechanical vibratory stimuli to cells. The microdevice makes it possible to modify the surface condition for cell seeding and culture cells with different conditions of mechanical vibratory stimulation in each actuator because PZT actuators can be controlled independently by applying sine wave AC voltages with different amplitude and frequency. Therefore, the cell culture microdevice allows screening and optimization of the influencing factors of multiple mechanical vibratory stimuli (amplitude, frequency, etc.). Figure 2 shows a PZT actuator consisting of multiple layers (Pt/Ti/PZT/Pt/Ti/SiO_2) formed on a Si_3N_4 diaphragm, which was fabricated by an anisotropic wet etching using KOH solution, were successfully patterned and structured.

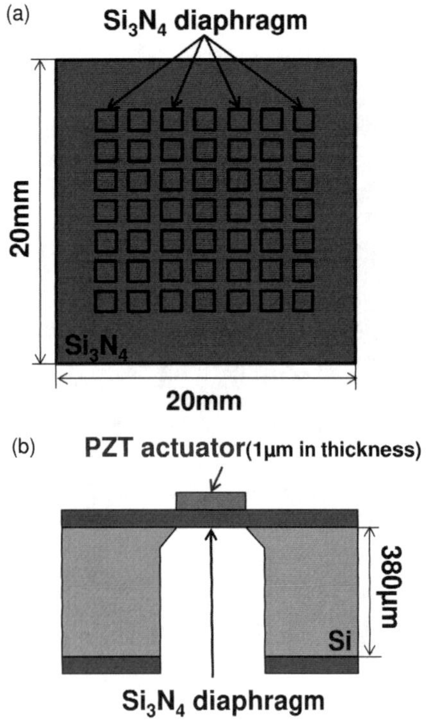

Figure 1. Schematic diagram of cell culture microdevice actuated by PZT thin film; (a) Plan view of device, (b) Cross-sectional view.

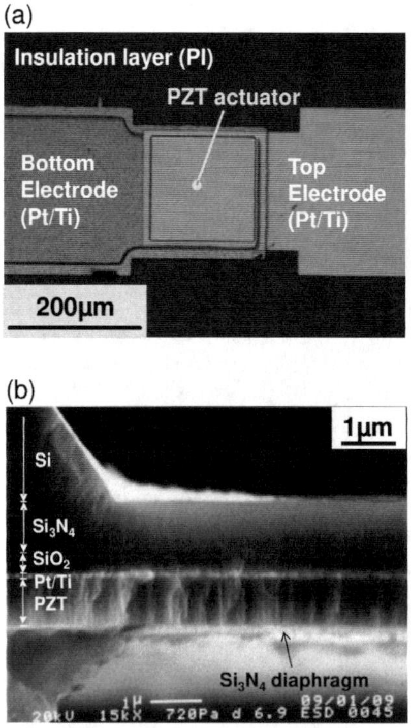

Figure 2. Fabricated microdevice; (a) Plan view of PZT actuator, (b) Cross-sectional view of PZT actuator.

3. Characteristics of cell culture microdevice

Figure 3 shows a typical result of time dependence of the diaphragm displacement of a PZT actuator with a size of 200×200 μm^2 measured by laser interferometer (Canon DS-80) which has a high frequency resolution of 500 kHz. The resulting displacement was estimated to be approximately 52 nm at an applied sine wave AC voltage of ±10 V at a drive frequency of 10 kHz. Although the diaphragm displacement was found to be disagreed with the applied AC voltage, it is considerable that the deformation of the diaphragm was restricted by the insulation layer of polyimide between top and bottom electrodes.

In order to improve drive performance of the cell culture microdeivce, poling treatment was executed at electric field of -150 kV/cm at 70 °C for 30 min, where the bottom electrode was connected to electrical ground. Figure 4 shows the diaphragm displacements of three PZT actuators, which were used for cell culture experiment of delivering mechanical vibratory stimuli to cells in this paper, with size of 200 × 200 μm^2 before and after poling. The average displacements of three diaphragms before the poling treatment were estimated to be approximately 25 nm (±5.6 nm) at an applied sine wave AC voltage of ±10 V at a drive frequency of 10 kHz. On the other hand, the average displacements were increased to 50 nm (±1.0 nm) after the poling treatment. As a result, the vibration characteristic of the PZT actuator was improved by using an electric field in poling process, because direction of polarization was reoriented by such a strong electrical field applied between top and bottom electrodes during the poling process. However, large displacement of diaphragm was not obtained. This was caused by the crystalline of the PZT thin film fabricated. Therefore, it is necessary to optimize the fabrication process.

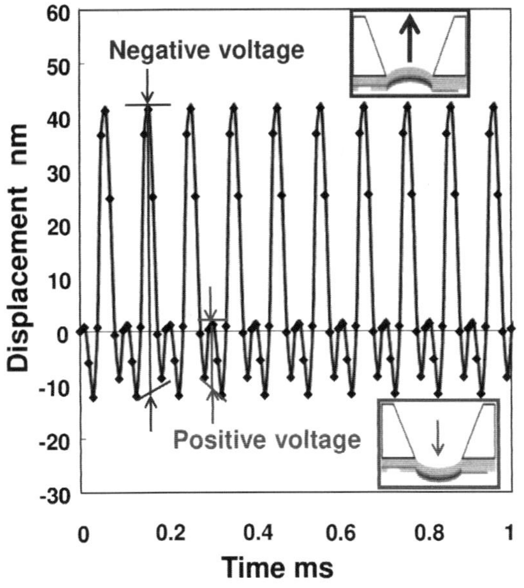

Figure 3. Diaphragm displacement of PZT actuator with 200×200 μm^2 diaphragm size as a function of time.

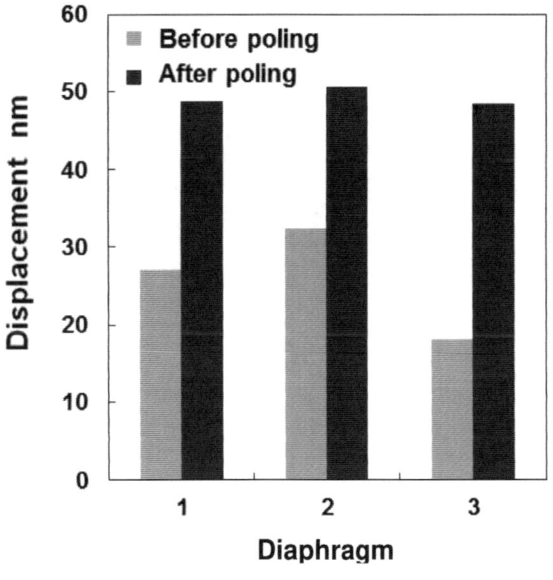

Figure 4. Effect of poling treatment on diaphragm displacement of PZT actuator.

4. Delivering mechanical vibratory stimuli to cultured cells

In order to demonstrate capabilities of cell culture with the microdevice for on-chip regulation of cell functions, a feasibility study on delivering mechanical vibratory stimuli to Mouse Embryonic Fibroblast (MEF) cells was executed. In this experiment, KOH-etched cavities surrounded by four (111) sidewalls were used as microchambers in order to introduce cells to be cultured. In cell-seeding process, a frame made of polydimethylsiloxane (PDMS) was mounted on the device for storing culture solution, as shown in Figure 5. After culture solution of 300 μL Dulbecco's Modified Eagle Medium (DMEM) was dropped into the PDMS frame, MEF cells (concentration of cell suspension: 1.0×10^5 cells/mL, volume: 100 μL) were seeded. After 5 minutes for settling the MEF cells, mechanical vibratory stimulation with displacement of 50 nm at a drive frequency of 10 kHz for 1h was applied. The MEF cells were cultured at 37 °C in 5 % CO_2 for 48 h after the stimulation. Figure 7 shows optical micrographs of the MEF cells just after seeding into the microchamber, after 24 h cultivation and after 48 h. Immobilization treatment of MEF cells cultured for 48 h by using 2.5% glutaraldehyde was carried out. After immobilization treatment, dewatering process with dehydrated ethanol was performed in order to observe the MEF cell adhesion. As a result, 2/13 (=15.4%) of the diaphragm without delivering the mechanical vibratory stimulation was adhered by MEF cells. On the other hand, 3/3 (=100%) of that with delivering the simulation were adhered by MEF cells. It was found to be applicable to our application. Figure 8 shows a SEM microphotograph of diaphragm without delivering the simulation. A lot of MEF cells adhered to Si (111) sidewalls of the microchamber were observed. This result was considered that the different condition of the substrate surface affected the adhesion of the MEF cells because cells prefer to adhere to a substrate depending on its surface energy [4,5].

Figure 5. PDMS frame mounted on the device for storing culture solution.

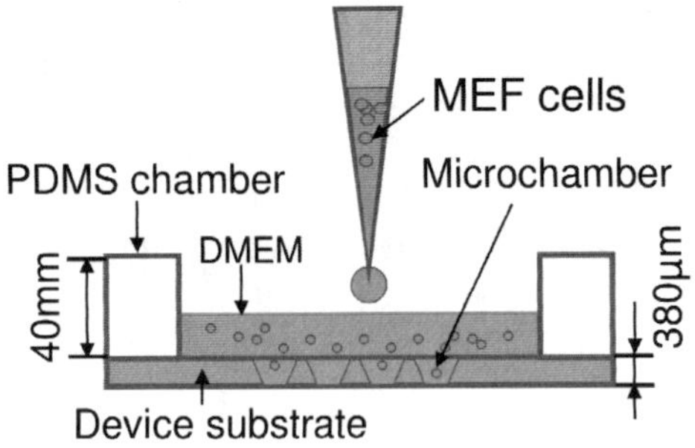

Figure 6. Schematic of feasibility study on cell culture for delivering mechanical vibratory stimuli to MEF cells.

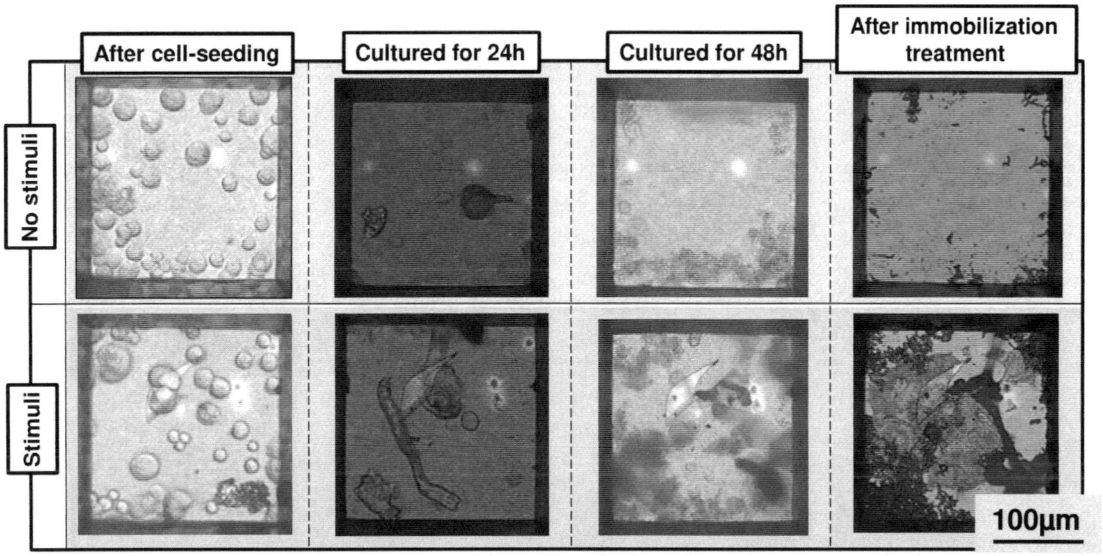

Figure 7. Time-series photographs of MEF cells adhered on Si_3N_4 diaphragm; Upper row is without mechanical vibratory stimulation, Lower row is with stimulation of applied displacement of 50 nm at a drive frequency 10 kHz for 1 h.

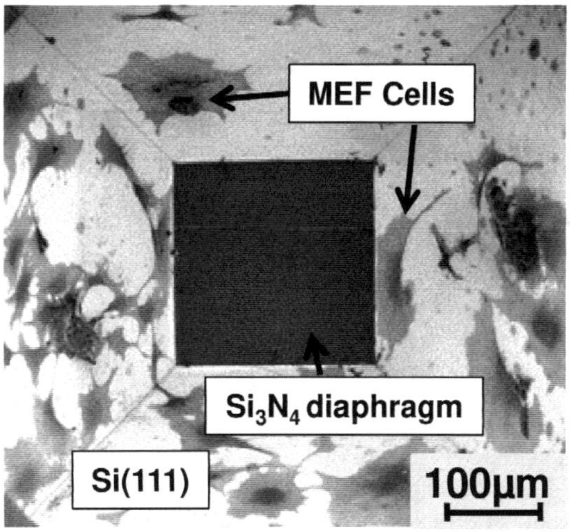

Figure 8. Adhesion of MEF cells in the microchamber.

5. Modification of substrate surface using parylene-C

It is desirable that substrate surface is homogeneous because substrate surface effects to cell adhesion, as shown in Figure 8. Therefore, depositing of parylene-C was performed in order to obtain homogeneous substrate surface, because parylene-C is well-known as one of biocompatible materials and modification of substrate surface is easily possible [6]. Therefore, three substrates under three conditions of substrate surfaces were prepared for evaluation of cell adhesion. First, Si_3N_4 membrane was deposited on Si substrate in low-pressure CVD (LPCVD) process after SPM treatment (H_2O_2:H_2SO_4=1:3) and rinsing with deionized water. The water contact angle of the substrate was 0°. The substrate was named as "SiN(0°)". Second, parylene-C with 1 μm in thickness was deposited on Si substrate. The water contact angle of the as-deposited parylene-C surface was 95° (named as

"PC(95°)"). Third, Si substrate after parylene-C deposition was treated by O_2-plasma. In the O_2-plasma treatment process, flow rate of O_2 was 50 sccm, RF power was 50W, pressure during the process was 380 Pa, and process time was 30 s. The water contact angle of the O_2-plasma treated parylene-C surface was 25° (named as "PC(25°)").

Figure 9 shows a SEM photograph of HeLa cells adhered to the O_2-plasma treated parylene-C surface. It can be confirmed the good adhesion of HeLa cells. Figure 10 shows adhesion areas of HeLa cells on each substrate surface. The adhesion areas of the HeLa cells were calculated from binary images of obtained optical photographs by using a photo-editing software (pickmap 1.0.0.4). As a result, the substrates with Si_3N_4 surface ("SiN(0°)") and O_2-plasma treated parylene-C surface ("PC(25 °)") were good for cell adhesion. The adhesion area of as-deposited parylene-C surface ("PC(95°)") was extremely low. Consequently, it was found that O_2-plasma treated parylene-C process could be applicable for obtaining homogeneous surface of cell culture microdevice shown in Figure 1.

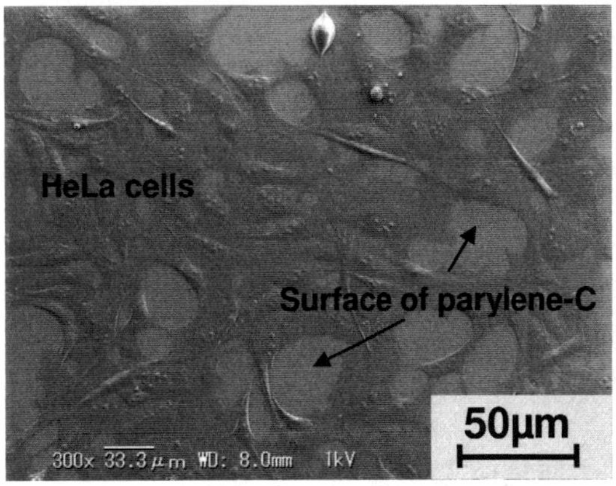

Figure 9. Adhesion of HeLa cells on O_2-plasma treated parylene-C surface.

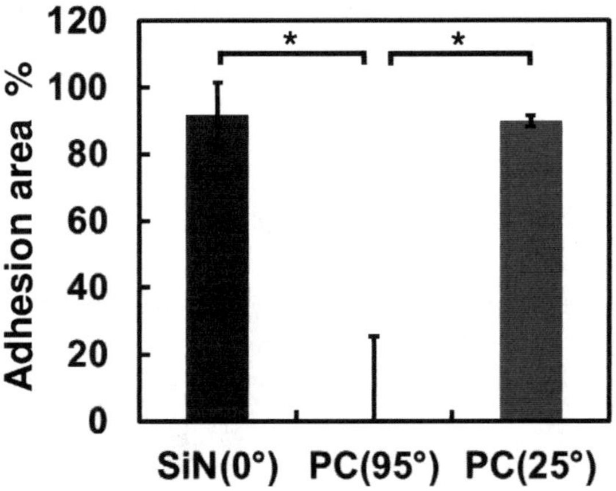

Figure 10. Adhesion area of HeLa cells on three types of substrate surfaces.

6. Conclusions

The cell culture microdevice with KOH-etched cavities surrounded by four (111) sidewalls as microchambers in order to introduce cells to be cultured was fabricated in this paper. The vibration characteristic of the PZT actuator was improved by using an electric field -150 kV/cm at 70 °C for 30 min in poling process. A feasibility study on cell culture for delivering mechanical vibratory stimuli to cells at applied displacement of 50 nm at a drive frequency of 10 kHz for 1h revealed the microdevice could be applicable to the culture with actual biological cells for our purpose. In order to obtain a homogeneous substrate surface, deposition of parylene-C film was performed on Si substrates. As a result, it was found that O_2-plasma treated parylene-C process could be applicable for obtaining homogeneous surface of cell culture microdevice, because the adhesion area of HeLa cells on O_2-plasma treated parylene-C substrate was 90%. In future, a prototype of the cell culture microdevice proposed here will be appeared and effect of delivering mechanical vibratory stimuli to cells on cell functions will be investigated.

Acknowledgments
This work was partially supported by the Global COE Program "Frontiers of Intelligent Sensing" from the Ministry of Education, Culture, Sports, Science and Technology, Japan.

References

[1] Bashir R 2004 *Adv. Drug Deliv. Rev.* **56** 1565-1586
[2] Masuzawa T, Kusama J, Kimura T, Shibata T, Kishida A and Kimura T 2005 *Proc. 21st Ann. Conf. of Society of Life Support Tech.* 80 (in Japanese)
[3] Konno K, Kosawada T, Yamazaki H, Hozumi Y and Goto K 2008 J. *Biomed. Sci. Eng.* **3** 38-49
[4] Ranella A, Barberoglou M, Bakogianni S, Fotakis S, Stratakis E 2010 *Acta Biomater.* **6** 2711–2720, (2010)
[5] Khang D, Lu J, Yao C, Haberstroh K, Webster K 2008 *Biomaterials* **29** 970–983
[6] Wei L, Lakhtakia A, Roopnariane A , Ritty T 2010 *Mat. Sci. Eng. C-Biomim.* **30** 1252-1259

Journal of Physics: Conference Series **352** (2012) 012028 doi:10.1088/1742-6596/352/1/012028

Aqueous synthesis and transmission electron microscopy observation of seed-grown spherical ferrite nanoparticles

T Tanaka[1], M Tada[2], T Nakagawa[3], M Abe[4]

[1] Tokyo Bio R&D Center, Tamagawa Seiki Co., Ltd., 4259-3 Nagatsuta-cho, Midori-ku, Yokohama, Kanagawa 226-8510, Japan

[2] Department of Physical Electronics, Tokyo Institute of Technology, 2-12-1 Ookayama, Meguro-ku, Tokyo 152-8552, Japan

[3] Department of Management of Industry and Technology, Osaka University, 1-1 yamadaoka, Suita, Osaka 565-0871, Japan

[4] Materials & Structures Laboratory, Tokyo Institute of Technology, 4259 Nagatsuta-cho, Midori-ku, Yokohama, Kanagawa 226-8501, Japan

E-mail: toshiyuki-tanaka@tamagawa-seiki.co.jp

Abstract. Uniform-sized spherical iron ferrite nanoparticles grew on seed crystals in an aqueous solution containing sucrose. Using the seed crystals which were highly dispersed in acidic or alkaline seed-crystal suspension without relation to pH of the suspension, we widely controlled the particle diameter in the range 20–200 nm by changing the additive amount of the seed crystals. By transmission electron microscopy observation and X-ray diffraction analysis, it indicated that the particles were highly crystalline but not amorphous. Selected area diffraction patterns of the particles by using transmission electron microscope revealed that the particles were composed of one to several crystals. Thus we provided the evidence that the particles grew on clusters composed of one to several seed crystals to which those of several dozen seed crystals were disintegrated.

1. Introduction

Functional magnetic nanoparticles have been developed for biomedical applications such as bio-separation, drug delivery systems, magnetic resonance imaging and hyperthermia therapy [1,2]. Especially, iron ferrite (intermediate between Fe_3O_4 and γ-Fe_2O_3) nanoparticles have high potential for the biomedical applications due to high biocompatibility and corrosion resistance. To enhance the quantifiablility and reproducibility of the biomedical applications, magnetic nanoparticles should have spherical shape and uniform size, which endow the nanoparticles with high dispersibility and uniform surface-immobilization, magnetic and hydrodynamic properties. Moreover, the particle size should be optimally controlled between several dozen and several hundreds nanometers in accordance with each biomedical application to obtain high throughput and high sensitivity. In previous studies, uniform-sized spherical iron ferrite particles were synthesized by various chemical syntheses, which were however limited to ≤ 30 nm or ≥ 200 nm in diameter [3–9].

We have already developed a synthesis method of uniform-sized spherical iron ferrite nanoparticles in an aqueous solution containing a disaccharide and ferrite seed crystals [10]. We found that disaccharides enable spherical growth of the ferrite nanoparticles and seed growth makes the particle size distribution narrow. We have also widely controlled the particle diameter in the range 20–200 nm by changing the amount of ferrite seed crystals highly dispersed in a HNO_3 solution [11]. By the relation between the adding seed crystal amounts and the resultant spherical particle diameters, we theoretically considered that clusters composed of the several dozen seed crystals were disintegrated into smaller clusters of a few seed crystals and then the spherical ferrite nanoparticles grew on the smaller clusters.

In this paper, we report the effect of seed-crystal suspensions on seed growth of uniform-sized spherical iron ferrite nanoparticles. We also show crystal morphology of the spherical ferrite nanoparticles by transmission electron microscopy (TEM) observation and present an evidence to support that the spherical ferrite nanoparticles grew on the disintegrated seed-crystal clusters.

2. Experimental

Iron ferrite seed crystals were prepared by a coprecipitation method. A NaOH solution is added to an aqueous solution of $FeCl_2$+$2FeCl_3$. The prepared ferrite seed crystals were separated from the reaction solution by a magnet and then washed in distilled water several times. They were dispersed into a HNO_3 solution and a tetramethylammonium hydroxide ($N(CH_3)_4OH$) solution based on Massart's method [12]. Thus we obtained acidic and alkaline suspensions of the seed crystals. The concentration of the seed crystals in the suspensions was measured as Fe concentration by spectrophotometric method using 2,2'-bipyridyl [13].

Uniform-sized spherical iron ferrite nanoparticles were synthesized by the seed-growth method from an aqueous solution containing sucrose [11]. An aqueous solution containing NaOH (10 mmol), $NaNO_3$ (20 mmol), and sucrose (50 mmol) was heated to 70 °C, to which the acidic or the alkaline seed-crystal suspension (about 1–1000 μmol Fe) and a $FeCl_2$ solution (7.5 mmol) were sequentially added. The mixed reaction solution was adjusted to 100 mL. The mixed solution was kept at 70 °C for 3 h under stirring and N_2 gas bubbling. The grown ferrite nanoparticles were separated from the solution by a magnet and then washed in distilled water.

The hydrodynamic diameter and zeta potential were measured by a dynamic light scattering method. The crystal structure and morphology were assessed by TEM operating at 200 kV and X-ray diffraction (XRD) using a Cu-Kα radiation source. For TEM observation of crystal structure, we prepared observation samples by slicing the spherical ferrite nanoparticles, about 150 nm in size, embedded in resin.

3. Results and discussion

The diameter of the synthesized iron ferrite seed crystals had geometric average d_s = 8.8 nm and geometric standard deviation σ_s = 1.37 as measured from TEM images. Table 1 shows properties of the acidic and the alkaline suspensions of the seed crystals. The seed crystals were highly dispersed as small clusters with approximate 20 nm in hydrodynamic diameter in both suspensions. They had high positive and negative surface charges in the acidic and the alkaline suspension, respectively. These are because pHs of the suspensions are far from the Fe_3O_4 isoelectric point of 6.5–6.8 and NO_3^- and $N(CH_3)_4^+$ are low polarizing ions [12].

Table 1. Properties of the acidic and the alkaline suspensions of ferrite seed crystals.

Solution	pH	Hydrodynamic diameter (nm) (average ± S.D.)	Zeta potential (mV)
HNO_3	1.7	22.8 ± 5.4	42
$N(CH_3)_4OH$	12.8	23.0 ± 5.6	−43

We have already reported the result of seed-grown spherical iron ferrite nanoparticles synthesized by using the acidic seed-crystal suspension [11]. As the additive concentration of the seed crystals in the reaction solution m_s decreased from 11,500 to 11.5 μmol Fe/L, the geometric average diameter of the synthesized nanoparticles d increased from 19 to 209 nm and geometric standard deviation of the diameter σ narrowed from 1.38 to 1.11 (figure 1(a)–(c) and 2). Here m_s was calculated from the seed-crystal concentration in the suspensions, the additive volume of the seed-crystal suspension to the reaction solution and total volume of the reaction solution. The size distribution became narrowed by a large amount of the seed crystals, because it suppresses secondary nucleation. The relationship between d and m_s follows

$$d = (n^{1/3} d_s m^{1/3}) m_s^{-1/3} \tag{1}$$

where m is Fe concentraion supplied from green rust, intermediate product, for the ferrite growth in the reaction solution [14], d_s average diameter of the seed crystals and n average number of the seed crystals composing clusters on which the ferrite growth occurs. The d vs. m_s curve obtained for d_s = 8.8 nm and m = 50 mmol Fe/L was best fitted by equation (1) with a nominal value of n = 3.4 (figure 2). The value was considerably smaller than n = 17 which was calculated from the relation $n = (d_s'/d_s)^3$ where d_s' = 22.8 nm is the hydrodynamic diameter of the seed crystal clusters measured in the acidic suspension before the ferrite growth experiments (Table 1). This suggested that the original clusters composed of several dozen seed crystals in the suspension were disintegrated to those of a few seed crystals during the ferrite growth in the reaction solution.

Figure 1(d)–(f) and table 2 show spherical iron ferrite nanoparitcles grown on seed crystals by adding the alkaline ($N(CH_3)_4OH$) seed-crystal suspension. The particle diameter d also widely controlled from 212 to 22 nm as m_s increased from 11.3 to 11,300 μmol Fe/L (figure 2). The standard deviation σ also decreased from 1.30 to 1.15 as d increased from 22 to 116 nm. However, the largest nanoparticles (d = 212 nm) were accompanied by fine particles (figure 1(d)) similarly to those synthesized by using the acidic suspension (figure 1(a)). This is because that secondary nucleation could not be suppressed enough by the smallest amount of the seed crystals. The relationship between d and m_s was also fitted by equation (1) and n was nominally estimated to 4.6 seed crystals (figure 2). The value was also smaller than n = 18 calculated from the hydrodynamic diameter d_s' = 23.0 nm in the alkaline suspension. The clusters composed of seed crystals in the alkaline suspension were also inferred to be resolved to those of a few seed crystals during the ferrite growth. On the other hand, we have shown that particle diameter could not be controlled by using the large amount of seed crystals aggregated in alkaline NH_3 solution [11]. Thus, regardless of whether pH of the seed-crystal suspension is acidic or alkaline, size of the spherical nanoparticles is widely controlled by using the seed crystals highly dispersed in the suspensions.

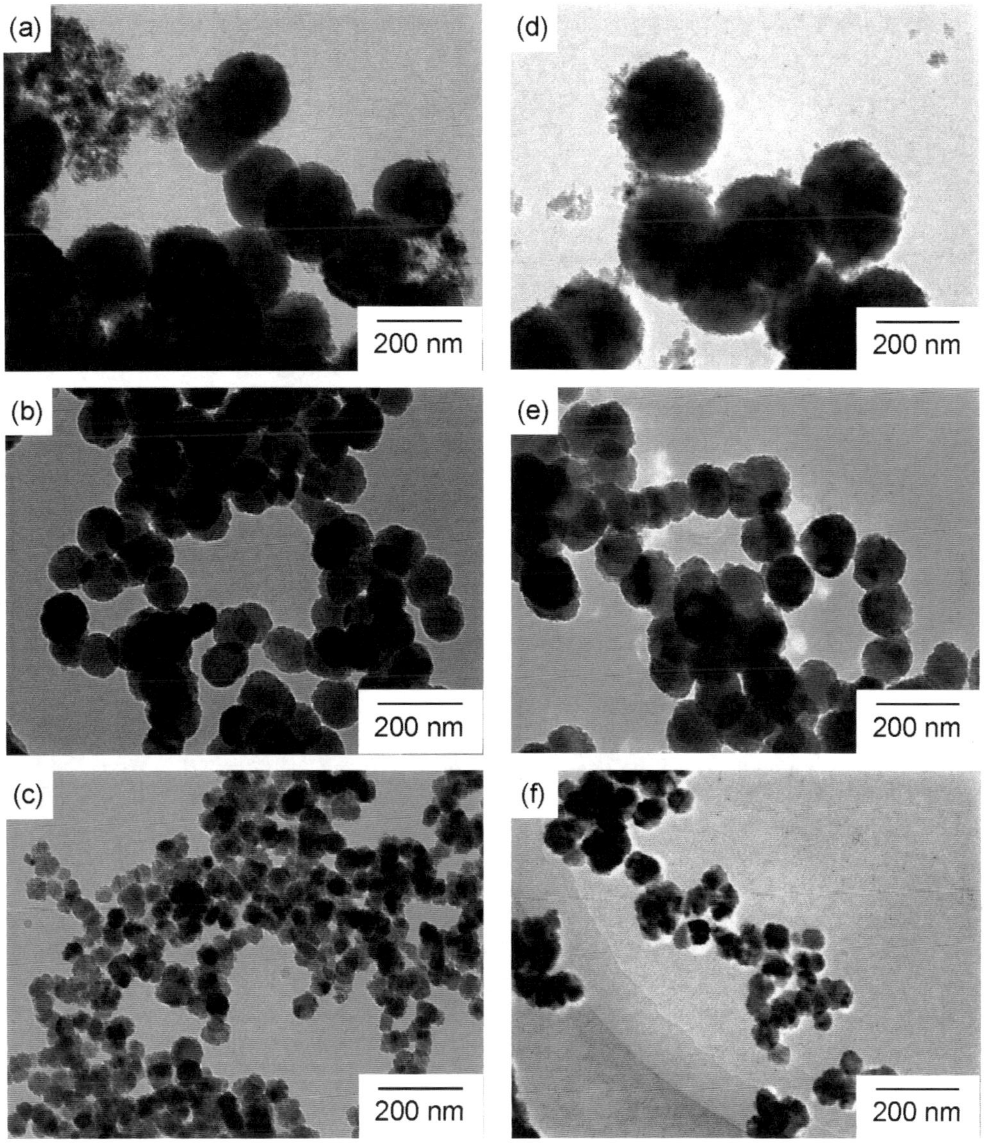

Figure 1. TEM images of spherical ferrite nanoparticles grown on seed crystals. Seed crystals were dispersed in HNO_3 solution ((a) 11.5, (b) 115 and (c) 1150 μmol Fe/L) and in $N(CH_3)_4OH$ solution ((d) 11.3, (e) 113 and (f) 1130 μmol Fe/L).

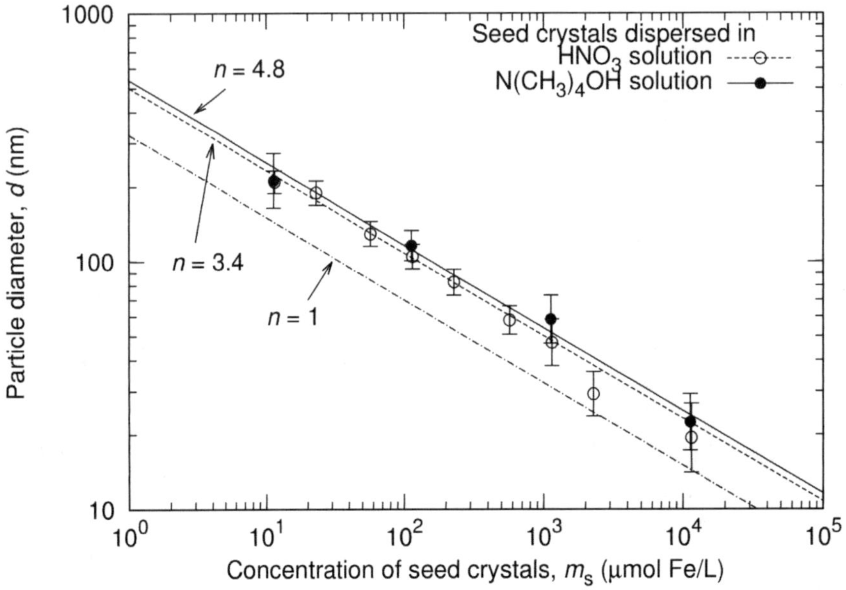

Figure 2. Relationship between diameter of seed-grown spherical ferrite nanoparticles d and additive concentration of seed crystals in reaction solution m_s. Curves are calculated from equation (1) with nominal average number of seed crystals n which compose clusters on which seed growth occurs. Chain line shows estimated curve of each nanoparticle grown on one seed crystal ($n = 1$).

Table 2. Relationship between additive concentration of seed crystals in reaction solution and diameter of spherical ferrite nanoparticles grown on the seed crystals measured from TEM images. The seed crystals were dispersed in alkaline suspension before particle synthesis.

Seed crystals	Spherical ferrite nanoparticles	
Concentration m_s (μmol Fe/L)	Geometric average diameter d (nm)	Geometric standard deviation σ
11.3	212 [a]	1.29 [a]
113	116	1.15
1130	58	1.25
11300	22	1.30

[a] Smaller particles in figure 1(d) were neglected.

Figure 3 show XRD profiles of seed-grown spherical ferrite nanoparticles whose diameter was controlled. All the nanoparticles had a single phase with a spinel structure and the diffraction spectrum became sharp as the particle diameter increased. As shown in figure 4, the uniform lattice fringe was observed at part of a seed-grown spherical ferrite nanoparticle although the nanoparticle had the concavo-convex surface. They clearly indicate that the spherical nanoparticles are highly crystalline but neither amorphous nor aggregate of smaller particles. The concavo-convex surface of the nanoparticles might be formed by growth suppression with weak adsorption of sucrose on the surfaces.

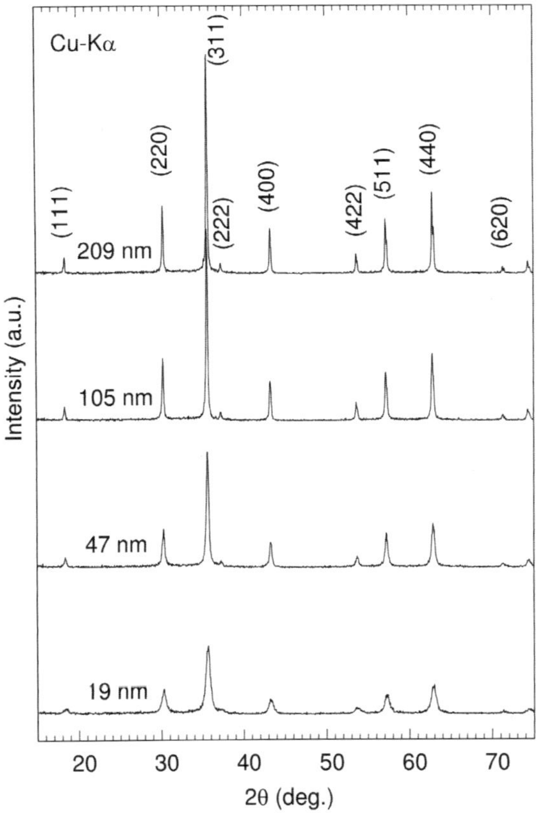

Figure 3. Powder XRD profiles of seed-grown uniform-sized spherical ferrite nanoparticles controlled their diameters by changing the additive concentration of seed crystals.

Figure 4. A high-resolution TEM image of part of a seed-grown spherical ferrite nanoparticle at an ultrahigh magnification.

Figure 5(a) and (b) shows dark- and bright-field TEM images of a seed-grown spherical ferrite nanoparticle. The dark-field image consists of several domains. The selected area diffraction (SAD) patterns of the particle at areas A–D, shown in figure 5(c)–(f), had clear diffraction spots and were different from one another. Their superposition gives a pattern similar to that taken on the whole area of the crystal (figure 5(g)). All spots are assigned as diffractions from crystal planes of spinel structure of ferrite (figure 5(h)). Thus it indicates that the nanoparticle is composed of several single crystals.

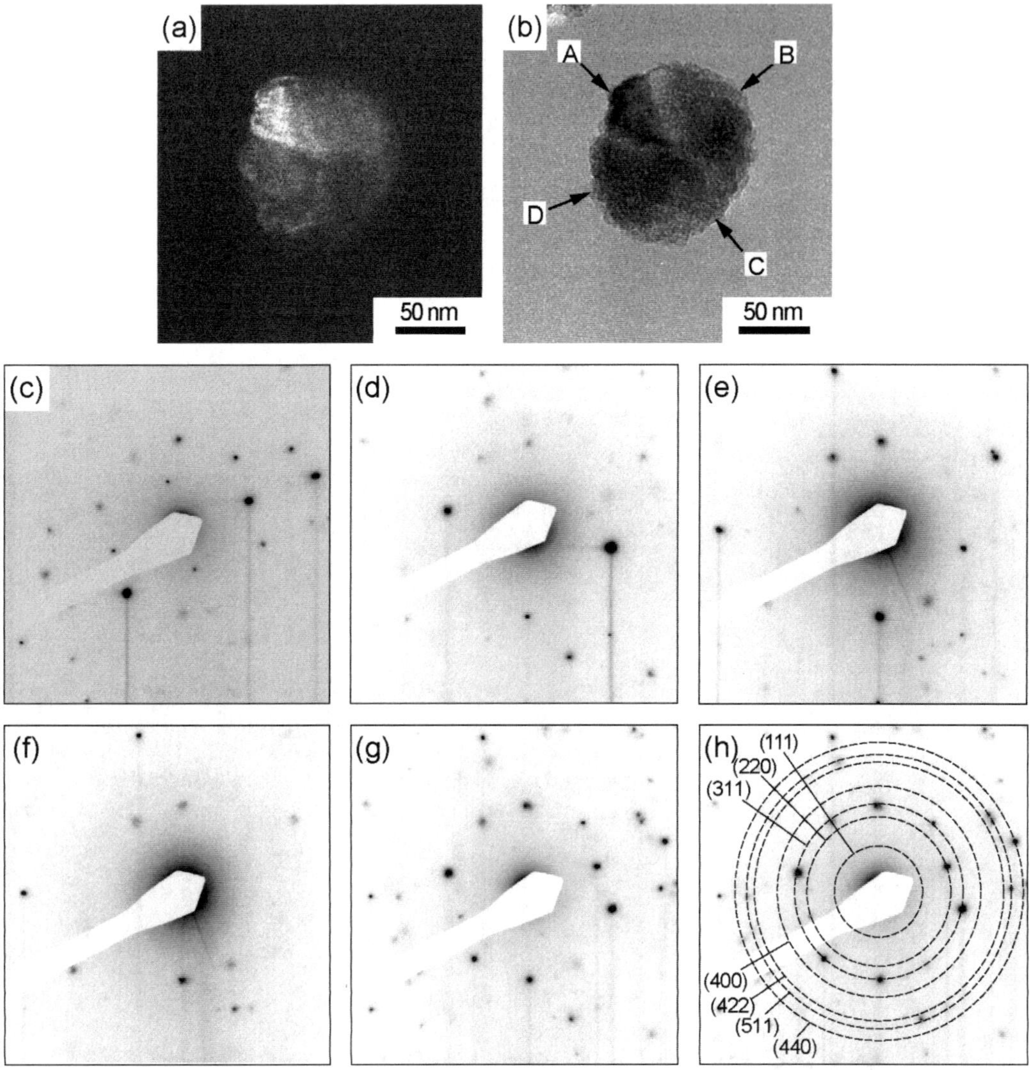

Figure 5. TEM images and electron diffraction patterns of a spherical ferrite nanoparticle composed of several crystals. (a), (b) Dark-field and bright-field TEM images. (c)–(f) SED patterns obtained at areas A–D. (g) Diffraction pattern of the whole particle. (h) Crystal planes responsible for the diffraction spots in figure (g).

As shown in figure 6(a), another spherical ferrite nanoparticle gave a dark-field TEM image with smooth brightness gradation over the whole range, which suggests that the particle is essentially single crystalline. This was confirmed because each of the SAD patterns taken at areas A–C in figure 6(b) was composed mainly of (002) diffraction spots arrayed on a line (figure 6(c)–(e) and (g)). Also each SAD pattern was similar to that taken on the whole area of the crystal (figure 6(f)).

Therefore TEM observation revealed that seed-grown spherical ferrite nanoparticles are composed of one to several single crystals. Since the seed crystals were single crystalline, ferrite growth occurred on clusters composed of one to several seed crystals as we expected. Thus we successfully proved that the clusters each composed of several dozen seed crystals were disintegrated to those composed of a few seed crystals during the ferrite growth.

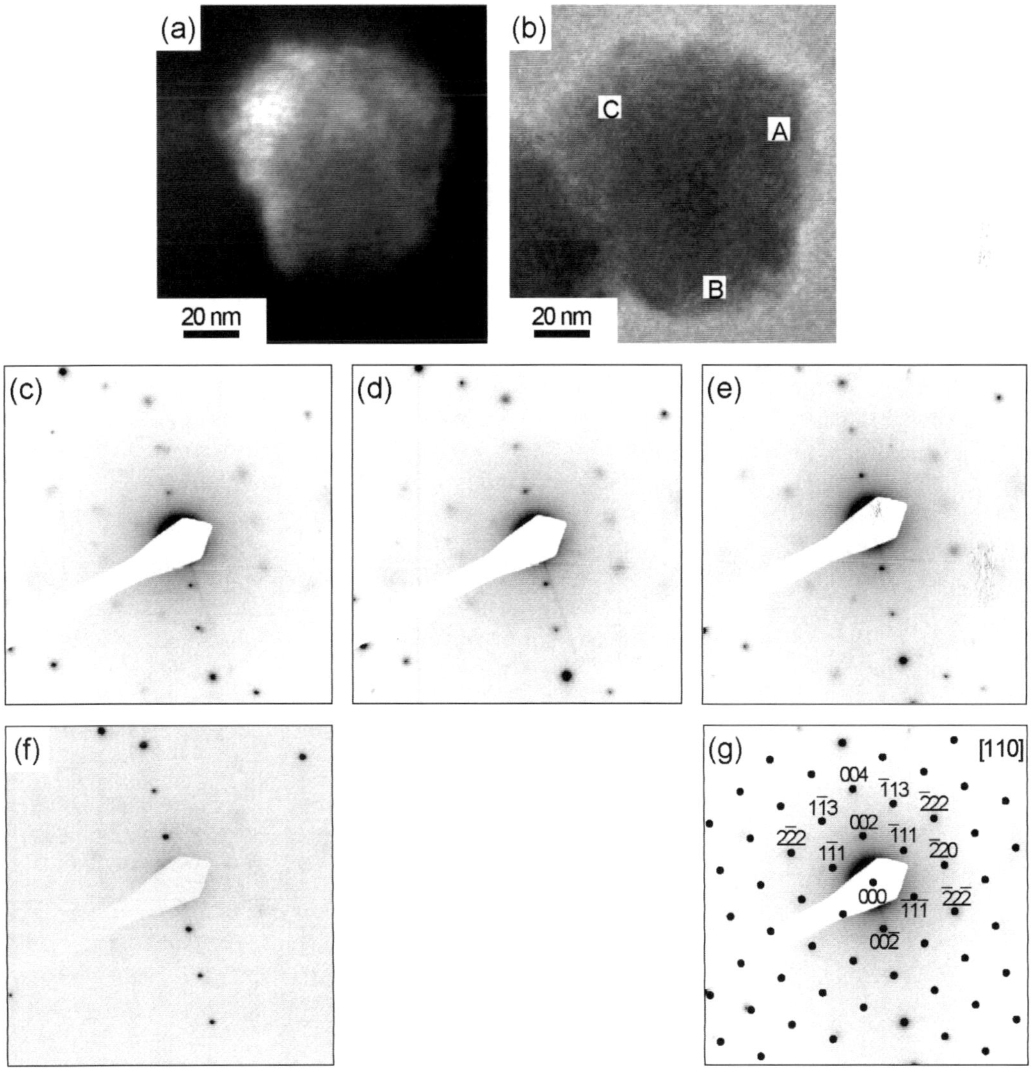

Figure 6. TEM images and electron diffraction patterns of a spherical ferrite nanoparticle which is single crystal. (a), (b) Dark-field and bright-field TEM images. (c)–(e) SED patterns obtained at areas A–C. (f) Diffraction pattern of the whole particle. (g) Crystal planes responsible for the diffraction spots in figure (e).

4. Conclusions

Using seed crystals highly dispersed in their acidic or alkaline suspension independently of the suspension pH, uniform-sized spherical ferrite nanoparticles grew on seed crystals and the particle diameter were controlled in the range 20–200 nm by changing the additive concentration of the seed crystals. TEM observation revealed that seed-grown spherical ferrite nanoparticles were highly crystalline and composed of one to several single crystals. Therefore we provided that the spherical ferrite nanoparticles grew on clusters composed of one to several seed crystals to which those of several dozen seed crystals resolved during the particle growth.

Acknowledgement

The authors are indebted to Mr. A. Genseki and Mr. J. Koki of Center for Advanced Materials Analysis, Tokyo Institute of Technology, for taking TEM micrograms.

References

[1] Shinkai M 2002 *J. Biosci. Bioeng.* **94** 606
[2] Pankhurst Q A, Connolly J, Jones S K and Dobson J 2003 *J. Phys.* D: *Appl. Phys.* **36** R167
[3] Kang Y S, Risbud S, Rabolt J F and Stroeve P 1996 *Chem. Mater.* **8** 2209
[4] Hyeon T, Lee S S, Park J, Chung Y and Na H B 2001 *J. Am. Chem. Soc.* **123** 12798
[5] Sun S and Zeng H 2002 *J. Am. Chem. Soc.* **124** 8204
[6] Sun S, Zeng H, Robinson D B, Raoux S, Rice P M, Wang S X and Li G 2004 *J. Am. Chem. Soc.* **126** 273
[7] Yu W W, Falkner J C, Yavuz C T and Colvin VL 2004 *Chem. Commun.* **2004** 2306
[8] Deng H, Li X, Peng Q, Wang X, Chen J and Li Y 2005 *Angew. Chem. Int. Ed.* **44** 2782
[9] Sugimoto T, Matijević E 1980 *J. Colloid Interface Sci.* **74** 227
[10] Tanaka T, Shimazu R, Nagai H, Tada M, Nakagawa T, Sandhu A, Handa H and Abe M 2009 *J. Magn. Magn. Mater.* **321** 1417
[11] Tanaka T, Nagai H, Tada M, Nakagawa T, Sandhu A, Tamaura Y, Handa H and Abe M 2009 *J. Mater. Res.* **24** 2051
[12] Massart R 1981 *IEEE Trans. Magn.* **17** 1247
[13] Iwasaki I, Katsura T, Ozawa T, Mashima M, Hamamura H and Iwasaki B 1960 *Bull. Volcanol. Soc. Jpn., Ser.* 2 **5** 75
[14] Tanaka T, Tada M, Nakagawa T, Tamaura Y and Abe M 2008 *J. Jpn. Soc. Powder Powder Metallurgy* **55** 831

Circular dichroism in optical second harmonic generated in reflection from chiral G-shaped metamaterials

Evgeniy Mamonov[1], Irina Kolmychek[1], Tatiana Murzina[1], Anton Maydykovsky[1], Oleg Aktsipetrov[1], Ventsislav Valev[2], Thierry Verbiest[2], Alejandro Silhanek[3] and Victor Moshchalkov[3]

[1] Department of Physics, Moscow State University, 119991 Moscow, Russia

[2] Molecular Electronics and Photonics, INPAC, Katholieke Universiteit Leuven, Celestijnenlaan 200 D, B-3001 Leuven, Belgium

[3] Nanoscale Superconductivity and Magnetism, Pulsed Fields Group, INPAC, Katholieke Universiteit Leuven, Celestijnenlaan 200 D, B-3001 Leuven, Belgium

E-mail: mamonov@shg.ru

Abstract. Influence of chirality on the optical second harmonic generated from planar array of *G*-shaped metamaterials is studied. Circular dichroism of these nanostructures manifests itself via different efficiency of left and right circularly polarized second harmonic that is observed for the samples of different handedness. This difference allows to distinguish between the two enantiomers.

1. Introduction

Metamaterials are artificial structures that possess unique properties not existing in natural materials [1], [2]. Optical planar metamaterials can demonstrate such intriguing effects as negative refraction [3] or asymmetric transmission [4]. In case of planar chiral metamaterials optical properties observed previously only for 3D chiral media can be observed, for example, circular dichroism [5] and circular conversion dichroism [6]. The shape of a metamaterial influences much the interaction of the electromagnetic field with the structure by causing specific field distribution in it. The symmetry of a metamaterial can bring about new features of this interaction and determines the existence of such effects as circular conversion dichroism and some other.

Second harmonic generation (SHG) is a perspective probe of 2D chiral metamaterials first of all because of an extremely high sensitivity of the SHG process to the symmetry of the structures [7] . It has been demonstrated that local field distribution strongly influences the parameters of the SHG [8], [9]. It is proven that SHG measurements can distinguish the chirality of the samples: namely, azimuthal dependencies of the SHG intensity have been demonstrated to be different for the two enantiomers composed of *G*- and mirror *G*-structures [10]. A strong field localization in such structures has been detected using confocal SHG measurements that turned out to be the main mechanism of the azimuthal dependence of the SHG intensity[11].

Linear optical effects such as changing the ellipticity of light and polarization plane rotation under the reflection from 2D chiral metamaterials have been studied in [12], while there has been no similar studies of the nonlinear response of such metamaterials. Here we report the observation of circular dichroism in optical second harmonic generation from chiral *G*-shaped nanostructures under their excitation by linearly polarized fundamental beam.

2. Samples and experimental setup

Samples consist of periodic array of G-shaped nanostructures, made of 25 nm thick Au deposited onto SiO_2/Si substrate. Distance between two structures is 200 nm and linewidth of gold is also 200 nm, period of the structure is 1.2 μm. The whole sample consists of 3333x3333 structures. *G*-shaped structure was prepared by electron-beam lithography on an area of 1x1 μm^2 [13]. There were two types of samples: single- and double-periodic.

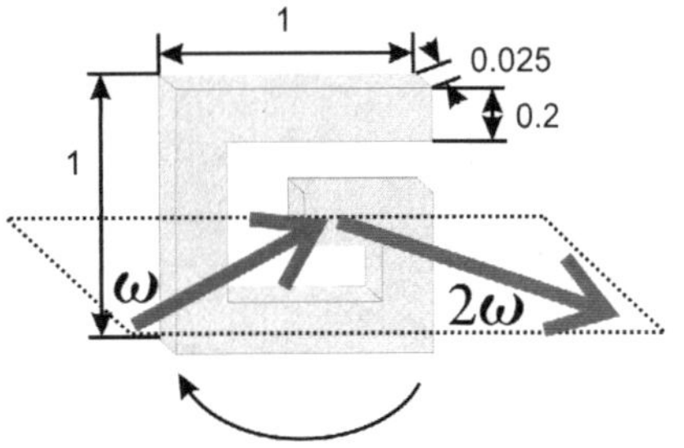

Figure 1. General view of single G-shaped structure and incidence plane location. Round arrow stands for direction of azimuthal rotation of the sample. Sample sizes are given in microns

For the experiments we used the output of a Ti:Sapphire laser at 780 nm with the mean power of 150 mW, pulse width of 70 fs and the repetition rate of 80 MHz. The pump beam was focused onto a spot of 50 μm in diameter. The angle of incidence of the pump beam was 45°. Reflected SHG radiation was spectrally selected by BG39 Schott color filters, passed through a diaphragm with the angular aperture of 5° and was detected by a photomultiplier. SHG polarization was controlled by Babinet-Soleil compensator and a polarizer. Linearly polarized pump beam was used. SHG polarization for studied samples was elliptical. The azimuthal angle $\psi = 0°$ corresponds to the case when the plane of incidence is parallel to one of the sides of the sample (for single-periodic structures the orientation of the plane of incidence is shown on the inset in Fig. 3).

3. Experimental results

3.1. Double-periodic samples

Figure 2 shows the dependencies of the intensity of the left and right circularly polarized SHG, $I_{2\omega}$ reflected from *G*- and mirror-*G*-shaped structures on the azimuthal angle, ψ. The p-polarized fundamental radiation was used. All the graphs reveal the four-fold symmetry. It can be seen that the efficiency of *right* and *left* circularly polarized SHG, e.g. the sign of the third Stokes parameter $S_3 = I_{right} - I_{left}$ [14], is different for mirror structures. Here the subscripts correspond to the different circular polarizations. This result demonstrate that the SHG probe allows for distinguishing of the enantiomers with different handedness.

It is worth noting that there is no such difference in the linear optical response of *G*-shaped structures. Polarization of reflected from studied structures light at fundamental beam is also elliptical. Figure 3 shows the azimuthal angular dependencies of the intensity of the left and right circularly polarized light, I_ω, reflected from *G*- and mirror-*G*-shaped structures for the p-polarized fundamental beam. The maxima of intensity of left circularly polarized light correspond to the minima of right circularly polarized light, while the sign of ellipticity (or direction of electric field vector rotation) of the radiation is the same for two mirror structures. This property may be caused by specific absorption of circularly polarized light by this structures because different ellipticity of the reflected light can be observed in two enantiomers of such symmetry for linearly polarized light [15].

Journal of Physics: Conference Series **352** (2012) 012029 doi:10.1088/1742-6596/352/1/012029

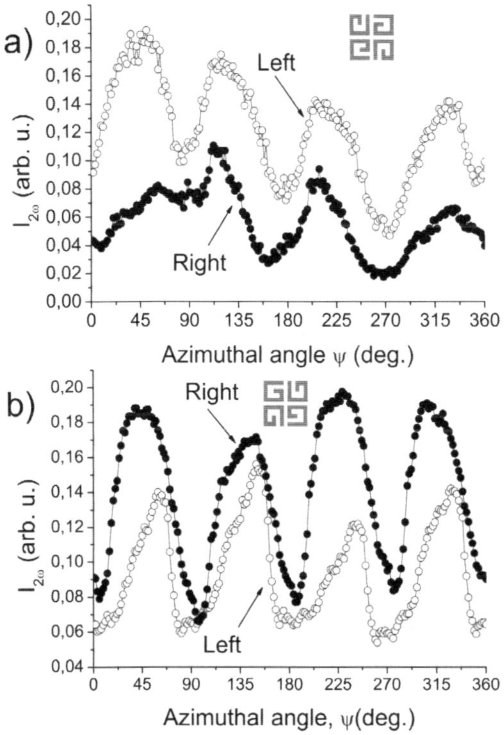

Figure 2. Circularly polarized SHG intensity dependence on azimuthal angle for double-periodic structures, a) is for mirror-G-shaped structures, b) is for G-shaped structure. Fundamental beam was p-polarized

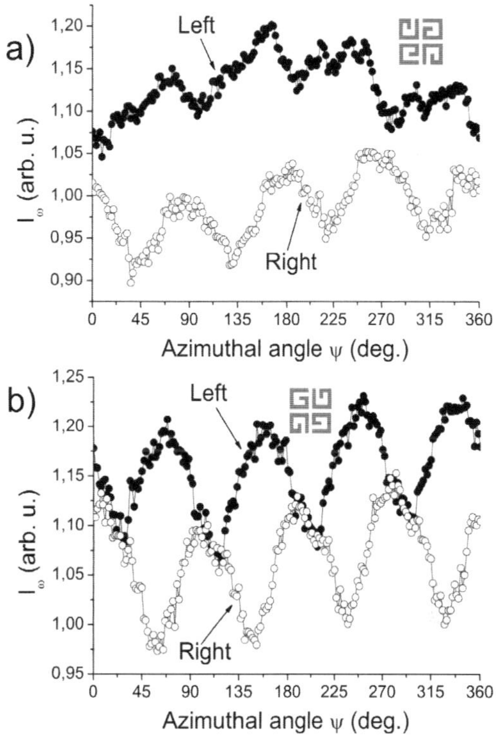

Figure 3. Intensity dependence of circularly polarized light reflected from the structure on azimuthal angle for double-periodic structures, a) is for mirror-G-shaped structures, b) is for G-shaped structure. Fundamental beam was p-polarized

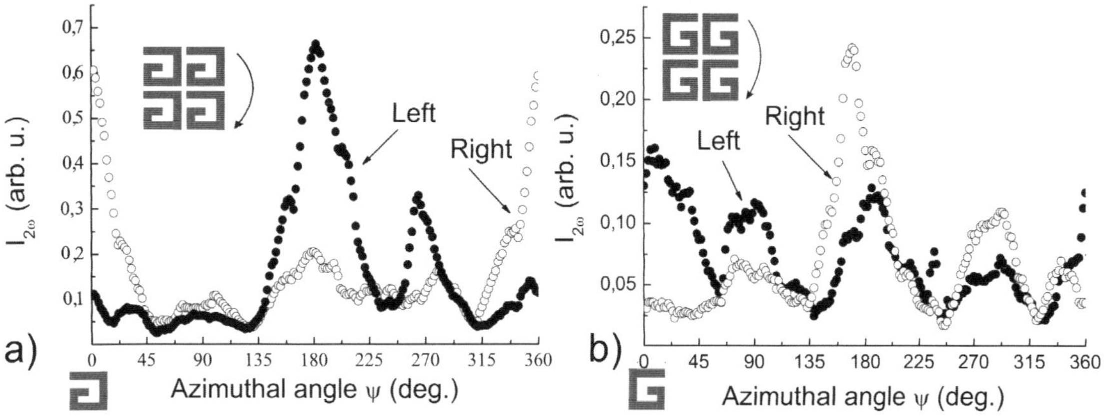

Figure 4. Circularly polarized SHG intensity dependence on azimuthal angle for single-periodic structures, a) is for mirror-G-shaped structures, b) is for G-shaped structure. Fundamental beam was s-polarized

3.2. Single-periodic samples

The anisotropy of circularly polarized SHG from one-periodic metamaterials is shown in Fig. 4. The maximum SHG efficiency is observed as the fundamental electric field is parallel to the one of the sides of the G element, i.e. at the azimuthal angles $\psi = 0°, 90°, 180°, 270°$. An interesting feature of the dependencies shown in Fig. 4 is the different efficiency of the *left* and *right* circularly polarized SHG. The highest maximum of the *right* circularly polarized SHG (open circles) for the case of G-shaped metamaterials (top panel) corresponds to $\psi = 180°$, while the highest maximum of the *left* circularly polarized SHG (filled circles) is attained at $\psi = 0°$. The opposite situation is observed in mirror-G metamaterials (bottom panel): the highest efficiency of *right* circular SHG is observed at $\psi = 0°$ (open circles) and the highest *left* circular SHG at $\psi = 180°$ (filled circles). These differences are attributed evidently to the chirality of the samples when the handedness of metamaterials leads to different anisotropy of circularly polarized nonlinear-optical response.

4. Discussion

This effect can be explained in terms of local field amplification at the fundamental or SHG wavelength. It is well known [9] that the second harmonic intensity can be described as follows: $I_{2\omega}(\psi) \propto (\hat{\chi}^{(2)})^2 < L^2_{2\omega}(\psi) >< L^4_{\omega}(\psi) > I(\omega)$, where $L_\omega, L_{2\omega}$ are the local field factors at the corresponding wavelengths, $I(\omega)$ is the pump intensity and the brackets denote the statistical averaging over the laser spot area. In our resent paper [11] we showed that there is a strong field localization on the fundamental frequency in such G-shaped metamaterials. Thus there are SHG hotspots [13] which can be considered as the main SHG sources that determine the SHG azimuthal dependencies from the whole structure. Mutual arrangement of the SHG hotspots can influence the phase difference between p- and s-polarized SHG and thus give rize to predominantly right or left circularly polarized SHG.

For azimuthal angles $\psi = 0°$ and $\psi = 180°$ of single-periodic G-shaped structures (Fig. 3) the phase difference is maximal. For double-periodic structures the sign of the phase difference remains nearly constant due to the arrangement of all the G-elements, which influences much the field distribution[16] in the structure and consequently the SHG properties.

5. Conclusions

In conclusion, circular dichroism in SHG from G-shaped chiral nanostructures is observed for linearly polarized fundamental beam, while no such effect is detected in the linear reflectivity. SHG hotspots interaction is considered as possible mechanism of the observed nonlinear-optical effect.

6. References

[1] Pendry J 2004 *Contemporary Physics* **45** 191–202
[2] Shalaev V M 2007 *Nat Photon* **1** 41–48
[3] Kim E, Shen Y R, Wu W, Ponizovskaya E, Yu Z, Bratkovsky A M, Wang S Y and Williams R S 2007 **91** 173105 ISSN 00036951
[4] Fedotov V A, Schwanecke A S, Zheludev N I, Khardikov V V and Prosvirnin S L 2007 *Nano Letters* **7** 1996–1999
[5] Decker M, Klein M W, Wegener M and Linden S 2007 *Opt. Lett.* **32** 856–858
[6] Fedotov V A, Mladyonov P L, Prosvirnin S L, Rogacheva A V, Chen Y and Zheludev N I 2006 *Phys. Rev. Lett.* **97**(16) 167401
[7] Klein M W, Wegener M, Feth N and Linden S 2007 *Opt. Express* **15** 5238–5247
[8] Aktsipetrov O A, Murzina T V, Kim E M, Kapra R V, Fedyanin A A, Inoue M, Kravets A F, Kuznetsova S V, Ivanchenko M V and Lifshits V G 2005 *J. Opt. Soc. Am. B* **22** 138–147
[9] Hubert C, Billot L, Adam P M, Bachelot R, Royer P, Grand J, Gindre D, Dorkenoo K D and Fort A 2007 **90** 181105 ISSN 00036951

[10] Valev V K, Silhanek A V, Verellen N, Gillijns W, Van Dorpe P, Aktsipetrov O A, Vandenbosch G A E, Moshchalkov V V and Verbiest T 2010 *Phys. Rev. Lett.* **104**(12) 127401

[11] Mamonov E A, Murzina T V, Kolmychek I A, Maydykovsky A I, Valev V K, Silhanek A V, Ponizovskaya E, Bratkovsky A, Verbiest T, Moshchalkov V V and Aktsipetrov O A 2011 *Opt. Lett.* **36** 3681–3683

[12] Papakostas A, Potts A, Bagnall D M, Prosvirnin S L, Coles H J and Zheludev N I 2003 *Phys. Rev. Lett.* **90**(10) 107404

[13] Valev V K, Zheng X, Biris C, Silhanek A, Volskiy V, Clercq B D, Aktsipetrov O A, Ameloot M, Panoiu N C, Vandenbosch G A E and Moshchalkov V V 2011 *Opt. Mater. Express* **1** 36–45

[14] Nee S M F 1999 *The Measurement, Instrumentation and Sensors Handbook* ed Webster J G (Boca Raton: CRC and IEEE)

[15] Kang M, Chen J, Cui H X, Li Y and Wang H T 2011 *Opt. Express* **19** 8347–8356

[16] Valev V K, Silhanek A V, Smisdom N, Clercq B D, Gillijns W, Aktsipetrov O A, Ameloot M, Moshchalkov V V and Verbiest T 2010 *Opt. Express* **18** 8286–8293

Journal of Physics: Conference Series **352** (2012) 012030 doi:10.1088/1742-6596/352/1/012030

Flow Regulator Powered by the Stalk of *Vorticella convarallia*

N. Matsumoto*, M. Nagai, Y. Hayasaka, M. Oguri, T. Kawashima, and T. Shibata

Department of Mechanical Engineering, Toyohashi University of Technology, Toyohashi, 441-8580, JAPAN

E-mail: n-matsumoto@mems.me.tut.ac.jp

Abstract. We propose flow regulation by the stalk of *Vorticella convarallia*. The motion of stalk can be controlled by introducing Ca^{2+} and chelator solutions alternatively. The fluid channel is narrowed by fully extended *V. convarallia*, which results in increasing fluidic resistance and reducing flow rate. The flow regulator consists of a microchannel and *Vorticella*. The microchannel has two layers. The top layer contains a fluid channel where *Vorticella* is stored to control the flow rate. The bottom layer includes a control channel which enables control of the fluid channel by an external compressed air supply. After cells were introduced to the device at 1.0×10^3 cells/μL, eight of 18 chambers contained more than a single cell. Four cells of *Vorticella* were cultured in one chamber at a maximum. The cells were alive for two days after the injection. We evaluated the change in mean flow rate through the fluidic channel. The flow rate without *V. convarallia* increased linearly from 5.5nL/min to 1.1μL/min with applied pressure from 4 to 18kPa. The switching time of solutions measured two seconds at 10 kPa. The response time for the device was estimated to be approximately four seconds after the partial removal of cell membrane.

1. Introduction

MEMS (Microelectromechanical system) devices have proven to be valuable in several subject areas. MEMS based physical sensors have contributed on improving the machinery performance. However MEMS based devices suffer from operation under water and dependency of an external driving circuit. They normally require a large power supply, which leads to poor portability. Biological actuators and chemical sensitive hydrogel have been developed to remove the power supply, because they obtain their driving energy from a solution. There is wide range of biological actuators available in micro scale such as cardiomyocytes [(1)-(2)] or *Mycoplasma* [(3)]. Those bio actuators consist of energy conversion circuits and sensing elements. Their energy conversion circuits convert chemical energy such as ATP to mechanical work [(4)]. A flow regulator equipped with a hydrogel actuator successfully regulated a flow through a micro fluid channel. [(5)] The hydrogel actuator suffered from small displacement and slow response to start the actuation. The displacement and response time of the biological actuator are superior to the hydrogel. For example, *Vorticella* can contract from 60μm to 20μm in 2s after partial removal of cell membrane. [(6)] Therefore, integration of microorganism to a microdevice has great potential for many applications.

We propose the flow regulator powered by stalk of *V. convallaria* (*Vorticella convallaria*). The device is an example of integration of biological actuator and artificial structure. *V. convallaria* is classified as stalked ciliate protozoan which inhabits at fresh water area (Figure 1). It consists of a filament like motive organ called stalk, a bell shaped zooid and an array of cilia located around the oral region. *V. convallaria* is normally anchored onto a substrate. The contraction of stalk completes in 7–8 ms. This sudden contraction is occurred by rapid release of Ca^{2+} produced within the zooid. The stalk returns to the original length in several seconds. The motion of stalk is controlled by partial removal of cell membrane with surfactant solution. After the treatment, it becomes permeable to external supply of Ca^{2+}. The motion of stalk can be controlled by varying concentration of Ca^{2+}. The flow rate regulation can be done by this particular movement of the stalk within the micro fluidic channel, which causes the change in fluidic resistance in the micro channel. The flow regulation is critical for the micro fluidic device, *e.g.* micro total analysis system, since precise control of injecting reagents is essential to obtain desired chemical reaction. In this paper, we fabricated the multilayer fluidic device for flow regulation and investigated the flow speed and the flow rate through the channel. We evaluated the stalk growth and distribution of *V. convallaria* cells in the fluidic channel. On-chip permeabilize treatment was also carried out.

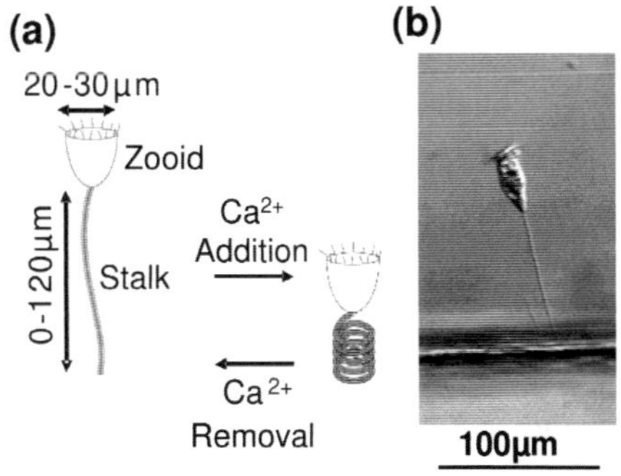

Figure 1. Schematic diagram of the motive organs of *V. convallaria.* a) Contraction and relaxation of stalk by injection and removal Ca^{2+} b) Image of *V. convallaria.*

2. Operational principal

The proposed micro flow regulator utilizes *V. convallaria* as biological actuator which changes its length by varying concentration of Ca^{2+}. Figure 2 shows the operational principal of proposed flow regulator. When *V. convallaria* is introduced to the fluid channel, it swims freely and adheres on the wall of channel spontaneously. Then the cell membrane is partially removed by surfactant solution. The stalk is contracted and extended by introduction of Ca^{2+} and chelator solution alternatively, which results in two forms. The relationship between three variables, flow rate, Q (m^3/s), applied pressure, P (Pa), and fluidic resistance, R (Pa s/m^3), is given by the following expression.

$$Q = \frac{P}{R} \quad 1)$$

Under the assumption that the applied pressure is constant, we can select two different flow rates. (a) A fully extended stalk causes clogging in the channel. The total fluid resistance in a single element is the summation of the fluid resistance of element, R, and the fluid resistance excreted by *V. convallaria,* ΔR. The flow rate is reduced to $P/(R+\Delta R)$. (b) The injection of Ca^{2+} makes the stalk contract. The fluid

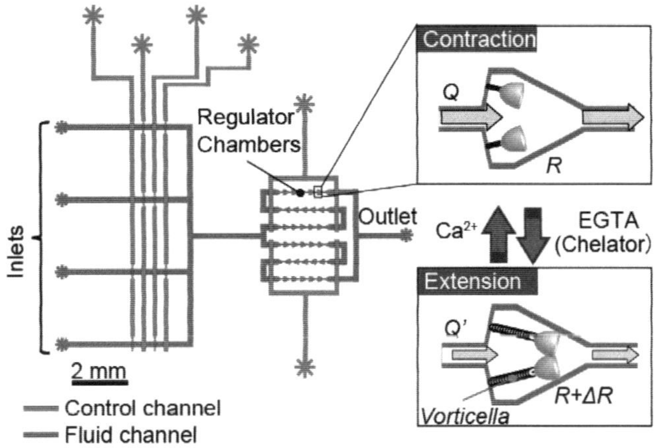

Figure 2. Schematic diagram of operational principle

resistance is reduced to *R*. The flow rate increases to *R/P*. The fluid resistance exerted in each element is enhanced and averaged by arranging 18 bottle neck shaped elements (the volume of single element: $300\times100\times28$ μm = 0.84 nL) in series for the fluid regulation effect to be greater. The flow regulator is equipped with two separate regulator sections. This allows us to carry out two different conditions for flow regulation in a single device by settling different concentration of *V. convallaria* on each section. The device consists of two layers of fluidic channel, fluid layer and control layer. There are three main elements in the device: 1) Regulator section on fluid channel to contain cells and control flow rate. 2) Pneumatic valves on control channels to select solutions. 3) Fluid channel with four inlets and one outlet. All reagents including Ca^{2+} and chelator solutions are delivered to the fluid channel with pneumatic valves by compressed air. The target solution of the current device is limited to water in order not to damage *V. convallaria*. A Wide variety of fluids can be used by making a partition and separating working fluid and analytic fluid.

3. Experimental method

3.1. Fabrication of multi-layer fluidic channel

We fabricated the multilayer fluidic channel to achieve fluid flow regulation. The device was fabricated by multilayer soft lithography technique [(7)] Polydimethylsiloxane (PDMS, Sylgard 184, Dow Corning) is used to cast the two channel layers. The upper and lower layers are fluid and control channels, respectively. The push-up valve structure is employed in our device[(8)]. The fluid channel is pushed upwards by the control channel via the valve membrane using compressed air.

Figure 3 shows a schematic diagram of the fabrication process. The fabrication is broken down in three main processes, (a) mould fabrication for fluid channel layer, (b) mould fabrication for control channel layer and (c) casting and bonding of channel layer. (a) We spin-coated a thick positive photoresist (AZ P4903) on silicon wafer with the thickness of 30 μm. The photoresist was patterned by standard photolithography. Upon development, patterned photoresist was reflowed by heating over the glass transition temperature, 110 °C for 15min to obtain semi elliptic cross sectional shape. Epoxy resin mould (Devcon ET 300, Illinois Tool Works Inc) was prepared by replica molding technique to enhance the durability of the mould. (b) The mould of control channel was made of thick negative photoresist, SU-8 3050. The photoresist was spun on silicon wafer to obtain the thickness of 30 μm. The mould was patterned by photolithography. (c) Two PDMS layers of control and fluid channels were fabricated individually. The base polymer and catalyst of PDMS, we use in this paper, were all mixed with the ratio on 10:1. Both moulds were treated with the vapor of Trichloro(1H,1H,2H,2H-perfluorooctyl)silane (Sigma-Aldrich) at 60 °C for 4min to improve the release of PDMS. Uncured PDMS was poured onto the epoxy resin mould and obtained about 1 cm thickness of fluid channel layer. For the control channel layer, the PDMS was spun at 1800 rpm for 60s. Both layers were cured at 80 °C for 1h. The fluid channel layer was peeled off from the mould. The access holes were punched using a biopsy punch (0.5mm diameter, Harris, Uni-Core, Ted Pella Inc, Redding, CA) and trimmed to shape. This layer was cleaned with isopropyl alcohol and dried with nitrogen gas. Both control and fluid layers were bonded after the corona discharge treatment with the corona treater, BD-20AC (Electro-Technic Products, Inc.) for 20 s. We aligned both layers under the microscope and baked the assembled device at 80 °C for 40 min. The assembled layer was peeled off from the mould and the access holes for the control channel were punched. The final device was made by bonding cover slip and the assembled layer by the same bonding technique.

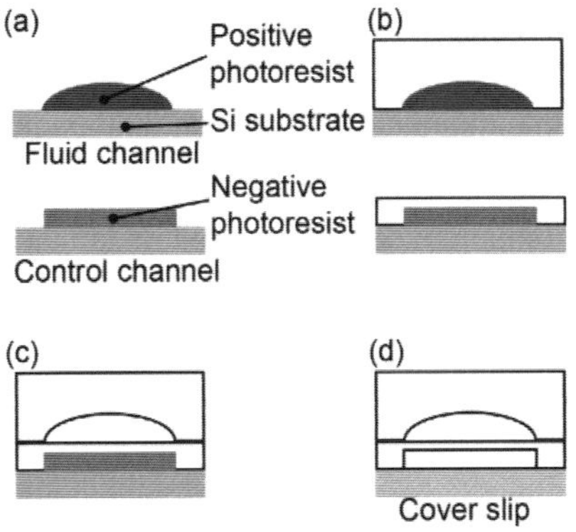

Figure 3. Device fabrication process

(a) Prepare the moulds for fluid and control channels (b) Block and film of PDMS was casted on the moulds and cured (c) Two layers of PDMS were aligned and bonded after corona discharge treatment (d)Assembled chip and cover slip were bonded with same manner as bonding of PDMS layers

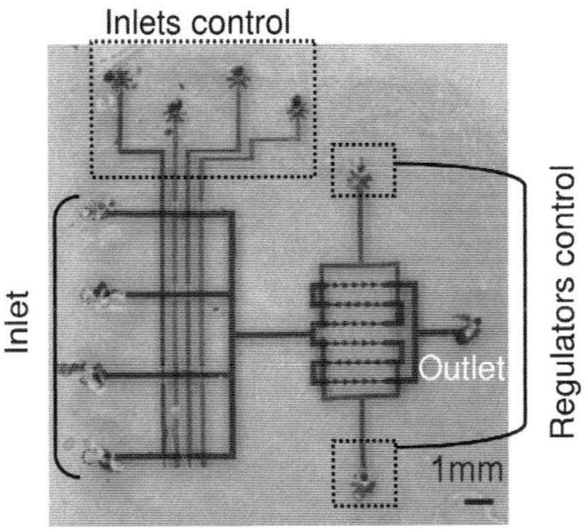

Figure 4. Picture of flow regulator

3.2. Device calibration

The manipulation of reagents (solution input and replacement) was calibrated using blue and red food dye solutions for flow visualization. The concentration of the dye solutions was adjusted to 10 mg/mL with deionized water. We monitored a flow though the fluid channel with the inverted microscope, Eclipse TE2000-U (Nikon Instruments Inc.) with objective lens of 10× and 20×. The flow rate was determined by measuring the mean flow velocity thorough a vinyl chloride tube with inner diameter of 3.2×10^2 μm connected to the exit. The mean velocity was obtained by travelling time of dye solution in unit length. The cross sectional area of the tube is $8.6\times10^4 \mu m^3$. The mean flow rate was obtained by the multiplication of the cross sectional area of the tube and the mean velocity.

3.3. Preparation and injection of V. convallaria

We prepared floating cells for injection. *V. convallaria* was cultured in the 1L flask containing the extraction of wheatgrass powder (Pines International Inc.) in deionised water. Then *V. convallaria* was detached from the flask surface by shaking the flask at 100 rpm for 24 h. The detached *V. convallaria* was transferred to a large Petri dish and incubated at 20 °C for 6 h to allow adhering to the bottom of dish surface. The adhered cells were rinsed with spring water twice and were scraped with a cell scraper. They were centrifuged at 3,000 g for 10 min to obtain concentrated suspension. The concentration of suspension was adjusted to be approximately 1,000 cells/μL. The suspension of *V. convallaria* cells was introduced to the fluid channel using a syringe. The cells were incubated at 20 °C for 48 h to allow adhering and stalk growth.

3.4. The treatment of V. convallaria in microchannel

The introduced cells were peameablized with a surfactant solution, 0.1%(w/w) saponin in a chelator solution (0.1M KCl, 4mM EGTA [Ethylene Glycol bis(2-aminoethylether) Tetraacetic Acid], 50mM Tris [Tris(hydroxymethyl)aminomethane]-HCl buffer, pH 7.0) for 5min. The surfactant solution was washed away with the chelator solution. The surfactant and chelator solutions were stained for visualization with red and blue food dyes, respectively. We observed the cell at 10 fps using the inverted microscope equipped with 10× and 20 × lenses and a CCD camera.

4. Result and discussion

4.1. Flow velocity and flow rate evaluation

We measured flow velocity and rate through the fluid channel of the fabricated device (Figure 4). The mean flow speed was obtained by measuring the traveling time and channel lengths from A to C and B to C. The lengths from the inlet 3 point A, the first chamber of flow regulator, point C is 7.7 mm (see the inset of figure 5). The length from the inlet 4, point B to point C is 9.8 mm. Figure 5 shows the result of fluid velocity measurement. The mean velocity increased steadily from 0.1 to 5.8 $\times 10^3$ μm/s with the applied pressure from 2 to 18 kPa. We measured the flow speed five times. Error distribution in figure 5 represents standard deviation. The slope of the flow speed was 3.2×10^2 μm/kPa·s. This agrees well with the trend in Poiseuille flow. From the velocity measurement, we also evaluated the possible device response time. As the travelling time of solution to the regulator was found to be about 2 s and the time taken for the contraction of peameablized stalk, the device response time could be roughly 4 s.

Figure 6 shows the result of flow rate measurement at the exit of the fluid channel as the function of applied pressure ranging from 2 to18kPa. We measured the flow rate five times. Error distribution in figure 6 represents standard deviation. At the inlet pressure of 6kPa and above, the flow rate increased linearly from 5.5nL/min to 1.1μL/min with the applied pressure from 4 to18kPa. The slope of flow rate was 0.5×10^2 nL/kPa·min. However, the for inlet pressure below 6kPa, the result did not follow the trend. This shows the existence of threshold pressure for flow at the exit of fluid channel. This phenomenon is currently under investigation.

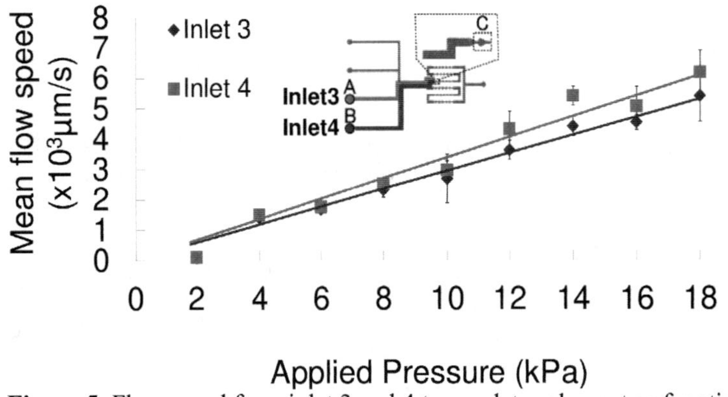

Figure 5: Flow speed from inlet 3 and 4 to regulator element as function of applied inlet pressure

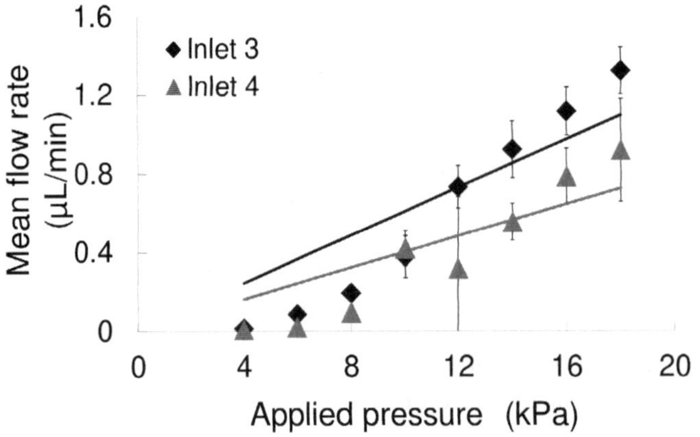

Figure 6. Flow rate at fluid channel exit as function of applied inlet pressure

4.2. The distribution of adhered V. convallaria and stalk growth within regulator section

We confirmed that 12 cells adhered to the surface two hours after cell introduction. Figure 7 shows the distribution of *V. convallaria* cells. Four cells adhered in a single chamber at maximum. The cells grew their stalks to 39±11 μm (n=12) at 48 hours of incubation (Figure 8). There was no detachment of the cells after seven days of incubation. However, the result of cell introduction was below the target values, 2–3 cells on each regulator chamber and the stalk length of 100 μm. Poor distribution of cell adhesion was caused by the property that the cells swim freely and adhere to their desired location. The cells adhered more to the inlet port than the chamber. The diameter of the inlet is approximately 0.5 mm, which is larger than the chamber. The number of cells adhered to the regulator chamber can be improved by confining them with additional control valves. The lack of nutrition in the suspension of mineral water may result in short length of stalk. The stalk growth can be improved by addition of the culture medium in the suspension.

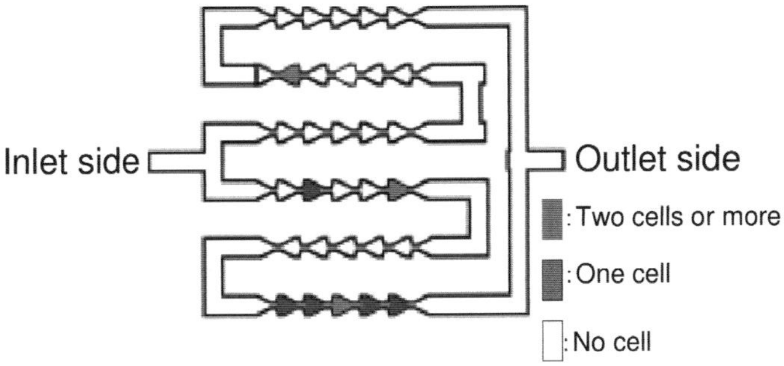

Figure 7. Distribution of cells within a regulator chamber 48 hour after the injection

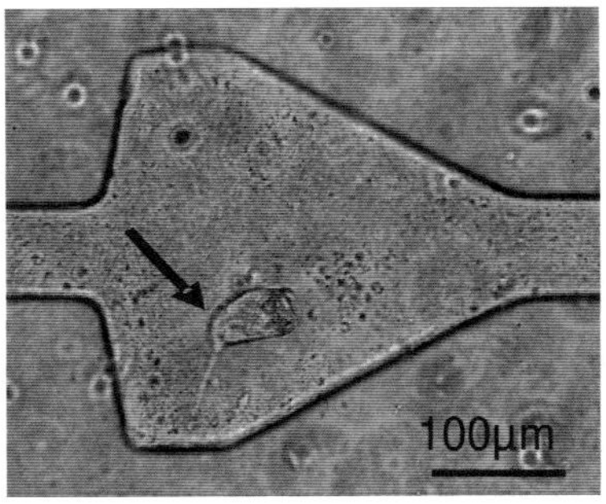

Figure 8. Image of *V. convallaria* after 48 hours of incubation

4.3. Partial cell membrane removal of adhered cell at regulator chamber

We observed the reduction of mechanical strength of the treated cell. This suggests the partial removal of cell membrane. Figure 9 shows the time progress images of chelator introduction to the regulator chamber. The zooid of the adhered cell was detached by the introduction of the chelator solution. As shown in Figure 9(a), the cell remained to regulator chamber at the introduction of the chelator. One second after the introduction, the zooid was detached and only the stalk remained attached (figure 9(b)). The treated cell could not withstand the fluid velocity at regulator chamber. When applied at 10 kPa, we estimate the flow velocity at around 3.3×10^3 μm/s. The inlet pressure should be reduced to prevent the zooid detachment after the permeabilize treatment. The inlet pressure of 6–8 kPa should be reasonable.

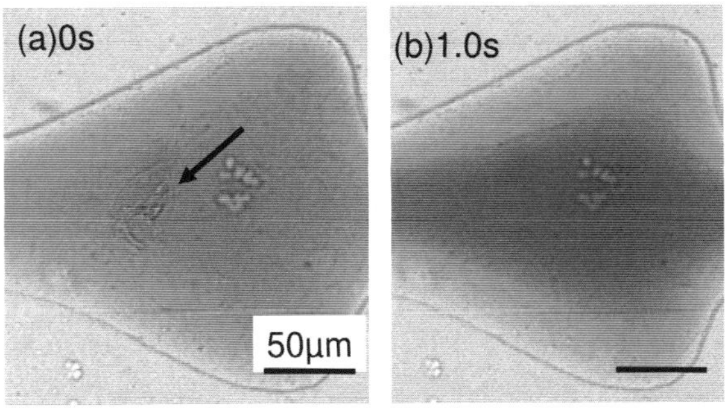

Figure 9. Time series images of the chelator solution injection at the inlet pressure of10 kPa. (a) At the instant of chelator solution introduction. (b) Detachment of the zooid was observed at 1s after the injection of solution.

5. Conclusion

We fabricated a multilayer fluidic channel, which is a platform of the proposed micro flow regulator. Introduction, stalk growth, adhesion and permeabilize treatment of *V. convallaria* were achieved on a chip. The stalk grew to 39±11 μm (n=12) after 48 hours of incubation. There were 12 cells in the entire regulator elements with the maximum of four cells in the single element. From the on-chip permeabilize treatment, the partial removal of cell membrane was successfully achieved by observing the detachment of zooid during the chelator solution injection. As for the future work, we will conduct the measurement of change in stalk length under varying concentration of Ca^{2+} and flow rate variation during stalk actuation. The stalk length will be improved by replacing mineral water to culture fluid. The number of adhered cell is planned to increase by actuating the control channel for regulator section in order to limit the movement of cells. The mechanism of flow regulation will also be investigated by visualizing the fluid flow with fluorescent particles.

Acknoledgement
This work was supported by Grant-in-Aid for Research Activity Start-up 22810012.

References

[1] Feinberg A, Feigel A, Shevkoplyas S, Sheehy S, Whitesides G and K Parker2007 *Science.* **317** pp 1366–1370
[2] Kim J, Park J, Yang S, Baek J, Kim B, Lee S, Yoon E, Chun K and Park S, 2007 *Lab Chip.* **7** pp 1504–1508
[3] Hiratsuka Y, Miyata M, Tada T and Uyeda T 2006 *Proc. Natl. Acad. Sci. U. S. A.* **103** pp 13618–13623.
[4] Fletcher D and Theriot J, 2004 *Phys Biol* **1** T1–T10.
[5] Baldi A, Gu Y, Loftness P, Siegel R and Ziaie B 2003 *J. Microelectromech. Syst.* **12** pp 613–621.
[6] Nagai M, Ryu S, Thorsen T, Matsudaira P and Fujita H 2010 *Lab Chip.* **10** pp 1574–1578
[7] Unger M , Chou H, Thorsen T, Scherer A, Quake S 2000 *Science.* **288** pp113–116
[8] Studer V, Hang G, Pandolfi A, Quake S 2004 *J. Appl. Phys.* **95** pp 393–398

Journal of Physics: Conference Series **352** (2012) 012031 doi:10.1088/1742-6596/352/1/012031

Biofouling of Cr-Nickel Spray Coated Films on Steel Surfaces

Kento Yoshida[1], Hideyuki Kanematsu[1], Daisuke Kuroda[1], Hajime Ikigai[1], Takeshi Kogo[1] and Seiji Yokoyama[2]

[1]Shiroko-cho, Suzuka, Mie, 510-0294, Japan. [2] Hibarigaoka 1-1, Tenpaku-cho, Toyohashi, Aichi, 441-8580, Japan

E-mail: kanemats@mse.suzuka-ct.ac.jp

Abstract. Nowadays, corrosion of metals brings us serious economic loss and it often reaches several percentage of GNP. Particularly the marine corrosion was serious and the counter measure was very hard to be established, since the number of factors is huge and complicated. One of the complicated factors in marine corrosion is biofouling. Biofouling was classified into two main categories, microfouling and macrofouling. The former is composed of biofilm formation mainly. Marine bacteria are attached to material surfaces, seeking for nutrition in oligotrophic environment and they excrete polysaccharide to form biofilm on metal surfaces. Then larger living matters are attached on the biofilms to develop biofouling on metal surfaces, which often lead loss and failures of metals in marine environments. From the viewpoint of corrosion protection and maintenance of marine structures, biofouling should be mitigated as much as possible. In this study, we applied spray coating to steels and investigated if chromium-nickel spray coating could mitigate the biofouling, being compared with the conventional aluminium-zinc spray coating in marine environments. The specimens used for this investigation are aluminium, zinc, aluminium-zinc, stacked chromium/nickel and those films were formed on carbon steel (JIS SS400). And the pores formed by spray coating were sealed by a commercial reagent for some specimens. All of those specimens were immersed into sea water located at Marina Kawage (854-3, Chisato, Tsu, Mie Prefecture) in Ise Bay for two weeks. The depth of the specimen was two meter from sea water surface and the distance was always kept constant, since they were suspended from the floating pier. The temperature in sea water changed from 10 to 15 degrees Celsius during the immersion test. The biofouling behavior was investigated by low vacuum SEM (Hitachi Miniscope TM1000) and X-ray fluorescent analysis. When the spray coated specimens with and without sealing agents were compared, the former showed higher antifouling properties generally. Aluminium-zinc alloy spray coated films had higher antifouling property. And the anti-property decreased in this order: Al-Zn alloy spray coating > Zinc spray coating > Aluminium spray coating > Stacked chromium/nickel spray coating. Aluminium and zinc spray coating has been evaluated high conventionally for anti-biofouling in marine environment. However, the Cr/Ni spray coating showed pretty high anti-fouling property.

1. Introduction

Corrosion of metallic materials brings us a lot of economic failures in our modern society. One of the estimation about them suggests that the failure reached several percentages of GDP in advanced countries usually[1]. Particularly, the corrosion in marine environment is the most remarkable to affect on economical failures, since it is very usual one. When biofilms form on material surfaces, oxygen concentration cells are established around biofilms, which leads to serious corrosion failures. With and without biofilm formation, the corrosion rate is generally very different. However, the precise evaluation and differentiation between usual corrosion and microbially influenced corrosion (MIC)[2]-[6] has been very hard, since the involvement of microorganism would be always very delicate and difficult to be fixed.

In this study, the effect of spray coating on the biofilm formation and MIC were investigated. Concretely speaking, various spray coated steel were immersed in marine environment and the biofouling behaviors were measured, evaluated and discussed.

2. Experimental

2.1. Specimens

Table 1 Spray coating conditions for various specimens.

coating materials	aluminum	zinc	aluminum-zinc	chromium	nickel
spray coating process	arc spray coating	arc spray coating	arc spray coating	plasma spray coating	plasma spray coating
voltage(V)	30	20	20	20	62
current(A)	125	50	50	50	500
gas pressure	5-6 kg/cm^2	5-6 kg/cm^2	5-6 kg/cm^2	Ar 75Psi, H_2 50Psi	Ar 75Psi, H_2 50Psi
spray coating distance	150mm	150mm	150mm	120mm	100mm

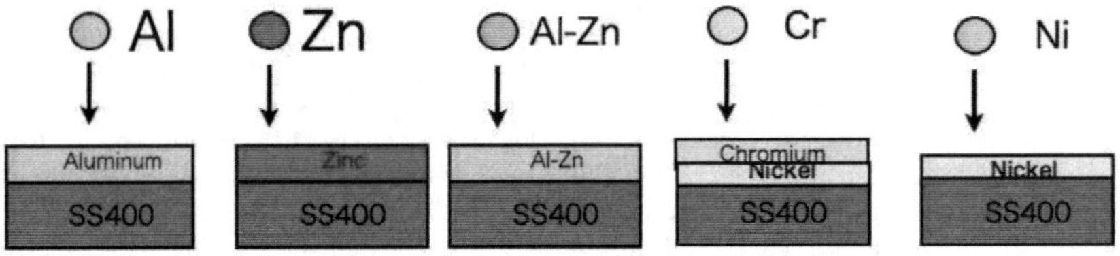

A general carbon steel (JIS SS400) was used as substrate. Aluminium, zinc, aluminium-zinc, chromium/nickel (Nickel layer was spray coated and then chromium layer was formed on it.) were spray coated on the substrate, respectively. The conditions for those spray coating processes were shown in the following table (Table 1). The figures below the table were schematic illustration of the spray coating process. The readers should note that chromium/nickel spray coating process was composed of two steps (nickel and chromium spray coating processes.)

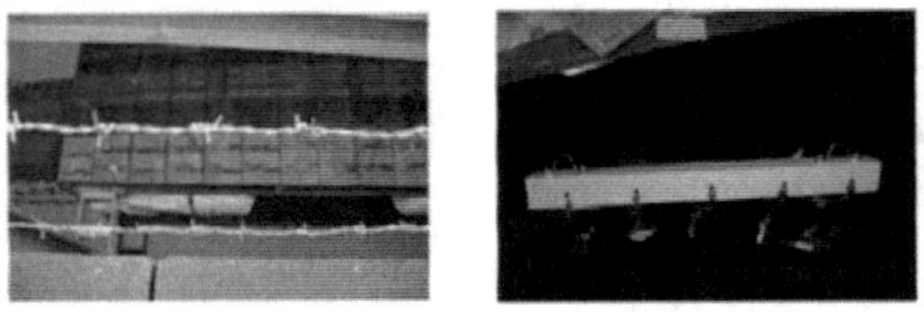

(1) floating dock (2) A Jig for immersion

Fig.1 Immersion tests in Ise Bay

After the spray coating process, the pores of spray-coated films were sealed by a commercial sealant, Permeate (D & D Co., 78 Sakura, Yokkaichi, Mie, 512-1211, Japan)[6],[7]. And those sealed specimens were compared to those without sealer.

2.2. Marine immersion tests

All of spray coated specimens were fixed to a jig for the immersion test and the jig with those specimens were immersed in sea water at Marina Kawage (854-3, Chisato, Tsu, Mie Prefecture, Japan) in Ise Bay. The jig with the specimens was hung from a floating dock and kept at the depth of 2m from the sea surface. The temperature of sea water was changed between 10 degrees Celsius and 15 degrees Celsius. The appearance of the jig and immersion site were shown in Fig.1.

2.3. Surface Observation

After immersion, specimens were taken out of the sea, washed by water and served to observations by low pressure Scanning Electron Microscopy (TM 1000, Miniscope, Hitachi). The acceleration voltage of 15kV was applied to the specimens and the reflection image was observed. In addition to that, fluorescent X-ray analysis (MESA 500W, Horiba Co.) was carried out to identify elements on specimens' surfaces before and after immersion.

3. Results and discussion

3.1. Low Pressure SEM Observation

The specimens were immersed in the sea for two weeks and observed by the low pressure SEM after the immersion. Some of the SEM photos were shown in Fig. 2 and Fig.3.

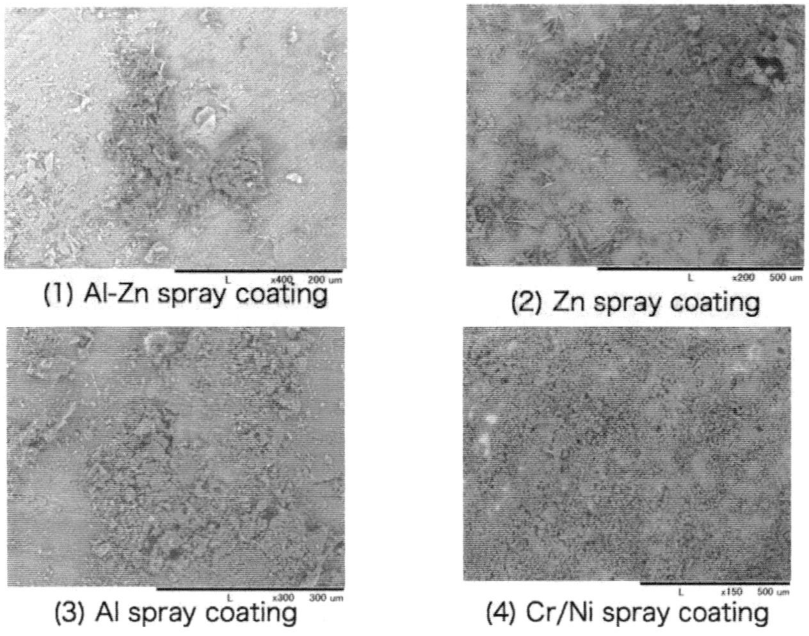

Fig.2 Low Pressure SEM Observation for spray coated specimens without sealer

In Fig.2, the results for the specimens with sealer were shown, while those without sealer in Fig.3. When biofouling occurred on the surface of specimen to some extent, the surface color usually changes and the fouling matters could be observed by naked eyes in most cases. As for both of sealed and non-sealed specimens, Al-Zn specimens were the hardest to have biofouling. The fouling was observed on the specimen's surface to some extent. However, there were few fouling places by naked eyes. Therefore, it could be judged to be the hardest for biofouling. The zinc coated specimen and aluminum coated specimens followed in this order. The Cr/Ni coated specimen was the easiest one for biofilms to form on the surfaces. For the Cr/Ni coated specimens, and red rusts were often observed. Since they were obviously iron oxides (mainly Fe_2O_3) formed on the specimens' surfaces

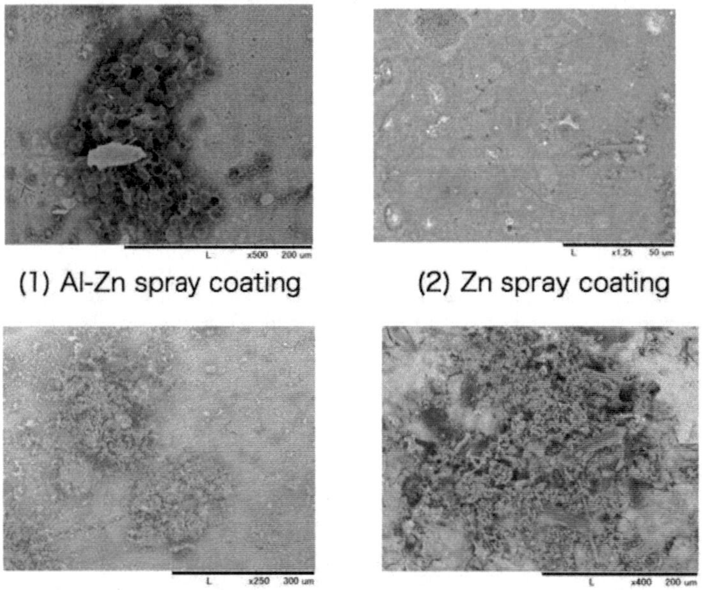

(1) Al-Zn spray coating (2) Zn spray coating

(3) Al spray coating (4) Cr/Ni spray coating

Fig.3 Low Pressure SEM Observation for spray coated specimens with sealer

Table 2 Fluorescent X-ray analyses for spray coated specimens without sealer

element	mass concentration(%)	
	before immersion	after immersic
Zn	59.86	85.03
Al	37.91	11.95
Si	0	1.46
Cl	0	1.19
Mg	1.92	

(1) Al-Zn

element	mass concentration(%)	
	before immersior	after immersior
Zn	99.38	95.14
Si	0	2.78
Cl	0	1.32

(2)Zn

element	mass concentration(%)	
	before immersion	after immersion
Al	98.35	77.85
Cl	0	13.52
Si	0	6.61
Fe	0	1.68
Mg	1.36	—

(3) Al

element	mass concentration(%)	
	before immersioı	after immersion
Cr	99.53	93.73
Ni	0	1.32
Fe	0	2.16
Si	0	2.06

(4) Cr/Ni

by the reaction between iron and oxygen or hydroxyl ion. For all of specimens, sealed coatings showed higher anti-corrosion resistance than non-sealed ones did. It indicates clearly that the sealer had a certain inhibition effect of biofilm formation and improved corrosion resistance to much extent.

3.2. Fluorescent X-ray analyses

Table 3 Fluorescent X-ray analyses for spray coated specimens with sealer

element	mass concentration(%)	
	before immersio	after immersion
Zn	55.34	90.35
Al	43.12	7.06
Si	0	1.51

(1) Al-Zn

element	mass concentration(%)	
	before immersion	after immersion
Zn	97.99	98.51
Al	0	1.07
Si	1.11	—

(2)Zn

element	mass concentration(%)	
	before immersio	after immersion
Al	99.81	93.27
Si	0	5.8

(3) Al

element	mass concentration(%)	
	before immersion	after immersion
Cr	96	94.87
Ni	1.45	1.12
Cl	0	1.26
Fe	0	1.13
Si	1.03	1.03

(4) Cr/Ni

All of the specimens were analyzed by a fluorescent X-ray apparatus. Table 2 and 3 show those results. The former corresponds to the results for the spray coated specimens without sealer, while the latter for those with sealer. According to the results, silicon was remarkable for al of the specimens. The change of the silicon contents was caused mainly by the incorporation of silicon from the sea water. Silicon compounds were incorporated into biofilm, dried in it after immersion and reflected in the results[8],[9]. Therefore, it could be an evaluation tool for the extent of biofilm formation. From the viewpoint, all specimens immersed into the sea showed the biofilm formation to some extent. On the other hand, chloride concentrations were pretty high for some specimens. It suggests that those specimens corroded remarkably. In this study, the strict and correct gravimetry was not carried out. However, the tendency was measured. According to the results, the weights for aluminum coated, zinc coated and chromium coated specimens decreased remarkably. Those metals are less noble than iron. Therefore, they dissolved significantly due to the galvanic cells between the metals and iron and the total weights decreased correspondingly. For Cr/Ni coated specimens, the red rust corresponding to iron oxide were observed on the specimens' surfaces. It suggests that the

Table 4 Final evaluation for all specimens.

Surface	evaluation	
better	1	Sealed Al-Zn
↑	2	non-sealed Al·
	3	sealed Zn
	4	non-sealed Zr
	5	sealed Al
	6	non-sealed Al
	7	sealed Cr/Ni
worse	8	non-sealed Cr

local galvanic cells established between the surface films and substrate also induced the corrosion and produced iron oxides on the surfaces of these layers.

From the viewpoint of gravimetry, silicon concentrations and observation by naked eyes, we concluded the fouling order among specimens for this experiment. The results were summarized in Table 4.

4. Conclusions

Various spray coated specimens with and without sealer (aluminum, zinc, aluminum-zinc and chromium/nickel coated specimens) were immersed into sea water and the biofouling behaviors were measured. The low pressure SEM observation and fluorescent X-ray analysis were used for the evaluation of biofilm formation. All of these results were investigated, discussed and summarized in Table 4. The order of inhibition effects against biofilm formation increased in the order shown in Table 4. The highest inhibition effect was shown by aluminum zinc spray coated film with sealer, while the worst one by chromium nickel spray coated specimen without sealer.

Acknowledgement

Kurashiki Borling Kiko Co. Ltd.(http://www.kbknet.co.jp/english/) and D&D Corporation (http://www.ddcorp.co.jp/company/index.html) gave us specimens and useful suggestions for this study. We authors appreciate them very highly. And this work was also supported financially by KAKENHI (#21360372)

References

[1] Flemming H.-C. 2009 *Marine and Industrial Biofouling* (Berlin Heidelberg, Germany: Springer Verlag)

[2] Hirai, N., Tanaka, T., Kanematsu, H., & Yamamoto, C. (2010). Fundamental study on the effect of EPS in biofilms on microbial influential corrosion. In C. and C. P. of M. in the O. and G. Industry (Ed.), (pp. 461-469). Houston, Texas, USA: 2010 MS&T'10.

[3] Kanematsu, H., Ikigai, H., Kikuchi, Y., & Oki, T. (2005). Electrochemical Stability of Hot Dip Galvanized Steel in an Acid Environment Cotaining Thiobacillus Ferrooxidans. *Transactions of the Institute of Metal Finishing*, *83*(4), 205-209.

[4] Kanematsu, H., Ikigai, H., & Yoshitake, M. (2009). Evaluation of Various Metallic Coatings on Steel to Mitigate Biofilm Formation. *International Journal of Molecular Science*, *10*(2), 559-571. doi:10.3390/ijms10020559

[5] Kanematsu, H., Ikegai, H., & Kuroda, D. (2011). Biofilm And Metallic Materials. *Rust Prevention And Control Japan*, *55*(10), 369-377.

[6] Kanematsu, H., Barry, D. M., McGrath, P., & Ohmori, A. 2004. *ITSC 2004*. Osaka, Japan: German Welding Society, ASM International Thermal Spray Society, Internationa Institute of Welding.

[7] Kanematsu, H., Murakami, K., & Nakata, K. (2009). *Sur/Fin 2009*. Louisville, Kentucky, USA: National Association of Surface Finishing (NASF)

[8] Satoshi Fukuzaki, Hideyuki Kanematsu, Hideo Itoh: Theory and Practice for Chemical Cleaning, Yoneda Shuppan Co. 2011

[9] Daisuke Kuroda, Hideyuki Kanematsu et al: Tetsu-to-Hagane (Journal of Iron and Steel Institute of Japan), in print.

Journal of Physics: Conference Series **352** (2012) 012032 doi:10.1088/1742-6596/352/1/012032

Electrochemical properties of arc-black and carbon nano-balloon as electrochemical capacitor electrodes

T Sato[1], Y Suda[1], H Uruno[1], H Takikawa[1], H Tanoue[1], H Ue[2], N Aoyagi[3], T Okawa[3] and K Shimizu[4]

[1] Department of Electrical and Electronic Information Engineering, Toyohashi University of Technology, 1-1 Hibarigaoka, Tempaku, Toyohashi, Aichi Japan
[2] Tokai Carbon Co., Ltd. 394-1 Subashiri, Oyama, Sunto, Shizuoka, Japan
[3] Daiken Chemical Co., Ltd. 2-7-19 Hanaten-Nishi, Joto, Osaka, Japan
[4] Shonan Plastic Mfg. Co., Ltd. (31-27 Daikan, Hiratsuka, Kanagawa Japan

E-mail: suda@ee.tut.ac.jp

Abstract. In this study, we used two types of carbon nanomaterials, arc-black (AcB) which has an amorphous structure and carbon nano-balloon (CNB) which has a graphitic structure as electrochemical capacitor electrodes. We made a coin electrode from these carbon materials and fabricated an electric double-layer capacitor (EDLC) that sandwiches a separator between the coin electrodes. On the other hand, RuO_2 was loaded on these carbon materials, and we fabricated a pseudo-capacitor that has an ion insertion mechanism into RuO_2. For comparison with these carbon materials, activated carbon (AC) was also used for a capacitor electrode. The electrochemical properties of all the capacitors were evaluated in 1M H_2SO_4 aqueous solution. As a result of EDLC performance, AcB electrode had a higher specific capacitance than AC electrode at a high scan rate ($\geq$ 100 mV/s). In the evaluation of pseudo-capacitor performance, RuO_2-loaded CNB electrode showed a high specific capacitance of 734 F/g per RuO_2 weight.

1. Introduction

Electrochemical capacitor (EC) has a high power density and a long lifetime. Then, EC is expected to be applicable to an electric vehicle. [1] ECs are distinguished by the difference of charging mechanism. There are two types of ECs: electric double-layer capacitors (EDLCs) and pseudo-capacitors (PCs). EDLC does not use a chemical reaction for charge/discharge and enables us to operate for a long cycle life. Commercial EDLCs employ inexpensive activated carbon (AC). On the other hand, PC, as represented by RuO_2 [2-6], has a different charge/discharge mechanism from EDLC, which is accompanied with a redox reaction and has a higher energy density than EDLC. However, Ru is expensive and Ru resources are limited.

Recently, novel carbon nanomaterials which have a high electric conductivity including carbon nanotube (CNT) [7-12], carbon nanofiber (CNF) [13-15], and graphene [16,17] are studied for EC. Fabrication of carbon nano-sheet is also studied for effective use of a high surface area of carbon materials. [12,16,17] Novel carbon nanomaterials have a lot of potential for EC application.

In this study, we used two types of carbon nanomaterials, arc-black (AcB) and carbon nano-balloon (CNB) which were developed in our laboratory, as EC electrodes and compared their electrochemical properties with activated carbon (AC). The applicability of AcB and CNB for EC electrode was examined by the following viewpoint: performance of EDLC electrode at a high voltage scan rate; and pseudo capacitance per RuO_2 weight.

2. Experimental

2.1. Electrochemical cell

The electrochemical cell used in this study is shown in Figure 1. The cell is made of two stainless steels and contains capacitor electrodes. When capacitor electrodes and liquid electrolyte are installed in the cell, the cell is evacuated by connecting with a rotary pump for degassing.
The dimension of coin-shaped electrode for this cell was 20 mm diameter and 1 mm thickness. Two coin-shaped electrodes were separated by double sheets of polypropylene (PP) (Nippon Kodoshi Corporation, MPF0830). This separator has a role to separate between anion and cation in an electrolyte. 1M H_2SO_4 was used as an electrolyte.

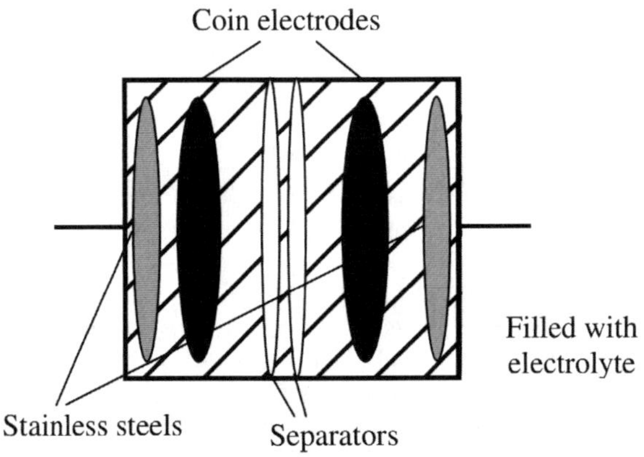

Figure 1. Schematic illustration of the electrochemical cell.

2.2. Fabrication of EDLC

2.2.1. Preparation of carbon nanomaterials for EDLC

AcB was prepared by employing an arc discharge of graphite rod electrodes [18]. Carbon atoms and molecules vaporized by an arc discharge were cooled down in the chamber due to the collision with N_2 gas molecules and deposited on the chamber wall as AcB. CNB was prepared by calcination of AcB in Ar gas ambient at 2500°C.

The specific surface areas of carbon nanomaterials were measured by the Brunauer-Emmett-Teller (BET) method. The resistivities of carbon nanomaterials were measured by the compressive electrical resistivity method [19]. A 300 mg carbon nanomaterial was compressed under a load of 3.75 MPa when measured.

2.2.2. Fabrication of EDLC coin electrodes

Each of the carbon nanomaterial used (AcB, CNB, or AC) was mixed with a conductive carbon black (Ketjenblack : EC-600JD, Shanghai Fuhua Co., Ltd.) and a binder (PTFE). The mass ratio of carbon nanomaterial, Ketjenblack, and PTFE was 8:1:1, respectively. A 200 mg quantity of mixed material was placed in a jig and pressed at 14 MPa for 30 min to form a coin-shaped electrode. Two coin-shaped electrodes were separated with double sheets of polypropylene separators.

2.3. Fabrication of PC

2.3.1. Surface functionalization of carbon nanomaterials

Prior to RuO_2 supporting, the carbon nanomaterials were oxidized in 15% H_2O_2 aqueous solution at 120°C for 2 h to add surface functional groups such as the hydroxyl group (–OH) and carboxyl group

(–COOH). The solution was then filtered and washed repeatedly with deionized water and dried in air at 120°C for 2 h.

2.3.2. Support of RuO_2 onto carbon nanomaterials
The functionalized carbon nanomaterials were firstly heated at 250°C for 2 h in air. A NH_4HCO_3 aqueous solution was added dropwise to an aqueous solution of $RuCl_3 \cdot xH_2O$ (280 mg) until the pH of the mixed solution reached 5 [20]. Then, the carbon nanomaterial was added to the solution. The total volume of the solution was adjusted to 50 ml. Then the solution was sonicated by an ultrasonic homogenizer (US-300T, Nihon Seiki) for 60 min. After sonication, the solution was filtered and washed repeatedly with deionized water and dried in air at 120°C for 2 h. The amounts of RuO_2 supported on the carbon nanomaterials were 20 wt.% each. A coin electrode for PC was prepared by the same procedure as that for EDLC. For EDLC and PC, H_2SO_4 aqueous solution with a concentration of 1 M was employed as an electrolyte.

2.4. Characterization of carbon nanomaterials and analyses of EC performances
Thermogravimetric analysis was carried out to measure the supported amount of metal oxides in still air at a heating rate of 10°C/min with a thermal analysis instrument (TG/DTA: Shimadzu, DTG-60). The morphologies and structures of RuO_2 were analyzed by transmission electron microscopy (TEM: JEOL, JEM-2100F) and scanning electron microscopy (SEM: HITACHI, S-4500), The cyclic voltammetric (CV) behavior of the electrodes was determined using electrochemical analyzer systems (Hokuto Denko, HZ-5000).

The specific capacitance C was calculated from the CV curve by the following formula [5]:

$$C = \frac{\int I \cdot dt}{2\Delta E} \quad (1)$$

where I is current, dt is time step, and ΔE is operating voltage. The specific capacitance per electrode weight C_m is given by the following formula:

$$C_m = \frac{C}{m} \quad (2)$$

The specific capacitance per one electrode C_s is given by the following formula:

$$C_s = 4C_m \quad (3)$$

The internal resistance of the EDLC electrode R was calculated by the following formula:

$$R = \frac{\Delta V}{I_{cd}} \quad (4)$$

where ΔV and I_{cd} are IR drop and discharge constant current, respectively. The charge/discharge current was 100 mA.

3. Results and discussion

3.1. Carbon nanomaterials used in this study
We used two types of carbon nanomaterial, arc-black (AcB) which has an amorphous structure and carbon nano-balloon (CNB) which has a graphitic structure as EC electrodes. Activated carbon (AC: YP-80F, Kuraray Chemical Co., Ltd.) was also used for comparison. Figure 2 shows SEM and TEM images of carbon nanomaterials. These properties are listed in Table 1. The AcB was diameter as large as CNB. The CNB resistance was lowest among all of carbon nanomaterials. The AC was largest specific surface area among three carbon nanomaterials.

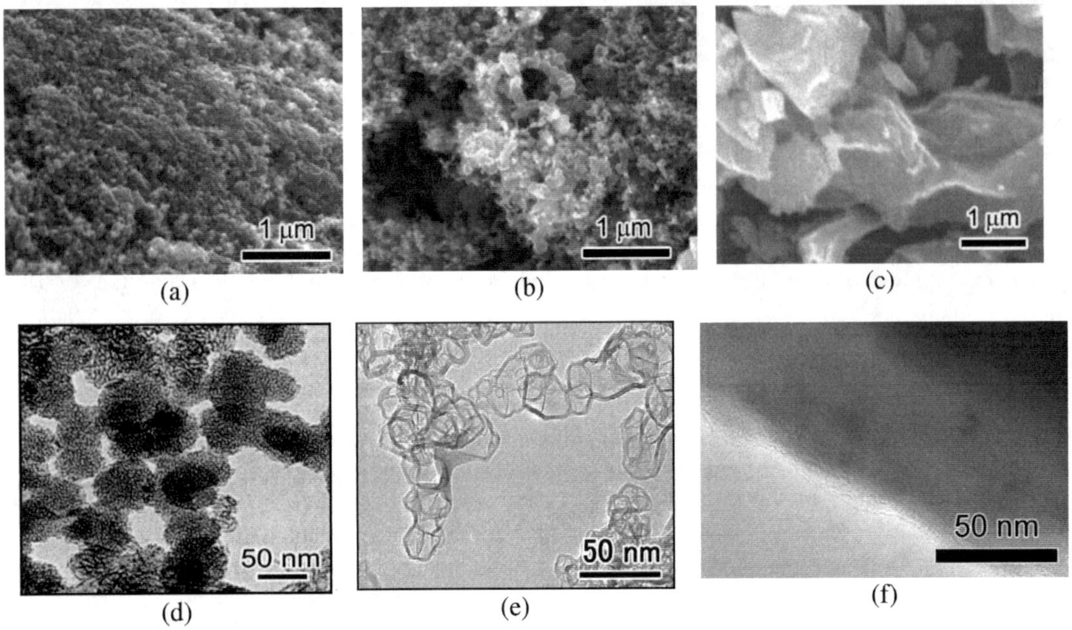

Figure 2. SEM images of carbon nanomaterials: (a)AcB, (b)CNB, and (c)AC, and TEM images of (d)AcB, (e)CNB and (f)AC.

Table 1. Characteristics of carbon nanomaterials

	Particle size (nm)	Surface area (m^2/g)	Resistivity ($\Omega\cdot$cm)
AcB	1-50	152	32
CNB	1-50	35	0.6
AC	1000-20000	2155	2.6

3.2. EDLC performance

Figure 3 shows the CV curves of EDLC electrodes of AcB, CNB and AC at 5 and 7000 cycle operations. Figure 4 shows cycle performance of these three electrodes. Almost no decrease in specific capacitances of all the electrodes was observed up to 7000 cycles. Figure 5 shows the CV curve of the EDLC electrodes at different voltage scan rates, 1 and 100 mV/s. Figure 6 shows the result of specific capacitances of the AcB, CNB, and AC electrodes at different voltage scan rates. In a lower scan rate region ($\leq$ 50 mV/s), the AC electrode had the highest specific capacitance. This is explained by that a high surface area of AC resulted in a lot of sites for electric double-layer in the AC electrode. In contrast, specific capacitance of the AC electrode becomes the lowest in a high voltage scan rate region ($\geq$100 mV/s). This result is still under consideration. However, Liu, et al has shown a similar experimental result where the specific capacitance of AC electrode significantly decreased at a high voltage scan rate [21]. They considered that the many channel is required for ion transportation as the voltage scan rate is increased. It is reasonable to say that the ion channel was not formed well in our AC electrode. Though the particle sizes of AcB and CNB are almost the same, the surface area of AcB is more than 4 times higher than that of CNB as shown in Table 1. This is the reason maybe that specific capacitance of the AcB electrode became the highest at higher voltage scan rate.

Figure 7 shows charge/discharge curves of the EDLCs composed of the carbon nanomaterials. The internal resistance of the EDLC electrode R in Figure 7 was calculated by formula (4). R of the AcB, CNB, and AC electrodes were 1.5, 1.2, and 1.5Ω, respectively. The CNB electrode had the lowest R, and this is correlated with the lowest resistivity of CNB among three carbon nanomaterials used in this study.

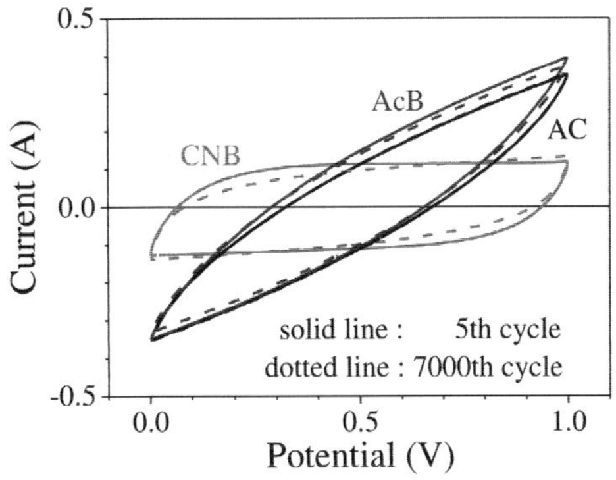

Figure 3. Cyclic voltammograms of AcB, CNB and AC EDLC electrodes at different cycle operations.

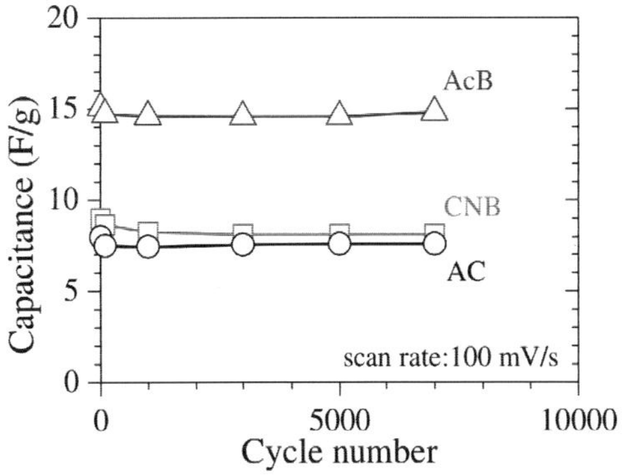

Figure 4. Cycle performances of AcB, CNB and AC EDLC electrodes.

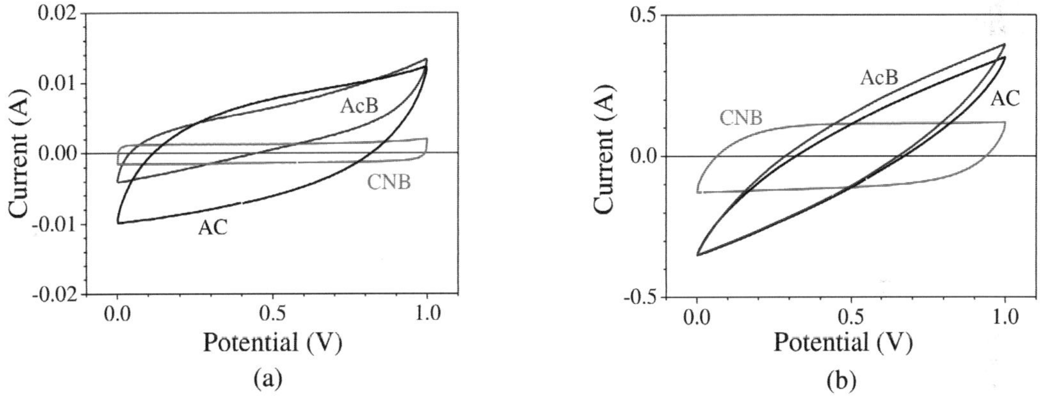

Figure 5. Cyclic voltammograms of AcB, CNB and AC EDLC electrodes at different voltage scan rates: (a)1 mV/s and (b)100 mV/s.

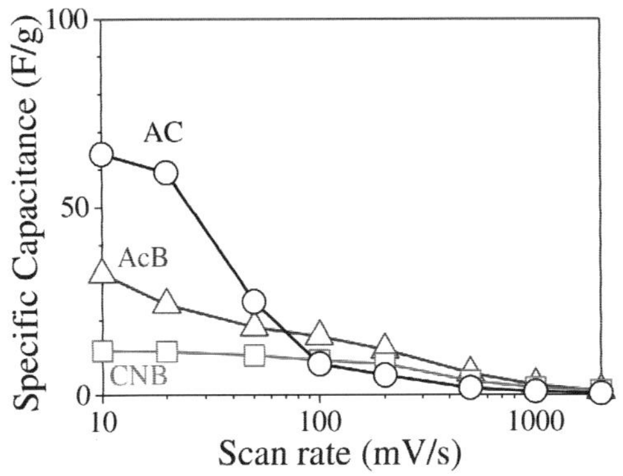

Figure 6. Specific capacitances of AcB, CNB and AC EDLC electrodes at different voltage scan rates.

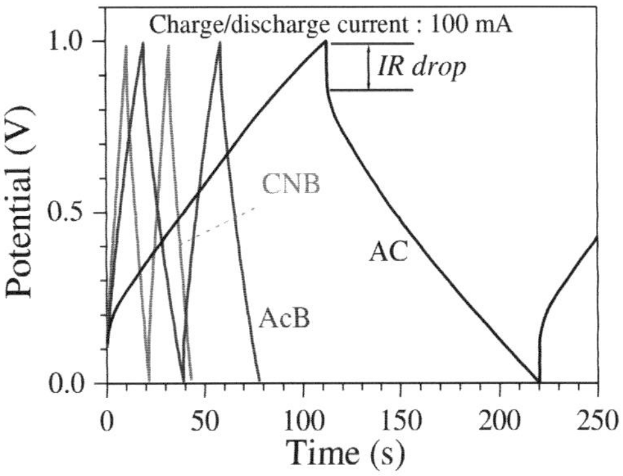

Figure 7. Charge/discharge curves of AcB, CNB and AC EDLC electrodes.

3.3. Characteristics of carbon nanomaterials as a support material of RuO_2 for PC

TEM images of the RuO_2-supported carbon nanomaterials are shown in Figure 8. The characteristics of RuO_2-supported carbon nanomaterials are listed in Table 2.

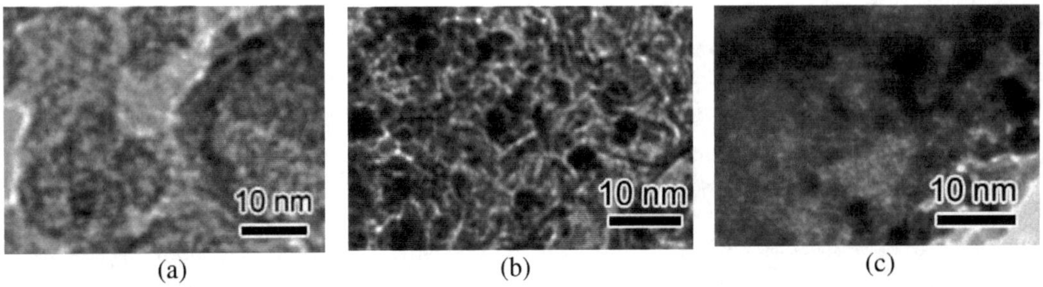

Figure 8. TEM images of RuO_2 supported carbon nanomaterials: (a)AcB, (b)CNB, and (c)AC.

Figure 9 shows the cyclic voltammograms of RuO_2-supported carbon nanomaterial electrodes. Specific capacitances of the RuO2/AcB, RuO2/CNB, and RuO2/AC electrodes are respectively calculated. then this became 178, 170, 225 F/g from the results in Figure 9 (a).

Subtracting the specific capacitance of carbon nanomaterials in RuO_2-supported carbon electrodes from that of the total capacitance, the specific capacitance of RuO_2 was calculated and listed in Table 3. Specific capacitances of the RuO_2/AcB and RuO_2/CNB were almost the same. The pseudo capacitance of the RuO_2-supported carbon electrodes was much higher than their electric double-layer capacitance. The RuO_2/CNB electrode showed the highest pseudo capacitance per the RuO_2 mass as shown in Table 3. This indicates that CNB is suitable for support material of RuO_2 in PC.

Figure 10 shows charge/discharge curves of the RuO_2/C in PC. The internal resistances R of RuO_2-supported AcB (RuO_2/AcB), CNB (RuO_2/CNB), and AC (RuO_2/AC) were 0.8, 0.6, and 0.6 Ω, respectively. Each R of the RuO_2/C shows respectively lower than that of carbon nanomaterial without RuO_2. These have an advantage in low output loss.

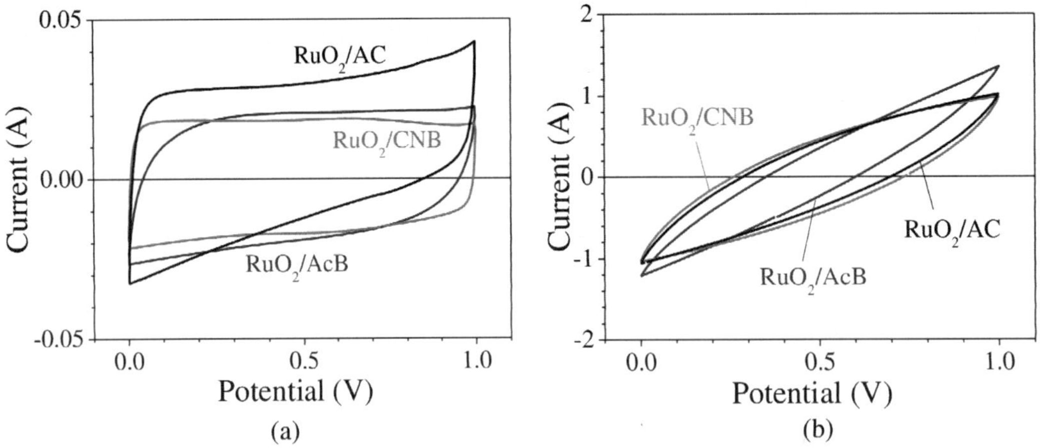

Figure 9. Cyclic voltammograms of RuO_2/C electrodes at different voltage scan rates: (a)1 mV/s and (b)100 mV/s.

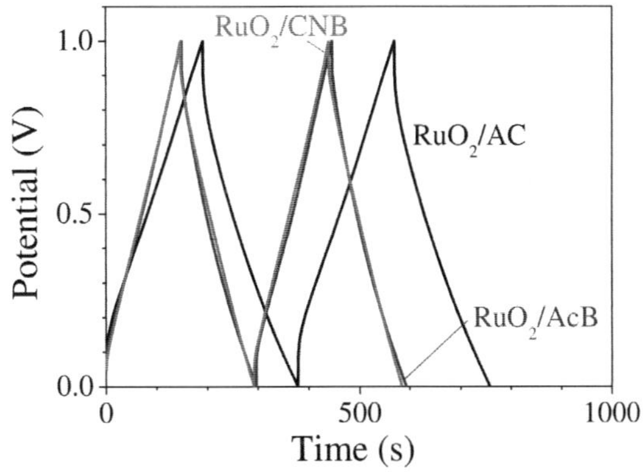

Figure 10. Charge/discharge curves of RuO_2/C electrodes.

Table 2. Characteristics of RuO_2/Carbon electrodes

	RuO_2 load (wt.%)	Particle size of RuO_2 (nm)
RuO_2/AcB	20.4	2.0
RuO_2/CNB	21.2	2.9
RuO_2/AC	20.8	2.5

Table 3. Specific capacitances of RuO_2/C electrodes

	Total specific capacitance (F/g)	component of carbon nanomaterials (F/g)	component of RuO_2 (F/g)
RuO_2/AcB	178	37	708
RuO_2/CNB	170	20	734
RuO_2/AC	225	101	721

4. Conclusions

AcB and CNB electrodes of EDLC were shown to operate for a long cycle life. The AcB electrode for EDLC indicated the highest specific capacitance among all the carbon nanomaterials in a high voltage scan rate region (≥ 100 mV/s). The RuO_2/CNB electrode showed the highest pseudo capacitance per the RuO_2 amount among all the RuO_2/C electrode. These results indicate that AcB and CNB are respectively suitable for EDLC electrode and support material of RuO_2 in PC.

5. References

[1] Kandalker S G, Gunjakar J L, and Lokhande C D 2008 *Applied Surface Science* **254** 5540-5544
[2] Hong J H, Han S, Hyeon T and Oh S M 2003 *J. Power Sources* **123** 79-85
[3] Yue F S, Feng W, Li Y B and Zhao H Y 2007 *New Carbon Materials* **22(1)** 53-58
[4] Kuratani K, Kiyobayashi T and Kuriyama N 2009 *J. Power Sources* **189** 1284-1291
[5] Hu C C and Huang Y H 2001 *Electrochimica Acta* **46** 3431-3444
[6] Ramania M, Harana B S, Whitea R E, Popova B N and Arsovb L 2001 *J. Power Sources* **93** 209-214
[7] Chen J H, Li W Z, Wang D Z, Yang S X, Wen J G and Ren Z F 2002 *Carbon* **40** 1193-1197

[8] Ma R Z, Liang J, Wei B Q, Zhang B, Xu C L and Wu D H 1999 *J. Power Sources* **84** 126-129
[9] Liu C G, Fang H T, Li F, Liu M and Cheng H M 2006 *Journal of Power Sources* **160** 758-761
[10] Lu W, Qu L, Henry K and Dai L 2009 *J. Power Sources* **189** 1270-1277
[11] Ye J S, Liu X, Cui H F, Zhang W D, Sheu F S and Lim T M 2005 *Electrochemistry Communications* **7** 249-255
[12] Y Honda et al 2008 *J. Power Sources* **185** 1580-1584
[13] Kim C, Park S H, Lee W J and Yang K S 2004 *Electrochimica Acta* **50** 877-881
[14] Seo M K and Park S J 2009 *Materials Science and Engineering* **B 164** 106-111
[15] Kim B H, Yang K S, Kim Y A, Kim Y J, An B and Oshida K 2011 *J. Power Sources* **196** 10496-10501
[16] Li Y, Zijll M V, Chiang S and Pan N 2011 *J. Power Sources* **196** 6003-6006
[17] Chen Y, Zhang X, Yu P and Ma Y 2010 *J.Power Sources* **195** 3031-3035
[18] Oke S et al 2009 *Chem. Engineering* **146** 434-438
[19] Ikeda T et al 2011 *Jpn. J. Appl. Phys.* **50** 01AF13
[20] Uruno H et al 2010 *IEEJ Trans. Fund. and Materials* **130-A(3)** 293-294
[21] Liu G, Kang F, Li B, Hang Z and Chuan X 2006 *J.Phys. and Chem. of Solids* **67** 1186-1189

Remote Sensing of Radiation Dose Rate by Customizing an Autonomous Robot

T Kobayashi[1], M Nakahara[1], K Morisato[1], T Takashina[2] and H Kanematsu[3]

[1] Department of Electronics & Control Engineering, Tsuyama National College of Technology, Okayama, 708-8509 Japan
[2] International Industry Academia Collaboration Division, Hiroshima University, Hiroshima, Hiroshima, 739-8511 Japan
[3] Department of Materials Science & Engineering, Suzuka National College of Technology, Mie, 510-0294 Japan

E-mail: t-koba@tsuyama-ct.ac.jp

Abstract. Distribution of radiation dose was measured by customizing an autonomous cleaning robot "Roomba" and a scintillation counter. The robot was used as a vehicle carrying the scintillation survey meter, and was additionally equipped with an H8 micro computer to remote-control the vehicle and to send measured data. The data obtained were arranged with position data, and then the distribution map of the radiation dose rate was produced. Manual, programmed and autonomous driving tests were conducted, and all performances were verified. That is, for each operational mode, the measurements both with moving and with discrete moving were tried in and outside of a room. Consequently, it has been confirmed that remote sensing of radiation dose rate is possible by customizing a robot on market.

1. Introduction

Robots used in hazardous environments have been studied and developed for nuclear industry, disaster relief and military usage for long time. Especially robots in nuclear industries have been applied for many purposes in practical operation [1]-[3]. In the fields of military and disaster relief, the most famous robot is "PackBot" produced by iRbot Co. More than 3000 PackBots are currently on station in Iraq and Afghanistan [4]. However each robot has special specification, namely for high level radiation or for travelling over the rubble and so on. Therefore these robots used in extreme environments cannot be used by anyone anywhere at any given time. Furthermore the price of PackBot is said to be 120 thousands dollars [5]. On the other hand, regarding home robot, more than 6 million sets of a cleaning robot "Roomba" have been sold so far [6], and the price has become affordable .

[1] t-koba@tsuyama-ct.ac.jp
[2] takashina@hiroshima-u.ac.jp
[3] kanemats@mse.suzuka-ct.ac.jp

In low radiation level area, for example, in Ningyo-toge Environmental Engineering Center of Japan Atomic Energy Agency and medical facilities, it is important to detect subtle indication in early stage by means of daily inspection and take measures before occurrence of unusual event in addition to a periodic examination. Some of the present authors have conducted a trial measuring of radiation dose rate distribution using equipment on the market without using specially developed equipment, and they have reported an example of the result where a cleaning robot was applied as a vehicle of a radiation survey meter [7].

After the earthquake, tsunami, and failure of cooling systems at Fukushima I Nuclear Power Plant concerning other nuclear facilities in Japan on March 11, 2011, decontamination work has been conducted, and ordinary life has been recovered. However, radiation survey is required in many places still now. The above mentioned method could be useful for investigating in school house and playground with radioactive contamination in terms of cost and convenience.

In the present paper, the measuring results of radiation dose rate and position in three modes; (1) remote manual, (2) remote programmed and (c) autonomous are reported using an additional micro computer board to handle the signal from external connecter of the cleaning robot and the scintillation counter.

2. Experimental

2.1. Equipment

Radiation survey meter used is a scintillation counter whose specifications are shown in Table 1. Geiger-Muller counter is usually used for radiation dose rate. In this study, however, the scintillation counter was applied so that higher sensitivity more than one order is required for achieve "continuous measurement method" where measurement is done with moving than for "discrete measurement method" where vehicle moves and stops to measure repeatedly. As the radiation source, a standard sample of Cesium (Cs) and a mantle mesh of lantern for camping which contains Thorium (Th) were used, and they were placed on the floor or in the test fields.

Table1. Specification of survey meter

Measuring ray	γ - ray
Measurable range	From background to 30.0 μSv/h
Scintillater	25.4φ×25.4mm NaI (T1)
Accuracy	Max. ±15%
Output range	DC +10mV (Full scale)
Time constant	3 seconds

A cleaning robot "Roomba" was used as the vehicle conveying the scintillation counter. Although Roomba is sold for cleaner, it is an open ware which anyone can access to the electric signals of the driving and sensing systems through an external connecter. Therefore organized activities like Packman game, in which many Roombas are handled by controlling the inner micro computer, have been held [8].

In this study, H8 micro computer board which is often used in electronics handicraft and robot Coclé was utilized for controlling the Roomba's movement and the data communication of the radiation dose rate measurement between the vehicle and a stationary personal computer. The system structure is shown in Fig.1.

Although the operation modes are "remote manual"," remote programmed" and "autonomous" as mentioned above, both "discrete measurement method" and "continuous measurement method" are available for the former two modes.

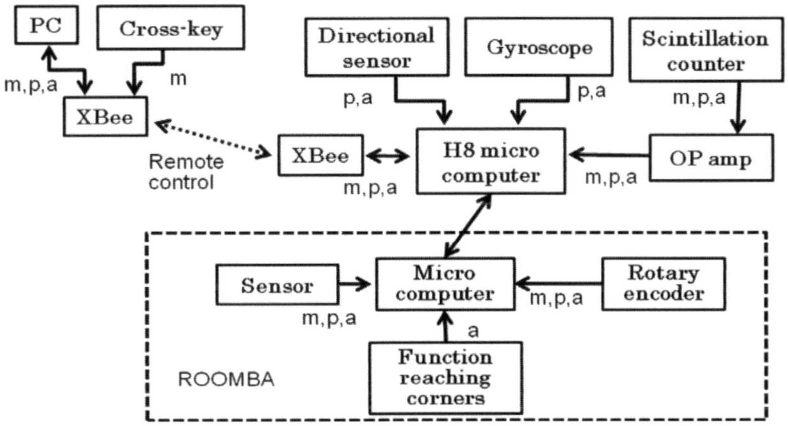

Figure1. Control system. Symbols "m", "p" and "a" represent manual, programmed and autonomous mode respectively.

2.2. Measuring mode

2.2.1 Remote manual. Measured field was the campus of Tsuyama National College of Technology, and the measurement was carried out by the remote manual mode with the discrete measurement method, that is, the vehicle moved by 10 m then stopped to measure for 3 seconds repeatedly.

2.2.2 Remote programmed. In the case of the remote programmed mode, it was assumed that the mode is used for the measurement in room or corridor in building. The measurement was carried out in the two cases which are "discrete method" running and stopping to measure for 3 seconds repeatedly and continuous method" programmed a moving route of the measurement.

2.2.3 Autonomous mode. Radiation sources placed in a room were measured with the original autonomous moving function mode.

3. Result and discussion

3.1. Simulation

In the "discrete method" since measuring is conducted in a stationary, precise value is expected. The "continuous method" provides a simple moving average (SMA) of radiation dose rate for last three seconds shown in Figure 2, that is the measured values are obtained in the following equation,

$$I_1(x_1) = \frac{1}{\Delta x}\int_{x_1-\Delta x}^{x_1} I(x)dx \qquad (1)$$

where I_1 is the simple moving average of radiation dose rate at a coordinates (x_1, 0, h), Δx is a distance travelling for a time constant Δt, $I(x)$ is the distribution of radiation dose rate at a coordinates (x, 0, h). When the vehicle travel with an speed of v, and Δx is written as $\Delta x = v\,\Delta t$.

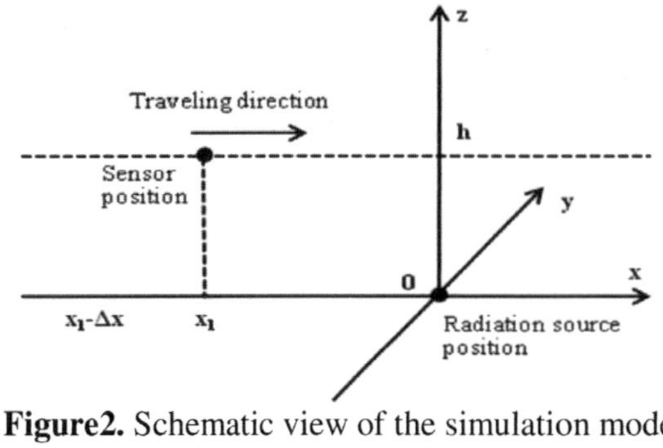

Figure2. Schematic view of the simulation model

Figure 3 shows examples of the distributions of radiation dose rate with the discrete method (v=0 mm/s) and the continuous method (v=10 to 200 mm/s) respectively. From the equation (1) and Figure 3, it is obvious that I_1 obtained at high travelling speed or longer moving average time is lower than that obtained in real case namely the discrete method. Therefore the obtained data in the continuous method needs to be considered in treatment.

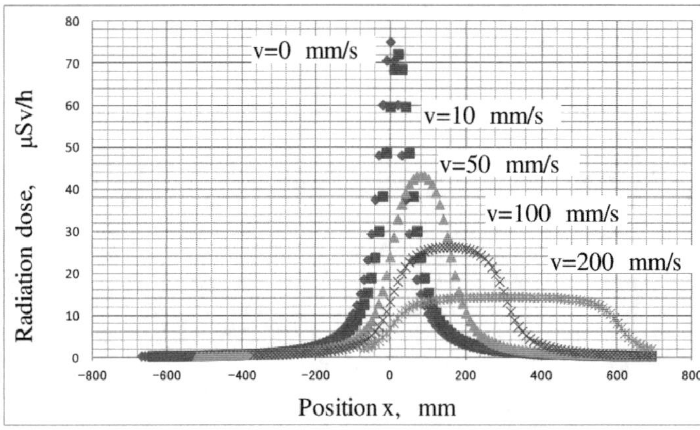

Figure 3. Simulation result of the radiation dose rate measured at some traveling speeds of the vehicle

3.2. Moving test
In all cases of manual, programmed and autonomous mode, planed driving was confirmed. The error factor of position coordinates mainly results from wheel slip, since position data are calculated from the encoder signal in driving unit which depends on the surface condition of the floor. In the present case, an error of around 3% in moving direction was obtained for a carpet floor. Further an error due to turning direction was investigated, and an error of 5% was obtained for a case where one wheel drove and the other wheel drove in the opposite spin direction.

3.3. Measuring the distribution of radiation dose rate
3.3.1 Remote manual. Radiation dose rate was measured in the discrete method, and a value of around 0.08 μ Sv/h was obtained. After that, the continuous method was also applied and almost same value as that in the discrete method was obtained because only natural radiation dose rate was measured.

3.3.2 Remote programmed. In Figure 4, a part of the measurement in the continuous method was shown. The indicated values at higher travelling speeds were lower than that obtained at lower

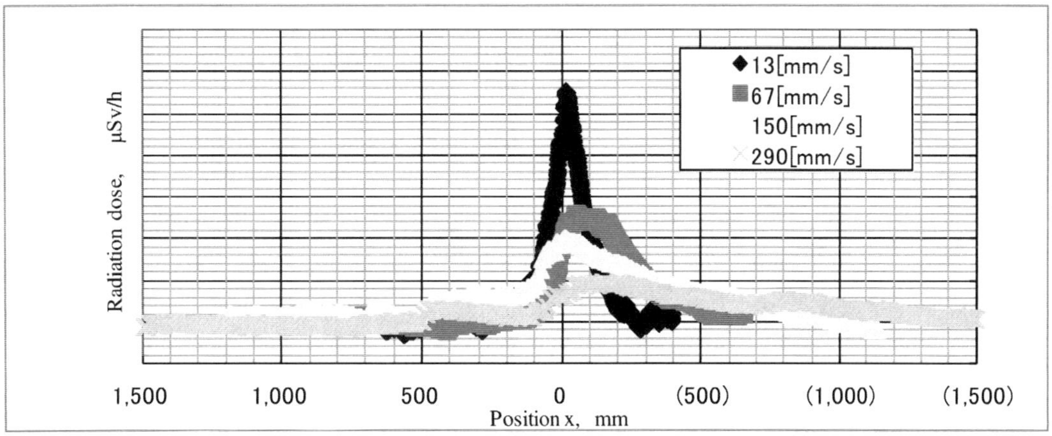

Figure 4. A part of the obtained result in continuous method with remote programmed mode

travelling speed. It is consistent with the simulation result. It is suggested that the difference was background due to natural radiation dose rate in a range from 0.07 to 0.1 μ Sv/h.

3.3.3 Autonomous mode. A test result measured in continuous method with the autonomous mode is shown in Figure 6. Radiation sources were placed in an experimental field shown in Figure 5. It was verified that a map of radiation dose rate distribution is successfully created by the customized system.

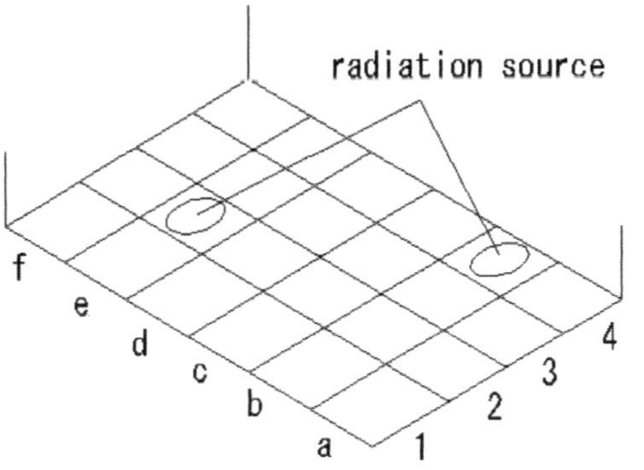

Figure 5. Overview of experimental field for the continuous measuring method

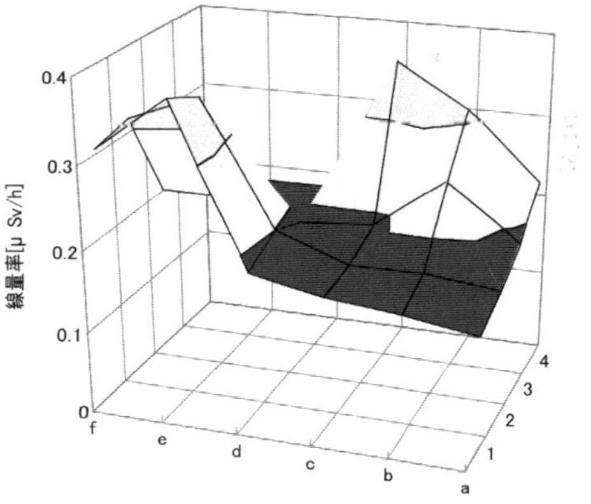

Figure 6. The distribution of radiation dose rate obtained by the continuous measuring method

4. Conclusion

An equipment and technique, measuring the distribution of radiation dose rate with moving by utilizing a cleaning robot on market, were studied and following results were obtained.

(1)Manual and programmed moving modes were additionally equipped to an autonomous cleaning robot which has an external signal connecter, and the moving function as a vehicle was confirmed.

(2)A scintillation counter was equipped with the travelling vehicle, then it was verified that the remote detection and monitoring of the radiation dose rate were capable in the discrete mode and continuous mode.
(3)It is shown that the position of the vehicle is detected by treating the signal from the travelling vehicle and the distribution of radiation dose rate is obtained.

Acknowledgements
A part of present project was financially supported by the Ministry of Education, Culture, Sports, Science and Technology.

References

[1] Ohmichi T 1992, Operation Robot under Radiation Environment, *Proc. Japan Welding Society*, vol.50, p.19-25

[2] Oka K 2003, Robot for the Nuclear Energy Disaster: Development of a Radiation-Proof Robot <Special Issue> Disaster Response Robotics, *Journal of the Japan Society of Mechanical Engineers*, vol.106, p.765-767.

[3] D'Annucci F and Lecour E 2010, Inspection of the Steam Generator Divider Plate, ICONE18-29457, *Proceedings of the 18th International Conference on Nuclear EngineeringICONE18, May 17-21, Xi'an, China*

[4] [http://www.webcitation.org/5xPANQOLV "iRobot Delivers 3,000th PackBot"]. Archived from the original on March 23, 2011. http://www.webcitation.org/5xPANQOLV.

[5][http://ja.wikipedia.org/wiki/%E3%83%91%E3%83%83%E3%82%AF%E3%83%9C%E3%83%83%E3%83%88]

[6] http://eetimes.jp/ee/articles/1109/15/news035.html

[7] Morisato K and Kobayashi T 2011 , *Proc. 41th the JSME students conference, Okayama, Feb.*

[8] http://pacman.elstonj.com/index.

Asia-Pacific Interdisciplinary Research Conference 2011 (AP-IRC 2011) IOP Publishing
Journal of Physics: Conference Series **352** (2012) 012034 doi:10.1088/1742-6596/352/1/012034

Optimisation of Flux Concentrator-Coil Design for Application in Magnetic Hyperthermia

M Kallumadil[1], M Abe[1,2], T Ueda[1] and H Handa[1,3]

[1]Solutions Research Laboratory, [2]Materials and Structures Laboratories, and [3]Department of Biological Information, Tokyo Institute of Technology, 4259 Nagatsuta-cho, Midori-ku, Yokohama 226-8503, Japan

mathew.kallumadil@gmail.com

Abstract. This study investigates some key optimisation aspects of applicator coils with ferrite-based flux concentrators for magnetic hyperthermia treatment. We found that the enhancement of the magnetic field flux from the coil with the aid of flux concentrators comes at a cost of increased inductance – which demands lower capacitance counterparts in a high-frequency resonance LC circuit – and higher AC resistance, which generates unwanted Joules heating. Positioning the coil at the application end of the flux concentrator increases the magnetic field strength dramatically, whilst reducing inductance and AC resistance. Preliminary experiments indicate that the increased magnetic field strength from flux concentrators of higher aspect ratio saturates above a certain concentrator-to-coil length ratio of around 15. Using a high degree of perforation of the core results in negligible difference in magnetic field progression. By optimising the relative coil-to-concentrator dimensions, AC resistance can be limited as proximity effects resulting from ferrite concentrator and coil interaction are reduced. The latter two findings enable new opportunities to improve the cooling efficiency.

1. Introduction

With a myriad of different types and pathophysiology, cancer and its treatment remains one of the most challenging healthcare challenges in the world. It is therefore not surprising that a majority of past and current research efforts are focused on the discovery and development of viable alternatives to the status quo of cancer treatment – namely surgery, radiotherapy or chemotherapy (or a combination of them). Many of the most promising new therapeutic technologies use magnetic nanoparticles for diagnostic purposes, for instance in contrast-enhanced magnetic resonance imaging (MRI), and/or for therapeutic use, by exploiting their unique properties to deliver remedial payloads (e.g. drugs, genes) or by delivering the therapy themselves, i.e. hyperthermia, the topic of this work. For a review on magnetic nanoparticles in medical applications, please refer to [1].

Only recently, magnetic hyperthermia has found its way to the hospital, with the first approved treatment centre at the Charité-Universitätsmedizin Berlin established and offering an alternative therapy to brain tumour patients (www.magforce.com).

During a typical magnetic hyperthermia therapy nanometre-sized magnetic particles are initially administered to the tumour site. A powerful alternating magnetic field is then used to trigger relaxation mechanisms in the magnetic nanoparticles, thereby releasing a possibly lethal dose of heat into the target volume [2]. Compared to thermal ablation, a process in which cells are necrotised at temperatures up to 56 °C, hyperthermia heat ranges only from 42-45 °C, where tumour cells are more

Published under licence by IOP Publishing Ltd

sensitive to heat than healthy tissue and can be forced to a programmed cell death, known as apoptosis [3].

Whereas many magnetic hyperthermia systems are designed to treat tumours located deep inside the body, our objective is to develop an applicator for the treatment of superficial targets, i.e. melanoma, and breast cancer at most. As such, we estimate that only a penetration depth of 1-3 cm for the magnetic field is needed, and based on requirements demanded by recent studies [4] we set a condition of achieving a frequency-field product of 8 MHzOe to heat gold-plated iron shots (1 mm in diameter) to 42.5 °C. In other words, for 500 kHz frequency, a field of 16 Oe is needed. In an ideal resonant circuit, frequency f is related to capacitance C and inductance L as $f = 1/(2\pi\sqrt{LC})$. Therefore, to achieve a certain frequency, increasing the inductance implies lowering the capacitance and vice versa. For high-current circuits (several 10s of Amps), however, only few manufacturers[1] offer high-power, high-frequency and low-valued capacitors at substantially higher prices. Therefore, we regarded better designed coils to have lower inductances.

Currently, during our preliminary phase, we employed a current of only 10 A, yet we anticipate to increase the input to about 20 A, if heat loss allows. For our purposes, we have already designed and built a control circuit using a complex programmable logic device (CPLD), a programmable control chip, to drive an LC resonant tank.

Additional to an LC tank, we propose the use of ferrite cores as flux concentrators in order to amplify the magnetic field. One downside of using cores is that various parameters of the coil configuration are shifted (e.g. inductance L, series resistance R_S), ultimately requiring new design considerations, such as the adjustment of the inductor-capacitor values in a given LC circuit and foremost the additional heat generation due to increased R_S. Heat is not only generated in the coil conductor but also in the ferrite core in high-frequency applications. In the case of our core material (see below for specifications), however, the core loss data provided by the manufacturer suggests at a 500 kHz frequency point, negligibly small heat is dissipated from the core at the above mentioned target field.

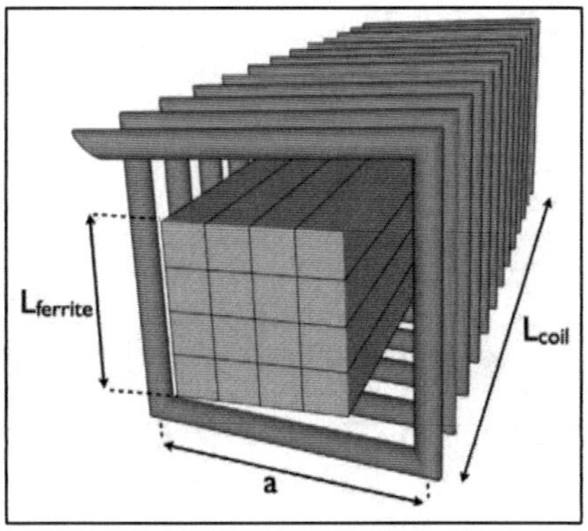

Figure 1 Diagram of experimental setup using a square coil and an array of 4x4 ferrite rods as flux concentrator.

In order to limit heat generation for an air-cooled system, we hoped to achieve a power dissipation of ~20 W maximum. According to Joule's Law, this means that for a current input of 10 A, an R_S of around 0.2 Ω or less could only be tolerated, which was around the noise level of our LCR meter. This challenge was exacerbated by skin and proximity effects from high frequencies, adjacent windings, the inclusion and proximity of the flux concentrator and the latter obstructing the airflow through the coil, respectively.

Here, it should be noted that the ferrite cores did not behave as ideal soft magnetic materials because they were used at near-saturation conditions, where magnetic domain movements essentially determine the dependence of the magnetic field on the current.

Therefore our studies are aimed at providing some guidelines on how to achieve design improvements for practical applications and non-ideal situations. This study is sectioned into four experiments: (i) the coil position with respect to the concentrator was tested, (ii) the relevance of concentrator aspect ratio was investigated, (iii) the ferrite flux concentrator was perforated and the different configurations were obtained and (iv) the dimensions of the concentrator relative to the coil were varied.

2. Experimental

[1] E.g. www.celem.com

To investigate potential design improvements, we constructed a number of different coils. All coils used 1.25 mm^2 (copper cross-section) wire with total OD=3 mm including insulation. The flux concentrator consisted of a collection of 120 mm-long, 5x5 mm square-faced ferrite core rods (PC-40, TDK® Corporation), as shown in figure 1. The coil specifications for each section are summarised in table 1, whereby a (round) solenoid as well as several square-faced coils were employed.

Magnetic field progression was measured using a gaussmeter (LakeShore® 455 DSP Gaussmeter), where both AC and DC values gave similar trends at a current input of 10 A. Only DC values are therefore presented in this work. Inductance and resistance were measured using an LCR meter (Hioki® LCR HiTESTER 3532-50). The former was constant across a frequency ranging from 100 kHz to 1 MHz and only relative values are provided here. Since the AC resistance data was recorded at around the noise floor of the LCR meter, the data were re-fitted with an averaging function for the reader.

Experiment	Coil Type	No. of turns	Pitch (mm)	Side length or ∅ (mm)
Coil Position	Round	6	3	30
Aspect Ratio	Square	7	3	35
Perforation	Square	36	3	35
Rel. Dimensions	Square	12	10	35

Table 1 Specifications of the coils used in the experiments.

3. Results & discussion

3.1. Coil positioning

In this setup, we altered the position of the coil with respect to a given flux concentrator consisting of 4 (= 2x2) ferrite rods. The progression of the magnetic field from the centre of the face of the flux concentrator is shown in figure 2(a). We discovered that positioning of the coil at the application end (in our experiment, that is the gaussmeter probe), enhanced the magnetic field by ca. 80% at 10 mm and 40% at 30 mm compared to a coil located in the middle of, or around the entire length of the core and even higher compared to a coil positioned at the opposite end.

This adjustment also reduced inductance considerably as compared to a central positioning. Using a 6-turn coil, we could indeed achieve 20 G at 10 mm, which satisfies our minimum requirement.

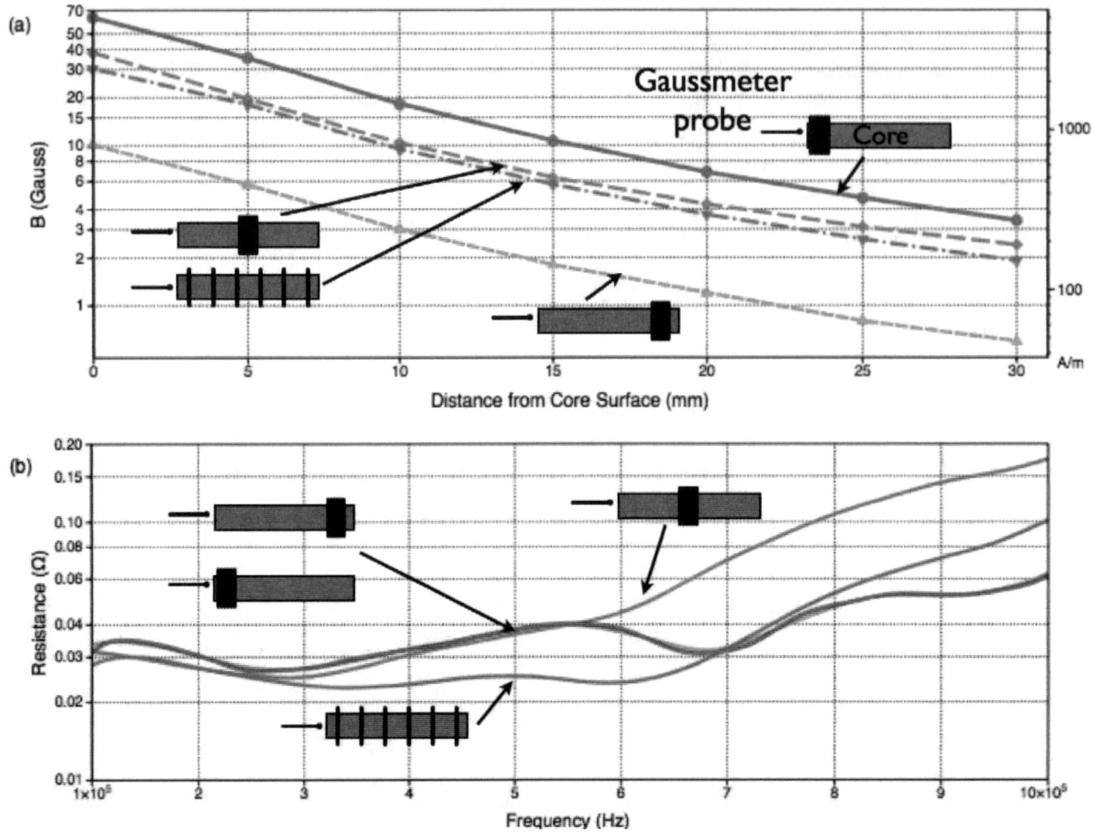

Figure 2 Coils of identical nominal conductor length were varied in positions, as well as pitch in the last case. Magnetic field (a) and AC resistance (b) for coils placed near, middle, far as well as for a fourth, loosely wound coil type from the application end (gaussmeter probe) are shown.

A coil with an identical nominal conductor length but wound at a pitch that covered the entire concentrator conserved flux leakage across the length of the concentrator; however, the number of turns per length was drastically reduced and resulted in substantially lower magnetic fields strength.

Resistances of the different configurations given in figure 3(b) showed that R_S was hardly affected by changing the coil positioning up to our target frequencies, unless wound loosely. For higher frequencies, a central positioning tended to increase more rapidly than other configurations.

3.2. Aspect Ratio

The aspect ratio in our case was defined as the ratio between the length of the flux concentrator and side length of the total ferrite face $AR=L_{ferrite}/a_{ferrite}$. Maintaining all other parameters constant, flux concentrators of varying lengths were inserted.

The data suggested that no significant improvement in magnetic field strength was seen above a certain aspect ratio – in our case above AR=6.2. This could be linked to the relative lengths of coil and concentrator, i.e. $L_{ferrite}/L_{coil}$. If we plot the magnetic field strength at 10 mm against $L_{ferrite}/L_{coil}$ (not shown) we find that the saturation occurs around a value of $L_{ferrite}/L_{coil}=15$, meaning that above a flux concentrator length that is 15 times longer than the coil length, no significant performance increase is observed.

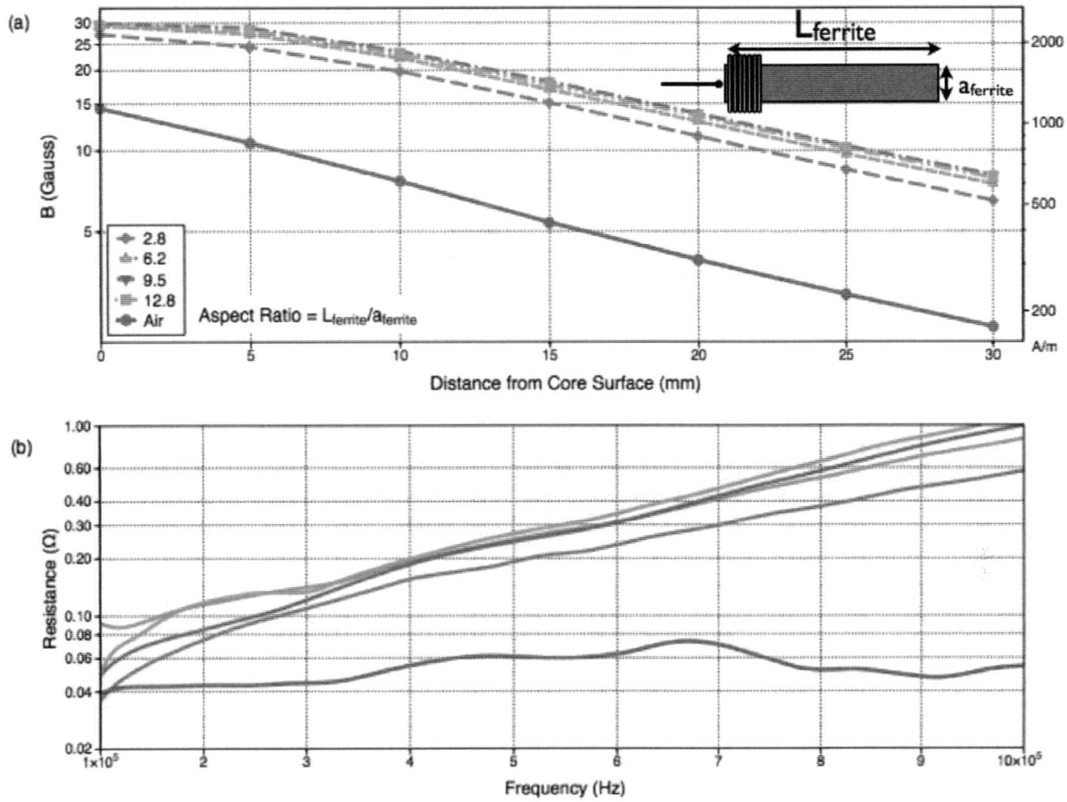

Figure 3 For a set coil, the aspect ratio of the flux concentrator was changed to compare performance differences. Magnetic field (a) and AC resistance (b) for ferrite cores ranging from 2.8 to 12.8 in aspect ratio.

Unsurprisingly, using shorter flux concentrators reduced the inductance and AC resistance substantially. The trade-off between loss of field strength (ca. 20% between AR=2.8 and AR above) and reduction of inductance (ca. 12%) and R_S (ca. 25%) meant that depending on the importance of a parameter of choice the design could be optimised.

With one turn more than the previous, this configuration provided a higher reach of our target field.

3.3. Concentrator perforation

In this experiment, we used a 36-turn, tightly-wound coil (hence higher inductance and AC resistance) to raise the measurement values above the noise floor. We arranged the ferrite rods in a number of configurations to explore the effects of interstitial gaps on the overall performance, as shown in figure 4. Perforation increased the effective aspect ratio and reduced the demagnetization effect, and therefore better outcome was expected for an *ideal* soft magnetic material.

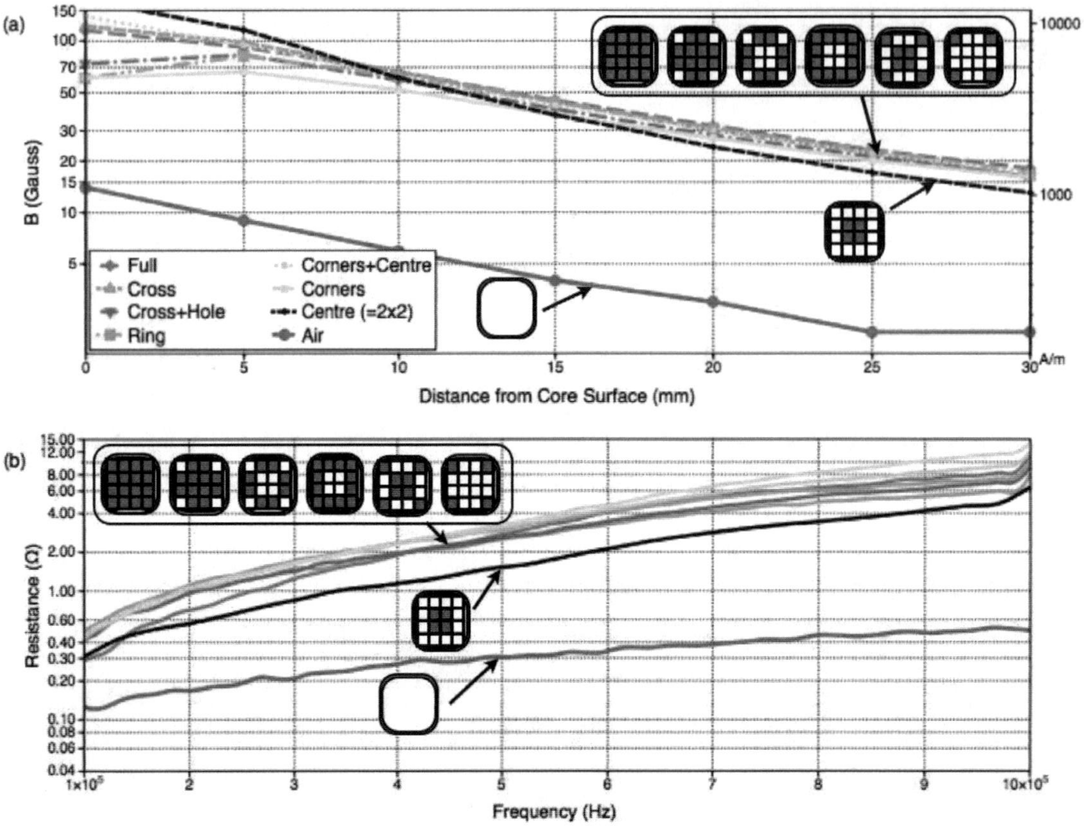

Figure 4 Magnetic field (a) and AC resistance (b) for ferrite cores of varying levels of perforation.

We discovered that the magnetic field was hardly affected, neither by the arrangement of the configuration nor by number of rods used. The exception occurs for only a single configuration (centre 2x2), which showed a significantly lower L and R_S, whilst delivering similar field progression as others. Note that this result was not dependent on the number of rods, as 4 rods placed on the corners gave a different result to 4 rods arranged in a 2x2 configuration. Although not with the 2x2 configuration, we could indeed achieve our target magnetic field at 30 mm with the other configurations.

3.4. Flux concentrator dimensions

The previous results prompted us to investigate the relationship between flux concentrator and coil dimensions. As exemplified in figure 5, for a given coil the concentrator dimensions were varied. Furthermore, for the 4x4 and 3x3 configurations, we also removed the centre rod(s).

We found that the difference in magnetic field strength between larger and smaller concentrators increased with distance between core and coil (figure 5(a)). In other words, whilst a large 4x4 configuration delivered an 8% higher magnetic field at 10 mm than the 2x2, it outperformed the latter by ca. 35% at 30 mm. The trade-off was rewarded with a 50% lower inductance and a lower trend in R_S (figure 7(b)). It is assumed that this was due to a lower proximity effect in the conductor as concentrator-coil distance was increased.

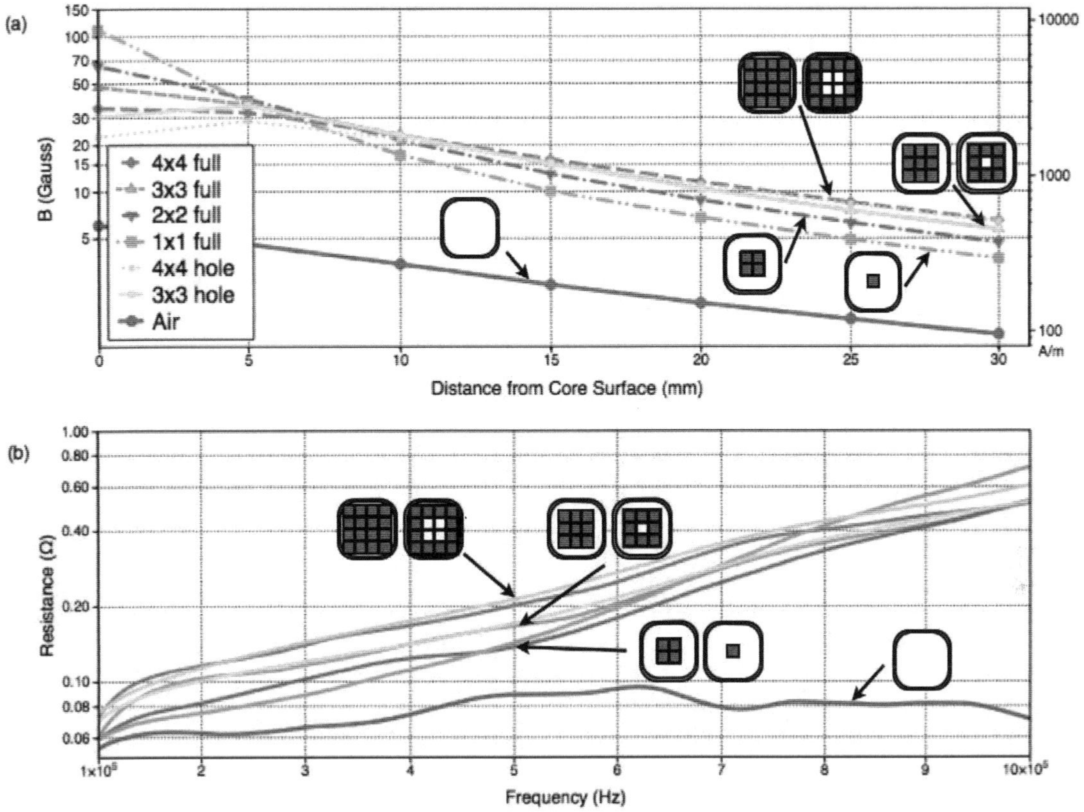

Figure 5 Magnetic field (a) and AC resistance (b) for ferrite cores with varying sizes, altering the space between the core and the coil.

The most astonishing results, however, were observed when comparing a 4x4 configuration with its counterpart with its central rods removed. All parameters between them can be regarded as identical. Thus, by inserting a hole (or possibly several) in a solid ferrite flux concentrator, the same qualitative performance were produced whilst enabling an opportunity for better cooling through the core of the coil.

4. Optimised configuration and conclusions

On the basis of these findings and referring to the target frequency-field product of 8MHzOe, we expect a 10-turn coil with a current of 15 A and a 'holed' 3x3 flux concentrator to be sufficient to meet the demands of our suggested therapy.

During this preliminary design phase, we were indeed able to achieve the required field strength at our minimum target distance only, partly due to a current of only 10 A. Furthermore, our "effective permeability" (i.e. the factor by which the magnetic field strength of an air coil could be improved by inserting the ferrite cores) was limited to ca. 10 or less, which indicated that the ferrite cores did not exhibit ideal soft magnetic properties when the relatively strong current (10 A) was applied.

However, we found that there are indeed several factors that may be considered to optimise an applicator. (1) Using the coil near the application end of the ferrite flux concentrator increased the magnetic field progression as well as lower the inductance. (2) A concentrator length of above 15 times the coil length did not increase the performance. (3) Using perforated flux concentrators had little influence on the magnetic field strength, but lowered inductance and resistance values. (4) Using a flux concentrator located centrally and with smaller dimensions (side length or diameter) resulted in a reduction in AC resistance. Perforating the centre of the flux concentrator, which had no detrimental effect on magnetic field strength, would also facilitate cooling.

Our next objectives include the measurement of real power dissipation from the inductor and actual power requirements for different flux concentrator configurations. Additionally, magnetic flux simulations are currently under way to confirm our measurements numerically and to explore more complex coil/concentrator designs.

Acknowledgements
The authors would like to thank Mr. Naoyoshi Sato from TDK® Corporation for supplying customised PC-40 ferrite rods to support this research. We further would like to thank Prof. Kitamoto for providing access to measurement systems that enabled this research and Prof. Nakagawa (Osaka University) for detailing the system requirements for a treatment.

References

[1] Pankhurst QA et al. 2003 *J. Phys. D: Appl. Phys.* **36** R167
[2] Jordan A et al. 1999 *J. Mag. Mag. Mat.* **201** 1-3 p413-419
[3] Saito H et al. 2008 *Cancer Science* **99** 4 805–809
[4] Nakagawa T and Abe M, *Digest 33rd Annual Meeting of Magn. Soc. Jap.*, (2009) **12** pD-6

Journal of Physics: Conference Series **352** (2012) 012035 doi:10.1088/1742-6596/352/1/012035

Development of high sensitive magnetic contaminant detection system using an HTS-rf-SQUID covered with HTS thin films

Y Hatsukade, R Kurosawa, Y Uchida, and S Tanaka
Department of Environmental & Life Sciences. Toyohashi University of Technology, 1-1 Hibarigaoka, Tenpaku-cho, Toyohashi, Aichi 441-8580, Japan

E-mail: hatukade@ens.tut.ac.jp

Abstract. For high sensitive detection of magnetic contaminant in electrode of lithium-ion battery, high-temperature superconductor (HTS) radio-frequency (rf) superconducting quantum interference devices (SQUIDs) based on a bicrystal $SrTiO_3$ (STO) substrate was designed, and fabricated employing $YBa_2Cu_3O_{7-x}$ thin films of about 200 nm in thickness. To improve characteristics such as effective area and 1/f noise profile of the SQUID, HTS thin films on normal STO substrates were overlapped on a wide superconducting weak link and/or a slit of the SQUID in flip-chip configuration. The noise profiles of the SQUID covered with the films on the respective positions were well improved compared to that of the bare SQUID. A magnetic contaminant detection system was developed employing the HTS-rf-SQUID covered with the films on both the positions. Using this system, a tungsten ball of 30 μm in diameter was successfully detected with a signal to noise ratio of about 14.

1. Introduction

Detection of ultra-small magnetic contaminants is a crucial issue for manufactures producing commercial products such as electrodes of lithium ion batteries. Although the industry is requiring a technology to identify the fine magnetic particles of less than 50 μm in diameter in the products, X-ray imaging and eddy current testing, which are commonly employed as the detection method, are unable to detect such small objects when the products are moving in the production lines. Therefore, a practical and high sensitive detection system for fine magnetic contaminants is demanded.

In this study, we designed and fabricated an $YBa_2Cu_3O_{7-x}$ (YBCO) high temperature superconductor (HTS) radio frequency (rf) superconducting quantum interference device (SQUID) with a rectangular hole based on a bicrystal $SrTiO_3$ (STO) substrate as a magnetic sensor [1, 2]. The HTS-rf-SQUID based on the single-layer YBCO thin-film technology and the bicrystal Josephson junction was employed to take advantage of the high magnetic sensitivity, high time resolution, the simple fabrication process, and easy-handling of the SQUID operated in liquid nitrogen. The characteristics of the HTS-rf-SQUIDs such as effective area and 1/f noise profile were improved by covering the SQUID partially with HTS thin films in a flip-chip configuration. We developed a magnetic contaminant detection system utilizing the HTS-rf-SQUID covered with the HTS films and the characteristics and the performance of the system were investigated.

2. Design, fabrication and evaluation of HTS-rf-SQUID

The pattern of the single-layer HTS rf SQUIDs based on the bicrystal STO substrate used in this study is shown in Figure 1. Concerning the HTS rf SQUID with an inductance L, resistance R, and critical current I_c, when it is operated with SQUID electronics with a tank circuit in the so-called adiabatic regime, where the resonance frequency ω_{rf} is much smaller than both the junction characteristic frequency $\omega_c = 2\pi RI_c/\Phi_0$ and the cut-off frequency of the SQUID ring $\omega_{cut} = 2\pi R/L$, and in the hysteretic mode, where the hysteretic parameter $\beta_{rf} = 2\pi LI_c/\Phi_0 > 1$ and the noise parameter $\Gamma = 2\pi k_B T/(I_c\Phi_0) << 1$, the following relation holds between the SQUID inductance L and the flux-to-voltage transfer function $dV/d\Phi$ with an appropriate quality factor Q of the tank circuit [3]:

$$\frac{dV}{d\phi} \approx \frac{\omega_{rf} L_T}{M} = \frac{\omega_{rf}}{k}\sqrt{\frac{L_T}{L}} \qquad (1)$$

where M and k are the mutual inductance and the coupling coefficient between L and L_T of the tank circuit. In order to achieve a lower flux noise level, $dV/d\Phi$ must be larger. $dV/d\Phi$ is inversely proportional to the root of L, while L is determined by the hole shape in the case of a washer SQUID ring [4]. In this study, we employed the rectangular shape of 800 x 50 μm^2 for the hole, since the SQUID inductance with a rectangular hole is smaller than that with a square hole with the same area. The outer dimension of the SQUID was 9 x 9 mm^2.

We fabricated the HTS-rf-SQUIDs based on the design shown in Figure 1. YBCO thin film with a thickness of about 200 nm was deposited on a bicrystal STO substrate with a miss-orientation angle of 30º. By conventional photolithography and ion-etching technologies, the SQUID with the bicrystal Josephson junctions of 4 μm in width was patterned. The SQUID had superconducting weak link with width of about 4 mm on the bicrystal grain boundary (GB) due to the SQUID pattern.

The fabricated HTS-rf-SQUID was operated with a SQUID electronics developed by Juelicher SQUID GmbH [5]. The SQUID was inductively coupled with a normal-conducting antenna from the electronics using the bicrystal STO substrates as the substrate resonator, which forms the tank circuit with the antenna [6]. The characteristics of the SQUID were measured in a magnetically shielded room (MSR) with a shielding factor of about 60 dB while cooling the SQUID in liquid nitrogen. The resonance frequencies ω_{rf} was about 680 MHz. The magnetic field noise profile of the SQUID is shown in Figure 2. In the measurements, the data was measured three times, and then averaged. The white flux and field noise levels $S_\Phi^{1/2}$ and $S_B^{1/2}$ at 1 kHz were about 52 $\mu\phi_0/Hz^{1/2}$ and 120 $fT/Hz^{1/2}$, respectively. The effective area A_{eff} of the SQUID was about 0.81. We note that the white field noise level was sufficiently low although the 1/f noise arose at about a few hundreds Hz and was relatively high compared to those of HTS-rf-SQUIDs with step-edge Josephson junctions [6, 7].

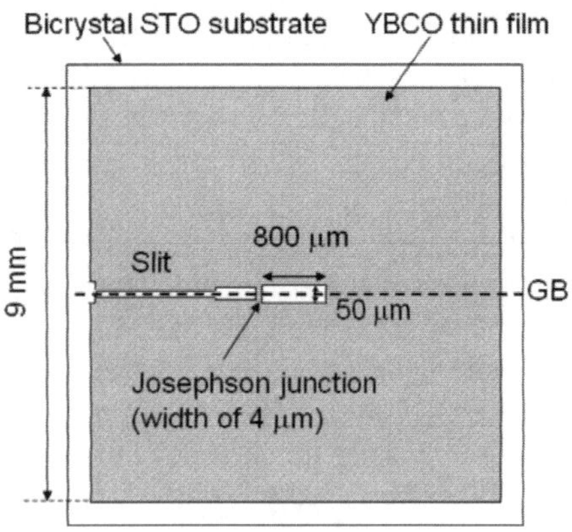

Figure 1. Design of HTS-rf-SQUID based on bicrystal STO substrate.

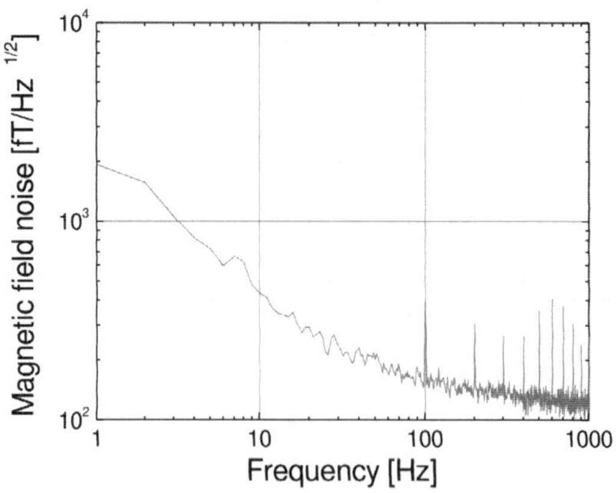

Figure 2. Magnetic field noise profile of HTS-rf-SQUID.

Journal of Physics: Conference Series **352** (2012) 012035 doi:10.1088/1742-6596/352/1/012035

3. HTS-rf-SQUID with HTS thin films in flip-chip configuration

In the SQUID pattern shown in Figure 1, the bicrystal GB runs across the washer SQUID ring to form the wide superconducting weak link, where penetration and movement of the flux vortices can easily occur [8]. For the application of the HTS-rf-SQUID to the magnetic contaminant detection, where the weak magnetic signal in the low frequency range less than 100 Hz must be measured, the reduction of the 1/f noise is strongly desirable. In order to prevent flux vortices from penetrating and moving in the wide weak link in the SQUID pattern, we fabricated HTS thin films to cover the weak link in a flip-chip configuration. YBCO thin film with the same thickness as the SQUID was deposited on a normal STO substrate without GB, and then rectangular pieces with an area of 4.1 x 2 mm^2 were cut out by a diamond saw. The cut YBCO films were positioned on the SQUID so that one of them covered the wide weak link while the other covered the slit of the SQUID. Both films were set carefully so as not to cover the SQUID hole. The positions of the films on the SQUID are indicated in the inset of Figure 3 with the dashed lines. The former film was expected to serve as a superconducting shield with its Meissner effect to avoid the penetration and motion of flux vortices in the weak link, while the latter was expected to serve as a superconducting shield to couple external magnetic flux to the SQUID hole, which would pass through the slit without the film, to increase the effective area. For adhesion between the films and the SQUID, we applied Apiezon grease between them, and pressed both the substrates so that the distance between the SQUID and the films became as small as possible. Polymer threads were used to reinforce the connection between the SQUID and the films tightly. We measured the characteristics such as the effective area A_{eff} and the field noise $S_B^{1/2}$ of the following four cases: the bare HTS-rf-SQUID, the SQUID with the HTS film on the GB, the SQUID with the HTS film on the slit, and the SQUID with both films on the respective positions. The measurements were carried out in the same MSR while cooling the SQUID with or without the HTS films in liquid nitrogen. The resonance frequency of the SQUID with the films slightly changed from the original value of 680 MHz of the bare SQUID. All the measurement data were obtained three times with different HTS films, and then averaged.

The field noise profiles of the SQUID with the HTS films on the GB and/or the slit are shown in Figure 3. The noise profile of the bare SQUID mentioned above is shown by the dotted line for comparison. The characteristics of all the SQUIDs with and without the films are summarized in Table I. The SQUID with the film on the GB shows superior noise characteristics to the others, especially in the low frequency range of less than 10 Hz. On the other hand, the noises from 10 to 100 Hz of the SQUIDs with the film on the slit and the films on both positions were better than that with only the film on the GB. This must be due to the fact that the effective areas A_{eff} of the formers were larger than that of the latter. The coverage of the slit contributed to increase the area by about 13 % as expected. In addition, it is noteworthy that the 1/f noise was also reduced compared to the bare SQUID. It is inferred that the density of the external flux must be higher around the edges on the slit due to focusing effect of a HTS thin film. The decrease in the 1/f noise should be because the coverage of the film on the slit contributed to the reduction of the flux trapping on the edges. We also note that the coverage of the GB also contributed to increase the effective area moderately. As shown in ref. 9, the increase in the number of flux trapped in a HTS film indicates the increase in non-superconducting area of the film. Therefore, we think that the possible removal of flux trapping in the GB by the coverage with the HTS film may serve to recover the actual effective area of the SQUID. Not as expected, the 1/f noise in the low frequency range of less than 10 Hz in the case of the SQUID with the two films was highest among the SQUIDs with the films, although the noise profile is better than that of the bare SQUID. In this case, the SQUID hole was left uncovered by the films. Therefore, the large 1/f noise with the coverage of both the GB and the slit may be due to the flux trapping at the edges of the SQUID hole, where the external flux density must be higher.

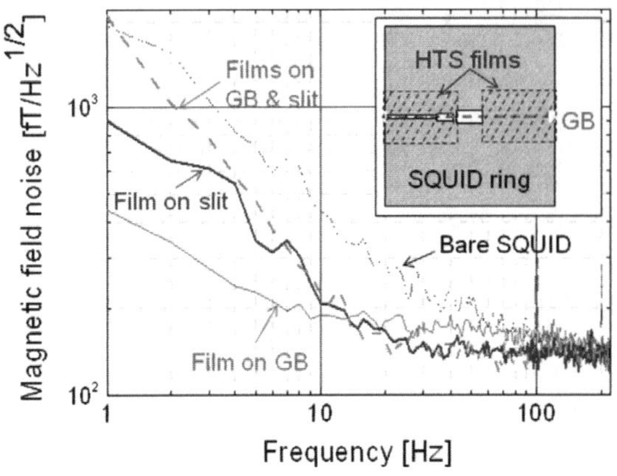

Figure 3. Field noise profiles of HTS-rf-SQUID with and without HTS thin films. The inset indicates the positions of the HTS films overlapping on grain boundary (GB) and slit in flip-chip configuration.

Table 1. Characteristics of HTS-rf-SQUID with and without HTS films on GB and slit at 77 K.

SQUID condition	A_{eff} [mm^2]	$S_B^{1/2}$ [fT/Hz$^{1/2}$] @ 1 Hz	$S_B^{1/2}$ [fT/Hz$^{1/2}$] @ 10 Hz	$S_B^{1/2}$ [fT/Hz$^{1/2}$] @ 100 Hz
Bare SQUID	0.81	1930	440	240
Film on GB	0.83	440	190	170
Film on slit	0.92	900	210	140
Films on GB & slit	0.93	2100	225	135

4. Magnetic contaminant system using HTS-rf-SQUID covered with HTS thin films

We developed a magnetic contaminant detection system with the HTS-rf-SQUID covered with the HTS films on both the GB and the slit utilizing the SQUID microscopic technique. A block diagram of the system is depicted in Figure 4 (a). The system consists of the covered HTS-rf-SQUID in the microscopic cryostat, SQUID electronics, electromagnetic and magnetic shielded cases, permanent magnet, belt conveyers, filters and PC. In this system, an object under test including a fine magnetic contaminant is set on the belt conveyer, and then magnetized in the vertical direction by the permanent magnet of 0.25 T. After the magnetization, the object is conveyed under the SQUID, which was installed in the cryostat and is cooled at about 77 K conductively, at a velocity of 100 mm/s. The remnant magnetic signal from the magnetized contaminant is measured by the SQUID, and the SQUID output voltage is recorded through a high-pass filter with a cut-off frequency of 0.5 Hz and a low-pass filter with a cut-off frequency of 20 Hz in the PC. The schematic figure of the SQUID mounted in the cryostat is illustrated in Figure 4 (b). The SQUID with the films was fixed with silver past on the sapphire rod, which was connected to a copper tank filled with liquid nitrogen. In order to reduce the liftoff distance, the normal STO substrates with original thickness of 0.5 mm were manually filed such that the thickness became to be about 0.25 mm. A sapphire window with thickness of 0.5 mm separates the SQUID and the room temperature. The minimum liftoff distance between the SQUID to an object is estimated to be about 1 mm. Figure 5 shows the noise profile of the covered SQUID installed in the system. For comparison, the noise of the bare SQUID without the films in the system is shown together. The noise level of the covered SQUID between 6 Hz to 100 Hz was lower than that of the bare SQUID.

By this system, detection of fine tungsten balls with diameters of less than 100 μm was examined. Figure 6 (a) shows the experimental results of a tungsten ball of 100 μm in diameter measured with the bare SQUID and the covered SQUID with the liftoff distances of about 1.5 and 1.75 mm, respectively. Due to the larger effective area, the signal measured with the covered SQUID was about 10 % larger than that measured with the bare SQUID, even though the liftoff distance of the former was a bit longer than that of the latter. The smallest tungsten ball of 30 μm in diameter was successfully detected with a signal to noise ratio (SNR) of about 14. Figure 7 (b) shows the relationship between diameter of the sample tungsten ball and signal amplitude measured from the samples. As shown, the signal amplitude is approximately proportional to the cubic of the sample diameter. Since the maximum peak-to-peak noise amplitude of the system was about 0.5 pT, it is demonstrated that a tungsten ball of diameter of about 15 μm can be detectable by the system with the SNR of at least 2.

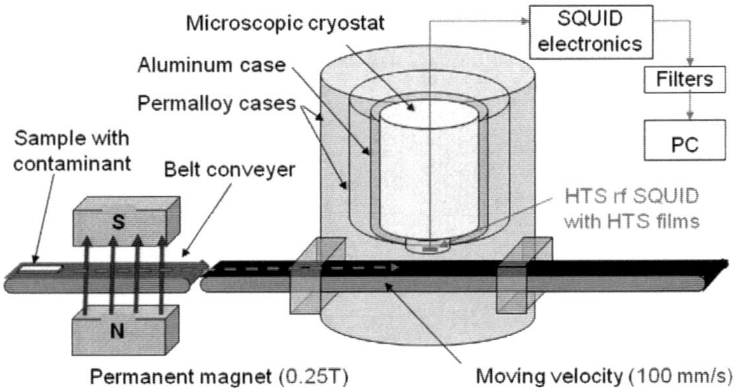

(a)

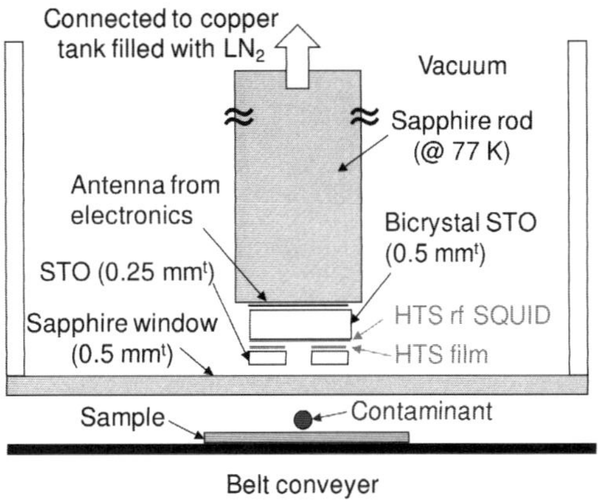

(b)

Figure 4. (a) Block diagram of magnetic contaminant detection system consisting of HTS-rf-SQUID in microscopic cryostat, electronics, magnetic and electromagnetic cases, belt conveyers, permanent magnet, filters, and PC.

(b) HTS-rf-SQUID covered with HTS films installed in microscopic cryostat.

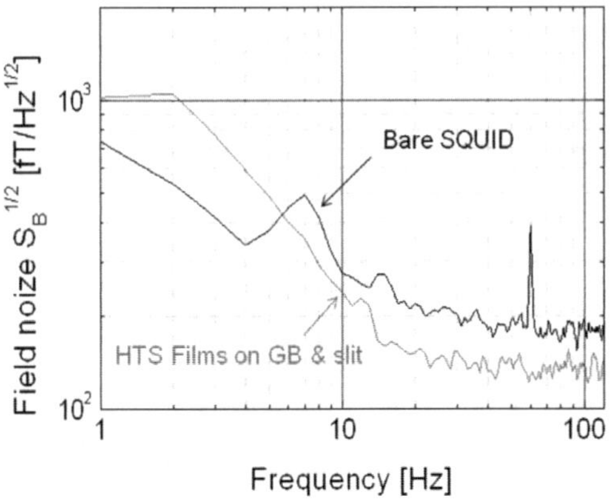

Figure 5. Field noise profile of HTS-rf-SQUID with and without HTS thin films installed in the magnetic contaminant detection system.

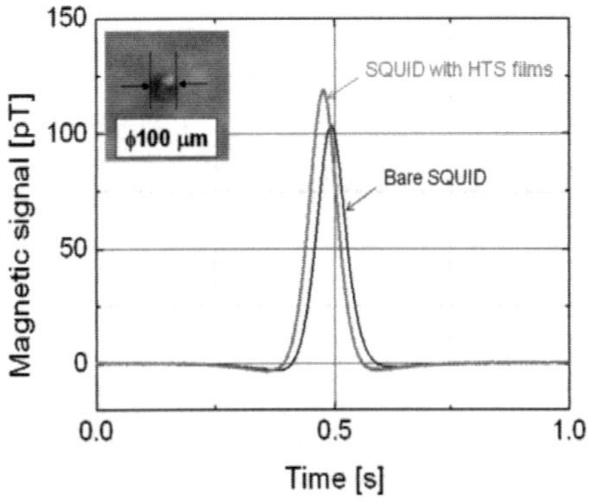

(a)

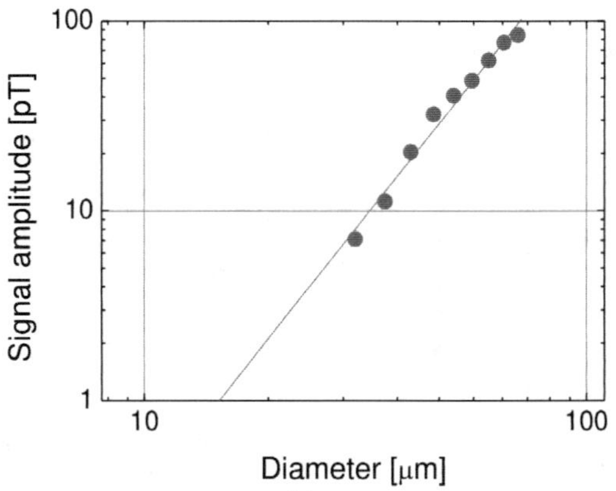

(b)

Figure 6. (a) Comparison of magnetic signals from tungsten ball of 100 μm in diameter measured with and without HTS films. (b) Signal amplitude as a function of diameter of tungsten ball.

5. Conclusion

The YBCO HTS-rf-SQUID was designed and fabricated on the bicrystal STO substrate. By covering the GB and/or the slit of the SQUID with the films on the normal STO substrates in the flip-chip configuration, the effective area and the 1/f noise profile of the SQUID were successfully improved. With the two films on both the GB and the slit, the SQUID had not only the largest effective area of 0.93 mm^2, but also relatively large noise of 1 – 2 pT/Hz$^{1/2}$ at 1 Hz. By installing the covered SQUID in the magnetic contaminant detection system, it is successfully demonstrated that the system has a potential to detect a tungsten ball of 15 μm in diameter with the SNR of at least 2.

Acknowledgements

This work is partially supported by "The Knowledge Hub" of Aichi, The Priority Research Project.

References

[1] Takemoto M, Akai T, Kitamura Y, Hatsukade Y and Tanaka S 2011 *IEEE Trans. Appl. Supercond.* **21** 522
[2] Chen J -C, Chen K -L, Yang H -C, Wu C -H and Horng H-E 2007 *Appl. Phys. Lett.* **90** 153504
[3] J. Clarke and A. I. Braginski (Eds.) 2004 *The SQUID Handbook Vol. I* (Wiley, Weinheim,) p 14
[4] Ketchen M B 1991 *IEEE Trans. Magn.* **27** 2916
[5] http://www.jsquid.com/index.htm/
[6] Zhang Y, Schubert J, Wolters N, Banzet M, Zander W and Krause H –J 2002 *Physica C* **372-376** 282
[7] Zhang Y, Yi H R, Schubert J, Zander W, Krause H -J, Bousack H and Braginski A I 1999 *IEEE Trans. Appl. Supercond.* **9** 3396
[8] Hirano S, Oyama H, Matsuda M, Morooka T, Nakayama S and Kuriki S 2001 *IEEE Trans. Appl. Supercond.* **11** 925
[9] Straub R, Keil S, Kleiner R and Koelle D 2001 *Appl. Phys. Lett.* **78** 3645

Journal of Physics: Conference Series **352** (2012) 012036 doi:10.1088/1742-6596/352/1/012036

Detection System for Metallic Contaminants by Eight-Channel SQUIDs

Y Kitamura[1], Y Hatsukade[1], S Tanaka[1], T Ohtani[2] and S Suzuki[2]
[1]Toyohashi University of Technology, 1-1 Tempaku-cho Toyohashi, Aichi 441-8580, Japan
[2]Advanced Food Tech Co., Ltd ., 333-9 Hamaike Nishimiyuki-cho Toyohashi, Aichi 441-8117, Japan

E-mail:y083813@ edu.imc.tut.ac.jp

Abstract. We developed a magnetic contaminant detection system for industrial products such as electrode foils of lithium ion battery employing eight high-T_c SQUID gradiometers. The system was based on pre-magnetization of a contaminant in an object under test by means of permanent magnets of 0.5 T, which magnetization direction was horizontal, in order to suppress the edge effect from the object composed of magnetic material. The object was conveyed to pass under the eight-channel gradiometer array, in which a pair of four gradiometers was aligned in two rows to cover target foils of several tens mm in width. The magnetization from the contaminant in the object was detected by the gradiometers of averaged flux white noise level of 25 $\mu\phi_0/Hz^{1/2}$. In case that an iron ball passed just under one gradiometer, an iron ball of about ϕ30 µm in diameter was successfully detected with a signal to noise ratio (S/N) of 5. From measurement results using an iron ball of about 100 µm in diameter, it was demonstrated that the system had a detectable range of 70 mm in width. There results suggest that the system is a promising tool for the quality control of lithium ion batteries.

1. Introduction

Recently, in the fields of leading-edge industry and the food processing, fine metallic contaminants mixed in the products cause the serious problems. Therefore, the metallic contaminant detection process becomes indispensable on the production lines. In such detection on the lines, precise detection technology is demanded. As the conventional methods, induction coil or x -ray computed tomography scanner have been employed. However, the induction coils are lacking in the sensitivity, and the x -ray computed tomography scanners are lacking in the time resolution. Thus, it is difficult by them to detect the metallic contaminants of less than 100 µm in diameter on the product lines [1-4]. However, there is an urgent need for precision industry products such as lithium-ion electrodes, in which 10µm–order size magnetic contaminants cause some faults. Furthermore, the matrix of industrial products can be magnetized and magnetic signals from the matrix are sometimes sufficiently large to mask signals from contaminants in many cases. Thus, we have developed a detection system employing high-T_c SQUID gradiometers and horizontal magnetization of such magnetic products in a longitudinal direction [5-7]. Since this system was the prototype which used a few SQUIDs, the detectable area was restricted. In order to detect the electrode with a width of several tens mm on the production line, it is necessary to arrange more sensors to cover the full width of the electrode. In this work, we developed a multi-channel contaminant detection system, in which eight SQUIDs are arranged two-dimensionally. The system characteristics were evaluated.

2. Detection system for metallic contaminants

2.1. Principle of detection of contaminant

A principle of the detection of the contaminant is shown in Figure 1. The contaminant detection method using SQUIDs require to magnetize a specimen with a permanent magnet to detect remnant magnetization of the contaminant. Since lithium ion electrodes often contain magnetic materials, they can be also magnetized. In order to distinguish the magnetization of the specimen and that of the contaminant, we employed the horizontal magnetization method in this work. By using this method, the magnetization field from the specimen appears only from the both ends of it as shown in Figures 1. On the other hand, the contaminant generates a local magnetization field near by the contaminant. By detecting the local magnetic field by a SQUID gradiometer, the signals from the matrix and that from the central contaminant are separable as shown in Figure 2.

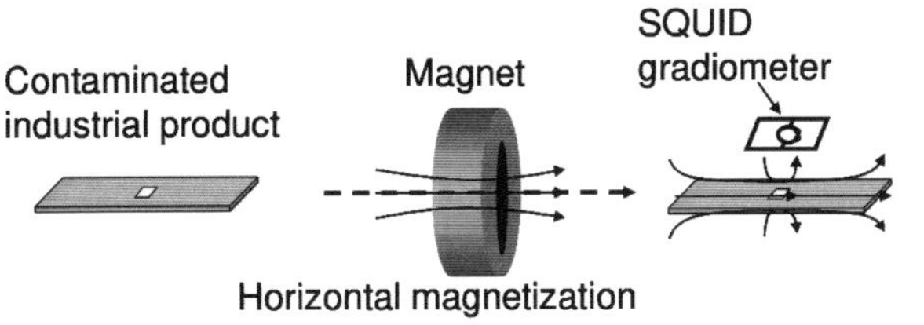

Figure 1. Diagram of detection method for fine metallic contaminant.

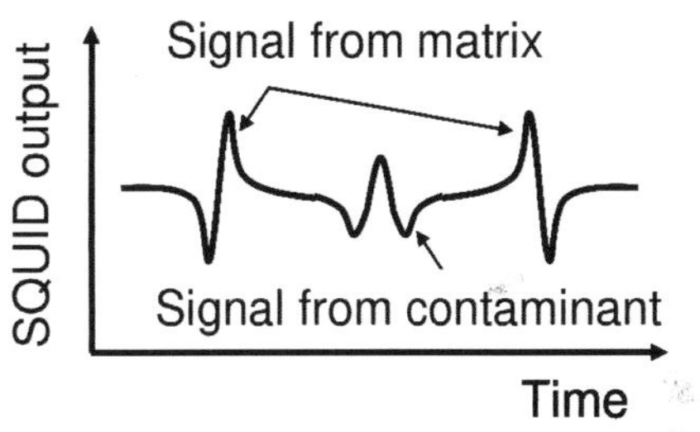

Figure 2. Signal from metallic contaminant in magnetic matrix measured by high-T_c SQUID gradiometer.

2.2. Detection system

Figure 3 shows the multi-channel high-T_c SQUID detection system for metallic contaminants, which we have developed. In the system, microscopic cryostat, in which Eight-channel SQUID array was mounted, is installed in the double-layer magnetic shield for reduction of environmental magnetic noise. In addition to the cryostat, the system is composed of flux locked loop (FLL) circuit [8] for eight channels, and a belt conveyor for movement of a specimen and a PC to store data. The dimensions (length × width × height) of the entire system are 2000 mm × 750 mm × 1550 mm. This contaminant detection system is aimed to be used in manufacturing process of lithium ion electrodes and the likes, thus the belt conveyor conveys a sample continuously at the speed of about 100 mm/sec. Eight high-T_c SQUID gradiometers were mounted at the tips of the sapphire rods using silver paint as adhesive. The rods are connected to a copper tank in the cryostat. As shown in Figure 4, the SQUIDs were arranged with intervals of 28 mm in the lateral direction and 18 mm in the longitudinal direction between the centres of the gradiometers for inspection of a full width of the lithium-ion electrode foil of about 70 mm. By arranging the SQUIDs in two rows, the second series of the 4 SQUIDs can cover the area between the front series of the 4 SQUIDs. Liquid nitrogen is poured in the tank in the cryostat and the SQUID gradiometers are cooled at near 77 K through the sapphire rods. The averaged white noise level of all the SQUID gradiometers was 25 $\mu\phi_0/Hz^{1/2}$ at 100 Hz. The magnetized specimen passes under the eight SQUIDs, and the remnant magnetization from the contaminant is measured. We record the outputs of the SQUIDs with a low pass filter with the cut-off frequency of 20 Hz and high pass filter with the cut-off frequency of 0.2 Hz.

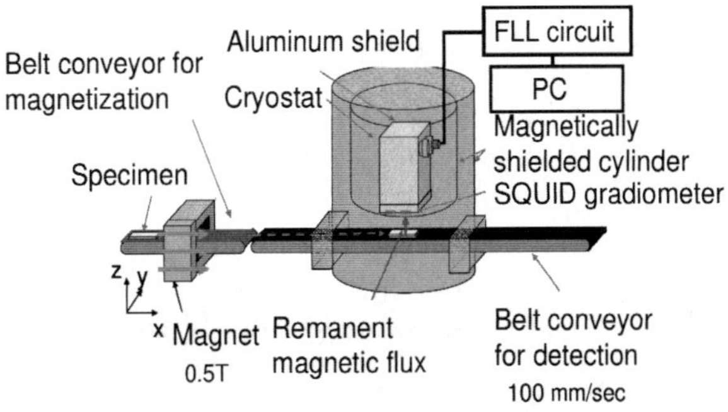

Figure 3. Block diagram of contaminant detection system.

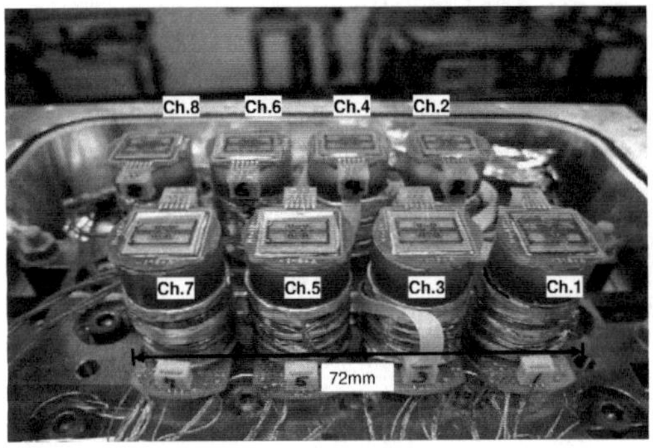

Figure 4. Arrangement of SQUIDs on tips of sapphire rods.

3. Characterization of system

3.1. Detection of contaminant on magnetic material

An electrode foil of a lithium-ion battery that we target is often made of magnetic materials. Therefore, we prepared a brass plate specimen as a magnetic material, on which an iron ball of 75 μm in diameter was fixed .The size of the brass plate used as a magnetic matrix is 100 mm× 10 mm × 1 mm. The brass specimen was inspected by the SQUID system. The sample was magnetized horizontally by the permanent magnet of about 0.5 T, and it was transported under the SQUID array by the belt conveyer. The remnant magnetization from the edges of the brass specimen was measured at 7.35 s and 8.37 s when each edge of the specimen passed under the SQUID, which positioned nearest to the specimen, as shown in Figure 5. On the other hand, a larger signal was detected at 7.85 s when the contaminant passed under the SQUID. From the result, it was demonstrated that it is possible to discriminate the contaminant signal from the matrix signal by this system.

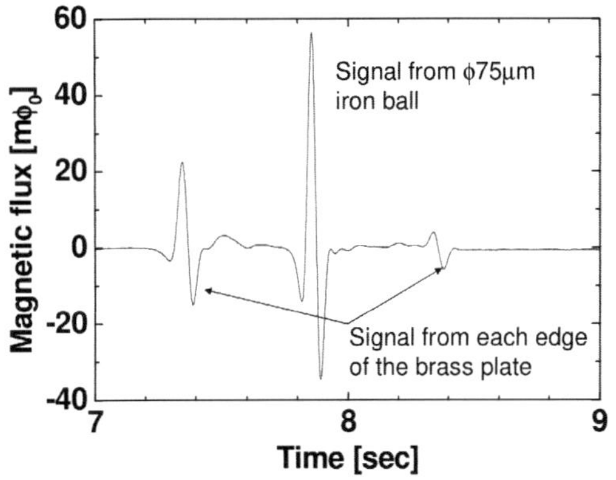

Figure 5. Detection result of a contaminant on magnetic material.

3.2. Relationship between diameter and signal of contaminant

Magnetic signals from contaminants with different diameters were measured with the developed contaminant detection system. Fine iron (S50C) balls with diameters of 27-95 μm were prepared as contaminants. Each iron ball was set on a glass plate. The iron balls on the glass plates were conveyed to pass just under one SQUID gradiometer of the array. The lift-off distance between the contaminant and the SQUID array was about 2-2.5 mm. The results are shown in Figure 6. In this experiment, one of the eight SQUID gradiometers was used. The relationship between measured signal amplitude and diameter of contaminant is shown in Figure 7. As shown in the figure, signal amplitude of contaminant was approximately proportional to the square of diameter of contaminant. The averaged peak-to-peak noise in the measurements was about 1.5 mϕ_0. Actually, a signal to noise ratio (S/N) necessary for proper detection of contaminant is 3 or more. From the result in Figure 7, it is suggested that the system can detect an iron ball of 20 μm in diameter with the S/N of 3. In this study, when an iron ball was in the center of a brass plate, the signal was successfully separated from the magnetization signals of the edges of the plate as shown in Figure 5. The averaged noise level except for the signal of the edges of the plate was 1.5 mϕ_0. From the result in Figure 7, since the signal strength from an iron ball of about 30 μm in diameter was 10 mϕ_0, it was shown that S/N was more than 3. Under these conditions, it is considered that measurement of ϕ20 μm of an iron ball on a brass plate is possible. We used brass plates as an object. However, we are planning to use electrode foil of lithium ion battery in stead of brass plates from now on. We used brass plates as an object. However, we are planning to use electrode foil of lithium ion battery in stead of brass plates from now on.

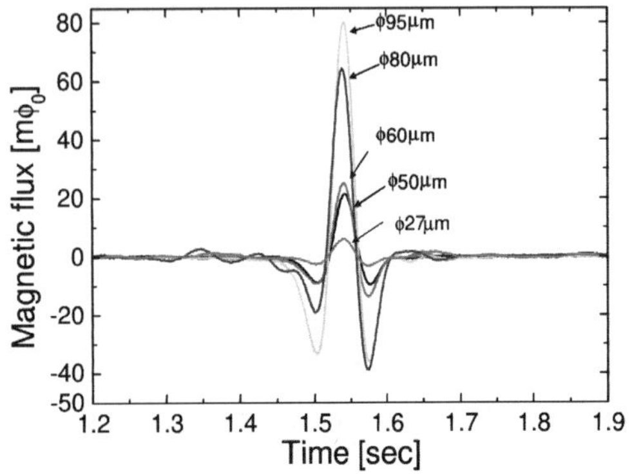

Figure 6. Measurement results of iron balls with different diameters.

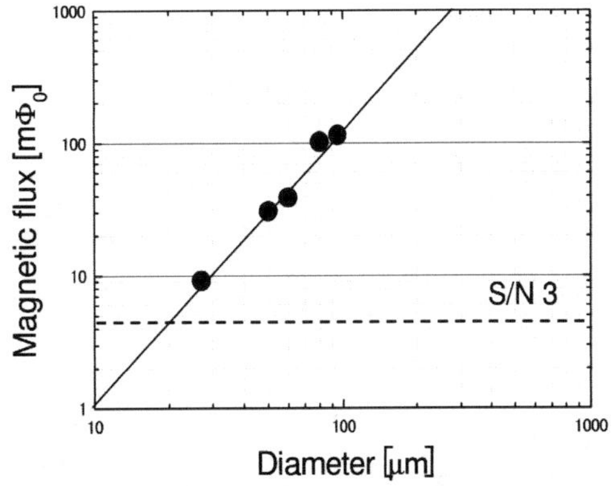

Figure 7. Peak-to-peak amplitude as a function of diameter of iron bal

3.3. Detectable area of system

In order to evaluate the sensitive width of the eight gradiometers, we prepared iron (S50C) balls with a diameter of 95 μm. The lift-off distance between the iron piece and the SQUID gradiometer was 2 mm. The position of the ball placed on the conveyor was changed in steps of 1 mm along the direction perpendicular to the conveyor motion until the entire range of (-38 mm to 46 mm) was covered. The object was magnetized in the same manner and the remnant signal from the iron ball was simultaneously detected by the eight SQUID gradiometers. The peak-to-peak values from Ch.1 and Ch.8 are plotted as a function of the offset distance in Figure 8. The position of each peak is where each sensor is located. The detectable width range of one SQUID gradiometer is estimated to be about 15 mm from Figure 8. By adding all the signals from the eight gradiometers, the valleys between the adjacent gradiometers in the signal outputs can be compensated as shown in Figure 8 by a dotted line. The noise level obtained from time traces was approximately 2 mϕ_0. If the detection width was defined as a range where the signal-to-noise ratio (S/N) was greater than 3 (corresponding to a signal level of 6 mϕ_0), the detection width of eight SQUID gradiometers was 72mm.

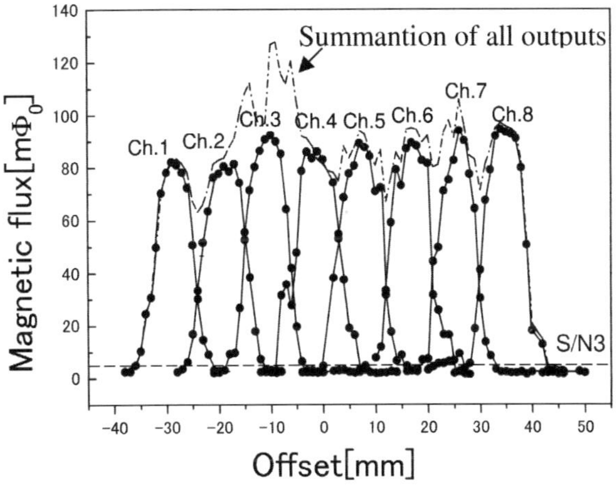

Figure 8. Distribution of peak-to-peak values from an iron ball measured by eight gradiometers. The size of the iron ball was 95 μm in diameter. The detectable width of the eight SQUID gradiometers was 72 mm if the detectable width was defined as a range where the signal-to-noise ratio (S/N) is greater than 3.

4. Conclusion

We developed the multi-channel contaminant detection system using eight channel high-T_c SQUID gradiometer array and the horizontal magnetization method for the industrial products such as the lithium ion electrodes. It was shown that this system has a capability to detect an iron ball of about 20 μm in diameter with the S/N of 3 or more at a lift-off of about 2 mm. It was demonstrated that the detectable range on the belt conveyer was bout 72mm in the case of the iron ball of 95 μm in diameter, thanks to the two-dimensionally arranged SQUID array. The iron ball with a diameter of 75 μm on the brass plate was detected with enough S/N while separating the signal from the ball from the signals from the plate by use of the gradiometers and the horizontal magnetization. The success of the eight-channel system proposed in this paper encourages us to apply this system to a production line on site.

References

[1] Nagaishi T, Ota H, Nishi K, Kuwa K, Fujita T and Tanaka S 2007 *IEEE Trans. Appl. Supercond.* **17** 800-3

[2] Tanaka S, Fujita H, Hatsukade Y, Nagaishi T, Nishi K, Ota H, Otani T and Suzuki S 2006 *Supercond. Sci. Technol.* **19** 280-3

[3] Krause H J, Panaitov G I, Wolter N, Lomparski D, Zander W, Zhang Y, Oberdoerffer E, Wollersheim D and Wilke W 2005 *IEEE Trans. Appl. Supercond.* **15** 729-32

[4] Bick M, Sullivan P, Tilbrook D L, Du J, Thorn B, Binks R, Sharman C, Leslie K E, Hinsch A, Macrae K and Foley C P. A 2005 *Supercond. Sci. Technol.* **18** 346-51

[5] Tanaka S, Akai T, Hatsukade Y, Ohtani T, Ikeda Y, Suzuki S and Tanabe K 2009 *IEEE Trans. Magn.* **45** 4510-13

[6] Tanaka S, Akai T, Hatsukade Y, Otani T, and Suzuki S *IEEE Trans. Appl. Supercond.* 2009 **19** 882-85

[7] Tanaka S, Akai T, Kitamura Y, Hatsukade Y, Otani T and Suzuki S 2011 *IEEE Trans. Appl. Supercond.* **21** 432-35

[8] J. Clarke and A. I. Braginski (Eds.) 2004 *The SQUID Handbook Vol. I* (Wiley, Weinheim,) p 11

Image Identification Based on Color and Luminance Information by Using an Optical Correlator

H. Kuboyama[1*], K. Moriyama [1], S. Arai [1], M. Fukuda [1], M. Kato [2], T. Kawaguchi [2], S. Yamamoto[3] and M. Inoue [1]

[1]Department of Electrical and Electronic Information Engineering, Toyohashi University of Technology, 1-1 Hibarigaoka, Tempaku-cho, Toyohashi, Aichi 441-8580, Japan
[2]PaPaLaB Ltd., 3-1-7 Wachiyama, Naka-ku Hamamatsu, Shizuoka 433-8123, Japan
[3]Hamamatsu University School of Medicine, 1-20-1 Handayama, Hamamatsu, Shizuoka 431-3192, Japan

E-mail: fukuda@ee.tut.ac.jp

Abstract. A technique for distinguishing real-world images has been developed using a joint transform correlator. The real-world images, taken with a digital camera, are converted into luminance-intensity histograms and x-y chromaticity diagrams in the correlator system. These two-dimensional patterns are displayed on a spatial light modulator and are distinguished using the correlator system. The proposed identification techniques could distinguish an object image similar to the reference image of an apple among various fruits and vegetables, and could also distinguish tumor cell images from other cell images using the correlator system. These experimental results demonstrate the feasibility of using an optical correlator for the identification of complex real-world images containing large volumes of information.

1. Introduction

The optical correlator is generally used for pattern recognition, and involves the use of an optical Fourier transform technique. For facial recognition and fingerprint recognition systems, several techniques for distinguishing the pattern information have been reported [1-2]. However, image identification using pattern information is only applicable to images having shape information, like faces and fingerprints. Also, if the image pattern is expanded or rotated, pattern recognition is markedly degraded in the optical correlator. In other words, image identification based on pattern information lacks flexibility, and has difficulty in distinguishing real-world images containing large volumes of information. Accordingly, we have investigated the influence of statistical color and luminance information on image recognition to address the weaknesses of the optical correlator.

The purpose of this study is to distinguish real-world images with techniques based on the statistical color and luminance information of the images on the optical correlator. In this paper, the details of these techniques and the experimental results for image identification using the developed techniques are described. The characteristics and architecture of the optical correlator are described in section 2. In section 3, the techniques for using the statistical color and luminance information of the image are described. In section 4, the experimental results are described and discussed. The results obtained in this study are summarized in section 5.

2. Joint transform correlator

Two main configurations are used to perform spatial correlation using optical systems: the VanderLugt correlator (VLC) [3] and the joint transform correlator (JTC) [4]. Compared to the VLC, the JTC has a disadvantage in that the processing speed is low. However, the JTC has the important advantage that images can be updated in real time. The JTC has two further advantages with regard to the hardware: compact size and a simple optical system without the requirement for complicated adjustments to the optical axis, and it is therefore smaller than the VLC. Consequently, a compact "slot-in type" optical correlator was developed previously [5].

2.1. Experimental set-up

The optical system of a JTC is shown in Fig. 1. A laser diode operating at a wavelength of 658 nm is used as a coherent source. A spatial light modulator (SLM) and a CMOS sensor are used as the system input plane and output plane, respectively. The light emitted from the laser diode is collimated with a lens having a focal length, *f*, of 100 mm. The collimated light is transformed from elliptically polarized light to linearly polarized light by the polarizer, with a polarization angle of 45° toward the horizontal direction, and the transformed light illuminates the SLM displaying the reference pattern (the letter "C") and the retrieved pattern (the letter "C") at a distance of $\pm d$ in the x-direction from the central part of the SLM (Fig. 1a). The reflected light from the SLM is phase-modulated according to the grey-scale of the input image and converted to an intensity-modulated signal with the polarizer set at a polarization angle of 135° from the horizontal axis. The intensity-modulated light is guided to the CMOS sensor by a lens with *f* of 120 mm. A Fourier-transformed pattern of the input image therefore appears on the CMOS sensor. The Fourier-transformed pattern is recorded as a power spectrum on the CMOS sensor. The recorded power-spectrum pattern is fed back to the SLM, and a similar process to that described above is repeated. The Fourier-transformed pattern of the power spectrum pattern is finally formed on the CMOS sensor. The cross-correlation spots, from which the optical intensity is expressed with the cross-correlation function of the retrieved and reference patterns, appear at a distance of $\pm 2d$ in the x-axis from the central part of the CMOS sensor (Fig. 1b).

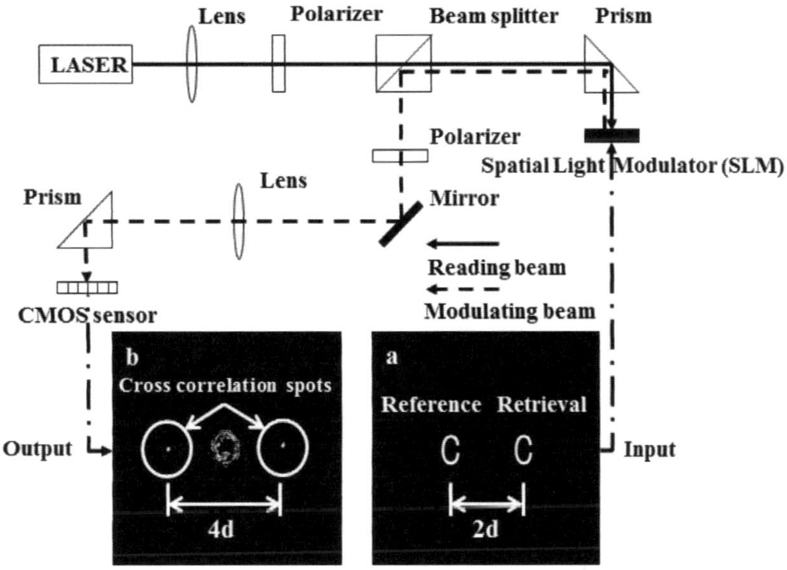

Figure 1. Block diagram of joint-transform correlator. Images (a) and (b) are the input and output images.

3. Color and luminance identification techniques

The optical correlator is generally applied to pattern recognition. Electrically addressed liquid crystal (LC) modulators are ordinarily used in optical correlators as the spatial modulators to input reference and retrieved data to the optical system. The input data are displayed with on-off (or black and white) pixel data on the modulator. The input data are therefore two-dimensional (2D) data, and displaying multi-dimensional data such as color information is difficult. Therefore, to distinguish images with the color and luminance information of the images on the optical correlator, this color and luminance information must be transformed into 2D patterns. Also, the "statistical" color and luminance information of the images is transformed into 2D patterns because the "statistical" information of the images scarcely depends on structural variation of the images, like an image expansion or rotation. The x-y chromaticity diagrams and luminance histograms were used to transform the statistical color and luminance information of the images into 2D patterns. The details of these techniques are described in sections 3.1 and 3.2.

3.1. Color identification technique

To transform the statistical color information into 2D patterns, a color specification system was used. The main color specification systems are the Munsell color system based on a color appearance system and the CIE (Commission Internationale de l'Eclairage) standard colorimetric system based on a color mixing system. Because the CIE standard colorimetric system has high precision in displaying optional colors when compared with the Munsell color system, the CIE system was used in this study. In the CIE standard colorimetric system, the techniques for transforming the color information of images into a chromaticity diagram have been reported in detail [6]. In these techniques, an x-y chromaticity diagram was used to transform the grey values of red, green, and blue included in each pixel of the images to a value given with X-Y-Z values and maps the value on to the color chart. In other words, the color information of the images can be transformed into a 2D pattern on the x-y chromaticity diagram. Accordingly, we have previously distinguished the color information of the images by pattern matching the color pattern on the x-y chromaticity diagram [7]. The procedure that is used to transform the color information of all of the image pixels into the x-y chromaticity diagram is shown in the flowchart (Fig. 2).

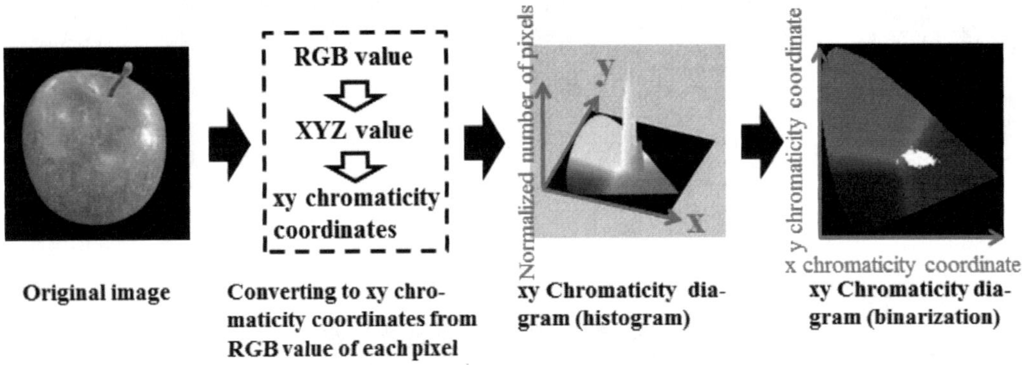

Figure 2. Flowchart for making x-y chromaticity diagrams.

3.2. Luminance identification technique

The luminance information of the images has been used for image texture analysis. In previous studies of texture analysis, some techniques that have been used to transform the statistical luminance information of images into 2D patterns have been reported [8]. From these techniques, we selected the luminance histogram. When a person sees the surfaces of an object, he or she can distinguish textures, such as glossiness and transparency, as well as color. In these textures, the glossiness and lightness that the person perceives has a high correlation with the skewness of the luminance histogram [9]. A luminance histogram is a 2D pattern where the x-axis and y-axis represent the luminance and the number of pixels, respectively. Accordingly, we have previously distinguished the texture information (luminance information) of the images by pattern matching the shape of the luminance histogram [10]. The procedure for deducing the luminance information of all of the image pixels for the luminance histograms is shown in the flowchart below (Fig. 3).

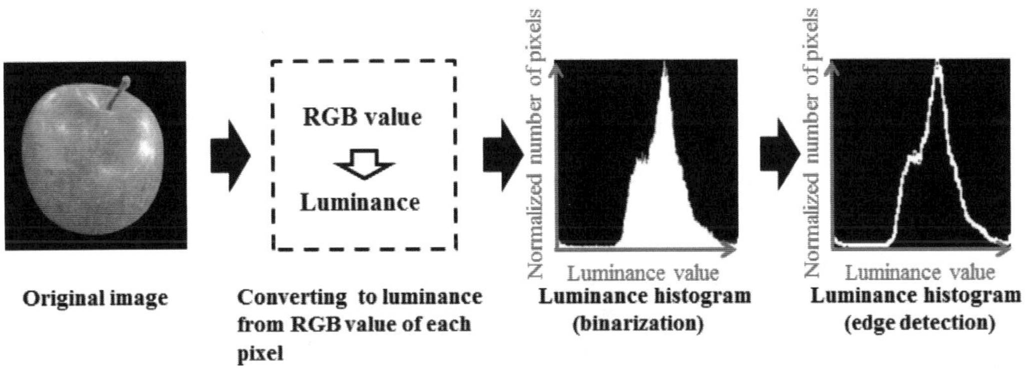

Figure 3. Flowchart for making luminance histograms.

4. Identification experiments

4.1. Images for identification experiments

The images of fruits and vegetables and the normal cell and tumor cell images taken by a digital camera were used as the real-world images. These images were distinguished with the identification techniques mentioned above on the JTC system. The details of these images are described in sections 4.1.1 and 4.1.2.

4.1.1. *Object images.*

The object images of fruits and vegetables are shown in Fig. 4. In the experiment, the images of the fruits and vegetables, which generally contain many different kinds of object information, were used (Fig. 4). Nine kinds of fruit and vegetables were photographed using the digital camera, the object patterns were extracted from the images taken, and the extracted object patterns were pasted into an image composed of 512×512 pixels (Fig. 4). In the experiment, 10 images of 9 kinds of fruit and vegetables were used.

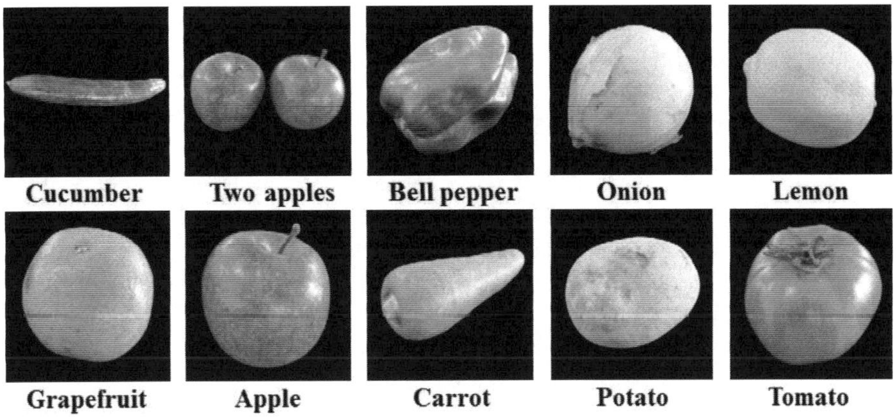

Figure 4. Images of fruit and vegetables (size of images: 512x512 pixels).

4.1.2. *Cell images.*

A whole image of a brain tumor cell of a rat taken using a digital camera in a microscope is shown in Fig. 5. The central part of Fig.5 is the tumor cell, and the surrounding part is a normal cell. Figure 6 shows the images taken of the cell of Fig. 5 by the digital camera under a microscope at 400 times magnification (Figures 6(a) and (b) are the normal and tumor cell images.). These images are composed of 2048×1536 pixels. The tumor cells have a high affinity to dye and show hyperplasia when compared to normal cells [11]. In the experiment, 30 images of normal and tumor cells, taken from the cell of Fig. 5 using the microscope with 400 times magnification, were used. In these 30 images, 15 images are the normal cell images and the other images are the tumor cell images.

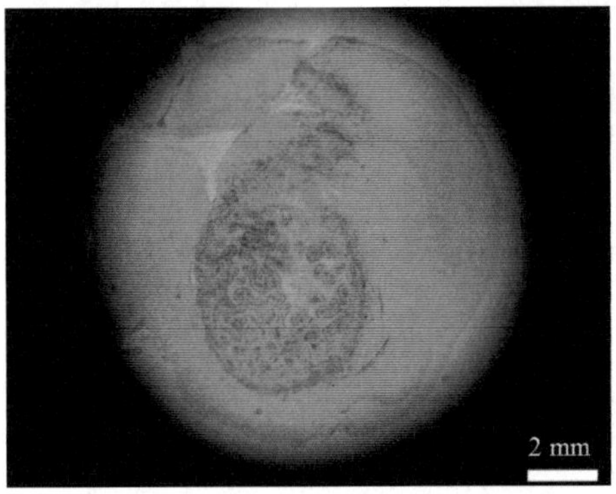

Figure 5. Whole image of a brain tumor cell of a rat (size of image: 2048×1536 pixels).

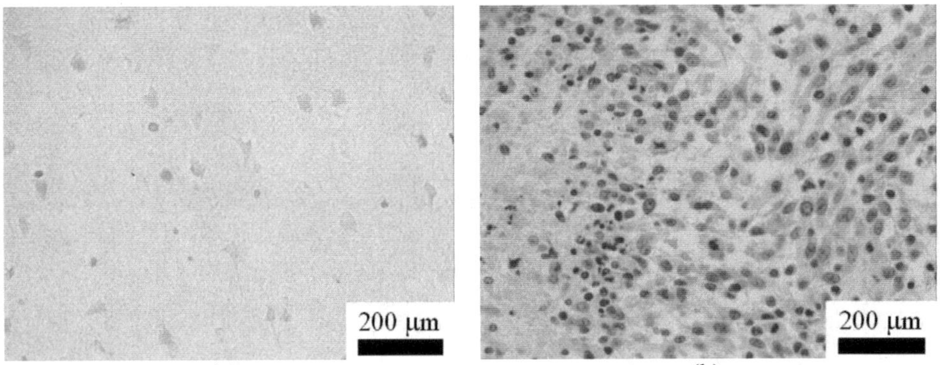

(a) (b)

Figure 6. (a) Normal cell image; (b) Tumor cell image (size of images: 2048×1536 pixels).

4.2. Identification experiment on object image

4.2.1. *Identification experiment with color identification technique.*

The color information from the object images of Fig. 4 was transformed into 2D color patterns on the color chart (Fig. 7). In Fig. 7, the color patterns of each object are distributed on each color region of the color chart (for example, the color patterns of an apple and a tomato are distributed on the red region of the color chart.). To calculate the correlation between the 2D color patterns on the JTC system, the image that placed the 2D color patterns in the same plane was given as the input image (Fig. 8), and was displayed on the SLM of the JTC system. In the experiment, the color pattern of the apple shown in Fig. 7 was used as the reference, and various object images were distinguished. Figure 9 shows examples of patterns from similar images, and shows that the intensity of the cross-correlation spot was high when the retrieval pattern coincided with the reference. The cross-correlation spot is expressed as the cross-correlation function of the retrieval and reference patterns, and the intensity of the cross-correlation spot depends on the similarity of the 2D patterns. For this reason, we assume that the intensity of the cross-correlation spot is equivalent to the similarity of the patterns.

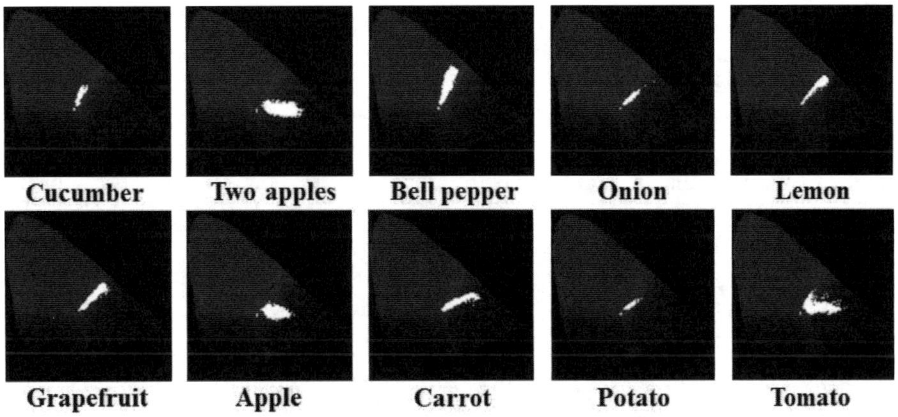

Figure 7. XY chromaticity diagrams of object images (size of images: 128×128 pixels).

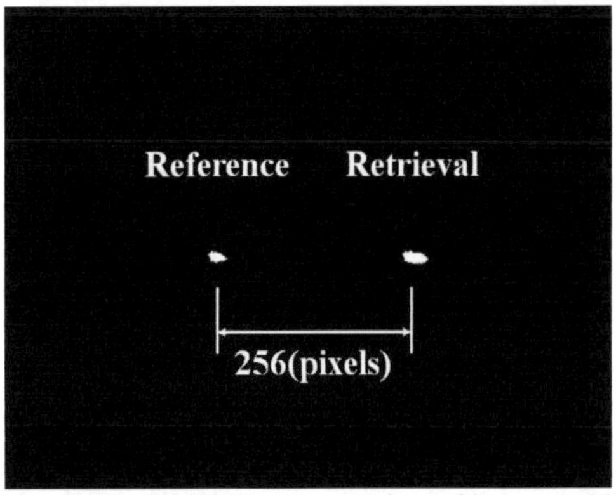

Figure 8. Input image to the JTC system (size of images: 800×600 pixels).

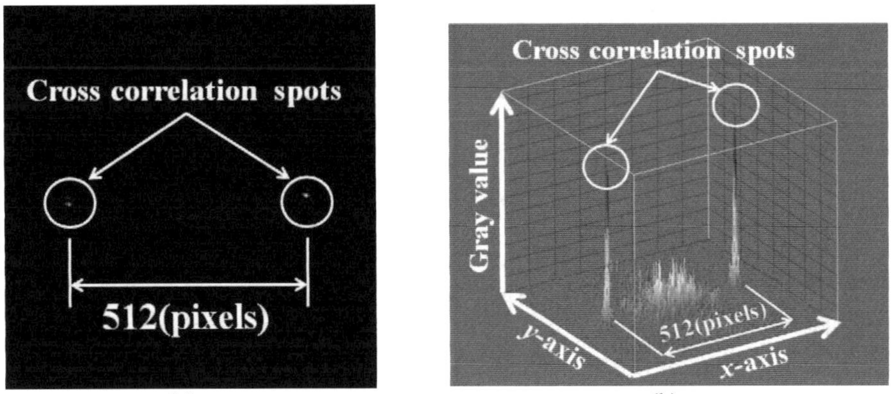

(a) (b)

Figure 9. (a) is the output image from the JTC system when the retrieval pattern coincided with the reference, and (b) is the three-dimensional image of (a).

4.2.2. *Identification experiment with luminance identification technique.*

The luminance information from the object images of Fig. 4 was transformed into 2D histograms corresponding to the luminance intensity (Fig. 10). In Fig. 10, the luminance histograms for a potato, onion, and carrot (which had matte surfaces) do not have a tail towards the high luminance direction. In contrast, the luminance histograms for the other objects (which had glossy surfaces) have a tail towards the high luminance direction.

The difference in luminance histogram among the samples depends on the glossiness of the sample surface. If surface is smooth (or rough), specular (or diffuse) reflection is generated at the surface in corresponding to the smoothness (or roughness). Here, diffuse reflection occurs over the surface of sample, while specular reflection is partially generated at the surface. Consequently, the luminance histograms of samples with glossy surface have high luminance components of low frequency and low/medium luminance components of high frequency. This results in the tail in the histogram of luminance intensity for glossy samples except for onion, carrot, and potato in Fig. 10.

In the experiment, the luminance histogram of the apple shown in Fig. 10 was used as the reference, and various object images were distinguished.

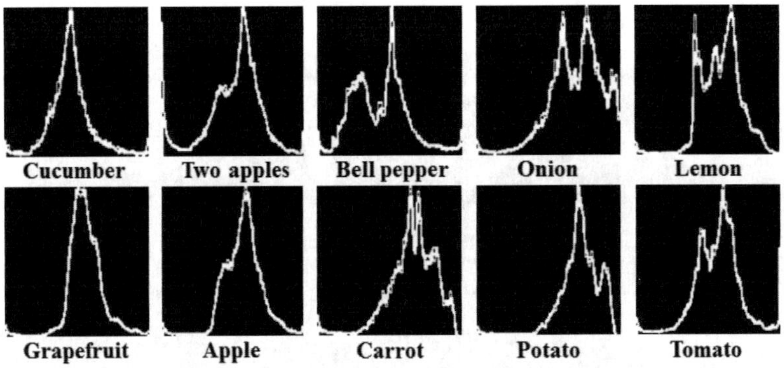

Figure 10. Luminance histograms of object images (size of images: 128×128 pixels).

4.2.3. *The results of the identification experiment with the color identification technique.*

The 2D color patterns of Fig. 7 were distinguished on the JTC system, and the results are shown in Fig. 11. The vertical axis in Fig. 11 was standardized using the intensity of the cross-correlation spot, obtained by using the same pattern as the retrieval and reference patterns. The object images were arranged according to the intensity of their cross-correlation spots along the horizontal axis of Fig. 11. In Fig. 11, the intensity of the cross-correlation spot for the tomato (i.e., a red object like the apple), the two apples and the carrot (i.e., a reddish object) was high, and that for the other objects was low. However, although the color pattern of the apple was used as the reference, the intensity of the cross-correlation spot for the tomato was higher than the intensity of the cross-correlation spot for the two apples. Therefore, this technique could not distinguish the color information of the images perfectly.

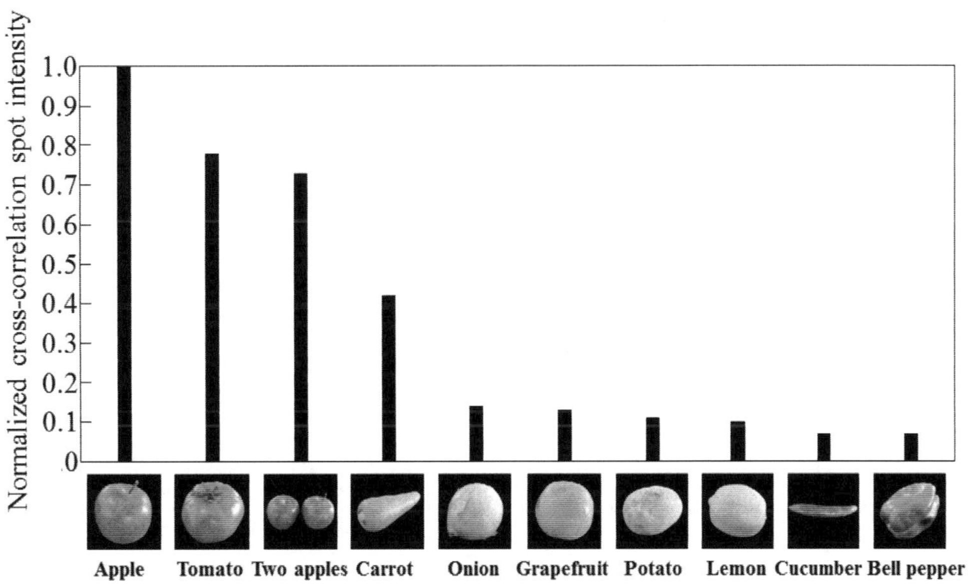

Figure 11. Results of object image identification with color identification technique.

4.2.4. *The results of the identification experiment with the luminance identification technique.*

The 2D histograms of Fig. 10 were also distinguished on the JTC system, and the results are shown in Fig. 12. In Fig. 12, the intensity of the cross-correlation spot for the two apples, the tomato, the lemon and the grapefruit (i.e., objects with a glossy surface like the apple) was high, and that for the potato, the onion, and the carrot (i.e., objects with matte surfaces) was low. However, although the green pepper and the cucumber have a glossy surface, the intensity of the cross-correlation spot for these objects was low. Therefore, this technique also could not perfectly distinguish the luminance (texture) information of the images.

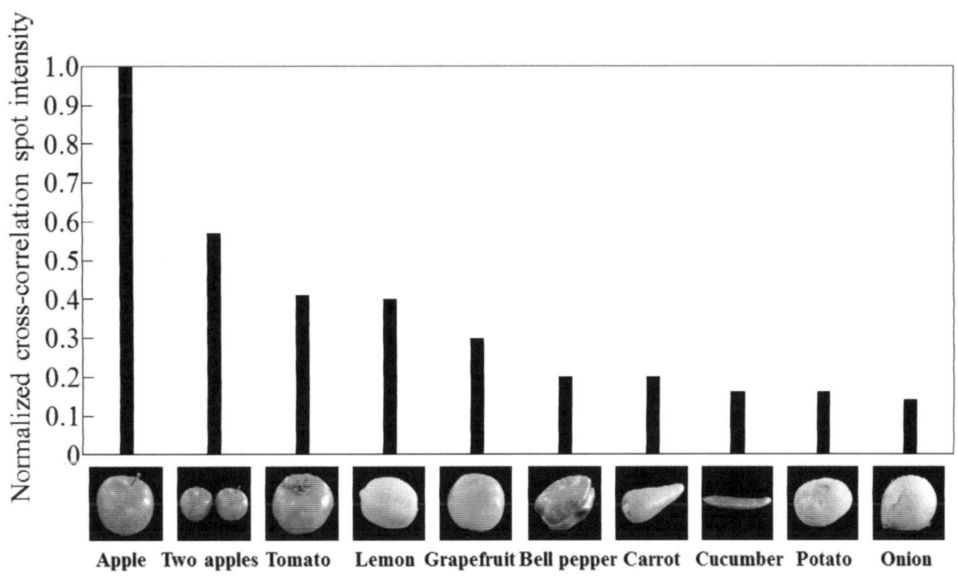

Figure 12. Result of object image identification with luminance identification technique.

4.2.5. *Comparison of proposed and previous identification techniques.*

The results of object image identification with color and luminance information, and that with pattern information were compared. The identification results found with the color and luminance information are shown in Fig. 13 as the black-filled bar. The vertical axis of Fig. 13 shows the mean value of the normalized cross-correlation spot intensity for each object of Fig. 11 and Fig. 12. Meanwhile, the identification results found with pattern information are described below. In facial recognition and fingerprint recognition systems, several techniques for distinguishing the pattern information have already been reported. One of these techniques, facial recognition with edge detection (which emphasizes the outline of the shape), has been developed, and has achieved recognition results with high accuracy [1]. Edge detection was therefore used for the identification technique with pattern information, and the object images were distinguished on the JTC system by using patterns that emphasized the outline of the object pattern (Fig. 14). The identification results with pattern information are shown in Fig. 13 as the white-filled bar.

According to the identification results with pattern information in Fig. 13, the intensities of the cross-correlation spots for the tomato, the onion, and the grapefruit (i.e., circular shapes like the apple) were high; while that for the cucumber and the carrot (i.e., slender shapes) was low. However, the intensity of the cross-correlation spot for the two apples (i.e., circular shapes) was low, because the object size of the two apples was not equal to the object size of the single apple. This technique for distinguishing shapes was therefore ineffective for images with different object sizes. According to the identification results with color and luminance information in Fig. 13, the intensity of the cross-correlation spot for the two apples and the tomato were high; and that for the other objects was low. Accordingly, although the object size of the two apples was not equal to the object size of the apple, the identification technique with the color and luminance information could distinguish the object image as being similar to the reference image of an apple.

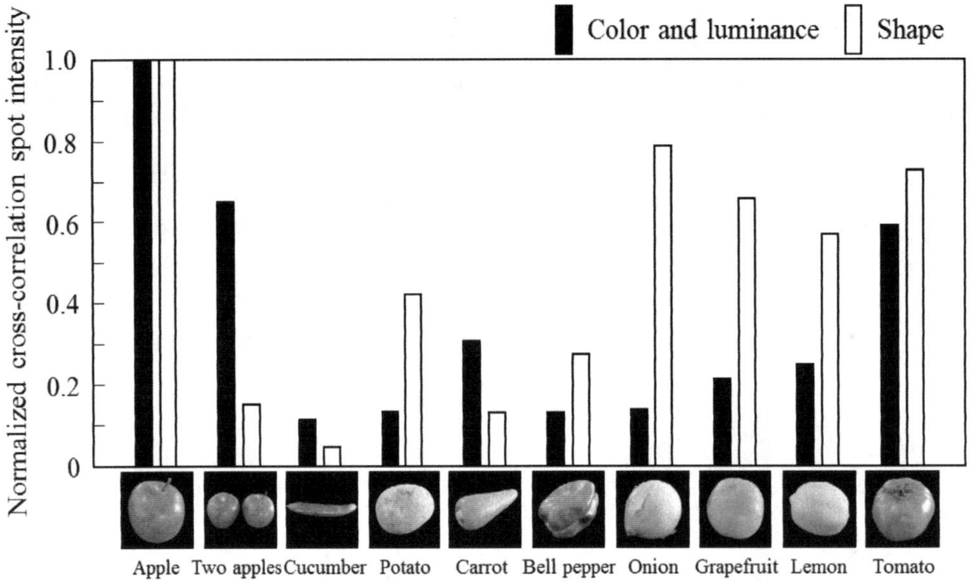

Figure 13. Result of object image identification with pattern, color, and luminance identification techniques.

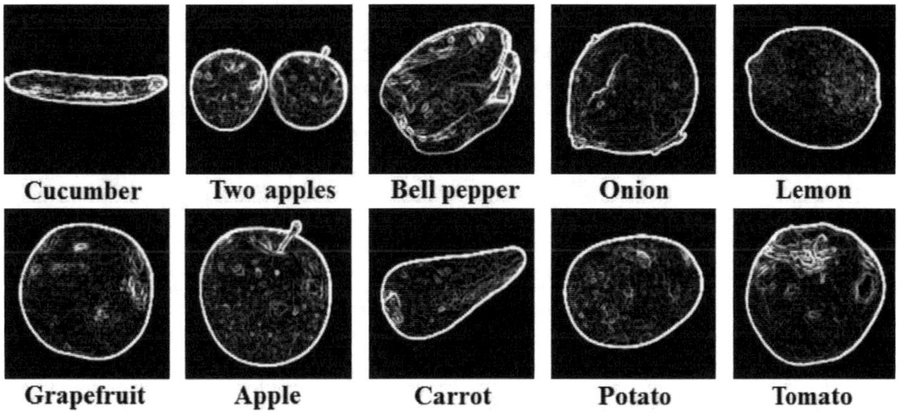

Figure 14. Outlines of fruits and vegetables (size of images: 128×128 pixels).

4.3. Identification experiment for cell images

4.3.1. *Identification experiment with color identification technique.*

Image processing was applied to the color patterns converted from the cell images to enhance the accuracy of the cell image identification. These image-processing steps are described below. The color patterns converted from Figs. 6(a) and (b) are shown in Figs. 15(a) and (b) (size of images: 128×128 pixels). As shown in Fig. 15, the color patterns converted from the tumor cell images were larger than those from normal cell images, and the color patterns converted from all cell images were distributed only on the blue region of the color chart. The reason for this is that the tumor cells have a higher affinity to dye when compared to the normal cells [11]. These color patterns were thus enlarged to emphasize the differences in the shapes of each color pattern between normal and tumor cells. Also, the enlarged color patterns were given edge detection image processing to enhance the sensitivity of the pattern matching. Figs. 16(a) and (b) (size of images: 128×128 pixels) are the color patterns that resulted from this image processing of the color patterns of Figs. 15(a) and (b). In the experiment, the color pattern of the arbitrarily selected tumor cell image was used as the reference, and the cell images were distinguished using the JTC system.

(a) (b)

Figure 15. Color patterns of (a) normal and (b) tumor cells.

(a) (b)

Figure 16. Color patterns following image processing of (a) normal and (b) tumor cells.

4.3.2. *Identification experiment with luminance identification technique.*

Some image processing was also provided for the luminance histograms converted from the cell images to enhance the accuracy of the cell image identification. These image-processing steps are described below. The luminance histograms converted from Figs. 6(a) and (b) are shown in Fig. 17(a) and (b) (size of images: 128×128 pixels). As shown in Fig. 17, the luminance histogram converted from the tumor cell image increased in the low luminance range, as compared to the normal cell image. The reason for this is that the tumor cells show hyperplasia when compared to normal cells [11]. Accordingly, the vertical axis of the luminance histograms was expressed logarithmically to emphasize the differences between the shapes of the histograms for the normal and tumor cells. Also, the luminance histograms for the high luminance range were removed to distinguish using only the low luminance range histograms. Figs. 18(a) and (b) (size of images: 128×128 pixels) are the luminance histograms that resulted from the above image processing of the luminance histograms of Figs. 17(a) and (b). In the experiment, the luminance histogram of the arbitrarily selected tumor cell image was again used as the reference, and the cell images were distinguished using the JTC system.

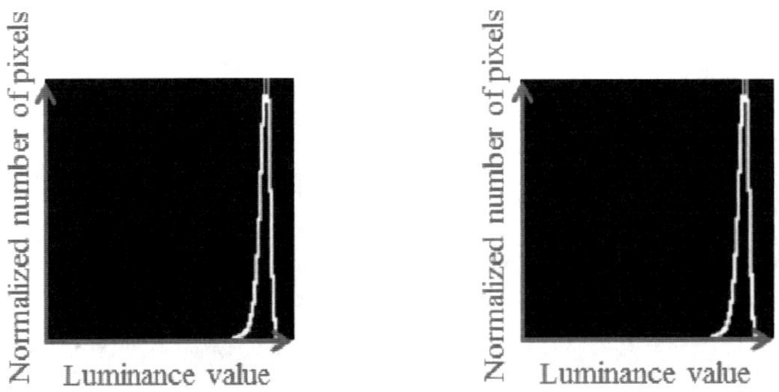

(a) (b)

Figure 17. Luminance histograms of (a) normal and (b) tumor cells.

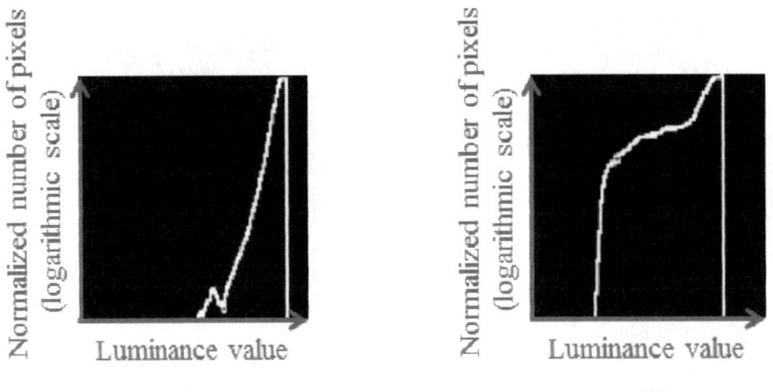

(a) (b)

Figure 18. Luminance histograms after image processing of (a) normal and (b) tumor cells.

4.3.3. *The results of identification experiments with color identification technique.*

The 2D color patterns converted from the cell images were distinguished using the JTC system, and the results are shown in Fig. 19. The vertical axis in Fig. 19 was standardized by the intensity of the cross-correlation spot, obtained by using the same pattern as the retrieval and reference patterns. The horizontal axis of Fig. 19 denotes the numbers of the cell images; numbers 1–15 are the normal cell images, and numbers 16–30 are the tumor cell images. Tumor cell image no. 19 was used as the reference image. In Fig. 19, the intensities of the cross-correlation spots for the normal cells were low, while the intensities for the tumor cells were high. The dispersion of the intensity of Fig. 19 is described as a normal distribution in Fig. 20. The vertical axis in Fig. 20 was standardized using the maximum of the probability density. In Fig. 20, the two distributions had similar standard deviations, and these distributions were well separated. However, a little part of the two distributions overlapped. The spot intensity at the crossover point of the two distributions was used as the threshold between the normal cells and the tumor cells. When the threshold was 0.49, the false acceptance rate (FAR) and the false rejection rate (FRR) were approximately 2.0% and 2.0%, respectively.

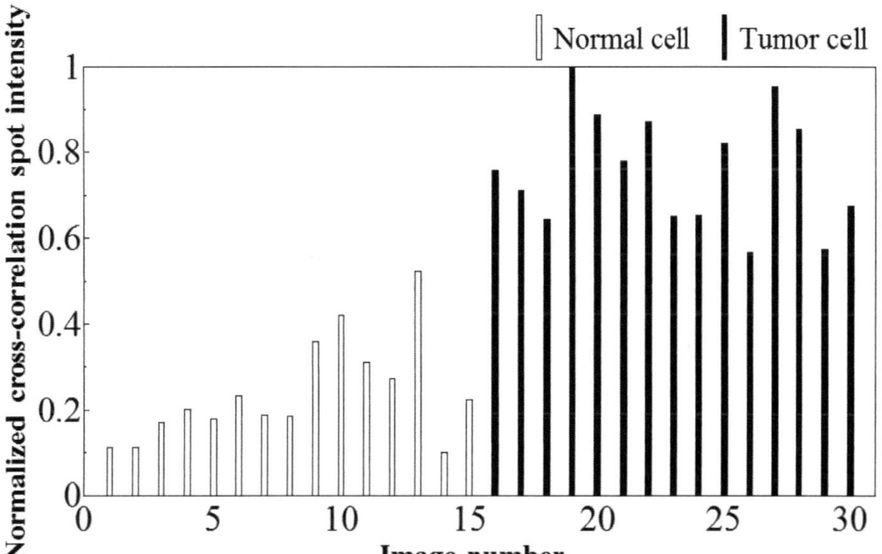

Figure 19. Results of cell image identification with the color identification technique.

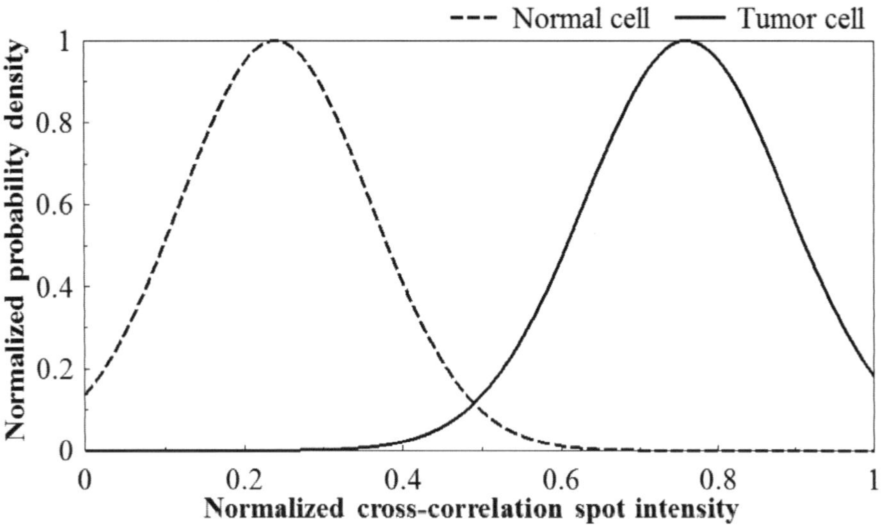

Figure 20. Normal distribution of the intensity of the cross-correlation spot.

4.3.4. *The results of identification experiments with luminance identification technique.*

The two-dimensional histograms converted from the cell images were distinguished using the JTC system, and the results are shown in Fig. 21. In Fig. 21, the intensities of the cross-correlation spots for the normal cells were low, and the intensities for the tumor cells were also comparatively low. The histogram for tumor cell image nos. 19, 28, and 29 are shown in Fig. 22. In Fig. 22, the histogram of no. 29 decreased in the low luminance range when compared with the other histograms, because the cell density of cell image no. 29 was low compared with the other cell images (Fig. 23). This means that a histogram in the low luminance range depends on the cell density of the cell image. Accordingly, the tumor cell images where the cell density was small compared to the reference image had low cross-correlation spot intensity. The dispersion of the intensity of Fig. 21 is described as a normal distribution in Fig. 24. In Fig. 24, the standard deviation of the normal distribution for the tumor cells was larger than that for the normal cells, and large parts of these distributions overlapped. When the threshold between the normal cells and the tumor cells was 0.22, FAR and FRR were approximately 33.1% and 22.3%, respectively.

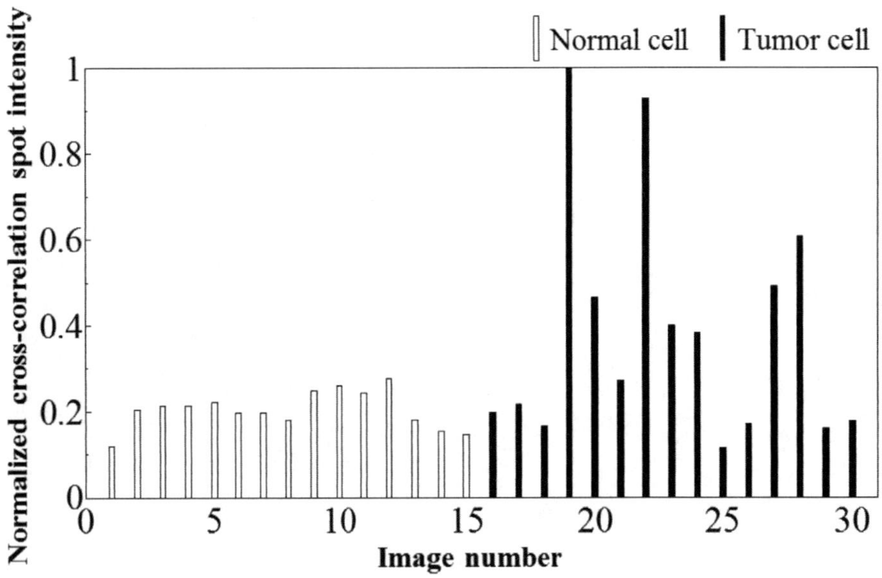

Figure 21. Results of cell image identification with the luminance identification technique.

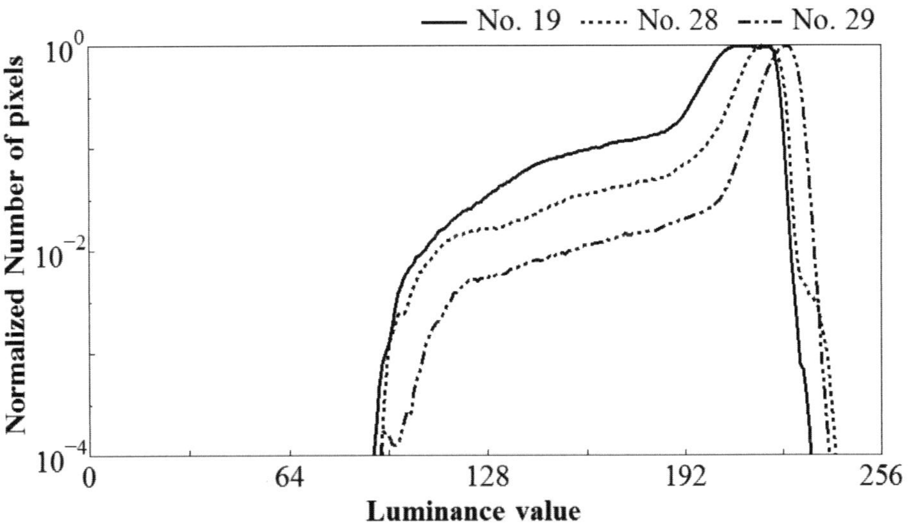

Figure 22. Luminance histograms converted from tumor cell images.

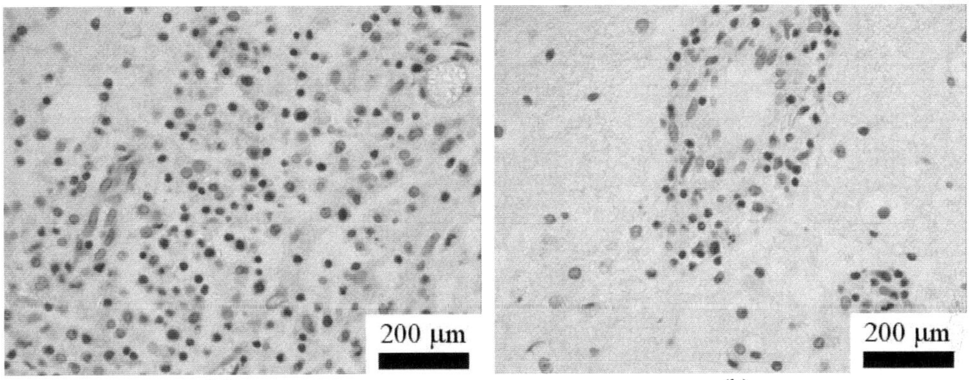

(a) (b)

Figure 23. (a) Tumor cell image no. 28. (b) Tumor cell image no. 29 (size of images: 2048×1536 pixels).

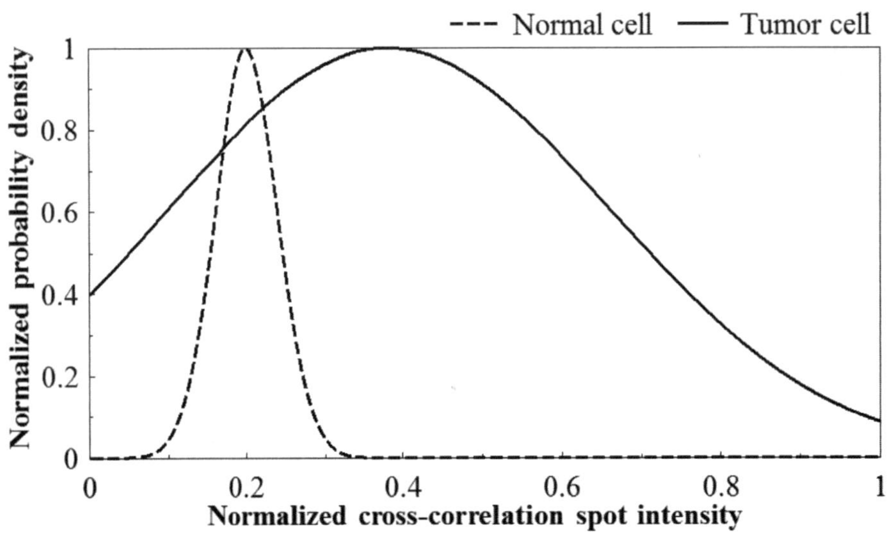

Figure 24. Normal distribution of the intensity of the cross-correlation spot.

4.3.5. *The results of identification experiments with color and luminance identification techniques.*

The results of cell image identification with both the color and the luminance information are shown in Fig. 25. The vertical axis of Fig. 25 shows the mean value of the normalized cross-correlation spot intensity for each cell image of Fig. 19 and Fig. 21. In Fig. 25, the intensities of the cross-correlation spots for the tumor cells were high as compared with the intensities for the normal cells. The dispersion of the intensity of Fig. 25 is described as a normal distribution in Fig. 26. In Fig. 26, the standard deviation of the normal distribution of the tumor cells was large compared with that of the normal cells, and part of these distributions overlapped. When the threshold between the normal cells and the tumor cells was 0.315, FAR and FRR were approximately 9.3% and 9.2%, respectively.

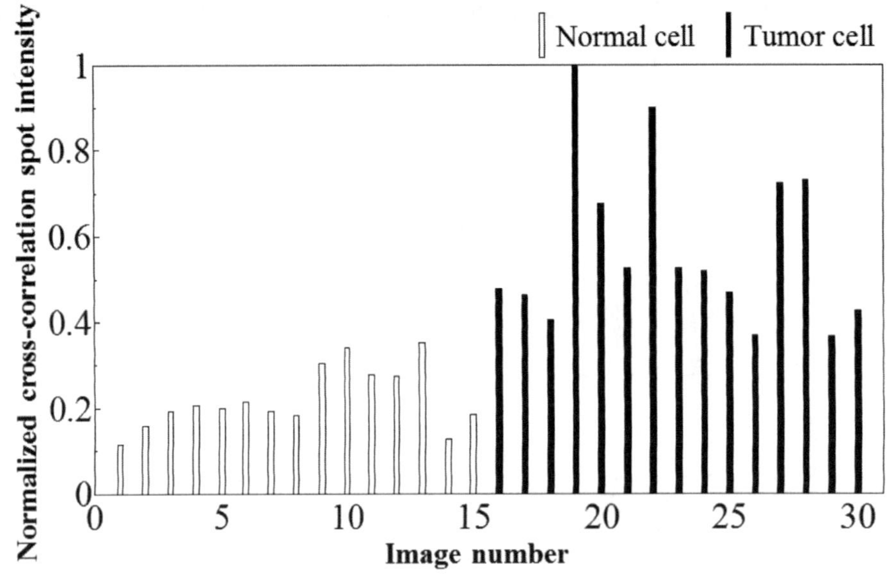

Figure 25. Results of cell image identification with color and luminance identification techniques.

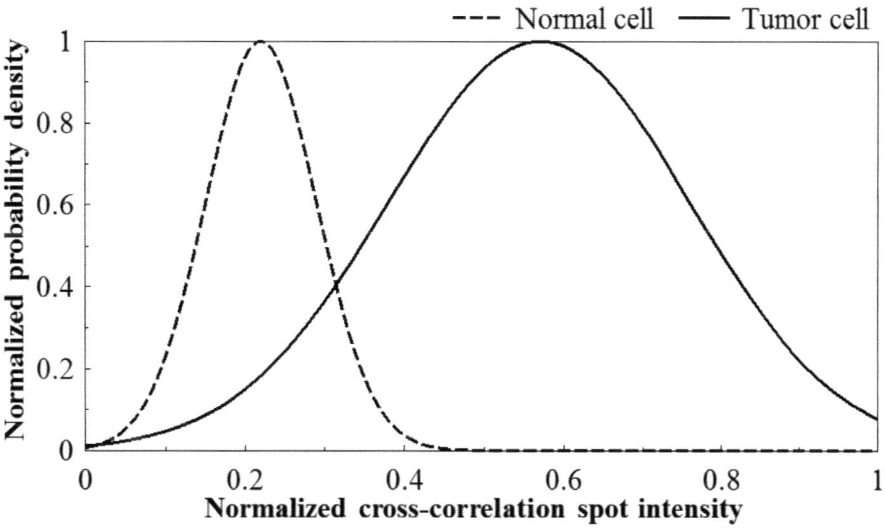

Figure 26. Normal distribution of the intensity of the cross-correlation spot.

5. Summary

A technique for distinguishing real-world images has been developed using a joint transform correlator. The statistical color and luminance information in the various real-world images was converted from digital images into 2D patterns and then distinguished using the optical correlator. The identification technique using the statistical color and luminance information could distinguish an object image similar to the reference image of an apple, unlike the identification technique using shape information, and could also distinguish tumor cell images from other cell images with high accuracy (FAR of 9.3% and FRR of 9.2% were achieved). These experimental results demonstrate the feasibility of using an optical correlator for the identification of complex real-world images containing large volumes of information.

Acknowledgment

This work was partially supported by the Hamamatsu Cluster of the Knowledge Cluster Project of the Ministry of Education, Culture, Sports, Science and Technology, Japan.

References

[1] Watanabe E, Ichikawa T, Akiyama R and Kodate K 2008 *Japanese J. Appl. Phys.* **47** 5964–5967
[2] Kobayashi Y and Toyoda H 1999 *Opt. Eng.* **38** 1205–1210
[3] VanderLugt A 1964 *IEEE Trans. Inform. Theory* **IT-10** 139–145
[4] Weaver C S and Goodman J W 1966 *Appl. Opt.* **5** 1248–1249
[5] Yamamoto S, Kuboyama H, Arai S, Yamaguchi K, Fukuda M, Kato M, Kawaguchi T and Inoue M 2010 *Proc. SPIE* vol 7723 pp 77230B–1–77230B–8
[6] Ota N 2001 *Color Dynamics* (Tokyo: Tokyo Denki University Press) pp 63–148
[7] Fukuda M, Yamamoto S, Kato M, Kawaguchi T, Aono M, Sekino H and Inoue M 2009 *Proc. SPIE* vol 7358 pp 73581B–1–73581B–8
[8] Haralick M R, Shanmugan K and Dinstein I H 1973 *IEEE Trans. Syst. Man Cybern.* **SMC 3** 610–621
[9] Motoyoshi I, Nishida S, Lavanya S and Edward H A 2007 *Nature* **447** 206–209
[10] Kuboyama H, Arai S, Yamaguchi K and Fukuda M 2010 *Moc conf.* pp 150–151
[11] Alberts B, Johnson A, Lewis J, Raff M, Roberts K and Walter P 1995 *Molecular Biology of the Cell* (Tokyo: Newton Press) pp 1255–1273

Journal of Physics: Conference Series **352** (2012) 012038 doi:10.1088/1742-6596/352/1/012038

Observation of Biological Tissues Using Common Path Optical Coherence Tomography with Gold Coated Conical Tip Lens Fiber

K. Taguchi, J. Sugiyama, M. Totsuka and S. Imanaka

Department of Photonics, Ritsumeikan University, Kusatsu, Shiga, 525-8577, Japan
E-mail: taguchi@se.ritsumei.ac.jp

Abstract. In this paper, we proposed a high lateral resolution common-path Fourier domain optical coherence tomography(OCT) system with the use of a chemically etched single mode fiber. In our experiments, single mode optical fiber for 1310nm was used for preparing the tapered tips. Our system used a conical microlens that was chemically etched by selective chemical etching technique using an etching solution of buffered hydrofluoric acid (BHF). From experimental results, we verified that our proposed optical coherence tomography system could operate as a common-path Fourier domain OCT system and conical tip lens fiber was very useful for a high lateral resolution common-path Fourier domain OCT system. Furthermore, we could observe a surface of paramecium bursaria and symbiotic chlorella in the paramecium bursaria using gold coated conical-tip fiber in the water.

1. Introduction

Optical Coherence Tomography is an imaging system which can observe multiple sample's surface and inside with rapidly as real time microscopic optical-imaging. In this research, we proposed Common Path OCT (CPOCT) for the samples not only in the air but also in high refractive index media such as water. Due to the elimination of the alignment process between the microlens and fiber end, there has been recent interest in developing in-built fiber probes for OCT applications. The advantage of a monolithic conical microlens is only in simple fabrication, but also in generating the back reflected reference signal and providing a capability for focusing the light for imaging without the addition of optical elements.

Axial resolution depends on only the center wavelength and bandwidth of the light sources, so we set up a goal that is to improve lateral resolution using SLD with a center wavelength of 1310nm and bandwidth of 70nm. It is needed to narrow down the beam diameter for improving lateral resolution. For that reason, using an optical lens was commonly way. However, there were problems that it made OCT system be bigger and a beam diameter spread very immediately. Therefore we proposed a conical microlens that was chemically etched in the core at the fiber tip[1][2]. Furthermore, we tried to use CPOCT in high refractive index media, like blood vessel identification. CPOCT using conical-tip could not be functioning in high refractive index media such as water because the value of back-reflected light field from the conical tip was too low. For this reason, we had to set reference intensity at a high level by coating the tip with a thin layer of gold. From experimental results, we verified that conical-tip fiber coated with a thin layer of gold could be operated as a fiber probe of

CPOCT with high lateral resolution. Furthermore, we could observe a surface of paramecium bursaria and symbiotic chlorella in the paramecium bursaria using gold coated conical-tip fiber in the water.

2. Chemically etched single mode fiber tip

A fiber can be sharpened by utilizing the difference in etching rate between its core and cladding when immersed in an etching solution of buffered hydrofluoric acid (BHF), a mixture of 50% weight hydrofluoric acid (HF) and 40% aqueous solution of ammonium fluoride (NH_4F).

Fig. 1. SEM images of optical fiber axicon microlens. The volume ratio of NH_4F:HF is 2:1. (a) The apex angle is 126deg. (b) Tilted SEM image of (a).

To make an axicon lenses, we prepared etching solutions. They were mixtures of NH_4F versus HF by volume ratio of 2:1. A cleaved fiber was immersed in the 2:1 BHF etching solution for 30min. The fiber used for fabrication was a standard single-mode fiber with an MFD of 9.5μm at λ=1310nm. Then axicon microlens was formed at the cleaved fiber. Fabricated axicons had apex angles of 126deg for 2:1 solution as shown in Fig. 1. We also studied the laser beam profiles emanating from the fiber probe by objective-lens imaging experiments.

Figure 2 shows two dimensional intensity profiles at the beam waist for (a) cleaved fiber and (b) 126deg axicon respectively. The central spot size is ~9.5μm for cleaved fiber and ~3.2μm for 126deg axicon with 1310 nm wavelength light-source. Table 1 shows the beam diameter as a function of the distance from beam waist(Z=0μm). From these results, we could verify that beam diameter emerging from axicon fiber was smaller than that emerging from cleaved fiber.

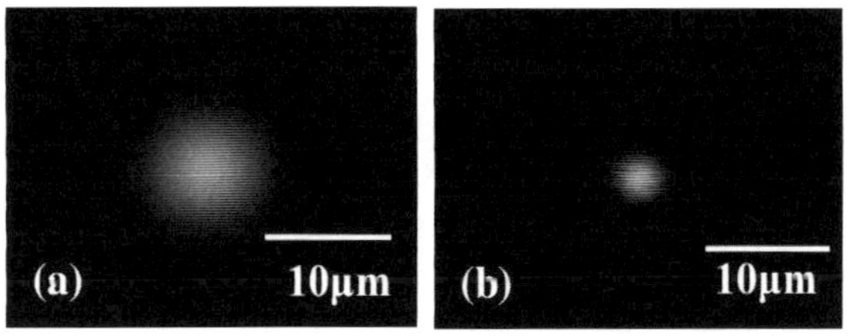

Fig. 2. CCD camera images of field distribution at the beam waist for (a) cleaved fiber and (b) 126 deg axicon fiber, respectively.

Journal of Physics: Conference Series **352** (2012) 012038 doi:10.1088/1742-6596/352/1/012038

Table 1. Measured beam diameter as a function of the distance from beam waist.

fiber probe	Z=0μm	Z=10μm	Z=30μm
cleaved fiber	9.5μm	9.8μm	10.3μm
axicon fiber	3.2μm	3.7μm	5.9μm

beam waist: Z=0μm

3. Experimental setup

The experimental setup of the common-path FD-OCT setup is shown in Fig. 3. The low coherence light source was a superluminescent diode (SLD) (Denselight Semiconductors) with a centre wavelength of 1310nm and spectra width of 70nm. A commercial spectrometer (B&W Tek Inc.) was used. The optical power, from the conical-tip fiber probe, incident on the sample was measured using a photo-detector (PDA10CS, Thorlabs Inc.) to be approximately 1mW. The motorized translation stage (KT-LS13-M, Zaber Technologies Inc.) provided the lateral scanning for the B-scan operation. A custom program was written with Labview version 8.2 (National Instruments) to operate and collect data from this experimental CPOCT system.

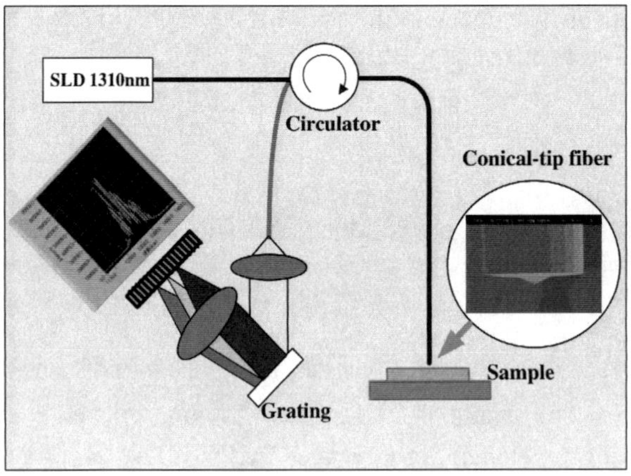

Fig. 3. Schematic of the in-fiber CPOCT setup incorporating the conical-tip fiber probe.

4. Experimental results

The lateral resolution of the 126deg conical-tip fiber was measured using an USAF resolution target card (Edmund Optics). The target card is made up of different rectangular gratings with different widths. For example, the line pairs for elements E2 and E6 of group 6 are 13.9μm and 8.8μm respectively. This implies that the corresponding thickness of each line in E2 is 6.95μm and the corresponding thickness of each line in E6 is 4.40μm. The scanning was carried out at a lateral step size of 0.1μm and with the fiber end positioned 30μm away from the target card. A cleaved fiber, using the same experimental setup as in Fig. 3, was used for comparison. From our previous work, it was found that the 126deg conical-tip fiber was able to clearly resolve the lines in element E6 of group 6 whereas the cleaved fiber was barely able to. From these results, we can verify that 126deg axicon fiber has higher resolution than that of the cleaved fiber. We also investigated the lateral resolution of the system at 100μm and 250μm away from the target card. Experimental results were shown in Fig. 4 and Table 2.

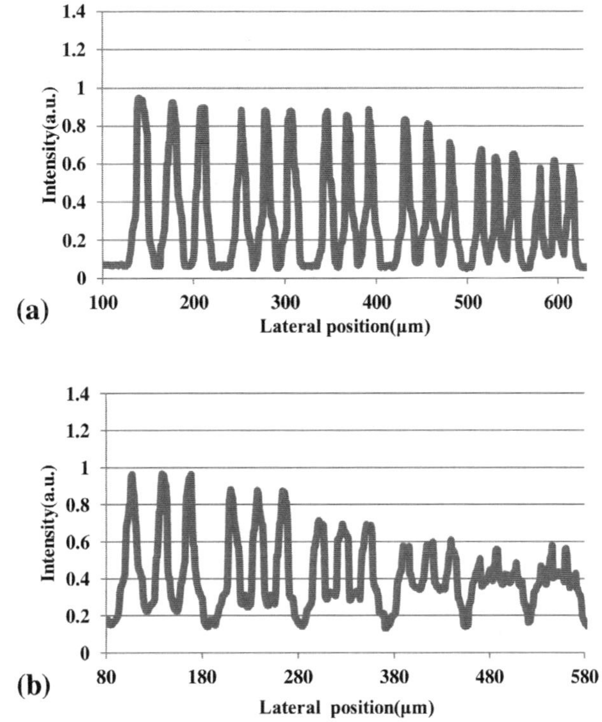

Fig. 4. Lateral plot of a single scan OCT data using 126deg axicon fiber. (a)100μm away from the target card(group 5). (b) 250μm away from the target card(group 4).

Table 2. Measured lateral resolution of 126deg conical-tip fiber

Distance D_z	30μm	100μm	250μm	600μm
Resolution	4.4μm	8.77μm	12.4μm	16.0μm

***Dz*: Distance between fiber probe and target card**

Table 2 shows the measured lateral resolution as a function of the distance between fiber tip and target card. Generally the depth range is limited by the wavenumber spacing between pixels of the spectrometer. So, we also investigated the lateral resolution of the system at the extreme end of our depth range (0.6mm from the fiber tip) where we found that the lateral resolution of our 126deg fiber decreased to ~16μm. According to these results, we could verify that chemically etched axicon fiber was useful for a high lateral resolution common-path Fourier domain OCT system. To demonstrate the imaging capability of the conical-tip fiber through the common-path technique, a two dimensional cross sectional scan of a sample of leaf was performed. Figure 5 depicts a cross-section OCT tomogram of the leaf sample obtained with the conical-tip fiber. The step-size set on the translation stage used for the scanning was 2μm step. The cell wall layer making up the epidermis layer is clearly evident from the rest of the section. In addition the vascular bundle, which plays the vital role of transporting nutrients to the surrounding tissue, can also be observed in the tomogram.

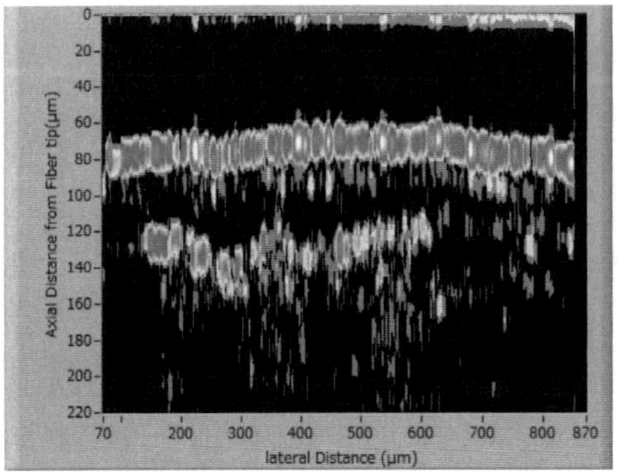

Fig. 5. A cross-section OCT tomogram of the leaf sample using the conical-tip fiber in the CPOCT setup.

Furthermore, we tried to use CPOCT in high refractive index media, like blood vessel identification. CPOCT using conical-tip could not be functioning in high refractive index media such as water because the value of back-reflected light field from the conical tip was too low. For this reason, we had to set reference intensity at a high level by coating the tip with a thin layer of gold. Figure 6 shows the measured signal to noise ratio (SNR) as a function of axial distance between the conical-tip fiber end and mirror for 126deg axicon. The SNR was obtained by using the peak signal of the mirror surface in water solution, after inverse Fourier transform, over the noise floor. The SNR > 30dB is achievable at an axial distance of 0.5mm from the fiber tip.

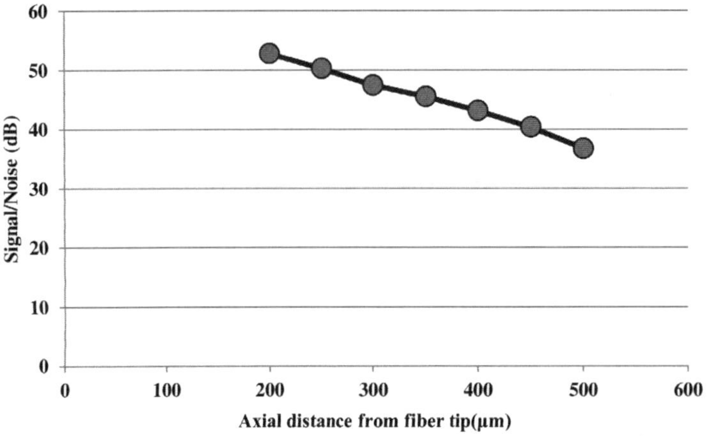

Fig. 6. Measured signal to noise ratio with the 126deg gold coated conical-tip fiber for different axial position between the fiber tip and mirror in water.

Figure 7(a) depicts the bright-field image of the paramecium bursaria. The dotted line indicates the scan path. We can see the symbiotic Chlorella in paramecium bursaria. We pretreated the paramecium using NiCl2 solution, in which paramecia stopped the action of cilia. So the paramecium bursaria did not move. A cross-section OCT tomogram of the paramecium bursaria was shown in Fig. 7(b). We could observe a surface of paramecium bursaria and symbiotic chlorella in the paramecium bursaria using gold coated conical-tip fiber in the water. From these experimental results, we verified that conical-tip fiber coated with a thin layer of gold could be operated as a fiber probe of CPOCT with high lateral resolution.

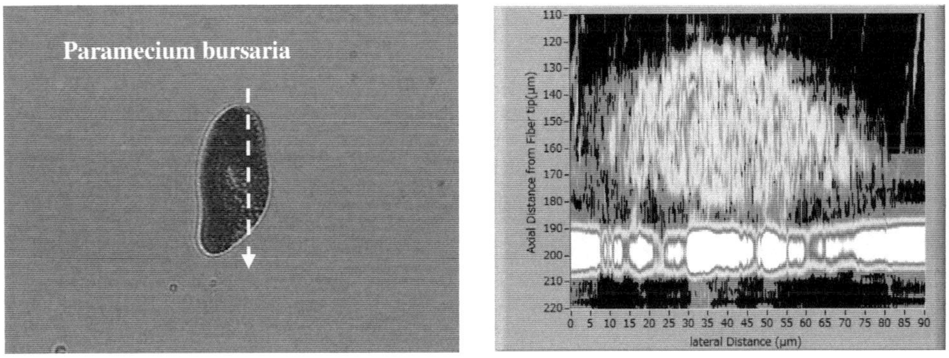

Fig. 7. (a)Image of paramecium bursaria with NiCl2 solution. The paramecia stop the action of cilia. (b) A cross-section OCT tomogram of the paramecium bursaria using the conical-tip fiber in the CPOCT setup.

5. Conclusions

In this paper, we proposed a high lateral resolution common-path Fourier domain OCT system with the use of a chemically etched single mode fiber. The conical-tip fiber could be readily fabricated via a selective-chemical etching technique at low cost and with high reproducibility. From experimental results, we verified that our proposed optical coherence tomography system could operate as a common-path Fourier domain OCT system and conical tip lens fiber was very useful for a high lateral resolution common-path Fourier domain OCT system. Furthermore, we could observe a cross-section OCT tomogram of the paramecium bursaria using the gold coated conical-tip fiber in the CPOCT setup.

References

[1] X. Tsampoula1, K. Taguchi, T. Čižmár, V Garces-Chavez, N. Ma, S. Mohanty, K.Mohanty, F. Gunn-Moore, K. Dholakia, Fibre based cellular transfection, OPTICS EXPRESS. **16** (2008) 17007-17013.

[2] K. M. Tan, M. Mazilu, T. H. Chow, W. M. Lee, K. Taguchi, B. K. Ng, W. Sibbett, C. S. Herrington, C.T. A. Brown, K. Dholakia, In-fiber common-path optical coherence tomography using a conical-tip fiber, Opt. Express. **17**, No.4(2009) 2375-2384.

Journal of Physics: Conference Series **352** (2012) 012039 doi:10.1088/1742-6596/352/1/012039

Three-Dimensional Optical Trapping for Cell Isolation Using Tapered Fiber Probe by Dynamic Chemical Etching

K. Taguchi, J. Okada, Y. Nomura and K. Tamura

Department of Photonics, Ritsumeikan University, Kusatsu, Shiga, 525-8577, Japan
E-mail: taguchi@se.ritsumei.ac.jp

Abstract. In this paper, chemically etched fiber probe was proposed for laser trapping and manipulation of cells. We fabricated tapered fiber probe by dynamic chemical etching technique. Three-Dimensional optical trap of a yeast cell dispersed in water solution could be formed by the fiber tip with 17deg tip. Optical forces were sufficient to move the yeast cell for trapping and manipulation. From these experimental results, it was found that our proposed tapered fiber tip was a promising tool for cell isolation.

1. Introduction

Ashkin *et al.* proposed the optical trapping of dielectric particles by a single-beam gradient force trap for the first time[1]. This laser beam trapping method is considered to be useful for manipulation of biological object because of noninvasive technology. Since then, this method was developed as an optical tweezers technology for various biological objects, such as viruses, bacteria and yeast cells, as well as for various dielectric particles. Its operation principle is based upon the electromagnetic momentum preservation. In recent years, a great deal of experimental and theoretical work has been done on the various types of laser beam trapping to manipulate microorganisms. We have already developed an optical trapping system with a single-mode lensed fiber probe [2]-[6]. This system is simpler and more flexible than conventional optical tweezers. Microsphere dispersed in liquid could be trapped by a single laser beam from an lensed optical fiber end inserted at an angle, and could be freely manipulated in 2-D plain synchronized to the lensed fiber probe as shown in Fig. 1.

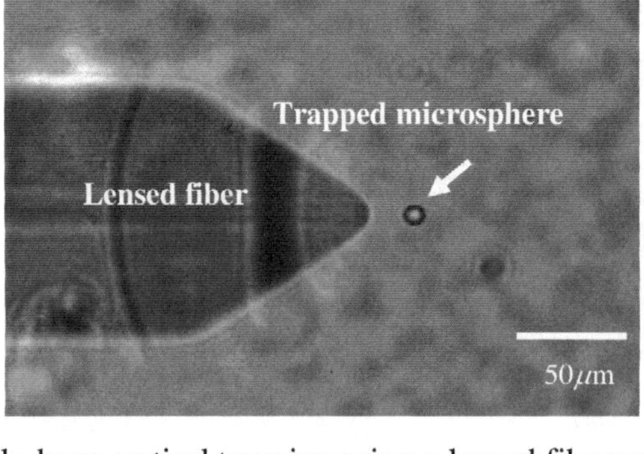

Fig. 1. Single-beam optical trapping using a lensed fiber probe(Top view).

However, these fiber optical tweezers cannot form a Three-Dimensional optical trap by using just single one optical fiber due to the weakly focused laser beam from the polished tapered fiber with spherical lens[2][3]. In this paper a novel tapered single fiber optic tweezers was proposed for cell manipulation and isolation. Optical fiber tips were fabricated by dynamic chemical etching method. From our experimental results, it was found that a Three-Dimensional optical trap of a yeast cell dispersed in water solution could be formed by the fiber tip with 17deg tip. Optical forces were sufficient to move the yeast cell for manipulation and isolation.

2. Fiber trapping system

Figure 2 shows the experimental set up for laser trapping with a tapered fiber probe inserted at an angle. The purpose of this paper is to show that it is possible to make efficient optical tweezers using a properly modified single mode fiber. The fiber was mounted on a linear actuator stage and immersed in water solution. In our previous work, it was found that the fiber insertion angle into a water solution was important parameter for fiber trapping and manipulation of a microsphere. A single-beam fiber optic trap was not created at all when the fiber insertion angle was less than 20 degrees. Therefore, in our experiment, a fiber probe was inserted at an angle of 50 degrees. In our experiments, a semiconductor laser module at 980nm was used. The output of laser light was coupled into an optical fiber which had an optical connector at the fiber end. The trapping fiber could be freely moved in the water solution by controlling the xyz manipulator. A microscope was used to observe the trapped objects and the trapping behavior was recorded on a VTR with a CCD camera.

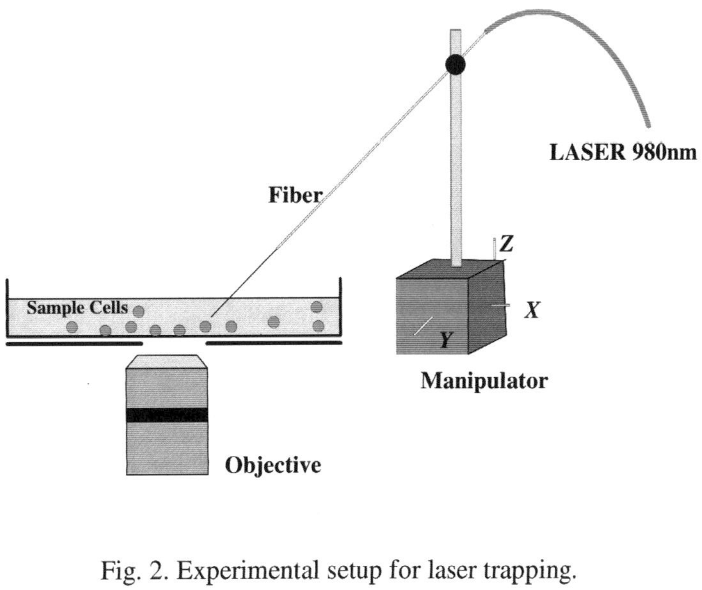

Fig. 2. Experimental setup for laser trapping.

This optical fiber trapping method has following merits: (i) optical trapping systems using optical fibers are simple and inexpensive. (ii) trapped objects can be moved easily and freely, synchronized to the trapping fiber. (iii) optical sources can be changed easily using optical connectors. (iv) the trapping point is easily noticeable, because the fiber end points out the focal point.

3. Tapered fiber probe

In this paper a novel tapered single fiber optic tweezers was proposed for cell manipulation and isolation. Optical fiber tips were fabricated by dynamic chemical etching method as shown in Fig. 3. The mechanically cleaved bare single mode fiber was dipped into Hydrofluoric (HF) acid containing a protective layer of Toluene at the top. By moving the fiber at variable speeds, a variety of tip shapes could be created. The motorized translation stage (KT-LS13-M, Zaber Technologies Inc.) was used for moving the fiber at variable speeds. A custom program was written with Labview version 8.2 (National Instruments) to operate this experimental system.

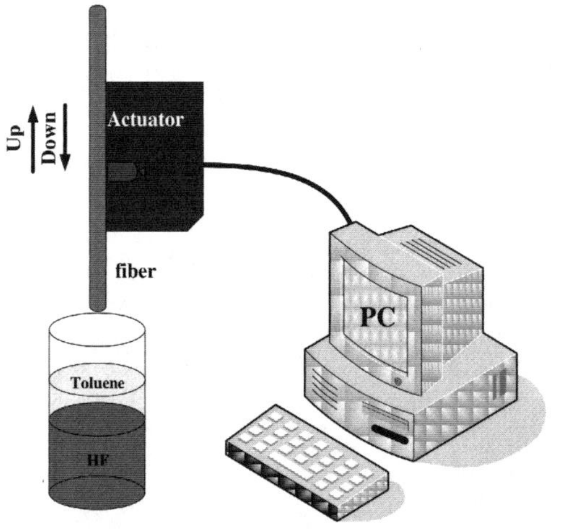

Fig. 3. Experimental setup for dynamic chemical etching.

Tip angle could be adjusted from 7deg to 37deg by dynamic chemical etching method. From our experimental results, it was found that a Three-Dimensional optical trap of yeast cell dispersed in water solution could be formed by the fiber tip with less than 17deg tip.

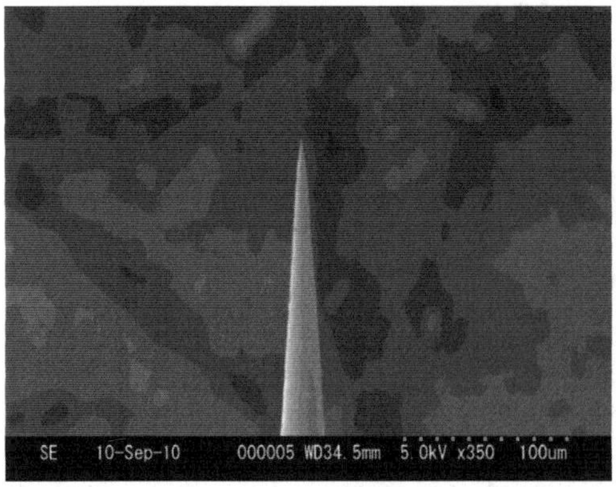

Fig. 4. SEM image of optical fiber tip by dynamic chemical etching method. The apex angle is 17deg.

A yeast cell was trapped at the taper tip as shown in Fig. 5(a). When the power was reduced, the trapped yeast cell was released from the optical trap at the fiber tip as shown in Fig. 5(b). So it was easy to trap and isolate cell, and we could see the clonal growth of the yeast cell trapped and isolated using 980nm laser. Furthermore we calculated the evanescent electromagnetic field in the vicinity of the fiber tip using FDTD method, and verified that a three-dimension gradient optical field was formed within the optical near field of the fiber probe. From these experimental and theoretical results, it was found that our proposed method was a promising tool for the isolation of microorganisms.

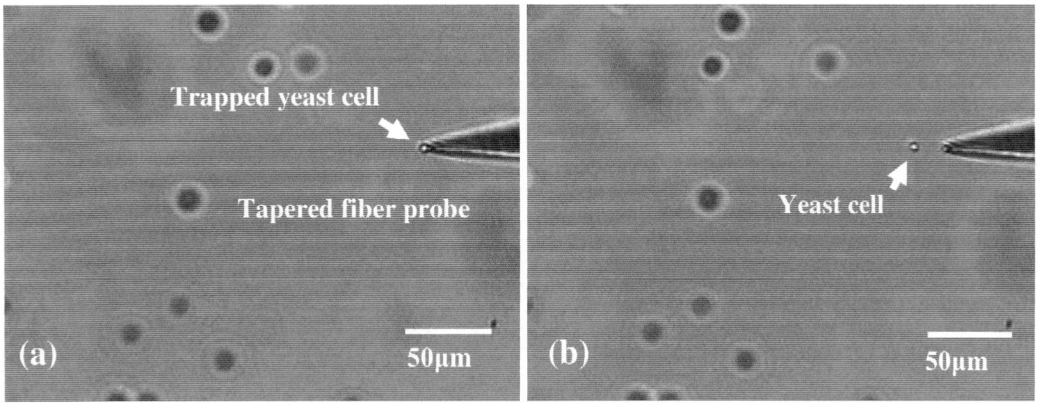

Fig. 5. Experimental results.

4. Conclusions

In this paper, chemically etched fiber probe was investigated for trapping. We fabricated optical fiber tips by dynamic chemical etching method. Tip angle could be adjusted from 7deg to 37deg by dynamic chemical etching. From our experimental results, it was found that a Three-Dimensional optical trap of yeast cell dispersed in water solution could be formed by the fiber tip with less than 17deg tip. Optical forces were sufficient to move the yeast cell for manipulation and isolation. From our experimental results, it was found that our proposed method was a promising tool for the isolation of microorganisms.

References

[1] A. Ashkin, J. M. Dziedzic, J. E. Bjorkholm, S. Chu, Observation of a single-beam gradient force optical trap for dielectric particles, Opt. Lett. **11** (1986) 288-290.

[2] K. Taguchi, K. Atsuta, T. Nakata, M. Ikeda, A Possible Manipulation of a Biological Cell by a Laser Beam Focused through Optical Fiber, IEICE TRANSACTION on Electronics, **E83-C**, No.4, (2000) 664-667.

[3] K. Taguchi, M. Tanaka, M. Ikeda, Optical Force on a Micro-sphere by the Laser Beam from a Lensed Optical Fiber Inserted at an Angle, OPTICAL REVIEW. **8**, No.3 (2001) 203-205.

[4] K. Taguchi, M. Tanaka, M. Ikeda, Investigation on the radius of hemispherical microlens of optical fiber end for three-dimensional trapping, Optical and Quantum Electronics. **34**, No. 10 (2002) 993-999.

[5] K. Taguchi, N. Watanabe, Single-beam optical fiber trap, Journal of Physics. **61** (2007) 1137–1141.

[6] K. Taguchi, M. Ikeda, Optical Rotational Manipulation System of a Biological Object Using Plural Optical Fibers and Optical Switch, Applied Electromagnetics and Mechanics, **25** (2007) 645-648.

Optical Manipulation of Symbiotic Chlorella in Paramecium Bursaria Using a Fiber Axicon Microlens

K. Taguchi, S. Hirota, H. Nakayama, D. Kunugihara and Y. Mihara

Department of Photonics, Ritsumeikan University, Kusatsu, Shiga, 525-8577, Japan
E-mail: taguchi@se.ritsumei.ac.jp

Abstract. In this paper, chemically etched axicon fiber was proposed for laser trapping of symbiotic chlorella from paramecium bursaria. We fabricated axicon micro lenses on a single-mode bare optical fiber by selective chemical etching technique. The laser beam from fiber axicon microlens was strongly focused and optical forces were sufficient to move a symbiotic chlorella. From experimental results, it was found that our proposed fiber axicon microlens was a promising tool for cell trapping without physical contact.

1. Introduction

The handling of individual cells is a key technique in cell engineering such as gene introduction, drug injection and clone technology. However, a wide variety of naturally occurring bacteria cannot be isolated by classical microbiological methods. More sophisticated electronic enumeration and sampling systems such as flow cytometry cannot prevent the formation of cell aggregates.

Ashkin *et al.* proposed the optical trapping of dielectric particles by a single-beam gradient force trap for the first time[1]. This laser beam trapping method is considered to be useful for manipulation of biological object because of noninvasive technology. Since then, this method was developed as an optical tweezers technology for various biological objects, such as viruses, bacteria and yeast cells, as well as for various dielectric particles. Its operation principle is based upon the electromagnetic momentum preservation. In recent years, a great deal of experimental and theoretical work has been done on the various types of laser beam trapping to manipulate microorganisms. We have already developed an optical trapping system with a single-mode lensed fiber probe [2]-[5]. This system is simpler and more flexible than conventional optical tweezers. Microsphere dispersed in liquid could be trapped by a single laser beam from an lensed optical fiber end inserted at an angle, and could be freely manipulated in 2-D plain synchronized to the lensed fiber probe.

In this paper, we investigated the manipulation of cells using fiber axicon microlens[6][7]. We fabricated axicon micro lenses on a single-mode bare optical fiber by selective chemical etching technique using an etching solution of buffered hydrofluoric acid(BHF), a mixture of 50% weight hydrofluoric acid (HF) and 40% aqueous solution of ammonium fluoride (NH4F). In our experiments, a semiconductor laser module at 980nm was used as a light source. From our experiment, we could verify that the laser beam from fiber axicon microlens was strongly focused and optical forces were sufficient to move a microorganisms and biological cells without physical contact. Furthermore we tried to trap and manipulate the symbiotic Chlorella in paramecium bursaria using a fiber axicon microlens. In our experiments, we used surfactant to flow out many symbiotic algae from paramecium bursaria. From these experimental results, it was found that our proposed fiber axicon microlens was a

Journal of Physics: Conference Series **352** (2012) 012040 doi:10.1088/1742-6596/352/1/012040

promising tool for the isolation of microorganisms which cannot be obtained in pure culture by conventional methods.

2. Chemically etched axicon fiber

We fabricated axicon microlenses on a single-mode bare optical fiber by using selective chemical etching technique. We also investigated the optical properties by measuring the beam profile of light emanating from the axicon microlens by objective lens imaging. A fiber can be sharpened by utilizing the difference in etching rate between its core and cladding when immersed in an etching solution of buffered hydrofluoric acid (BHF), a mixture of 50% weight hydrofluoric acid (HF) and 40% aqueous solution of ammonium fluoride (NH_4F).

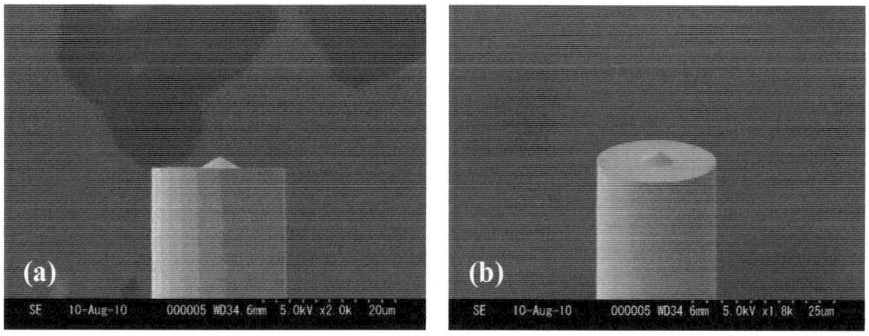

Fig. 1. SEM images of optical fiber axicon microlens. The volume ratio of NH_4F:HF is 2:1. (a) The apex angle is 114deg. (b) Tilted SEM image of (a).

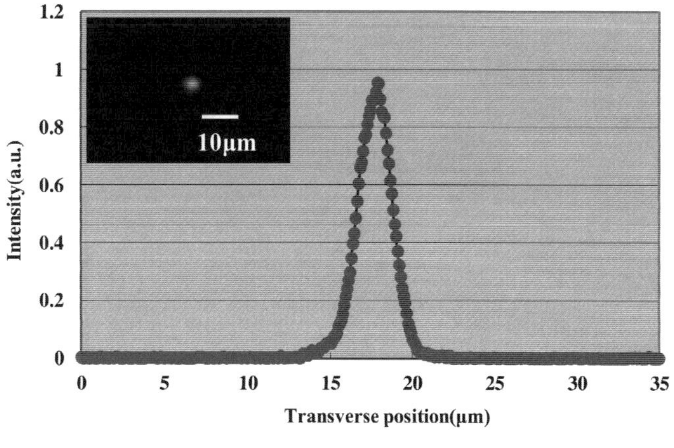

Fig. 2. Laser beam profile and CCD camera images of field distribution at the beam waist in liquid for 114deg axicon.

The Teflon beaker filled with BHF was warmed in a double boiler to a temperature of 35 degree Celsius. A cleaved fiber was immersed in the 2:1 etching solution for 35min. Then an axicon microlens was formed at the cleaved fiber. Fabricated axicons had apex angles of 114deg for 2:1 solutions as shown in Fig. 1. We also studied the laser beam profiles emanating from the axicon lens by objective-lens imaging experiments.

The smallest central spot size in liquid is ~4μm for 114deg axicon with 980 nm wavelength light-source. On the other hand, the fiber used for fabrication was a standard single-mode fiber with an MFD of 5.8μm at λ=980nm. From these results, we can say that fiber axicon microlens is useful tool for trapping and manipulation of an object.

3. Fiber trapping system

Figure 3 shows the experimental set up for laser trapping with a fiber axicon microlens inserted at an angle. The purpose of this paper is to show that it is possible to make efficient optical tweezers using a properly modified single mode fiber. The fiber was mounted on a linear actuator stage and immersed in water drop. In our previous work, it was found that the fiber insertion angle into a water drop was important parameter for fiber trapping and manipulation. A single-beam fiber optic trap was not created at all when the fiber insertion angle was less than 20 degrees. Therefore, in our experiment, a fiber probe was inserted at an angle of 50 degrees. In our experiments, a semiconductor laser module at 980nm was used. The output of laser light was coupled into an optical fiber which had an optical connector at the fiber end. The trapping fiber was attached to an xyz manipulator and the fiber was introduced into a water drop at an angle. The trapping fiber could be freely moved in the water drop by controlling the xyz manipulator. A microscope was used to observe the trapped objects and the trapping behavior was recorded on a VTR with a CCD camera.

This optical fiber trapping method has following merits: (i) optical trapping systems using optical fibers are simple and inexpensive. (ii) trapped objects can be moved easily and freely, synchronized to the trapping fiber. (iii) optical sources can be changed easily using optical connectors. (iv) the trapping point is easily noticeable, because the fiber axicon microlens points out the focal point.

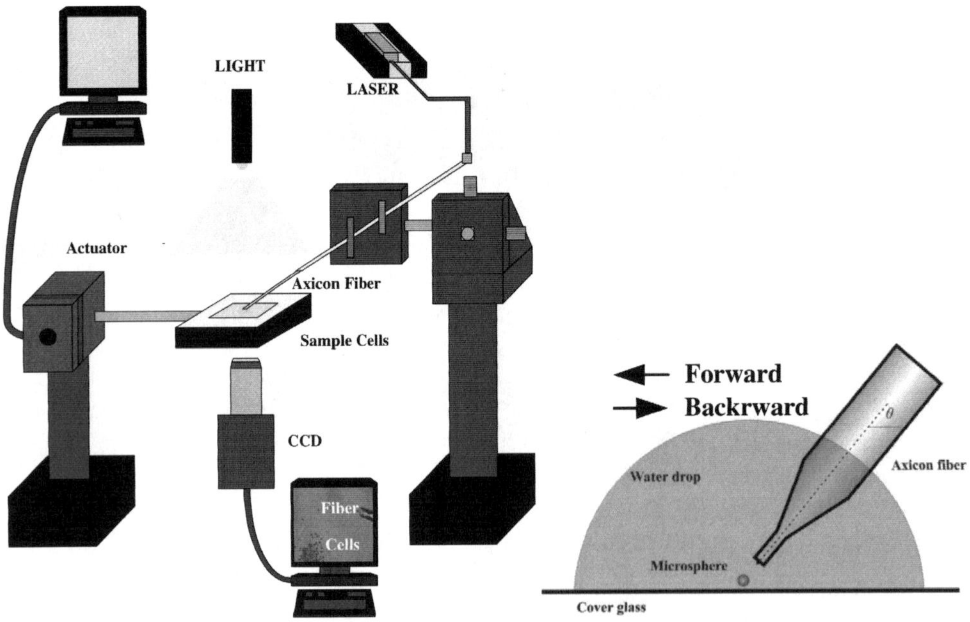

Fig. 3. Experimental setup for laser trapping.

From our experiment, we could verify that the laser beam from fiber axicon microlens was strongly focused and optical forces were sufficient to move a microsphere and biological cells without physical contact as shown in Fig. 4. In Fig. 4, a 2μm-diameter microsphere diapered in water solution was trapped near the fiber end. Under the trapped status, we could freely move the trapped microsphere to the forward and backward or right and left directions synchronized to the trapping fiber.

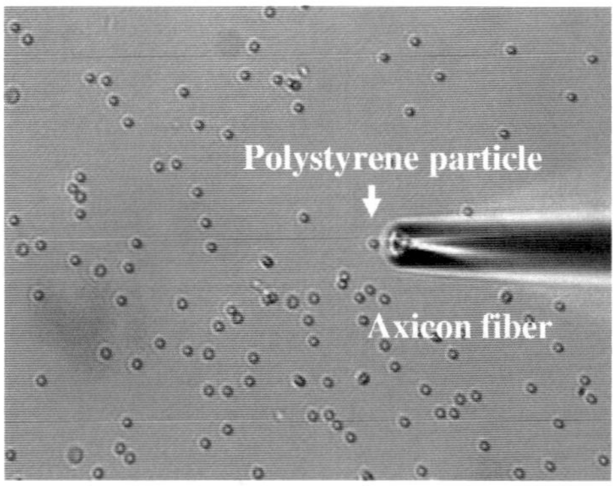

Fig. 4. Trapped microsphere using axicon fiber.

Next, we tried to trap and manipulate the symbiotic Chlorella in paramecium bursaria using a fiber axicon microlens. Many researchers study with a chlorella of the second and third-generation. But the chlorella of second and third-generations are not known an exact species of chlorella. So, the study become instability and unreliable. In our experiments, we used surface-active agents to flow out many symbiotic algae from paramecium bursaria as shown in Fig. 5.

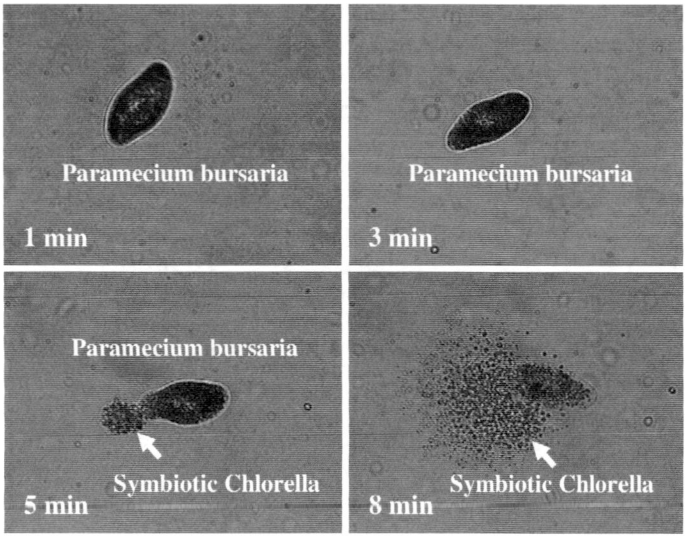

Fig. 5. Paramecium bursaria with surface-active agents.

According to the results in Fig. 5, after 5min we could see the symbiotic Chlorella, and we could trap the symbiotic Chlorella by the laser beam from axicon fiber as shown in Fig. 6(a). Furthermore, we could isolate the symbiotic Chlorella from paramecium bursaria as shown in Fig. 6(b).

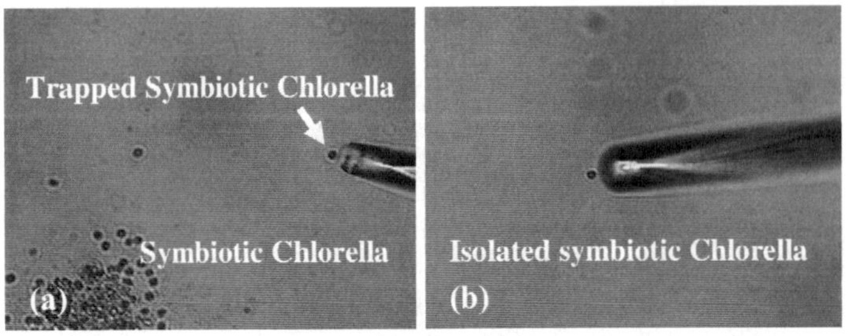

Fig. 6. (a)Symbiotic Chlorella trapped using axicon fiber. (b) Isolated symbiotic Chlorella.

Furthermore, we tried to investigate the cell tapping performance of this special designed fiber probe. A 980 nm laser diode with a power of 10mW and 20mW was used for the optical tweezers. Constant moving speed was achieved with a motorized stage (KT-LS13-M, Zaber Technologies Inc.). Maximum forward and backward (see Fig. 3.) trapping forces for the symbiotic Chlorella were investigated. When the moving speed increased, the drag forces increased proportionally and were equal to the optical force. When the Chlorella escaped from the trap, the current drag force was equal to the maximal trapping force. Experimental results are listed in Table 1. According to results, the forward trapping force was always larger than that obtained from backward trapping. This is because the scattering force coming from reflection always pushes the micro object to move in the direction of light propagation. Optical forces were sufficient to move the symbiotic Chlorella without physical contact.

Table 1. Maximum manipulation velocity for single fiber tweezers.

laser power from fiber end	**maximum forward speed**	**maximum backward speed**
10mW	**～ 60μm/s**	**～ 18μm/s**
20mW	**～110μm/s**	**～ 34μm/s**

4. Conclusions

In this paper, chemically etched axicon fiber was proposed for trapping. We fabricated axicon micro lenses on a single-mode bare optical fiber by selective chemical etching technique using an etching solution of buffered hydrofluoric acid(BHF), a mixture of 50% weight hydrofluoric acid (HF) and 40% aqueous solution of ammonium fluoride (NH_4F). From these experimental results, it was found that our proposed fiber axicon microlens was a promising tool for the isolation of microorganisms which cannot be obtained in pure culture by conventional methods.

References

[1] A. Ashkin, J. M. Dziedzic, J. E. Bjorkholm, S. Chu, Observation of a single-beam gradient force optical trap for dielectric particles, Opt. Lett. **11** (1986) 288-290.
[2] K. Taguchi, K. Atsuta, T. Nakata, M. Ikeda, A Possible Manipulation of a Biological Cell by a Laser Beam Focused through Optical Fiber, IEICE TRANSACTION on Electronics, **E83-C**, No.4, (2000) 664-667.
[3] K. Taguchi, M. Tanaka, M. Ikeda, Optical Force on a Micro-sphere by the Laser Beam from a Lensed Optical Fiber Inserted at an Angle, OPTICAL REVIEW. **8**, No.3 (2001) 203-205.
[4] K. Taguchi, M. Tanaka, M. Ikeda, Investigation on the radius of hemispherical microlens of optical fiber end for three-dimensional trapping, Optical and Quantum Electronics. **34**, No. 10 (2002) 993-999.
[5] K. Taguchi, N. Watanabe, Single-beam optical fiber trap, Journal of Physics. **61** (2007) 1137–1141.
[6] X. Tsampoula1, K. Taguchi, T. Čižmár, V Garces-Chavez, N. Ma, S. Mohanty, K.Mohanty, F. Gunn-Moore, K. Dholakia, Fibre based cellular transfection, OPTICS EXPRESS. **16** (2008) 17007-17013.
[7] K. M. Tan, M. Mazilu, T. H. Chow, W. M. Lee, K. Taguchi, B. K. Ng, W. Sibbett, C. S. Herrington, C.T. A. Brown, K. Dholakia, In-fiber common-path optical coherence tomography using a conical-tip fiber, Opt. Express. **17**, No.4(2009) 2375-2384.

Development of a cell microarray chip for detection of circulating tumor cells

S Yamamura[1], S Yatsushiro[1], K Abe[1], Y Baba[1,2] and M Kataoka[1]
[1]Health Research Institute, National Institute of Advanced Industrial Science and Technology (AIST), 2217-14 Hayashi-cho, Takamatsu, Kagawa 761-0395, Japan
[2]Department of Applied Chemistry, Graduate School of Engineering, Nagoya University, Furo-cho, Chikusa-ku, Nagoya 464-8603, Japan

E-mail: yamamura-s@aist.go.jp

Abstract. Detection of circulating tumor cells (CTCs) in the peripheral blood of metastatic cancer patients has clinical significance in earlier diagnosis of metastases. In this study, a novel cell microarray chip for accurate and rapid detection of tumor cells from human leukocytes was developed. The chip with 20,944 microchambers (105 μm diameter and 50 μm depth) was made from polystyrene, and the surface was rendered to hydrophilic by means of reactive-ion etching, which led to the formation of mono-layers of leukocytes on the microchambers. As the model of CTCs detection, we spiked human bronchioalveolar carcinoma (H1650) cells into human T lymphoblastoid leukemia (CEM) cells suspension and detected H1650 cells using the chip. A CEM suspension contained with H1650 cells was dispersed on the chip surface, followed by 10 min standing to allow the cells to settle down into the microchambers. About 30 CEM cells were accommodated in each microchamber, over 600,000 CEM cells in total being on a chip. We could detect 1 H1650 cell per 10^6 CEM cells on the microarray by staining with fluorescence-conjugated antibody (Anti-Cytokeratin) and cell membrane marker (DiD). Thus, this cell microarray chip has highly potential to be a novel tool of accurate and rapid detection of CTCs.

1. Introduction

In patients with metastatic cancer, there is growing evidence that the presence of circulating tumor cells (CTCs) in the blood may be an important indicator of the potential for prognosis in metastatic cancer. However, it is estimated that CTCs in the peripheral blood can occur at a frequency of approximately 1 CTC per 10^5–10^7 peripheral blood mononuclear cells in metastatic cancer patients[1]. Although some techniques for CTCs detection, PCR[2] and flow cytometry[3], are developed, they especially have a problem in sensitivity. Recently CTC separation methods, have been developed, capture cells with specific cell surface markers such as EpCAM by using magnetic beads or micro-structures coated with a monoclonal antibody. CellSearch system (VeridexTM, Warren, PA), a commercialized semi-automated technology, enriches whole blood for CTCs by adding ferrofluids coated with EpCAM-specific antibodies[4]. In addition, some microfluidic devices have been developed to detect CTCs from whole blood, and the devices enable easy and efficient separation of CTCs[5].

Although these EpCAM-based methods can detect low frequency of only EpCAM-positive CTCs from whole blood, they cannot achieve reproducible recovery of CTCs from all types of tumor.
In previous our study, we already developed a cell microarray chip for rapid and highly sensitive detection of antigen-specific single B-cells[6] and malaria-infected erythrocytes[7]. Here, we have developed a novel high-throughput screening and analysis system for CTCs allowing high sensitivity in a short time using a cell microarray chip made from polystyrene with over 20,000 individually addressable microchambers. Our cell microarray chip has been improved to allow the dispersion of a leukocytes suspension contained with cancer cells into the microchambers regularly, with the formation of a monolayer, and analysis with a microarray scanner for detection of the presence of cancer cells in leukocytes. The potential of a cell microarray chip system for the early diagnosis of metastases was shown, it allowing accurate detection of CTCs in a short time.

2. Experimental

2.1. Construction of a cell microarray chip

The cell microarray chip with 20,944 microchambers (105 μm diameter, 50 μm depth, and 300 μm pitch) was made from polystyrene by the Lithographie Galvanoformung Abformung (LIGA) process by Starlight Co. Ltd. (Osaka, Japan) (Figure 1)[6]. The polystyrene microarray chip was fabricated with a nickel mold by injection molding. The microarray chip has 112 (14 x 8) clusters of 187 microchambers, respectively. There are block numbers on the clusters for easy confirmation of cancer cells in leukocytes. Each microchamber comprises a frustum in shape. The cell microarray chip surface was rendered hydrophilic by means of reactive ion-etching (RIE) treatment using a SAMCO RIE system (SAMCO, Inc., Tokyo, Japan) for achieving leukocytes confinement in the microchambers. RIE exposure time controlled cell adhesion on the polystyrene chip surface. The effect of RIE exposure on the microarray chip surface was examined by measuring the contact angle of water on the chip surface using a contact-angle meter (Kyowa Interface Science Co., Ltd., Saitama, Japan)[6].

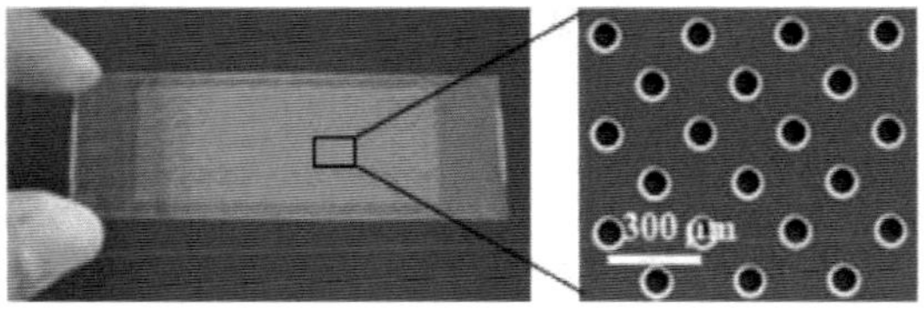

Figure 1. A real picture and SEM image of a cell microarray chip.

2.2. Detection of carcinoma cells on a cell microarray chip

For tightly confinement of leukocytes in the microchambers, T lymphoblastoid leukemia (CEM) cells in RPMI 1640 medium (pH 7.4) with Penicilline-Streptomycin, Fungizone and Fetal Bovine Serum (FBS) at cell concentration of 2.5 x10^6cells/ml, 5.0 x10^6cells/ml, 7.5 x10^6cells/ml, or 1.0x10^7cells/ml were dispersed on the cell microarray chip, followed by 15 min standing to allow the CEM cells to settle down enter into the microchambers under gravitational force. Microchambers were examined by light microscopy. For analysis of human bronchioalveolar carcinoma (H1650) cells in CEM cells suspension on the cell microarray chip, 0.0001%-0.1% H1650 cells were prepared with RPMI 1640 medium. Stock solutions of Anti-Cytokeratin (BD Biosciences, CA) and DiD (Life Technologies, Co., CA), which are Cytokeratin-specific fluorescence dye (Ex: 622 nm, Em: 645 nm) and cell membrane-specific fluorescence dye (Ex: 644 nm、Em: 665 nm), were added to CEM cells suspension spiked with H1650 cells to give a final concentration of 20 μM and 1 μM, respectively. The CEM cells suspension was dispersed manually on the cell microarray chip using a pipette, followed by 15 min standing to allow the CEM cells to settle down into the microchambers under gravitational force, and then staining with Anti-Cytokeratin and DiD. Then excess cells on the chip surface were removed by gentle washing with RPMI 1640 medium. The microarray chip was scanned with a confocal laser-based fluorescence microarray scanner, CRBIO IIe (Hitachi Software Engineering Co., Ltd., Tokyo, Japan). The system exhibits resolution of up to 10 μm and sensitivity of < 0.1 fluorescent molecule/μm^2, and is fitted with filters with emission wavelengths of 535 and 585 nm. The fluorescence intensity of individual CEM and H1650 cells was determined with DNASIS Array version 2.1 software (Hitachi Software Enginnering Co., Ltd), and the both cells that exhibited fluorescence intensities of ten times higher that of background were taken to be fluorescence-positive ones.

3. Results and Discussions

3.1. Dispersion of leukocytes on a cell microarray chip

To achieve the confinement of leukocytes in microchambers, the hydrophilicity on the microarray chip surface was optimized by means of RIE exposure[6,7]. It was observed that increase in the exposure time of RIE increases the hydrophilicity of the chip surface, which is inversely proportional to the decrease in the contact angle of the chip surface[6]. And also it was observed that the polystyrene microarray chip maintained a hydrophilicity character without alteration for at least one week. Eighty-seconds exposure gave appropriate hydrophilicity on the chip surface, and CEM cells could settle in each microchamber (data not shown). For the formation of a mono-layer of CEM cells on the bottom surface of the microchambers after washing, the cell microarray chip was designed to have 105 μm diameter microchambers of 50 μm in depth and a cone-shaped frustum. Because in previous our study, we already developed a cell microarray chip for the formation of monolayers of erythrocytes[7]. The CEM cells suspension is dropped on to the chip using a pipette, and then the CEM cells settle down under gravitational force and adhere to the chip surface with the formation of multi-layer. After washing of the chip surface using a pipette, CEM cells adhered to the bottom surface of each microchamber as a monolayer formation (Figure 2). It was observed that the number of CEM cells confined in the microchambers depended on the CEM cells concentration on the cell microarray chip, and CEM cells were tightly confined with over 7.5 x10^6 cells/ml. The number of confined CEM cells was determined to be 29 ± 2 per microchamber with over 7.5 x10^6 cells/ml. So, over 600,000 CEM cells are dispersed as a monolayer in the microchambers on a microarray chip. This cell microarray chip system has highly potential to be high-throughput screening and analysis of many kind of cells without any complicated equipments.

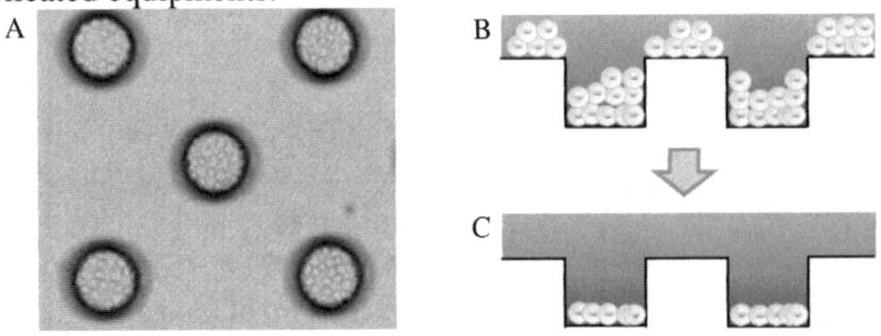

Figure 2. Light microscopic image of CEM cells on a cell microarray chip (A). Schematic cross-section images of CEM cells in microchambers before (B) and after (C) washing.

3.2. Detection of carcinoma cells on a cell microarray chip

CEM cells suspension spiked with carcinoma cells, H1650 was dispersed and stained with fluorescence-conjugated antibody (Anti-Cytokeratin) and cell membrane marker (DiD) on the cell microarray chip and then scanning images were obtained (Figure 3). Cytokeratin-positive cells were not observed among CEM cells as used a negative control. Cytokeratin-positive cells were observed in the microchambers with 0.01%, 0.001%, and 0.0001% H1650 cells, the total numbers of fluorescence-positive erythrocytes being as 99±2.6, 9.7±0.3 and 1.0±0.0 (n=3) in 184 microchamber areas, as determined with DNASIS Array software, respectively. The percentage of cancer cells was determined using the following formula: [(number of Cytokeratin-positive H1650/1,000,000 erythrocytes) x 100]. The percentages of H1650 were calculated to be 0.0099%, 0.00097% and 0.0001%, respectively. These calculated cancer cells ratios were well consistent with the practical ratios. Our cell microarray chip exhibits sensitivity (1 H1650 cell per 10^6 CEM cells) as high as that of reported microchip devices for CTCs detection, and only 15 min is required for the detection carcinoma cells without any complicated technical handling. Additionally, with this cell microarray chip, high accuracy was obtained because only cancer cells among mono-layered leukocytes were targeted on and detected with a confocal fluorescence laser scanning system. Therefore, this cell microarray chip has highly potential to be a novel device of accurate and rapid detection of CTCs and early diagnosis system of metastatic cancer.

Figure 3. Detection image of a carcinoma cell on a cell microarray chip.

Acknowledgements
We appreciate the help of Dr. M. Kurokawa, Starlight Co., Ltd., Japan, for fabricating the microarray chips. We would also like to thank Assoc. Prof. T. Suzuki, Kagawa University, Japan, for the technical assistance in the treatment of the cell microarray chips. This research was supported by the Japan Society for the Promotion of Science (JSPS) through its "Funding Program for World-Leading Innovative R&D on Science and Technology (FIRST Program)", and by grants-in-aid (23310093, 22710124, and 23590165) from JSPS.

References

[1] Paterlini-Brechot P, Benali NL, 2007 *Cancer Let.* **253** 180-204
[2] Smith BM, Slade MJ, English J, Graham H, Luchtenborg M, Sinnett HD, Cross NC, Coombes RC, 2000 *J. Clin. Oncol.* **18**, 1432–1439
[3] Moreno JG, O'Hara SM, Gross S, Doyle G, Fritsche H, Gomella LG, Terstappen LW, 2001 *Urology* **58,** 386–392
[4] Riethdorf S, Fritsche H, Muller V, Rau T, Schindlbeck C, Rack B, Janni W, Coith C, Beck K, Janicke F, Jackson S, Gornet T, Cristofanilli M, Pantel K, 2007 *Clin. Cancer Res.* **13**, 920–928
[5] Nagrath S, Sequist LV, Maheswaran S, Bell DW, Irimia D, Ulkus L, Smith MR, Kwak EL, Digumarthy S, Muzikansky A, Ryan P, Balis UJ, Tompkins RG, Haber DA, Toner M, 2007 *Nature* **450**, 1235–123
[6] Yamamura S, Kishi H, Tokimitsu Y, Kondo S, Honda R, Ramachandra Rao S, Omori M., Tamiya E, Muraguchi A, 2005 *Anal. Chem.* **77**, 8050 -8056
[7] Yatsushiro S, Yamamura S, Yamaguchi Y., Shinohara Y., Tamiya E., Horii T., Baba Y., Kataoka M., 2010 *PloS One* **5**, e13179

High-affinity RNA aptamers to C-reactive protein (CRP): newly developed pre-elution methods for aptamer selection

N Orito[1], S Umekage[1], K Sato[2], S Kawauchi[2], H Tanaka[2], E Sakai[1], T Tanaka[1], and Y Kikuchi*[1]

[1]Department of Environmental and Life Sciences, Toyohashi University of Technology, Tempaku-cho, Toyohashi 441-8580, Japan
[2] Mitsubishi Chemical Medience Co., Shibaura, Minato-ku, Tokyo 108-8559, Japan
kikuchi@tut.jp

Abstract. We have developed a modified SELEX (systematic evolution of ligands by exponential enrichment) method to obtain RNA aptamers with high affinity to C-reactive protein (CRP). CRP is a clinical biomarker present in plasma, the level of which increases in response to infections and noninfectious inflammation. The CRP level is also an important prognostic indicator in patients with several syndromes. At present, CRP content in blood is measured immunochemically using antibodies. To develop a more sensitive method using RNA aptamers, we have attempted to obtain high-affinity RNA aptamers to CRP. We succeeded in obtaining an RNA aptamer with high affinity to CRP using a CRP-immobilized Sepharose column and pre-elution procedure. Pre-elution is a method that removes the weak binding portion from a selected RNA population by washing for a short time with buffer containing CRP. By surface plasmon-resonance (SPR) analysis, the affinity constant of this aptamer for CRP was calculated to be K_D = 2.25×10^{-9} (M). The secondary structure, contact sites with CRP protein, and application of this aptamer will be described.

Introduction

RNA aptamers are biologically functional RNA molecules with specific binding abilities to target molecules, such as proteins and small biomolecules [1 – 3]. The aptamers are usually artificially selected from randomized RNA pools (combinatorial library) by several rounds of selective amplification. This selection procedure is called SELEX (systematic evolution of ligands by exponential enrichment) [2]. Many aptamers for different molecules have been reported [2,4,5]. We have also previously reported an RNA aptamer that specifically inhibits the highly active bacterial protease subtilisin [6]. Aptamers are also of interest as unique tools in diagnostics and therapeutics, because their action is similar to that of antibodies. An RNA aptamer has already been developed as an RNA drug for inhibition of macular degradation by specifically targeting the vascular endothelial growth factor [7].

C-reactive protein (CRP) is a clinical biomarker present in plasma, and the level of CRP increases in response to infections and noninfectious inflammation [8]. Initially, CRP was thought to only be a marker of vascular inflammation, but a recent study indicated that it is also important in atherogenesis [9]. The CRP level is also an important prognostic indicator in patients with acute coronary syndrome. At present, CRP content in blood is measured immunochemically using antibodies [10]. To develop a more sensitive RNA aptamer-based method, we have attempted to obtain a high-affinity RNA aptamer to CRP.

Here, we report the development of an RNA aptamer with high affinity to CRP using a newly developed method involving pre-elution. The RNA aptamer to CRP obtained here showed much higher affinity to CRP than antibodies. The secondary structure of the aptamer and contact sites with CRP protein are also reported.

Materials and methods

Materials

C-reactive protein (recombinant, CRP) was purchased from Oriental Yeast Co. CNBr-activated Sepharose 4B was purchased from Pharmacia. T7 RNA polymerase and RNasin (human placenta ribonuclease inhibitor) were from Toyobo. RNase-free DNase I (RQ-DNase) was purchased from Promega. All other enzymes and chemicals used in this study were purchased from commercial sources. CRP-immobilized Sepharose 4B was prepared by coupling of CNBr-activated Sepharose 4B with CRP according to the supplier's instructions. The CRP content in the gel was 1.2 mg/mL of gel. The CRP-immobilized Sepharose 4B gel was washed several times and stored in binding buffer consisting of 20 mM Tris-HCl (pH 7.5), 140 mM NaCl, 2 mM $CaCl_2$, and 1 mM $MgCl_2$ at 4°C.

Synthetic DNAs

The template DNA used for synthesis of the initial random RNA population was constructed with the following oligonucleotides: Template FR (5'-TCCGCATACGAGCACTGAGT(N_{50})GTACTACGGATCGAAGTCCC*TATAGTGA*-

3'), Template SL (5'-TCCGCATACGAGCACTGAGT(N_{12})GGCGGGCC(N_{12})GGCCCGCC(N_{12})GTACTACGGATCGAAGTCCC*TATAGTGA*-3') (N = G, A, T, or C), Primer 1 (5'-*TAATACGACTCACTATA*GGGACTTCGATCCG-3'), and Primer 2 (5'-TCCGCATACGAGCACTGAGT-3'). Template FR contained 50 nucleotides of randomized sequence (N_{50}) flanked by fixed sequences and part of the T7 RNA polymerase promoter sequence (shown in italics) at the 3'-terminus. Template SL contained a predetermined stem structure of high G+C content consisting of the sequences 5'-GGCGGGCC-3' and 5'-GGCCCGCC-3' (underlined). This stem structure had a random loop (N_{12}), and this stem-loop structure was flanked by two random sequences (N_{12}) containing fixed sequences the same as that of template FR. Primer 1 had the whole T7 promoter sequence and a sequence complementary to the 3'-terminus of Template FR or Template SL. Primer 2 had the same sequence as the 5'-terminal fixed region of Template FR or Template SL.

Preparation of RNA pools

RNAs were synthesized by *in vitro* transcription using T7 RNA polymerase. First, to prepare template double-stranded DNA (dsDNA), PCR-like reaction (filling in of nucleotides into the single-stranded regions) was performed using 50 pmol of Template FR (or 50 pmol of Template SL), 100 pmol of Primer 1, and *Taq* polymerase (Toyobo). Seven cycles of temperature shifts (denaturation, hybridization, and elongation) were performed. This reaction was calculated to yield approximately 3×10^{13} different dsDNA molecules. Using ten times the amount of dsDNA (3×10^{14} different dsDNA molecules) from Template FR or SL as a template, the first RNA pool was prepared from the reaction mixture containing T7 Buffer (Toyobo), 0.5 mM each rNTP, 1 units/μL T7 RNA polymerase, and 0.27 units/μL RNasin in a total volume of 300 μL. The mixture was incubated at 37°C for 90 min. The DNA template was removed by RQ-DNase I treatment and the RNA was purified by denaturing 10% PAGE (polyacrylamide gel electrophoresis). After this purification, the RNAs obtained with Template FR and Template SL were mixed and this preparation was used as the first generation of the RNA pool.

In vitro *selection*

The selection of RNA was performed using a CRP-immobilized Sepharose 4B column (0.08 – 0.5 mL gel bed volume = 0.1 – 0.6 mg of total CRP protein immobilized; see Table 1). The column was equilibrated with binding buffer consisting of 20 mM

Tris-HCl (pH 7.5), 140 mM NaCl, 2 mM $CaCl_2$, and 1 mM $MgCl_2$. The first RNA preparation ($6{\times}10^{14}$ molecules; see Materials and methods) was dissolved in 100 μL of binding buffer and heated at 65°C for 10 min, then incubated for 30 min at room temperature. To remove the Sepharose 4B-binding population, the RNA solution was first poured into the pre-column containing only 0.5 mL of Sepharose 4B gel. After application of the RNA preparation, 3 mL of the binding buffer was then added to the pre-column to obtain the eluate. This pre-column treatment was performed only in the first three selection rounds of SELEX (Table 1). The eluate was then poured into the CRP-immobilized Sepharose 4B column (0.5 mL of gel containing 0.6 mg immobilized CRP; see Table 1). The column was then washed three times with 5 mL of the binding buffer. The total volume of this washing buffer was 30 column volumes. For elution of the CRP-binding population, 1 mL of the elution buffer (binding buffer containing 10 μM CRP) was added to the column and incubated for 30 min at room temperature. This elution was repeated and the RNA preparation was collected by ethanol precipitation. This RNA preparation was reverse transcribed and the cDNA was then amplified by PCR. Reverse transcription of selected RNAs was performed using Primer 2 and a Thermo Script RT kit (Invitrogen). The resulting cDNA was amplified by PCR using Primers 1 and 2.

Binding assay using disk-filters

RNA aptamers were first radiolabeled internally by *in vitro* transcription using [α-^{32}P] UTP as described [11]. The labeled RNA was dissolved in binding buffer. The RNA solution was incubated at 65°C for 10 min and then at room temperature for 30 min to allow proper folding. The RNA was mixed with various concentrations (see Figs. 1 and 3) of CRP protein solution and incubated at 25°C for 30 min. The final concentration of RNA in the binding reaction mixture was 40 nM. The mixture was filtrated *in vacuo* through a disk-filter (MF-membrane filter with 0.45 μm pore size; Millipore), and the filter was washed with binding buffer. After drying, the radioactivity on the filter was measured using a liquid scintillation spectrophotometer (Aloka).

Molecular biological methods

DNA manipulation, cloning techniques, and nucleotide sequencing were performed as described by Sambrook and Russell [12]. Terminal labeling with [^{32}P], secondary structure analyses, and gel electrophoresis of RNAs were performed as described by Kikuchi *et al.* [13, 14].

Results and discussion

Effects of pre-elution by target ligand on SELEX

The selection of RNA was performed using a CRP-immobilized Sepharose 4B column. After the tenth round of selection without pre-elution as described in Materials and methods, we found that the RNA preparation did not have any significant binding ability to CRP in the binding assay using the disk-filter method (Fig. 1). This indicated that the binding ability of the selected RNA preparation to the target may be too weak to detect by this disk-filter method, even if the RNA preparation was trapped by the affinity column during the selection rounds.

To remove the weak binding portion from the RNA population, on and after the sixth selective round (Table 1), we introduced the pre-elution step just before the true elution step. In the pre-elution step, 1 mL of the elution buffer (10 μM CRP-containing binding buffer) was added to the column and incubated for a short time (0.5 – 15 min). These pre-elution steps were repeated several times in some selective rounds (see Table 1). The composition of the pre-elution buffer was the same as the elution buffer. A total of 19 rounds of selective amplification were performed (Table 1). The pre-elution steps were very effective to obtain the strong binding fraction (see below).

Figure 1 shows the results from the disk-filter assay of RNA populations from some selective rounds. The selected RNA populations were labeled internally by *in vitro* transcription of the dsDNA preparations from the tenth, fourteenth, and nineteenth rounds of selection using [α-^{32}P] UTP as described in Materials and methods. The RNA preparation (10000 – 100000 cpm) was dissolved in 100 μL of binding buffer, heated at 65°C for 10 min, and incubated for 30 min at room temperature to allow appropriate folding. The RNA preparation was then mixed with the CRP solution of various concentrations as shown in Fig. 1. Values of binding are expressed as % of total input RNA. The RNAs from the tenth, fourteenth and nineteenth rounds of selection with pre-elution step showed dose-dependent binding. First, the RNA preparation from tenth round containing pre-elution steps at 6th to 10th rounds (see Table 1) clearly showed the binding abilities, although none of the RNA preparation from tenth round without pre-elution step showed the ability (compare open squares with closed squares in Fig. 1). The RNA preparation from the nineteenth round of selection showed extremely high binding ability. Fig. 1 shows that the binding ability increased especially after the fourteenth round. During the fourteenth and seventeenth rounds, we adopted four pre-elution steps for 10 min (Table 1). The incubation time of the pre-elution of the last two rounds (the 18th and 19th) was extended to 15 min. Pre-elution was performed under very stringent selection conditions indicating that this was a very effective means of obtaining RNA aptamers with high binding ability.

Table 1. Selection conditions of RNA populations. All selection rounds (19 rounds) using CRP protein-immobilized Sepharose column are shown. **Pre-column, +**: in first 3 rounds, RNA populations were poured into free Sepharose column (no CRP) to remove Sepharose-binding portions. **CRP in column**: total amount of CRP immobilized on Sepharose gel used. **Wash**: volume of washing buffer and frequency are shown. **Pre-elution:** frequency and time (min, in parentheses) of pre-elution using 1 mL of elution buffer are shown. For example, "2 (5)" in 8th and 9th rounds indicates that two 5-min incubations with elution buffer were performed. See text for full description.

Round	**Pre-column**	**CRP in column [mg]**	**Wash**	**Pre-elution (min)**
1	+	**0.6**	**5 mL × 3**	-
2	+	**0.6**	**5 mL × 3**	-
3	+	**0.6**	**5 mL × 3**	-
4	-	**0.3**	**5 mL × 3**	-
5	-	**0.3**	**5 mL × 10**	-
6	-	**0.3**	**5 mL × 10**	**1 (0.5)**
7	-	**0.3**	**5 mL × 10**	**1 (1)**
8	-	**0.2**	**5 mL × 10**	**2 (5)**
9	-	**0.2**	**5 mL × 10**	**2 (5)**
10	-	**0.1**	**5 mL × 10**	**3 (10)**
11	-	**0.3**	**5 mL × 10**	**3 (10)**
12	-	**0.3**	**5 mL × 10**	**3 (10)**
13	-	**0.3**	**5 mL × 10**	**3 (10)**
14	-	**0.3**	**5 mL × 10**	**4 (10)**
15	-	**0.3**	**5 mL × 10**	**4 (10)**
16	-	**0.3**	**5 mL × 10**	**4 (10)**
17	-	**0.3**	**5 mL × 10**	**4 (10)**
18	-	**0.3**	**5 mL × 10**	**2 (15)**
19	-	**0.3**	**5 mL × 10**	**2 (15)**

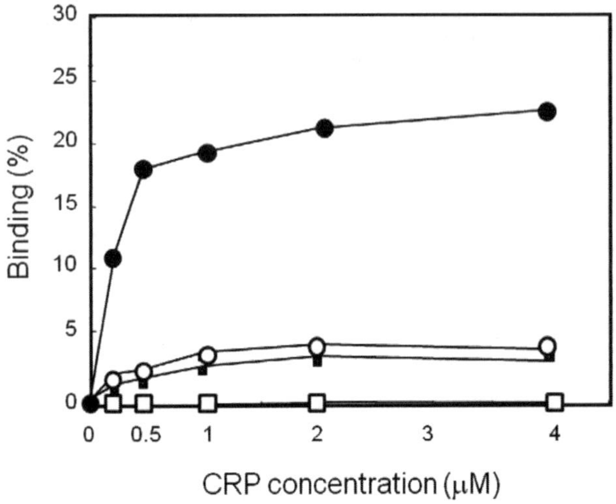

Figure 1. Binding assay of selected RNA populations using disk-filter. Binding abilities of RNA populations of several rounds of selection were assayed using increasing concentrations of CRP. Closed circles, RNA from 19th round; open circles, from 14th round; closed squares, from 10th round; open squares, from round 0, tRNA, and from 10th round without the pre-elution. See text for full description.

CRP1-1, an RNA aptamer with high affinity to CRP

The selected RNAs from the nineteenth round were reverse transcribed, and resulting cDNAs were amplified by PCR and subcloned into the plasmid vector pGEM-3Z (Promega). We arbitrarily chose 110 clones and sequenced the inserts of these plasmids. All clones sequenced originated from Template FR, although the first populations contained RNAs originating from equal amounts of Templates FR and SL (see Materials and methods). Template SL was designed to take predetermined stem-loop structures in the population. Although the stem-loop structures were thought to provide an advantage for binding ability selection, there were no such structures in the clones sequenced. This suggested that the stem-loop structures, even if they are dominant in the selection steps, were disadvantageous in the RT-PCR or transcription steps, because the stem with high G+C content may cause elongation pauses during reactions with reverse transcriptase or T7 RNA polymerase. If we try another SELEX starting with only Template SL, it is possible that another aptamer containing this stem-loop may be obtained, although we have yet to perform such experiments.

Of 110 clones sequenced, 87 (79%) had the same sequence shown in Fig. 2A, which we call CRP1-1. Thirteen of the 110 clones (12%) had the sequence shown in Fig. 2B as CRP1-2. Other clones had completely different sequences and lower binding capabilities compared to CRP1-1 and CRP1-2 (data not shown). We have confirmed the binding abilities of the aptamers CRP1-1 and CRP1-2 using disk-filter assay. As shown in Fig. 3, CRP1-1 had higher binding ability to C-reactive protein than CRP1-2. Therefore, we examined CRP1-1 in more detail.

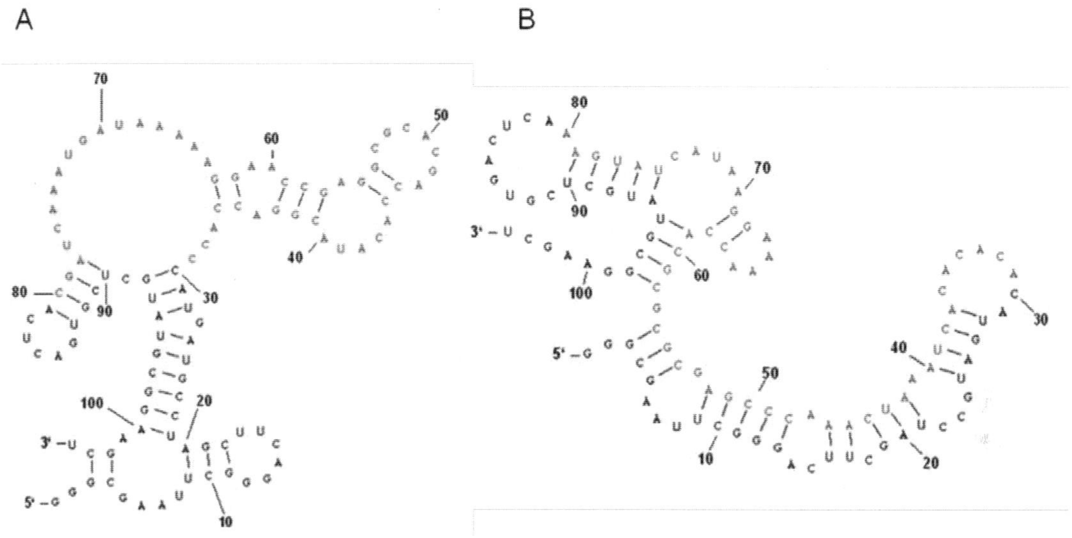

Figure 2. Nucleotide sequences and secondary structures of the RNA aptamers CRP1-1 (A) and CRP1-2 (B). The secondary structures were generated by computer. Colored regions (red and green letters) originated from random region of starting template and the black regions are constant regions used as primer binding sites for amplification.

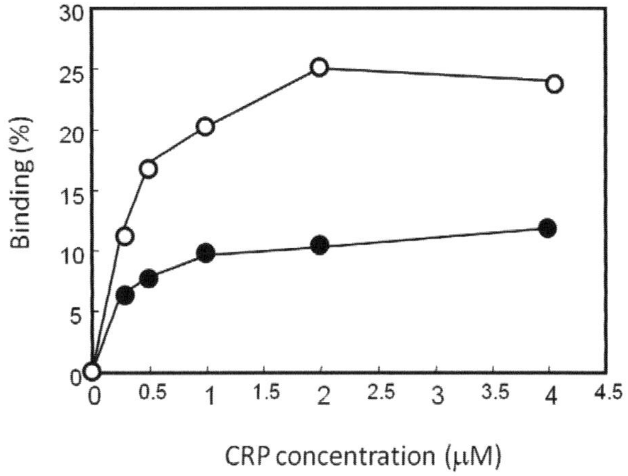

Figure 3. Binding of the RNA aptamers CRP1-1 and CRP1-2. Binding abilities of the aptamers CRP1-1 (open circles) and CRP1-2 (closed circles) to increasing concentrations of CRP protein were assayed.

SPR analysis of CRP1-1

Surface plasmon-resonance (SPR) analysis using Biacore biosensor was performed. Biacore Model X was used throughout this study. First, the 3'-end of CRP1-1 was biotinylated and immobilized on the surface of a streptavidin-immobilized chip. For kinetic analysis, CRP protein solutions of several concentrations (0.25 – 4 μg/mL) were loaded onto the CRP1-1 immobilized surface of a sensor chip at a flow rate of 20 μL/min for 5 min in the association phase. Then, the running buffer was injected at the same flow rate for 5 min in the dissociation phase. Figure 4 shows the sensorgrams of CRP protein binding. The k_{on}, k_{off}, and K_D values were calculated to be 1.6×10^5 ($M^{-1}s^{-1}$), 3.6×10^{-4} (s^{-1}), and 2.25×10^{-9} (M), respectively. The CRP RNA aptamer reported previously had a binding constant of 125 nM [15]. The binding constants of standard anti-CRP antibodies were reported to be 440 – 550 nM [16]. The binding constant of our CRP1-1 (2.25 nM) has almost two orders of magnitude lower (higher affinity) than those of the reported RNA aptamer and antibodies. CRP1-1 constructed here exceeds the binding constant of standard anti-CRP antibodies.

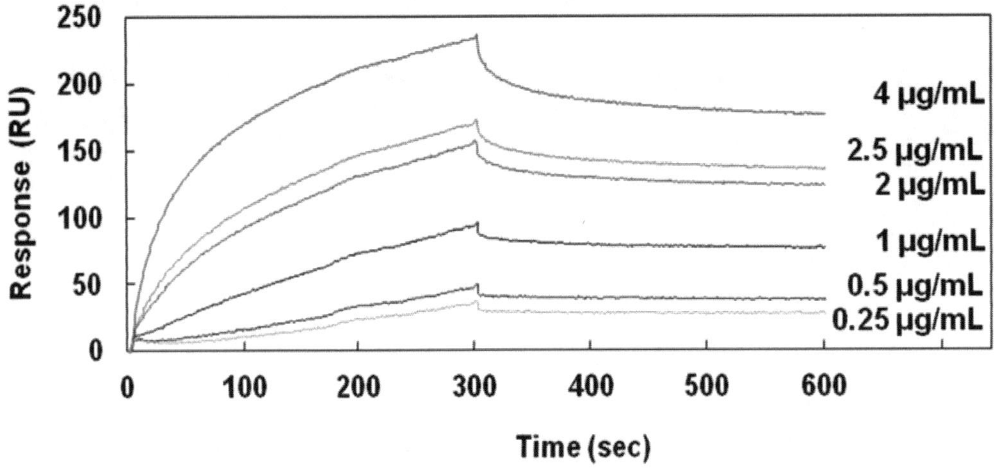

Figure 4. Binding kinetics of CRP protein against the RNA aptamer CRP1-1. Sensorgrams clearly showed the affinity of CRP protein to the CRP1-1. CRP protein at each indicated concentration (0.25 – 4 μg/mL) was loaded onto CRP1-1-immobilized surface of the sensor chip. Biacore X (GE Healthcare) was used.

This experiment indicated that $CaCl_2$ in the buffer is indispensable for binding. Usually, the binding buffer contained 2 mM $CaCl_2$. When $CaCl_2$ was removed from the binding buffer, no binding was detected by SPR analysis. These observations indicated that CRP1-1 recognizes the pentamer of CRP protein, because $CaCl_2$ was required to form pentamers of CRP [17]. Alternatively, CRP1-1 aptamer may require $CaCl_2$ for binding itself, irrespective of pentamer formation.

Secondary structure of CRP1-1

The secondary structure of CRP1-1 was first generated by computer using GENETYX-MAC 8.0 software (Fig. 2). This secondary structure was confirmed enzymatically using single-strand-specific nuclease S_1 and double-strand-specific ribonuclease (RNase) V_1. RNAs radioactively labeled at the 5'- or 3'-end under native conditions were used as the substrates for these enzymes. The enzymatic reaction products were analyzed by denaturing PAGE (Fig. 5). Based on the results of this experiment, the cleaved positions (the bands in Fig. 5) were marked in Fig. 6. There was good agreement between these experimental results (Fig. 5) and the computer-generated secondary structure (Figs. 2A and 6), indicating that the computer-generated secondary structure was mostly correct.

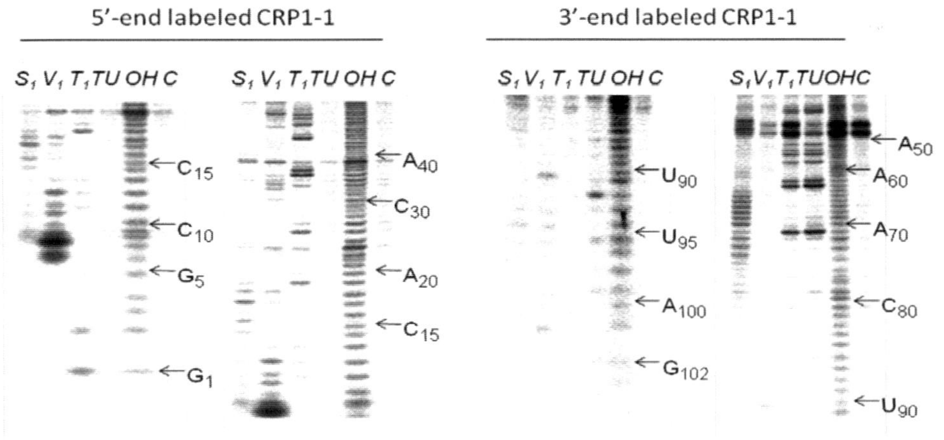

Figure 5. Analysis of the secondary structure of CRP1-1. Samples of the native form of CRP1-1 (5'- or 3'-end radiolabeled) were partially digested by nuclease S_1 (single-strand-specific nuclease), ribonuclease V_1 (RNase V_1, double-strand-specific RNase), and RNase T_1 (for G-mapping). The products were analyzed by denaturing PAGE. Autoradiograms are shown. S_1, nuclease S_1; V_1, RNase V_1; T_1, RNase T_1; TU, RNase T_1 in the presence of 4 M urea; OH, products of alkaline digestion; C, no enzyme control. Positions of some nucleotides in the CRP1-1 sequence are shown on the right sides of autoradiograms.

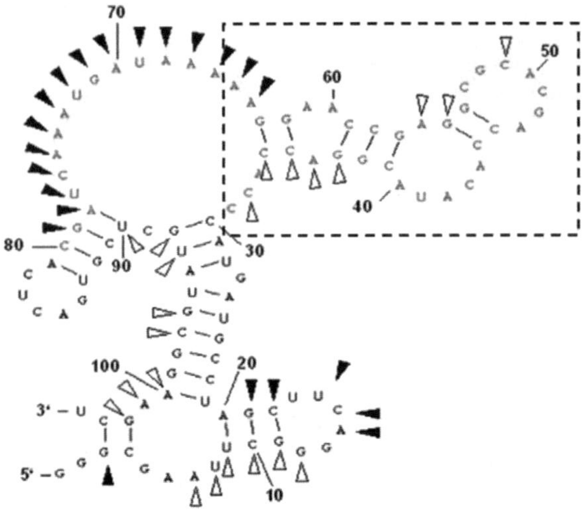

Figure 6. Secondary structure of CRP1-1 and the region that makes contact with CRP protein. Sites sensitive to nucleases derived from the results of Figure 5 are indicated by arrowheads. Closed arrowhead, the site cleaved by nuclease S_1; open arrowhead, the site cleaved by RNase V_1. Square shown by dotted lines indicates the putative region of contact with CRP protein. See text for full description.

The area in CRP1-1 in direct contacted with CRP protein

Using the footprinting technique, we searched for the area in the CRP1-1 sequence that makes contact with CRP protein. An experiment similar to that shown in Figure 5 was performed but in the presence of CRP protein for nuclease reactions. The intensity of the cleavage signals (bands) from nucleotide number 31 to 65 decreased significantly in the presence of CRP protein in the nuclease reaction mixture (data not shown). This suggested that the region from nucleotide 31 to 65 may make direct contact with CRP protein. This region is indicated by a square (enclosed by dotted lines) in Fig. 6.

Based on this result, we examined whether this region alone is sufficient for binding ability to CRP protein. We synthesized a small RNA containing only this region from nucleotide 31 to 65, but it showed no binding ability to CRP protein. These observations suggested that the region of contact in the CRP1-1 aptamer to CRP protein is indeed from nucleotide 31 to 65, but that other regions are also important for binding or for retaining some higher order structure necessary for binding.

Conclusions

We obtained an RNA aptamer, CRP1-1, with high affinity to CRP protein by the newly developed pre-elution method. The dissociation constant of CRP1-1 to CRP protein was calculated to be $K_D = 2.25\times10^{-9}$ (M) by SPR analysis using a Biacore biosensor. This value is almost two orders of magnitude lower (higher affinity) than that of the previously reported CRP RNA aptamer. The value was derived from the k_{on} value of 1.6×10^5 ($M^{-1}s^{-1}$) and the k_{off} value of 3.6×10^{-4} (s^{-1}). Until the tenth round of selection, we used gradually decreasing concentrations of CRP protein in the CRP-immobilized column (Table 1). This may have contributed to obtaining a high k_{on} value of selected populations. The pre-elution method described here may contribute to obtaining a favorable k_{off} value of CRP1-1 by removing a population easily released from the CRP-immobilized column.

Various methods for detection of CRP using RNA aptamers with low affinity have been reported [15, 18, 19]. These methods may be improved by use of an RNA aptamer with high affinity.

References

[1] Ellington AD and Szostak JW 1990 *Nature* **346**, 818

[2] Tuerk C and Gold L 1990 *Science* **249**, 505

[3] Gold L, Polisky B, Uhlenbeck O and Yarus M 1995 *Ann. Rev. Biochem.* **64**, 763

[4] Jenison RD, Gill SC, Pardi A and Polisky B 1994 *Science* **263**, 1425

[5] Bock LC, Griffin LC, Latham, JA, Vermaas EH and Toole JJ 1992 *Nature* **355**, 564

[6] Takeno H, Yamamoto S, Tanaka T, Sakano Y and Kikuchi Y 1999 *J. Biochem.* **125**, 1115

[7] Ng EWM, Shima DT, Calias P, Cunningham ET Jr, Guyer DR and Adamis AP 2006 *Nature Rev. Drug Discov.* **5**, 123

[8] Oliveira EB, Gotschlich EC and Liu TY 1977 *Proc. Natl. Acad. Sci. USA* **74**, 3148

[9] Clearfield MB 2005 *J. Am. Osteopath. Assoc.* **105**, 409

[10] Wu TL, Tsao KC, Chang CPY, Li CN, Sun CF and Wu JT 2002 *Clin. Chim. Acta* **322**, 163

[11] Kikuchi Y and Sasaki N 1991 *Nucleic Acids Res.* **19**, 6751

[12] Sambrook J and Russell DW 2001 *Molecular Cloning: A Laboratory Manual, Third Edition*, Cold Spring Harbor Laboratory, Cold Spring Harbor, NY

[13] Kikuchi Y, Ando Y and Shiba T 1986 *Nature* **323**, 824

[14] Kikuchi Y, Sasaki, N and Ando-Yamagami Y 1990 *Proc. Natl. Acad. Sci. USA* **87**, 8105

[15] Bini A, Centi S, Tombelli S, Minunni M and Mascini M 2008 *Anal. Bioanal. Chem.*

390, 1077

[16] Lin S, Lee CK, Lin YH, Lee SY, Sheu BC, Tsai JC and Hsu SM 2006 *Biosens. Bioelectron.* **22**, 715

[17] Pepys MB and Hirschfield GM 2003 *J. Clin. Invest.* **111**, 1805

[18] Pultar J, Sauer U, Domnanich P and Preininger C 2009 *Biosens. Bioelectron.* **24**, 1456

[19] Centi S, Sanmartin LB, Tombelli S, Palchetti I and Mascini M 2009 *Electroanalysis* **21**, 1309

Crystal Structure of an L-Carnitine Complex with Pyrogallol[4]arene

I. Fujisawa, D. Takeuchi, Y. Kitamura, R. Okamoto, and K. Aoki

Department of Environmental and Life Sciences, Toyohashi University of Technology, 1-1 Hibarigaoka, Tempaku-cho, Toyohashi 441-8580, Japan

ifujisawa@ens.tut.ac.jp

Abstract. L-Carnitine is essential for the transport of long-chain fatty acids from cytosol into mitochondria for generating metabolic energy. The survey of crystal structures of carnitine-containing proteins in the Protein Data Bank reveals that carnitine can take several conformations with the quarternary trimethylammonium terminal being always bound to aromatic residues through cation–π interactions in acyltransferases or carnitine-binding proteins. In order to demonstrate the importance of cation–π interaction as a carnitine recognition mechanism in the artificial receptor–ligand system that mimics the carnitine-binding sites, we have determined the crystal structure of a complex formed between L-carnitine and pyrogallol[4]arene (pyrogallol cyclic tetramer: PCT) as a carnitine receptor, 2PCT·2(L-carnitine)·4EtOH. There form two crystallographically independent monomeric [PCT·L-carnitine] substructures, which further form an obliquely arranged capsule-like dimeric $[PCT·L\text{-carnitine}]_2$ structure through a pair of O–H (PCT)···O (L-carnitine) hydrogen bonds. This is the first report of PCT complex with chiral molecules. In each of the two monomeric [PCT·L-carnitine] substructures, the L-carnitine molecule takes the elongated form with an intramolecular hydrogen bond between the hydroxyl group and the carboxylate oxygen, and the cationic trimethylammonium moiety is incorporated into the cavity of the bowl-shaped PCT molecule through cation–π interactions. These features are similar to those at the D-carnitine-binding site in the crystal structure of the glycine betaine/carnitine/choline-binding protein complex.

1. Introduction

Carnitine is needed for the transport of long-chain fatty acids through mitochondrial membrane; long-chain acyl-CoA synthesized in cytosol can't be transported as such into mitochondria but must be once converted to acyl-carnitine by carnitineacyltransferase I using carnitine, and incorporated acyl-carnitine in mitochondria is subsequently catalyzed by carnitineacyltransferase II to give acyl-CoA again, followed by β-oxidation [1, 2]. Because β-oxidation in mitochondria is an important metabolism pathway of fatty acids, the research of the molecules targeting acyltransferases is proceeded in expectation of the application for diet foods or supplements.

Crystal structures of three kinds of protein–carnitine complexes are reported in the Protein Data Bank. In the acetyltransferase–L-carnitine complex [3], the carboxylate group of the L-carnitine ligand forms hydrogen bonds with tyrosine, serine and threonine residues in the protein, the hydroxyl group forms a hydrogen bond with the histidine residue, and the quaternary trimethylammonium group forms a cation–π interaction with the phenylalanine residue, stabilizing the bent structure of the carnitine molecule. In the glycine betaine/carnitine/choline-binding protein–D-carnitine complex [4], the carboxylate group of D-carnitine forms hydrogen bonds with glutamine and asparagine residues in the protein and the trimethylammonium cation forms cation–π interactions with surrounding four tyrosine residues, stabilizing the elongated structure of carnitine with an intramolecular hydrogen bond formed between the hydroxyl group and the carboxylate oxygen. The carnitine molecule takes different conformations depending on each binding protein, but in all cases, the cation–π interaction between the trimethylammonium cation in carnitine molecule and aromatic rings in proteins (C···aromatic ring < 4.1 Å) is observed. This suggests the common importance of the cation–π interaction in the carnitine–protein recognition and binding.

In order to demonstrate the importance of the cation–π interaction as a ligand recognition mechanism, we have constructed artificial receptor–ligand systems that mimic interactions at ligand-binding sites between the quaternary alkylammonium moiety of the bioactive ligands, involving acetylcholine, and the aromatic ring(s) of the proteins [5-9]. Most recently, as such a system, we have reported [10] crystal structures of the complexes formed between resorcin[4]arene or tetramethylated resorcin[4]arene and L-carnitine. We report here the crystal structure of an L-carnitine complex with pyrogallol[4]arene (PCT) as a carnitine receptor, showing that cation–π interaction formed between the trimethylammonium moiety of L-carnitine and aromatic rings of PCT and that the L-carnitine conformation is similar to that in the glycine betaine/carnitine/choline-binding protein complex. Seven structures of PCT complexes have already been reported, but all have an inversion center, and consist of only achiral molecules. The present structure has no inversion center, because chiral L-carnitine is included.

2. Materials and Methods

PCT (**1**) host compound was synthesized by an analogous method [11] for RCT. The host solution and a L-carnitine solution were prepared by dissolving the host compound (700 mg) in ethanol (42 mL) and L-carnitine (1.0 g) in water (10 mL), respectively. The host solution (0.1 mL), L-carnitine solution (2 mL) and water (0.4 mL) were mixed and allowed to stand at room temperature to give dark-orange crystals after 5 days. Molecular formula was determined by X-ray analysis.

X-ray intensity data were measured on a Rigaku AFC-7R diffractometer (at the Instrument Center of the Institute for Molecular Science in Okazaki) with a rotating anode generator and a Mercury CCD camera, using graphite-monochromated Mo Kα radiation (λ(Mo Kα) = 0.7169 Å) at 295 K. Data reduction, the cell refinement, and semi-empirical absorption corrections were performed with the program CrystalClear [12]. The structure was solved by direct methods and refined by full-matrix least-squares techniques on F_o^2, minimizing $\Sigma w(|F_o| - |F_c|)^2$, with the programs SHELXS and SHELXL [13], respectively, on the platform and graphic software Yadokari-XG [14]. No attempt was made to locate H atoms. Crystal data for the PCT complex, 2(**1**)·2(L-carnitine)·4EtOH: $C_{94}H_{132}N_2O_{34}$, fw = 1834.04, triclinic, space group $P1$, a = 12.7778(9). b = 12.9479(10), c = 15.0200(15) Å, α = 89.832(7), β = 82.108(5), γ = 69.720(4)°, V = 2306.2(3) Å^3, Z = 1, D_{calcd} = 1.240 g cm^{-3}, μ(Mo Kα) = 0.149 cm^{-1}, F(000) = 914. All non-H atoms were refined with anisotropic temperature factors. Final R = 0.089 and R_w = 0.230 (for 10318 reflections with $I > 2\sigma(I)$ out of 14100 unique reflections in the range $2 < 2\theta < 55°$) and GOF = 1.55. Crystallographic data have been deposited with the Cambridge Crystallographic Data Center as supplementary publication No. CCDC-809923. Copies of the data can be obtained free of charge from the Cambridge, CB2 1EZ, UK; Fax: +44 1223 336033; e-mail: deposit@ccdc.cam.ac.uk.

3. Results and Discussion

In the asymmetric unit, two PCT (PCT1 and PCT2), two carnitine (Car1 and Car2) and four solvent ethanol molecules are included. As shown in Figure 1, there form two crystallographically independent monomeric PCT·L-carnitine substructures, [PCT1·Car1] and [PCT2·Car2], which further form an obliquely arranged [PCT·L-carnitine]$_2$ dimeric structure through an inter-substructure O–H (PCT1)···O (Car2) hydrogen bond (O(1)···O(28) = 2.951(8) Å). In each of the two substructures, the cationic trimethylammonium end of the carnitine molecule is incorporated into the cavity of a bowl-shaped PCT molecule through cation–π interactions. In addition, in the [PCT1·Car1] substructure, the carboxylate end of Car1 forms an intra-substructure hydrogen bond with the phenolic oxygen of PCT1 (O(10)···O(25) = 2.908(8) Å). Car1 and Car2 ligands take a fully extended backbone (N–C–C–C–C) conformation with an intramolecular O–H (hydroxyl)···O (carboxylate) hydrogen bond (O(27) ···O(25) = 2.881(7) Å for Car1 and O(30)···O(29) = 2.667(7) Å for Car2). These observations are in consistent with a thermodynamic solution study [15] showing that L-carnitine was strongly bound to PCT with K = 18,000 M^{-1} and all of the functional groups (trimethylammonium, carboxylate, and hydroxyl) and the extended hydrocarbon chain are assumed to contribute to the affinity enhancement through cation–π, hydrogen bonding, ion-dipole, and van der Waals interactions (see Figure 2 in Ref. 15).

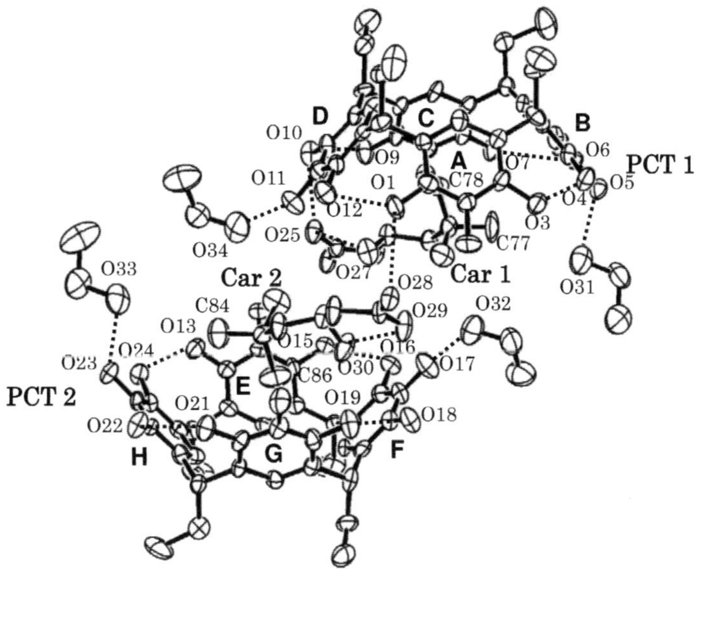

Figure 1. Molecular structure of the [**1**·L-carnitine]$_2$ dimer. Distances (Å) from the carbon atoms of the trimethylammonium group of carnitine to the π-centroids of **1** (< 4.1 Å): for Car1–PCT1, C78···ring A = 3.72, C77···ring B = 3.76, C78···ring C = 3.63, C78···ring D = 3.62; for Car2–PCT2, C86···ring E = 3.62, C86···ring F = 3.74, C86···ring G = 3.63, and C84···ring H = 3.76. Broken lines denote hydrogen bonds.

All crystal structures of PCT complexes registered in Cambridge Structural Database have a crystallographic inversion center which generates upside-down arranged two PCT molecules in the crystal lattice to construct a capsule or a capsule-like supramolecule. The present PCT complex has no inversion center because the chiral L-carnitine molecule is included but, interestingly, a capsular-like supramolecule is formed here in a similar manner to the known PCT complex structures. As shown in Figure 2, though the whole backbone shapes in the two substructures are similar to each other, each hydrogen bond network is not the same. Closer observations reveal that carboxylate of

Car1 forms hydrogen bonds to three PCT molecules, including the pairing PCT molecule, while carboxyl group of Car2 forms hydrogen bonds to three PCT molecules, not including the pairing PCT, but including the opposite side substructure's PCT molecule.

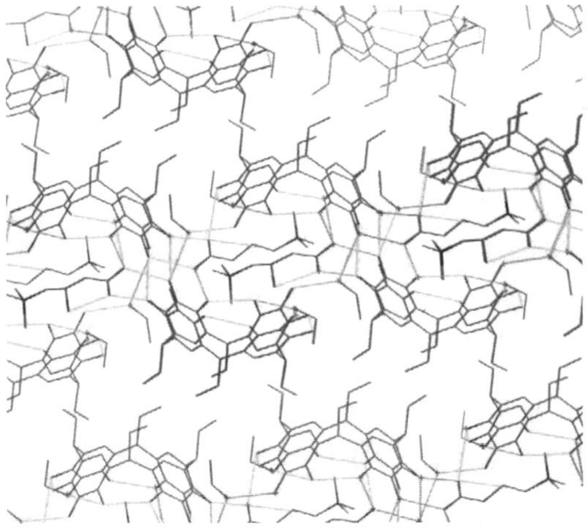

Figure 2. Crystal packing of the [**1**·L-carnitine]$_2$ dimer viewed from *a*-axis direction of the unit cell. PCT, carnitine and ethanol molecules construct hydrogen bonding networks expanding laterally, which layers pile up, forming the hydrophobic layers. Light blue lines denote hydrogen bonds.

Though the complex was prepared under a mixture of water and ethanol, only ethanol molecules were located in the crystal structure. Oxygen atoms of all the four ethanol molecules construct hydrogen bonding networks with PCT molecules, and the methyl groups direct toward hydrophobic layers made by lateral array of PCT molecules, where two ethanol molecules have C–H···π interactions in the distance less than 4.1 Å with aromatic rings of PCT. These well fitted ethanol molecules may contribute to the stabilization of the crystal structure and a small R-factor, 8.9%, compared with resorcin[4]arene or tetramethylated resorcin[4]arene complexes with carnitine, which showed higher *R* values, 11.7 or 13.1%, respectively [10].

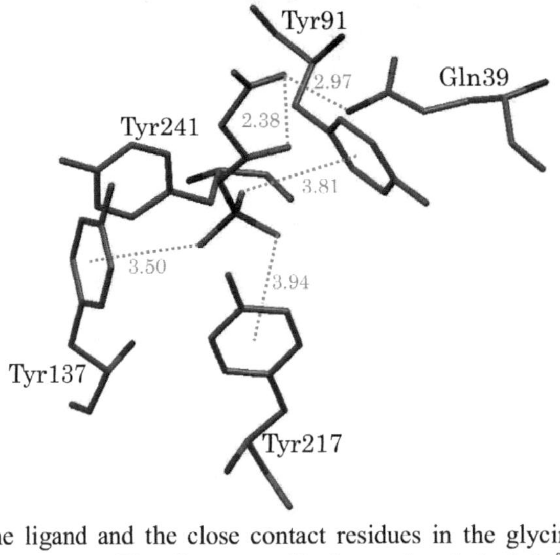

Figure 3. The D-carnitine ligand and the close contact residues in the glycine betaine/carnitine/choline-binding protein complex structure. The distances of hydrogen bonds (< 3.0 Å) and cation–π interactions (< 4.1 Å) are denoted.

The carbon atoms of the trimethylammonium moiety of Car1 and Car2 molecules are in distances to form cation–π interactions to aromatic rings of PCT1 and PCT2, as noted above (Figure 1). Also in the structure of the glycine betaine/carnitine/choline-binding protein complex [4], aromatic rings of four tyrosine residues surround the trimethylammonium end of carnitine (Figure 3). Differences around trimethylammonium cation between the present complex and the protein complex are angles between a carnitine molecule and four aromatic rings, and the scale of the cavity made by four aromatic rings. The angles between the vector made of a carbon atom of carboxylate and a nitrogen atom and the plane comprised of bottom carbon atoms of four aromatic rings of the PCT complex and the protein complex are almost perpendicular and about 20°, respectively. The distances between the two aromatic rings which are opposite to each other at the upper part of the cavity are 8–9 Å for both the PCT complex and the protein complex, while at the bottom part of the cavity, the corresponding distances are about 5 and 8–9 Å, respectively. The bottom of the PCT molecule is too narrow for the carnitine molecule to insert enough its cationic end into the cavity, resulting in the different spatial arrangement of cation–π interactions, comparing with that in the protein complex. To reach the precise imitation of the carnitine binding structure and to design more appropriate carnitine-receptor molecules that mimic the carnitine binding site in the glycine betaine/carnitine/choline-binding protein, these points should be improved utilizing different-shaped synthetic receptors [16] which have linkers with different length. The carboxylate oxygen atoms of Car1 and Car2 form hydrogen bonds to three PCT molecules and Car1 and Car2 have intramolecular hydrogen bonds. These interactions stabilize the elongated structure of Car1 and Car2, which is similar to that of the glycine betaine/carnitine/choline-binding protein complex but different from the partially folded structure in the RCT–L-carnitine complex [10] (Figure 4).

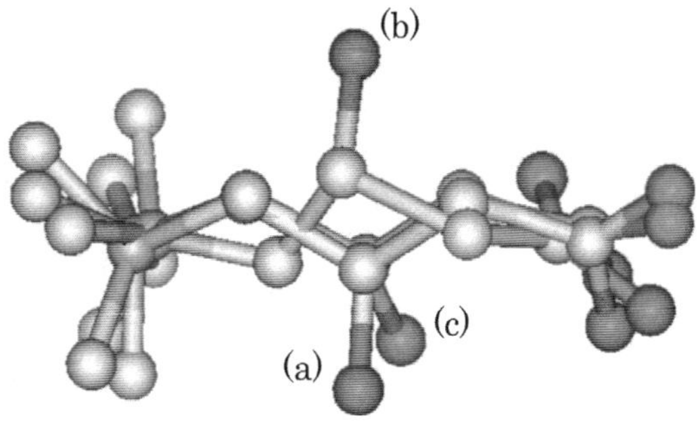

Figure 4. Comparison of three carnitine structures of the PCT (a), RCT (b) [10], or glycine betaine/carnitine/choline-binding protein (c) [4] complexes. One of the carnitine models of the PCT and RCT complexes are superimposed to that of the glycine betaine/carnitine/choline-binding protein complex with least-square calculation by the program MIFit [17] to match five backbone atoms from a nitrogen atom to a carbon atom of carboxylate, with r.m.s. distances 0.19 and 0.72 (Å), respectively.

4. Conclusion

This study provides an additional X-ray example that shows the existence of cation–π interaction between the quaternary trimethylammonium moiety of the biologically important L-carnitine ligand and aromatic rings in the artificial receptor–ligand system that mimics the carnitine binding sites in acyltransferases or carnitine-transporters. The extended structures of carnitine ligands are similar to that observed in the glycine betaine/carnitine/choline-binding protein complex [4].

References

[1] L. L. Bieber, *Ann. Rev. Biochem.*, **57**, 261 (1988).
[2] Friedman S. and Fraenkel G., Reversible acetylation of carnitine., *Arch. Biochem. Biophys.*, **51**, 491 (1955).
[3] G. Jogl, L. Tong, *Cell*, **112**, 113 (2003).
[4] Y. Du, W.W. Shi, Y.X. He, Y.H. Yang, C.Z. Zhou, and Y. Chen, *Biochem. J.*, **436**, 283 (2011).
[5] K. Murayama and K. Aoki, *Chem. Commun.*, 119 (1997).
[6] K. Murayama and K. Aoki, *Chem. Lett.*, 301 (1998).
[7] K. Aoki, K. Murayama and H. Nishiyama, *J. Chem. Soc. Chem. Commun.*, 2221 (1995).
[8] K. Murayama and K. Aoki, *Chem. Commun.*, 607 (1998).
[9] K. Aoki, T. Nagae, S. Yamaguchi and I. Fujisawa, *Bull. Chem. Soc. Jpn.*, **78**, 2066 (2005).
[10] I. Fujisawa, D. Takeuchi, R. Kato, K. Murayama, and K. Aoki, *Bull. Chem. Soc. Jpn.*, **84**, 1133 (2011).
[11] Y. Aoyama, Y. Tanaka and S. Sugahara, *J. Am. Chem. Soc.*, **111**, 5397 (1989).
[12] "CrystalClear: An integrated program for the collection and processing of area detector data," Rigaku Corporation (1997).
[13] G.M. Sheldrick, *Acta Cryst.*, **A64**, 112 (2008).
[14] C. Kabuto, S. Akine, T. Nemoto and E. Kwon, *J. Cryst. Soc. Jpn.*, **51**(3), 218 (2009).
[15] B. Schnatwinkel, M. V. Rekharsky, V. V. Borovkov, Y. Inoue, and J. Mattay, *Tetrahedr. Lett.*, **50**, 1374 (2009).
[16] J.C. Ma and D.A. Dougherty, Chem. Rev., **1997**, 97, 1303(1997)
[17] MIFit is a cross-platform interactive graphics application for molecular modeling, fitting, and refinement of protein structures from x-ray crystallography, originally created by Duncan McRee.

Journal of Physics: Conference Series **352** (2012) 012044 doi:10.1088/1742-6596/352/1/012044

Quantitative analysis of plasma interleiukin-6 by immunoassay on microchip

K Abe[1], Y Hashimoto[1], S Yatsushiro[1], S Yamamura[1], M Tanaka[1], T Ooie[1], Y Baba[1, 2], M Kataoka[1]

[1] Health Research Institute, National Institute of Advanced Industrial Science and Technology, 2217-14, Hayashi-cho, Takamatsu, Kagawa, 761-0395, Japan

[2] Department of Applied Chemistry, Graduate School of Engineering Nagoya University, Furo-cho, Chikusa-ku, Nagoya, 464-8603, Japan

E-mail : k-abe@aist.go.jp

Abstract. Sandwich enzyme-linked immunoassay (ELISA) is one of the most frequently employed assays for clinical diagnosis, since this enables the investigator to identify specific protein biomarkers. However, the conventional assay using a 96-well microtitration plate is time- and sample-consuming, and therefore is not suitable for rapid diagnosis. To overcome these drawbacks, we performed a sandwich ELISA on a microchip. We employed the piezoelectric inkjet printing for deposition and fixation of 1st antibody on the microchannnel surface (300 μm width and 100 μm depth). Model analyte was interleukin-6 (IL-6) which was one of the inflammatory cytokine. After blocking the microchannel, antigen, biotin-labeled 2nd antibody, and avidin-labeled peroxidase were infused into the microchannel and incubated for 20 min, 10 min, and 5 min, respectively. This assay could detect 2 pg/ml and quantitatively measure the range of 0-32 pg/ml. Liner regression analysis of plasma IL-6 concentration obtained by microchip and conventional methods exhibited a significant relationship (R^2=0.9964). This assay reduced the time for the antigen-antibody reaction to 1/6, and the consumption of samples and reagents to 1/50 compared with the conventional method. This assay enables us to determine plasma IL-6 with accuracy, high sensitivity, time saving ability, and low consumption of sample and reagents, and thus will be applicable to clinic diagnosis.

1. Introduction

Measurement of protein biomarkers in body fluids such as blood and urine is important to understand physiological and pathological state [1, 2]. The most widely used for protein analysis in clinic and experimental laboratories is the enzyme-linked immunosorbent assay (ELISA), due to their specificity and sensitivity. This fundamental principle is antigen-antibody specific binding where a capture antibody (1st antibody) immobilized on a solid substrate will react with its corresponding antigen, then which will react with an enzyme-labeled antibody (2nd antibody). ELISA is conventionally performed on disposable 96-well microtitration plates. However, this method consumes relatively large sample volumes (50-100 μl) and requires considerable incubation time (2 hr or more) [3].

Recently, the analysis of biomarkers at near patient which is called point-of-care testing (POCT) is a continuously expanding trend in the practice of laboratory diagnosis [4]. To overcome disadvantages of conventional ELISA method, miniaturization in microfluidic systems is widely used [5]. Microfluidic immunoassays could reduce analysis time, because increased surface area to volume ratios speeds up antigen-antibody reactions [6]. We developed sandwich ELISA on cyclic olefin copolymer (COC) microchip [7]. For this, we employed the piezoelectric inkjet printing for deposition and fixation of 1st antibody on microchannnel surface (300 μm width and 100 μm depth). We previously reported that carboxyterminal propeptide of type I procollagen (PICP), which was a biomarker for bone formation, could measure only 30-min incubation. Although the normal range of serum PICP is ng/ml level, the normal range of other biomarkers are various from pg/ml to μg/ml level. Therefore, more sensitive assay system is required.

In the present study, we selected interleukin-6 (IL-6) as a model analyte, because the normal plasma IL-6 level is less than 4 pg/ml. IL-6 is one of the inflammatory cytokine, and has been implicated in the pathogenesis of a number of inflammatory conditions, such as rheumatoid arthritis, cardiovascular disease [8]. To archive high sensitivity, we examined the concentration and the spotting pattern of 1st antibody on the microchannel. Furthermore, we examined the feasibility of determining plasma concentration of IL-6 in microchannel was comparable to that obtained by the conventional sandwich ELISA using the 96-well microtitration plates. Our new microchip method could detect 2 pg/ml of IL-6 and quantitatively measure the range of 0-32 pg/ml for 35-min incubation. We demonstrated possible application of microchip for monitoring plasma IL-6 level and its potential for POCT.

2. Materials and methods

2.1. Principle

Schematic illustrations of sandwich ELISA on microchip for the determination of plasma IL-6 are shown in Figure 1.

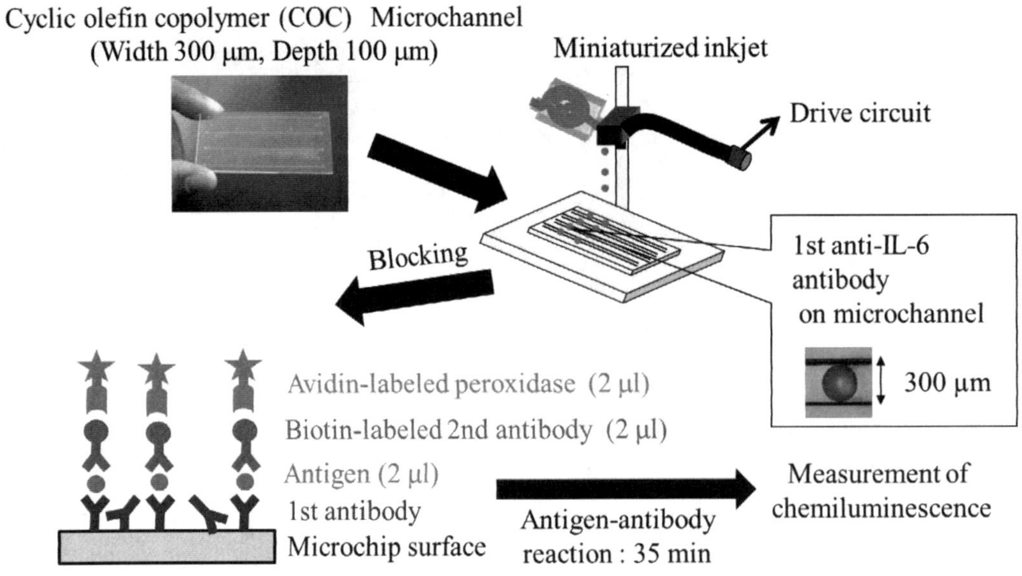

Figure 1. Schematic process for detection of IL-6 on a microchip

2.2. *Reagents and analytical procedure*
Human IL-6 ELISA MAX Deluxe was purchased from BioLegend (Sandiego, USA) and employed for the determination of the plasma concentration of IL-6 with a microtitration plate as the conventional method according to the supplied instructions.

2.3. *Analysis of IL-6 on the microchip*
For the sandwich ELISA on the microchip, we used the same monoclonal anti-IL-6 antibodies that were used in the Human IL-6 ELISA MAX Deluxe for 96-well of microtitration plates. For the deposition and fixation of 1st antibody in spotting buffer on the surface of the microchannel, a piezoelectric inkjet printing system (Cluster Technology Co., Ltd. Osaka, Japan) was employed. The volume of the ejected from the inkjet nozzle was 65 pl, and 60 drops were spotted onto the microchannel surface. After sealing of the microchip surface with a cover film (TOYO INK MFG. CO., LTD, Tokyo, Japan), blocking solution (Block Ace : DS Pharma Biomedical, Osaka, Japan) was infused into each microchannel, and the microchip was left undisturbed for 4 hr at room temperature. After washing of microchannel with 0.05% TritonX in PBS, 2 µl of antigen, biotin-labeled 2nd antibody, and avidin-labeled peroxidase were infused into the microchannel and incubated for 20 min, 10 min, and 5 min, respectively. After that, 2 µl of substrate solution (SuperSignal West Femto Maximum Sensitivity Substrate : Thermo Scientific, Rockford, IL, USA) for peroxidase reaction was infused into the microchannel. The chemiluminescence intensity of each spot was detected using ImageQuant™ LAS4000 (GE Healthcare Bio-Sciences, Uppsala, Sweden) and its integration value was analyzed using ImageQuant TL (GE Healthcare Bio-Sciences).

2.4. *Blood sample preparation*
Peripheral venous blood samples were collected from healthy subjects by standard venipuncture. All blood samples were collected into EDTA-coated evacuated blood container tubes (TERUMO, Tokyo, Japan), and then the plasma were obtained by centrifugation at 800 x g for 10 min. The plasma was then stored at -80℃ until used. When using the experiment of comparative analysis of the microchip method and the conventional microtitration plate method, purified IL-6 antigen (BioLegend) was added to the plasma, because the conventional microtitratin plate method could not accurately determine the healthy subjects under the detection limit (4 pg/ml).

2.5. *Ethics*
This study was approved by the institutional Review Board on using human derivatives for biomedical research of National Institute of Advanced Industrial Science and Technology, and the Ethical Committee of University of Tokushima. All subjects provided written informed consent for the collection of samples and subsequent analysis. No current external funding sources for this study.

3. Results and discussion

To archive high sensitivity in short incubation time, we examined the concentration of 1st antibody on the surface of the microchannel. Using high concentration of 1st antibody, we could detect 2 pg/ml of purified IL-6 solution in 35 min (Figure 2). Furthermore, the fixation area of 1st antibody was changed from 1 spot to 3 spots, luminescence intensity was increased 3 times (Figure 2).

The relationship between the concentrations of purified IL-6 in the range of 0-32 pg/ml and the luminescence intensity derived from avidin-labeled peroxidase was examined (Figure 3). Three spots of 1st anti-IL-6 antibody were deposited into each microchannel for triplicate determinations, and purified IL-6 solution (0-32 pg/ml), biotin-labeled 2nd anti-IL-6 antibody, avidin-labeled peroxidase were infused into each microchannel. The luminescence increased in a IL-6 concentration dependent manner. The mean (±SD) of luminescence intensity for the 3 different spots at each purified IL-6 concentration was obtained and plotted against the dose, giving a good relationship between the range of 0-32 pg/ml (R^2=0.9987, Figure 3). The detection limit was 2 pg/ml. This range is sufficient for clinical estimation of IL-6 concentration in the blood. These data indicated that this sandwich ELISA preformed on a microchip with aid of piezoelectric printing technology is suitable for the determination of plasma IL-6 concentrations.

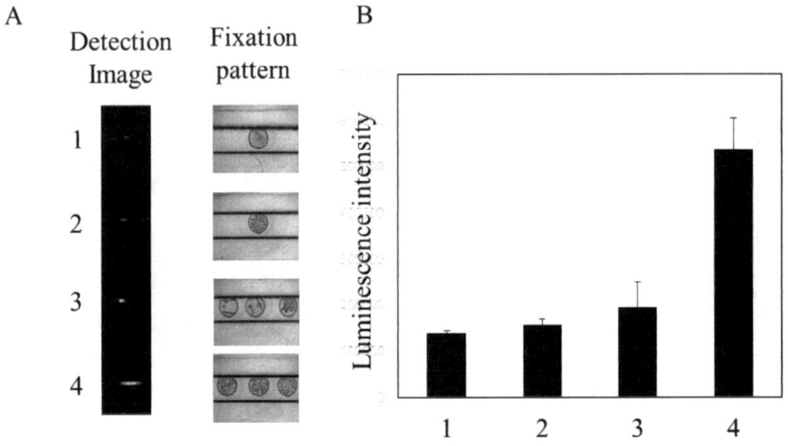

Figure 2. Effect of the concentration and fixation area of 1st anti-IL-6 antibody on the luminescence intensity. (1):0.2mg/ml, 60 droplets, 1 spot (2): 1 mg/ml, 60 droplets, 1 spot (3): 0.2mg/ml, 60 droplets, 3 spots (4) 1 mg/ml, 60 droplets, 3 spots (A) Detection images of 2 pg/ml of purified IL-6 antigen and fixation patterns of 1st anti-IL-6 antibody. (B) Determination of luminescence intensity obtained by the assay.

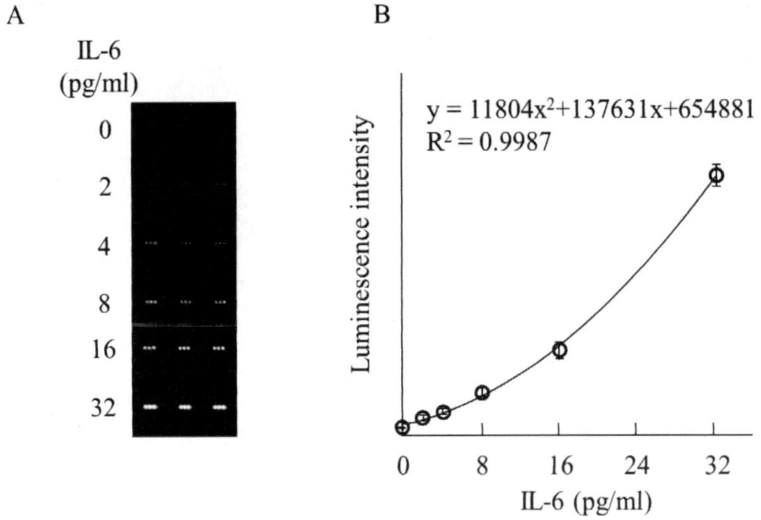

Figure 3. Detection of purified IL-6 by sandwich ELISA on a microchip.
Three deposits of 1st antibody were made into each microchannel, and 0 to 32 pg/ml IL-6, biotin-labeled 2nd antibody, and avidin-labeled peroxidase were then infused, allowing triplicate determination to be made in 1 channel. (A) The luminescence increased in IL-6 concentration-dependent manner. (B) Standard curve for the IL-6 concentration versus luminescence intensity obtained by the assay.

Plasma IL-6 concentration in 4 subjects were calculated by reference to the standard curve generated from data obtained from the sandwich ELISA on microchip. Each estimated plasma IL-6 concentration was compared with the results obtained by using the 96-well microtitration plate commonly used in clinical laboratories. As shown in Figure 4, linear regression analysis of these IL-6 concentration obtained by both methods revealed a significant relationship (R^2=0.9964, Figure 4). This result indicated that accurate determination of plasma IL-6 could be performed by use of our new methods.

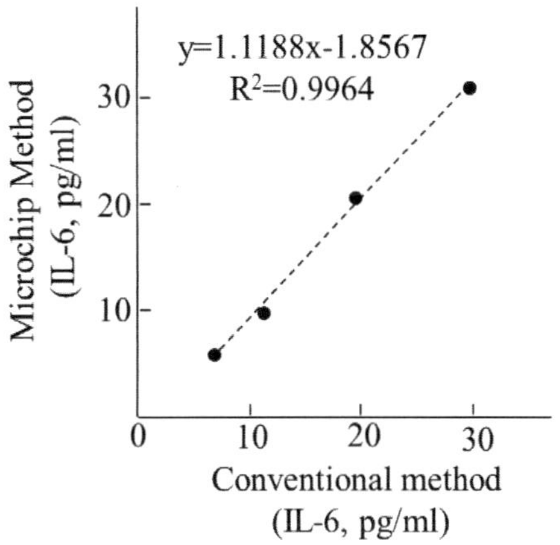

Figure 4. Comparative analysis of the values obtained with the microchip method and the conventional microtitration plate method. Linear regression analysis was used.

4. Conclusion

Sandwich ELISA preformed on a microchip could detect 2 pg/ml of IL-6 for 35-min incubation. This assay reduced the time for the antigen-antibody reaction to 1/6, and the consumption of sample and reagent to 1/50 compared with conventional method. Furthermore, this assay enables us to determine plasma IL-6 with accuracy, and will be applicable to clinic diagnosis.

Acknowledgments

This work was supported by Japan Society for the Promotion of Science, Grant-in-Aid for Scientific Research (KAKENHI) 23792494.

References

[1] Hu S, Loo JA, Wong DT. *Proteomics*. 2006, **6**, 6326-53.
[2] Pisitkun T, Johnstone R, Knepper MA. *Mol Cell Proteomics*. 2006, **5**,1760-71.
[3] Porstmann T, Kiessig ST. *J Immunol Methods*. 1992, **150**, 5-21.
[4] Junker R, Schlebusch H, Luppa PB. *Dtsch Arztebl Int*. 2010, **107**, 561-7.
[5] Ng AH, Uddayasankar U, Wheeler AR. *Anal Bioanal Chem*. 2010, **397**, 991-1007.
[6] Sato K, Yamanaka M, Takahashi H, Tokeshi M, Kimura H, Kitamori T. *Electrophoresis*. 2002,**23**,734-9.
[7] Yatsushiro S, Akamine R, Yamamura S, Hino M, Kajimoto K, Abe K, Abe H, Kido J, Tanaka M, Shinohara Y, Baba Y, Ooie T, Kataoka M. *PLoS One*. 2011, **6**, e18807.
[8] White UA, Stephens JM. *Curr Pharm Des*. 2011,**17**, 340-6.

Emotion Telepresence: Emotion Augmentation through Affective Haptics and Visual Stimuli

D. Tsetserukou[1], A. Neviarouskaya[2]
[1]Electronics-Inspired Interdisciplinary Research Institute, Toyohashi University of Technology, 1-1 Hibariga-Oka, Tenpaku-Cho, Toyohashi, Aichi 441-8581, Japan
[2]Department of Computer Science and Engineering, Toyohashi University of Technology

E-mail: tsetserukou@eiiris.tut.ac.jp

Abstract. The paper focuses on a novel concept of emotional telepresence. The iFeel_IM! system which is in the vanguard of this technology integrates 3D virtual world Second Life, intelligent component for automatic emotion recognition from text messages, and innovative affective haptic interfaces providing additional nonverbal communication channels through simulation of emotional feedback and social touch (physical co-presence). Users can not only exchange messages but also emotionally and physically feel the presence of the communication partner (e.g., family member, friend, or beloved person). The next prototype of the system will include the tablet computer. The user can realize haptic interaction with avatar, and thus influence its mood and emotion of the partner. The finger gesture language will be designed for communication with avatar. This will bring new level of immersion of on-line communication.

1. Introduction

Telepresence system allows a person to feel as if they were present at a place other than their true location. The sense of presence is provided with such stimuli as vision, hearing, sense of touch, etc. from a remote location [1][2][3].

Lobbard and Dutton concept can roughly be divided into two categories of presence: physical and social presence [4]. Physical presence refers to the sense of being physically located in mediated environment. While social presence implies the sense of social interaction with remotely located communication partner. We argue that the emotional telepresence, concept proposed by us, is the integral and one of the most important parts of feeling of being there.

Emotional telepresence, technology that lets users feel emotionally as if they were present and communicating at a remote physical location. The remote environment can be real, virtual, or augmented. Whereas conventional telepresence technologies focus on providing high-quality visual and audio feedback from the remote location to the user, emotional telepresence aims to enhance emotional immersion during online communication. In a nutshell, it ultimately aims to endow the online communicating partners with emotional experiences similar to those of face-to-face communication.

As the intersection of these categories, sense of co-presence combines the essential characteristics of emotional, physical, and social presence. The graphical representation of the relationship between all three main components is given in Figure 1.

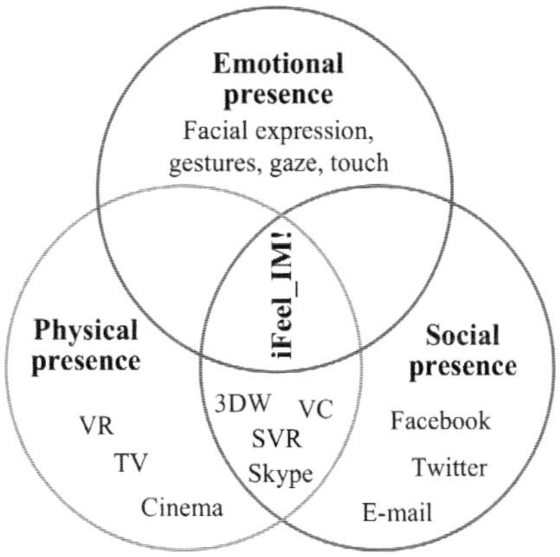

Figure 1. Relationship between physical, social, and emotional presence. Abbreviations are as follows: VR – virtual reality; 3DW – 3D worlds (e.g. SL, Sims); VC – video conferencing; SVR – shared virtual reality.

Conventional mediated systems usually:

- support only simple textual cues such as emoticons,
- lack visual emotional signals such as facial expressions and gestures,
- support only manual control of the expressiveness of graphical representations of users (avatars), and
- ignore such important social-communication channels as the sense of touch. Tactile interfaces could let users enhance their emotional-communication abilities by adding a new dimension to mobile communications.

Driven by the motivation to enhance social interactivity and emotionally immersive experience of real-time messaging, we proposed in the idea of reinforcing (intensifying) own feelings and reproducing (simulating) the emotions felt by the partner through specially designed system iFeel_IM!. The philosophy behind the iFeel_IM! (intelligent system for Feeling enhancement powered by affect sensitive Instant Messenger) is "*I feel* [therefore] *I am!*". The system employs haptic devices and visual stimulation to convey and augment the emotions experienced during online conversations [5]. It includes all three categories of the presence.

2. Architecture of the system

Figure 2 shows the structure of iFeel_IM! As you can see, the wearable part of the system is based on the human body and includes such parts as the heart, hands, abdomen, and sides.

iFeel_IM! stresses

- automatic sensing of emotions conveyed through text messages (artificial intelligence),
- visualization of the detected emotions through avatars in a virtual world,
- enhancement of the user's affective state, and
- reproduction of social touch through haptic stimulation in the real world.

We use Second Life as the communication platform. With Second Life, users can flexibly create their online identities (avatars) and play various animations (for example, facial expressions and gestures) of avatars by typing special abbreviations in a chat window.

We implement control of the conversation through EmoHeart, a Second Life object attached to an avatar's chest. EmoHeart communicates with the *Affect Analysis Model* (AAM), a system for textual affect sensing [6]. It also senses symbolic cues or keywords in the text that indicate a hug and generates a hugging visualization (that is, it triggers the related animation). iFeel_IM! stores the results from the AAM (the dominant emotion and intensity) and EmoHeart (the hug communicative function) along with the chat messages in a file on each user's local computer.

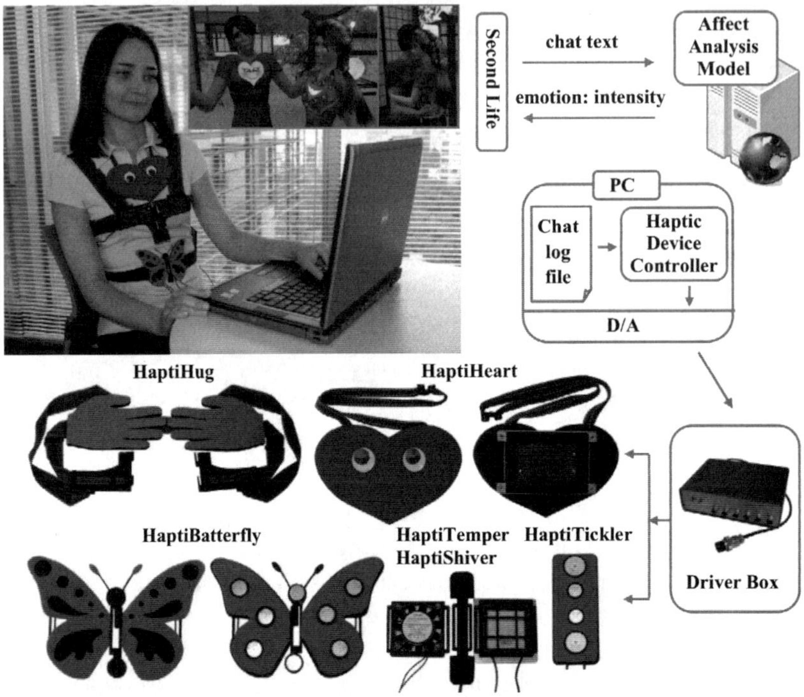

Figure 2. Architecture of the real-time communication system iFeel_IM!

The *haptic-device controller* analyzes the data in real time and generates control signals for the digital/analog converter, which then feeds control cues for the haptic devices to the *driver box*. On the basis of the transmitted signal, iFeel_IM! activates the user's corresponding haptic device. iFeel_IM! employs six haptic devices: HaptiHeart, HaptiHug, HaptiButterfly, HaptiTickler, HaptiTemper, and HaptiShiver.

3. User study

While demonstrating the iFeel_IM! system at the conferences, most of user reported that the haptic modalities (the simulated heartbeats, hugging, and tickling) produced highly realistic sensations. Subjects particularly enjoyed wearing HaptiHug; the simultaneous experience of the hugging animation and hugging sensation evoked surprise and joy. People who tried the system reported that the simulated heartbeats, hugging, and tickling were highly realistic.

The preliminary user study was carried out with six participants to evaluate the effectiveness of emotion elicitation and hug reproduction. The results revealed that the devices successfully generated the corresponding emotion (see Table 1, in which 100 percent stands for six positive replies). We assume that HaptiTemper would reinforce anger if it was activated along with HaptiHeart.

During the demonstration, we observed that HaptiButterfly and HaptiTickler induced most participants to smile and some of them to laugh. They expressed anxiety when HaptiHeart generated a fast, intense heartbeat pattern. The atmosphere between the participants and exhibitors was relaxed and joyful, indicating that iFeel_IM! successfully elicited emotions.

Table 1. Results of a six-user study evaluating the effectiveness of emotion elicitation and hug reproduction.

	Percentage of participants experiencing the emotion.			
Device	***Joy***	***Sadness***	***Anger***	***Fear***
HaptiHeart	—	83.3	66.7	100
HaptiButterfly	83.3	—	—	—
HaptiTickler	100	—	—	—
HaptiHug*	100	—	—	—

**100 percent of the HaptiHug users experienced social touch.*

4. Future work

The extensive user study on iFeel_IM! system will be conducted. We compare the iFeel_IM! system based on three different settings. The communication systems with emotional visual and haptic feedback, with only visual feedback, and without emotion enhancement (the textual communication) will be assessed. The following dimensions will be evaluated: (1) Interactivity, (2) Involvement (sense of co-presence), (3) Enjoyment, and (4) Overall satisfaction.

The novel robotic system with direct haptic avatar-mediated communication will be developed. The user wearing the affective haptic devises and holding tablet computer will interact with avatar by finger gestures (Figure 3(a)). The special gesture language will be designed to trigger the avatar (partner) emotions. The examples of basic touch gestures are given in Figure 3(b). User can hug the avatar by touching the surface with two fingers and bring them closer together. The heartbeat is generated through rapid touch of surface with fingertips. The time between consequent touches defines the frequency of the heartbeat. Tickling of avatar is driven by quick brushing of the surface with continuously engaging in the process fingertips. To make avatar feel afraid the user touches the surface by five fingertips continuously.

Figure 3. a) Tablet computer b) Basic touch gestures.

5. Discussion and conclusions

While developing the iFeel_IM! system (communication system with emotional telepresence), we attempted to bridge the gap between mediated and face-to-face communications by enabling and enriching the spectrum of senses such as vision and touch along with cognition and inner personal state.

In the paper we described the architecture of the iFeel_IM! and the development of a novel avatar-mediated communication system driven by touch gestures. The emotional brain of our system, Affect Analysis Model, can sense emotions and intensity with high accuracy. Haptic devices were designed with particular emphasis on natural and realistic representation of the physical stimuli, modular expandability, ergonomic human-friendly and emotional haptic design. User can perceive the intensive emotions during online communication, use desirable type of stimuli, comfortably wear and easily detach devices from torso.

Our iFeel_IM! demonstrations resulted in keen discussions of possible applications for future research. Examples include

- treating depression and anxiety (problematic emotional states),
- controlling and modulating moods on the basis of physiological signals,
- affective and collaborative games, and
- psychological testing.

We believe that iFeel_IM! could greatly enhance communication in online virtual environments that facilitate social contact. It could also greatly enhance social life in terms of both interpersonal relationships and the character of community.

6. References

[1] Minsky M. Telepresence: A Manifesto. *Omni magazine*. 44-52 (June 1980)
[2] Tachi S. Telexistence. *World Scientific Publishing Company*, Singapore (2009.September)
[3] Guizzo, E.: When my avatar went to work. *IEEE Spectrum*. 9, 24-30 (September 2010)
[4] Lombard, M., and Ditton, T.: At the heart of it all: the concept of presence. *Journal of Computer Mediated Communication*. 3(2) (1997)
[5] Tsetserukou D and Neviarouskaya A. iFeel_IM!: augmenting emotions during online communication. *IEEE Computer Graphics ad Applications*, 30(5), September/October 2010, pp. 72-80.
[6] Neviarouskaya A, Prendinger H, and Ishizuka M. Compositionality principle in recognition of fine-grained emotions from text. *Int. AAAI Conf. Weblogs and Social Media* (ICWSM 09), AAAI Press, pp. 278–281 (2009)

A Neurological Enigma: The Inborn Numerical Competence of Humans and Animals

Hans J. Gross

BEEgroup & Chair of Biochemistry, Biocenter, Julius-Maximilians-University, D-97074 Würzburg, Germany

hj.gross@biozentrum.uni-wuerzburg.de

Abstract. " Subitizing" means our ability to recognize and memorize object numbers precisely under conditions where counting is impossible. This is an inborn archaic process which was named after the Latin "*subito*" = *suddenly*, *immediately*, indicating that the objects in question are presented to test persons only for the fraction of a second in order to prevent counting. Sequential counting, however, is an outstanding cultural achievement of mankind and means to count "1, 2, 3, 4, 5, 6, 7, 8 …" without a limit. In contrast to inborn "subitizing", counting has to be trained, beginning in our early childhood with the help of our fingers. For humans we know since 140 years that we can "subitize" only up to 4 objects correctly and that mistakes occur from 5 objects on. Similar results have been obtained for a number of non-human vertebrates from salamanders to pigeons and dolphins. To our surprise, we have detected this inborn numerical competence for the first time in case of an invertebrate, the honeybee which recognizes and memorizes 3 to 4 objects under rigorous test conditions. This common ability of humans and honeybees to "subitize" up to 4 objects correctly and the miraculous but rare ability of persons with Savant syndrome to "subitize" more than hundred objects precisely raises a number of intriguing questions concerning the evolution and the significance of this biological enigma.

The origin and the evolution of human counting abilities and numerical competence in pre-historical times remains obscure [1]. It may have evolved from the ability to "subitize" [2] which means the ability to recognize small object numbers correctly without counting or when counting is prevented or impossible [3]. "Subitizing", the recognition and memorization of up to 3 or 4 objects is a rather universal ability. For humans it has been demonstrated in a famous experiment as early as 1873 by Jevons [3] that mistakes occur if 5 or more objects have to be estimated ("subitized") under conditions which do not allow regular, sequential counting. Table 1 summarizes the results of these experiments: Jevons [3] estimated within the fraction of a second the number of black beans which fell into a white box whenever he threw a handful of beans over the box. Only up to 4 beans were recognized correctly and an increasing number of mistakes occurred in case of 5 or more beans. These findings from 140 years ago about the "magical number *four*" have been discussed and reproduced in detail recently [4].

Similar results have been elaborated for many non-human vertebrates like birds, racoons, dolphins, monkeys, and salamanders and are summarized by Gross et al.[5]. The data from table 1 gave rise to a new hypothesis about the evolution of counting systems in ancient civilizations, namely that a new symbol for number 5 had to be introduced in order to avoid the need for counting object numbers beyond four [2].

Table 1. A summary of 1027 trials to determine the precision of numerical discrimination when sequential counting is prevented (modified, from ref. 3). The number of trials in which each real number was correctly or incorrectly guessed is shown. For example, number 9 was correctly guessed in 76 out of 122 trials, but in 24 cases it was mistaken for 8, in 2 cases for 7, in 18 cases for 10 and in 2 cases for 11. Note the tendency to over-estimate small numbers and to under-estimate large ones.

Estimated Numbers	Actual Numbers												
	3	4	5	6	7	8	9	10	11	12	13	14	15
3	**23**												
4		**65**											
5			**102**	7									
6			4	**120**	18								
7			1	20	**113**	30	2						
8					25	**76**	24	6	1				
9						28	**76**	37	11	1			
10						1	18	**46**	19	4			
11							2	16	**26**	17	7	2	
12								2	12	**19**	11	3	2
13										3	**6**	3	1
14										1	1	**4**	6
15											1	2	**2**
Totals	23	65	107	147	156	135	122	107	69	45	26	14	11
% correct Numbers	100	100	95.3	81.6	72.4	56.3	62.3	43	37.7	42.2	23	28.6	18.2

The archaic, inborn "subitizing", which cannot be significantly improved by training, and classical counting which has to be trained[6], beginning with the help of our fingers [7,8], are two different phenomena [9]. But what about the precise recognition of more than 4 objects, e.g., the 5 or 6 dots which a dice-player recognizes without counting? This is a clear case of pattern recognition due to the symmetric arrangement of the dots. Another case of precise recognition of numbers up to 9 without counting has been used in ancient Egypt (figure 1). The precise recognition of numbers 5 to 9 occurs by a combination of "subitizing" and a non-counting process which identifies 7 as two groups of 4 and 3, and 9 as three groups of 3 objects.

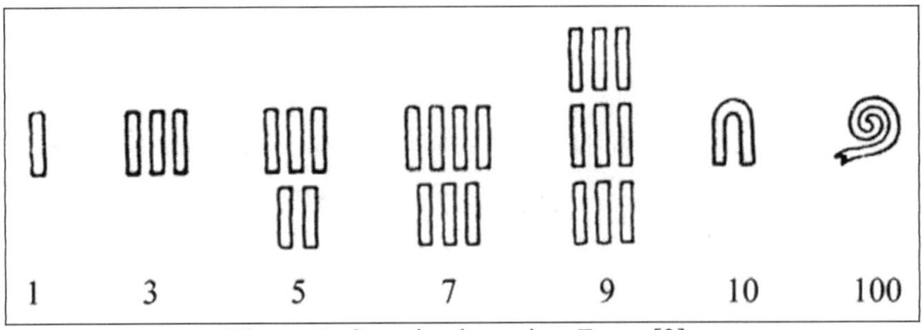

Figure 1. Counting in ancient Egypt [2].

Numerical competence has been documented for quite a number of non-human vertebrates (summarized in ref. 5). In principle those experiments rely on the fact that experimental animals prefer 2 pieces of food if they have the choice between an offer of one versus two pieces of a reward. If the alternative of two versus three pieces of food are offered, they prefer to take the three pieces, indicating that they understand the difference between 2 and 3 objects. However, in the case of three versus four pieces of reward the choice approaches 50:50, indicating that the limit of numerical competence has been reached.

Our recent discovery of numerical competence in an invertebrate, the honeybee, has been an unexpected surprise [5]. Using a rigorous experimental approach we found that honeybees recognize and memorize up to 3 and 4 objects for about 5 seconds during their flight through our delayed match-to-sample (DMTS) experimental apparatus (figure 2). In our experiments we were extremely careful to avoid any possibility for the bees of orientation via pattern recognition or olfactory or visual cues[5]. In this context a special case of a visual cue should be remembered, namely the "Clever Hans Phenomenon". This term is used to refer to inadvertent cuing of behaviours and of inadvertent minute signalling from the experimenter to the experimental animal. This phenomenon, which unfortunately has never been reproduced, goes back to a famous horse named "Clever Hans" which drew worldwide attention in Berlin from 1904 on. "Hans" performed simple mathematical calculations and answered questions by tapping the right number with his front hoof, using an alphabet in which letters were represented by numbers on a blackboard in front of him. In 1907 it was disclosed that the horse was an outstanding observer: It recognized microscopic changes in his master´s face when the correct number had been tapped [10]. Consequently, O. Koehler avoided strictly any direct or personal contact between the experimenter and his birds [11,12]. This precaution is missing in case of the "counting" and speaking grey parrots [13]. Such an experimental artifact due to visual interaction between experimenter and honeybee can be excluded in our study because bees are unable to realize such subtle details as reported for "Clever Hans".

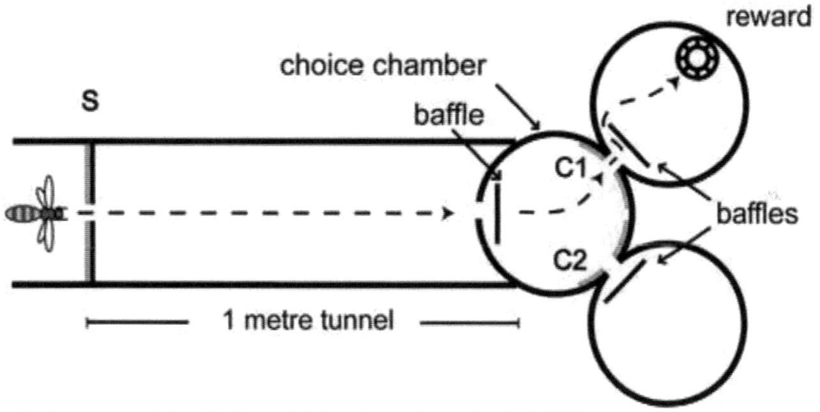

Figure 2. Layout of the Delayed Match-to-Sample (DMTS) experimental apparatus[5]. The bee encounters and flies through the initial sample pattern (S) before traversing a 1m long tunnel with a Perspex roof. There is a baffle behind the entrance of the decision chamber and baffles behind the entrances to the choice chambers. The baffles prevent the bees from experiencing the stimuli in the decision chamber until they had entered it, and from viewing the feeder from the decision chamber. Upon entering the choice chamber, the bee is presented with two choice patterns (C1 and C2), only one of which (C1 in this case) has the same number of dots as S. The bee must choose the matching pattern C1 in order to obtain a hidden reward of sugar solution.

Since bees certainly "subitize" and do not count as we count [9], this indicates that the honeybee with its one million brain cells has practically the same numerical competence for "subitizing" as humans with their 100 billion neurons. This surprising notion gives rise to some intriguing questions:
(a) What is the driving force which maintains such a similar numerical ability during the evolution of so enormously different species as humans and an insect, the honeybee? The answer may be that it has been and still is a matter of life and death for humans and for other vertebrates to estimate within the

fraction of a second whether 3, 4 or "many" lions are approaching, facilitating the decision to fight or to escape [14]. For honeybees the ability to remember up to 3 or 4 objects (trees, houses, other landmarks) may be useful for their orientation to find back to their hive when they return from distant foraging flights [5,14]. (b) Why is the human ability to "subitize" without mistake limited to 4 objects? This question deserves serious attention and discussion. At the very end of this essay a hypothesis will be discussed which proposes that our numerical competence is potentially unlimited, but suppressed. Two examples of rare deviations from the average human cognitive performance may serve to illustrate this idea.

First, we have to remember the fact that Down syndrome (trisomy 21) infants have an additional extra copy of chromosome 21 or of its long arm. It is assumed that altered expression patterns of chromosome 21 genes cause the retarded and incomplete maturation of their brain and, as a consequence, their cognitive dysfunction [15]. Children with this handicap are in general unable to "subitize" 2 to 4 objects [16], but instead count 1, 2, 3, 4, 5 …

Second, we have to discuss the opposite phenomenon, namely a variant of autism which is a mental condition of reduced ability to interact and to communicate with the outside world: the Savant syndrome [17]. In the following, I rely on and refer to the outstanding reviews of a world class expert for the Savant syndrome, D.A. Treffert [18,19]. Up to 10% of the autistic persons are so-called savants. They are able to perform difficult mathematical operations, prime number computing, calendar calculations, can reproduce whole books, even telephone books of big cities forward and backward and play many instruments or speak many languages without training. Typical for savants are their special skills, always in combination with a phenomenal memory. They often display an extreme variant of "subitizing". A well known example is the autistic Raymond in Barry Levinson´s famous movie "Rain Man" (1988) who immediately recognizes the number of 246 toothpicks when a package had dropped and who performs complex mathematical operations within seconds. A typical example for a task which a number-oriented savant will solve without delay is the following: "Multiply 937 with 791 and identify the two prime numbers neighbouring the result of the multiplication!" The result is presented in table 2.

Table 2. The Savant syndrome: Lightening calculation and prime number computing. The two prime numbers neighbouring 741167 are underlined.

937 x 791 = 741167

740693, 740711, 740713, 740717, 740737, 740749, 740801, 740849,
740891, 740893, 740897, 740903, 740923, 740939, 740951, 740969,
740989, 741001, 741007, 741011, 741031, 741043, 741053, 741061,
741071, 741077, 741079, 741101, 741119, 741121, 741127, 741131,
741137, <u>741163</u>, <u>741187</u>, 741193, 741227, 741229, 741233, 741253,
741283, 741337, 741341, 741343, 741347, 741373, 741401, 741409,
741413, 741431, 741457, 741467, 741469, 741473, 741479, 741491,
741493, 741509, 741541, 741547, 741563, 741569, 741593, 741599,
741641, 741661, 741667, 741677, 741679, 741683, 741691, 741709,

No doubt, many savants have the potential for lightening calculations [20], which allows them to perform any mathematical operations [21] and to identify thousands of prime numbers within seconds. They may see thousands of prime numbers with their inner eye without the need for lightening calculations. One of several other possibilities is that they identify prime numbers using the famous "Sieve of Eratosthenes", the function of which is explained in figure 3.

(1)	2	3	~~4~~	5	~~6~~	7	~~8~~	~~9~~	~~10~~	11	~~12~~	13
~~14~~	~~15~~	~~16~~	17	~~18~~	19	~~20~~	~~21~~	~~22~~	23	~~24~~	~~25~~	~~26~~
~~27~~	~~28~~	29	~~30~~	31	~~32~~	~~33~~	~~34~~	~~35~~	~~36~~	37	~~38~~	~~39~~
~~40~~	41	~~42~~	43	~~44~~	~~45~~	~~46~~	47	~~48~~	~~49~~	~~50~~	~~51~~	~~52~~
53	~~54~~	~~55~~	~~56~~	~~57~~	~~58~~	59	~~60~~	61	~~62~~	~~63~~	~~64~~	~~65~~
~~66~~	67	~~68~~	~~69~~	~~70~~	71	~~72~~	73	~~74~~	~~75~~	~~76~~	~~77~~	~~78~~
79	~~80~~	~~81~~	~~82~~	83	~~84~~	~~85~~	~~86~~	~~87~~	~~88~~	89	~~90~~	~~91~~
~~92~~	~~93~~	~~94~~	~~95~~	~~96~~	97	~~98~~	~~99~~	~~100~~	101	~~102~~	103	~~104~~
~~105~~	~~106~~	107	~~108~~	109	~~110~~	~~111~~	~~112~~	113	~~114~~	~~115~~	~~116~~	~~117~~
~~118~~	~~119~~	~~120~~	~~121~~	~~122~~	~~123~~	~~124~~	~~125~~	~~126~~	127	~~128~~	~~129~~	~~130~~

Figure 3. The "Sieve of Eratosthenes". Natural numbers are prime numbers if they are bigger than 1 and do not have divisors except 1 and itself. The identification of prime numbers, using the simple method of the Greek mathematician Eratosthenes (276/273 – 195 B.C.), here shown up to 130, works as follows: Cross out all multiples of 2, 3, 5, 7, 9, 11 and 13, and the remaining numbers, here shown in yellow, are prime numbers. This procedure can be extended into the millions: Just cross out all multiples of the other yellow prime numbers (17, 19, 23, 29, 37, 41, 43 …) and the remaining numbers are primes.

The extraordinary Savant syndrome, however, does not only occur in association with autism. Only about 50% of the savants suffer from an autistic disorder. The other 50% have some injury or impairment of their central nervous system, e.g., brain damage before, during or later after birth due to accidents or disease in infancy, childhood or adult live [18,19]. Due to brain damage this type of savant status can be associated with handicaps like a low IQ, mental retardation, blindness, stuttering or language impairments etc. An attractive but controversial hypothesis argues that damage and dysfunction of the dominant left half of the brain activates the pre-existing but dormant Savant syndrome abilities of the right hemisphere. As a consequence, this would mean that all of us are potential savants with an enormous inborn numerical competence and with a phenomenal memory - but that our left hemisphere inhibits this potential. For the evolution and survival of *Homo sapiens* this makes sense, since our species would not be fit for the challenges of our daily life if we all would be savants. It could in fact be that our inborn numerical competence to "subitize" up to 4 objects correctly is only a tiny fraction of our pre-existing but dormant potential to be a savant. The numerical capacity of animals, the surely unique numerical competence of savants and its possible relation to our limited inborn ability to "subitize", and its absence in Down syndrome patients, remains an exciting neurological enigma.

Acknowledgements
I thank Profs. Hildburg Beier, Holger Hoehn and Jürgen Tautz for critical reading of this manuscript, and Alexander Rais, Dr. Klara Azzami (Würzburg) and Markus Gross (Vienna) for competent help with the illustrations.

References

[1] Ifra G 1981 *Histoire Universelle des Chiffres* (Edition Seghers, Paris) English Issue: Ifra G *From One to Zero: A Universal History of Numbers*. Viking 1985 c/o Penguin Books Ltd., London. German Issue: Ifra G 1993 *Universalgeschichte der Zahlen* (Zweitausendeins)

[2] Gross H J 2011a Give me 5 ... The invention of number five in ancient civilizations – a consequence of our limited inborn numerical competence. *Commun Integr Biol* 4:1 62-63

[3] Jevons W S 1871 The power of numerical discrimination. *Nature* **3** 281-282

[4] Cowan N 2000 The magical number four in short-term memory: A reconsideration of mental storage capacity. *Behav Brain Sci* **24** 87-185

[5] Gross H J, Pahl M, Si A, Zhu H, Tautz J and Zhang S 2009 Number-based visual generalisation in the honeybee. *PLoS ONE* **4** (1): e4263. doi:10.1371/journal.pone.0004263

[6] Sarnecka B W and Carey S 2008 How counting represents number: What children must learn and when they learn it. *Cognition* **108** 662-674

[7] Ardila A 2010 On the evolution of calculation abilities. *Front Evol Neurosci* 2:7. doi: 10.3389/fnevo.2010.00007

[8] Rusconi E, Walsh V and Butterworth B 2005 Dexterity with numbers: rTMS over left angular gyrus disrupts finger gnosis and number processing. *Neurophysiologica* **43** 1609-1624

[9] Gross H J 2011b To bee or not to bee, this is the question ... The inborn numerical competence of humans and honeybees. *Commun Integr Biol* 4:5, 1-4

[10] Hediger HKP 1981 The Clever Hans Phenomenon from an Animal Psychologist´s Point of View. In The Clever Hans Phenomenon: Communication with Horses, Whales, Apes and People, TA Sebeok and R Rosenthal, Eds. *Annals NY Acad Sci* **364** 1-17

[11] Koehler O 1941 Vom Erlernen unbenannter Anzahlen bei Vögeln. *Naturwissenschaften* **29** 201-218

[12] Koehler O 1950 The ability of birds to `count`. *Bull Anim Behav Soc* **9** 41-45

[13] Pepperberg I M 2006 Grey parrot numerical competence: a review. *Anim Cogn* 377-391

[14] Gross H J 2012 The magical number four: A biological, historical and mythological enigma. *Commun Integr Biol*, 5:1, 1-2

[15] Lott I T and Dierssen M 2010 Cognitive deficits and associated neurological complications in individuals with Down´s syndrome. *Lancet Neurol* **9** 623-633

[16] Nye J, Fluck M and Buckley S 2001 Counting and cardinal understanding in children with Down syndrome and typically developing children. *Down syndrome Research and Practice* **7** 68-78

[17] www.savantsyndrome.com

[18] Treffert D A 2009 The savant syndrome: an extraordinary condition. A synopsis: past, present, future. *Phil. Trans. R. Soc.* **B 364**, 1351-1357

[19] Treffert DA 2006 Extraordinary people: understanding savant syndrome. Omaha, NE: iUniverse, Inc.

[20] Welling H 1994 Prime number identification in idiots savants: can they calculate them? *J Autism Dev Disord* **24** 199-207

[21] Dehaene S, Molko N, Cohen L and Wilson AJ 2004 Arithmetic of the Brain. *Curr Opin Neurobiol* **14** 218-224

Journal of Physics: Conference Series **352** (2012) 012047 doi:10.1088/1742-6596/352/1/012047

The "ouzo effect", recent developments and application to therapeutic drug carrying

Robert Botet

Laboratoire de Physique des Solides Bât.510, CNRS UMR8502 - Université Paris-Sud, Centre d'Orsay, F-91405 Orsay, France

E-mail: robert.botet@u-psud.fr

Abstract. This short review is about the spontaneous emulsification effect, aka the "ouzo effect". Under certain conditions, pouring a mixture of a totally water-miscible solvent and a hydrophobic oil into water, generates spontaneously nanometric droplets which are stable, even without surfactant. A basic example is anise-flavored aperitif, which is known from ages in South Europe and North Africa. Then, it is an amazingly old topic, potentially important in a number of applications – such as food additives, paints, cosmetic products or pharmaceutic drugs –, though the main mechanisms are yet essentially unexplained. This phenomenon is presently under intensive investigation using both microfluidic experiments and large-scale numerical simulations, through a CNRS project grouping four laboratories in France. This presentation will give an overview of the history, context and development of the ouzo effect, as well as recent advancements and ideas in the field. This unique effect is now related to two major streams of the scientific research, namely: nano-technology and bio-technology. Consequences in the latter domain is outlined.

1. Introduction

The "ouzo effect" [1] is a general method to emulsify various solutes (including polymers, oils, lipids, or pharmaceuticals) in aqueous media through a simple spontaneous emulsification route. The generality of the process makes it possible to produce large range of submicrometric particles, either solid or droplets. Moreover, by changing the solute content in the aqueous phase, as well as combining different solutes, it becomes possible to prepare nano-objects such as polymer capsules or fluorescent nanoparticles in a simple one-pot procedure. Maybe the best example to date is the technological breakthrough that the BASF Company accomplished by producing solid β-carotene dispersions that opened a new market for food additives in various dietary formulations [2].

Although the principle behind the ouzo effect has been known for a long time, many unsolved issues remain on the nature of the driving force that leads to metastable nanocolloids. In particular, the proper mechanisms occurring during the irreversible evolution to the metastable state are unclear and disagree with the classical theories of the nucleation.

We discuss below the state-of-the-art theoretical approaches of this phenomenon, and explain how a precise understanding of the ouzo effect could be important to generate new efficient forms of pharmaceutic drug-delivery.

2. The ouzo effect

Mixing two immiscible liquids results generally in phase separation and formation of two pure liquid phases. However, it is often desirable to obtain emulsion, that is a stable fine dispersion of one liquid into the other. This is of paramount importance for the chemical industry, especially in foods [3], paints and cosmetic products [4], but also for pharmaceutical drug delivery [5]. Indeed, stable drops protected with a shell, provide generally a safe way to carry active compounds in an aqueous fluid.

The ouzo effect is a special kind of emulsification process, which is currently under tremendous investigation because it does not require energy (thus interesting if one is concerned with cost issues) nor surfactant (concern with environmental issues). The process may have alternative names [6], such as: "spontaneous emulsification", "solvent shifting", "coacervation with addition of a nonsolvent", "solvent displacement process" or "nanoprecipitation", reflecting the successive discoveries of the same phenomenon by different teams, in different domains.

2.1. why the name : "ouzo effect"?

The ouzo effect is known from ages in anise-flavored alcoholic beverages. In Greece, the drink is called *ouzo* – hence the name –, while it is called *pastis* in France, *sambuca* in Italy, *raki* in Turkey, and *arak* in Eastern Mediterranean countries. Actually, all these beverages are made of anethole oil dispersed in alcohol. Anethole is an essential oil extracted by distillation from various anise plants which grow naturally in the Mediterranean regions, as well as south-east Asia (China and Vietnam).

The common special feature of all these beverages is their unique behavior during dilution with water. Adding a large amount of water causes indeed the clear liquor to turn instantly into opaque translucent milky-white color. This phenomenon is known as "louching", and it is the most spectacular feature of the physical process. It is due to the instantaneous formation of a number of small drops of about 1μm diameter, thus strongly scattering visible light.

2.2. a quick-and-dirty description of the "ouzo effect"

We can give here a very schematic description of the ouzo effect taking the example of the ouzo: at the very beginning of the process, the system is a ternary liquid system, with anethole oil and alcohol dispersed in water. Oil is almost insoluble in water though it is soluble in alcohol, while alcohol is soluble in water. Then, oil tends to be surrounded with alcohol molecules, and alcohol molecules are at ease when surrounded with the water fluid. Then, it looks natural that the system separates in two phases: drops of oil+alcohol, dispersed in water. However, the things are not as simple as that elementary explanation. Particularly, it fails to explain the size and stability of the droplets. If strict thermodynamic equilibrium was expected, the final system state should be indeed multiphasic.

Only recently, in 2003, this effect has been analyzed in great details by Vitale and Katz [1], who gave the naming "ouzo effect" for the phenomenon (and we keep here this terminology). In their paper, the authors drew and commented in details a phase diagram of the ouzo effect. Actually, it was not the first time that a phase diagram was shown for this kind of emulsification process [7, 8], however, Vitale and Katz were the first ones to put clearly the problem of relevance and meaning of this phase diagram. After this seminal paper, more complete phase diagrams have been drawn [9]. This approach strongly suggests that there exists a generic scenario – still not fully understood – of the ouzo effect. In other words, this process is not an 'accident' found with special compounds at special values of the parameters, but it is a new general process which should appear in many applications in definite range of the physico-chemical parameters.

Moreover, it appeared that the process was known and used well before this time, without synthetic idea about the generality of the phenomeon. For example, the term: "spontaneous emulsification" was introduced by Ruschak and Miller [10] in 1972 about a study on the ouzo effect in the ethanol-toluene-water system. Actually, a review of a number of examples of the ouzo effect described in scientific papers up to 2005, was recently completed by Ganachaud and Katz [6].

2.3. main features of the ouzo effect

The first and most natural question to ask is about the size of the solvent+solute drops in the dispersing phase. Amazingly, the experimental size distributions are found to be remarkably monodisperse [11]. However, one does not know for sure the reason why the drop-growing process stops, and why the definite size [12]. The typical size is found to range from 100nm [13] to a few micrometers in diameter, and it depends essentially on the ratio of excess solute to solvent [14] (i.e. subtracting the solute which remains dissolved in the final solvent/water solution from the initial amount of solute). In particular, it does not depend on stirring, pH, or ionic strength [15]. Also, such an emulsion appears to be stable for months [1].

Presently, most of the investigation efforts to understand the spontaneous emulsification, focus on the experimental phase diagram of ternary solutions as it is expected to give the key to control the entire process. Some parts of the diagram are well known, specially the binodal line [16], which separates the domain of parameters for which the equilibrated system is under a single homogeneous phase, from the domain for which several phases appear. The binodal line is found writing that the thermodynamic chemical potential of all solution components is equal in each phase. Another line of interest is the spinodal line [16], which separates the domain where the phase separation grows by nucleation, from the domain where the different pure phases appear spontaneously. This line is found as the boundary of thermodynamic instability of a solution to decomposition into multiple phases.

It has been shown clearly that the ouzo limit – which separates the domain where the system experiences spontaneous emulsion from the domain where the system evolves by nucleation or spinodal decomposition – is indeed different from both the spinodal line and the binodal line. Then, the common belief is that the ouzo limit cannot be explained by classical thermodynamics at equilibrium, but is directly related to the dynamics of appearance of metastable states. In this context, the very definition of 'phase diagram' becomes ambiguous, as this diagram should, in principle, be related to the only equilibrium states of the system. However, if we suppose that the phase diagram is still meaningful for metastable states too (at least these metastable states of very low energies), one can draw a general sketch as outlined on the figure 1.

2.4. Kinetic Monte-Carlo approach

Since classical thermodynamics requires the system at the equilibrium for the state variables be properly defined, such an approach of the ouzo effect looks questionable. Indeed, spontaneous emulsification process starts with strong off-equilibrium evolution, which makes direct application of the free energy concept hard to justify. This may be a reason why the various theoretical approaches of the ouzo effect is in quantitative disagreement with the experimental data. Actually, any correct theoretical description of the phenomenon should take into account the irreversible evolution of the system.

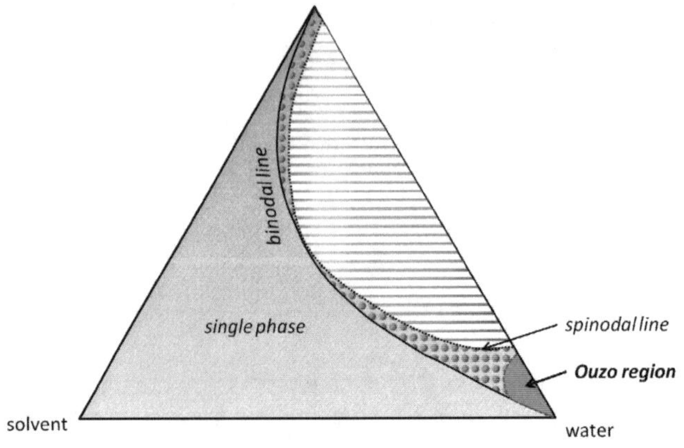

Figure 1. Generic phase diagram for solute-solvent-water ternary mixture, when solute is not soluble in water. The continuous line is the equilibrium binodal line that separates the single-phase region from the two-phase region. The dotted line is the spinodal line which separates the domain where the system evolves by nucleation (the domain with blue dots), and the domain where spontaneous phase separation occurs (the domain with yellow stripes). The ouzo phase (domain in pink color) is limited inside the region of the bi-phase equilibrium state, between the binodal and the spinodal lines. In this domain, the mixture decomposes into a monodisperse population of nanometric drops dispersed in an aqueous solution

Alternative theoretical approaches requires numerics. Moreover, numerical simulations are particularly useful to analyze short-time behaviors. They can be Molecular Dynamics or Monte-Carlo methods.

Molecular Dynamics [17] has the advantage of being quite realistic [18]: if we know precisely the interactions between the different species, one solves simultaneously all the equations of motions to find the complete evolution of the system. Of course, one has to pay the accuracy of the method with a huge amount of calculations necessary to realize a simulation. Then, the Molecular Dynamics method is restricted to short times and small systems.

The Kinetic Monte-Carlo method [19] does not need so serious computation cost. One can draw here the main lines for the specific case of the mixing of several liquids [20]. The system is approximated as an ensemble of molecules on the sites of a regular lattice. Then, each site has the same number, say z, of neighboring sites. The cubic 3D lattice is the most popular example, though use of other lattices can give more realistic visual impression.

2.4.1. exchange energy between two neighboring molecules

As we are dealing with a continuous fluid, all the sites of the lattice must be occupied with molecules. Moreover, a lattice site can be filled with one and only one molecule at a given time (steric exclusion). It means that one assumes that the molecules of different species are all of about the same volume. Moreover, neighboring molecules are assumed to be linked together with chemical bonds. The bond energy between a molecule A and a molecule B, is noted E_{AB}. We have then to estimate the difference of energy of the system when two molecules A and B exchange their positions.

A key remark is that all the changes of energy appear under combinations of the only reduced exchange energies:

$$W_{ij} = \frac{z}{k_B T}\left(E_{ij} - \frac{E_{ii} + E_{jj}}{2}\right) \tag{1}$$

in which z is the number of neighbors of a molecule (on the cubic lattice: $z = 6$, which is a reasonable value for real molecules), k_B is the Boltzmann constant ($k_B = 1.38\ 10^{-23}$ J.K^{-1}) and T the temperature in K. Indeed, suppose that the couple of neighboring molecules A and B undergoes position exchange. Let us denote i_k the sites neighbors to A and j_k the sites neighbors to B, before position exchange. Then, the change, ΔE, of the system energy during the move, writes:

$$\begin{aligned}\frac{\Delta E}{k_B T} &= \frac{1}{k_B T}\sum_{k=1}^{z}(E_{Bi_k} + E_{Aj_k} - E_{Ai_k} - E_{Bj_k})\\ &= \frac{1}{z}\sum_{k=1}^{z}(W_{Bi_k} + W_{Aj_k} - W_{Ai_k} - W_{Bj_k})\end{aligned}$$

Generally, values of the non-dimensional W_{ij} are positive as a consequence of Berthelot's geometric mean of the constants of the pure components [21]. For the ternary system made of three sorts of molecules A, B, C, one has to consider three exchange energies, namely: W_{AB}, W_{AC}, W_{BC}, and we have not to consider the exchange of molecules of the same sort (exchange of two neighboring molecules A, for example, does not change the state of the system).

2.4.2. probability per unit of time to exchange two neighboring molecules

A characteristic time, say τ_e is attached to the microscopic position exchange of two neighboring molecules. According to the Metropolis algorithm [22], the probability, $P_e(A, B)$, per unit of time for two molecules A and B, to exchange their position, is then:

$$\begin{aligned}P_e(A,B) &= \frac{1}{\tau_e} & \text{if } \Delta E \leq 0\\ P_e(A,B) &= \frac{1}{\tau_e}\ e^{-\Delta E/k_B T} & \text{if } \Delta E > 0\end{aligned} \tag{2}$$

where ΔE is the amount of energy required for the move. The probability of the reverse event (the exchange $(B, A) \rightarrow (A, B)$, all the rest of the system being identical), is such that the detailed balance condition:

$$\frac{P_e(A,B)}{P_e(B,A)} = e^{-\Delta E/k_B T}$$

is realized.

2.4.3. the Brownian motion of the clusters

A cluster is defined as an ensemble of molecules of the same sort, connected together through continuous sequence of neighboring bonds. The ability of the lattice approximation to define simply the clusters is a great advantage of the Monte Carlo approach (in the Molecular Dynamics method, the proper definition of the clusters is not that clear). Then, one can consider the Brownian translation and Brownian rotation of the clusters.

Brownian translation event by a lattice step occurs with the characteristic time τ_t depending on the size of the cluster and the viscosity of the liquid. Here we will choose a simple formula such as the Stokes formula: $\tau_t = \tau_{t0} R_\perp$, where $R_\perp$ is a typical cluster radius perpendicular to the move, and τ_{t0} the characteristic time for the Brownian motion of a single molecule. The liquid viscosity is supposed to be uniform throughout the system. Use of the Stokes drag formula is indeed a crude approximation. It requires in principle that the diffusing molecule keeps a compact shape, and that its own velocity is small. If we could define the Reynolds number at this scale, it should be of order 10^{-3} in water, which is a small number, then the Stokes drag formula is probably a good approximation.

Then, we define the probability, $P_t(\mathcal{C})$ of translation by a lattice step, of the cluster $\mathcal{C}$, in a way consistent with the notations of the previous section, that is:

$$P_t(\mathcal{C}) = \frac{1}{\tau_{t0}\, R_\perp} \tag{3}$$

and the translation occurs along one of the z possible directions allowed by the lattice structure. Note that the larger the cluster, the smaller its translational probability per unit of time. During the move, molecules of the fluid not belonging to the cluster, have to be displaced. A random rule mixing is used, that is these molecules are put randomly in the sites made vacant by the cluster move.

The case of the rotational motion is more difficult to handle on a lattice because of the geometric constraints. Indeed, a cluster can rotate by a finite angle (e.g.: $\pi/2$ on the cubic lattice). As we consider only small displacements, the point around which the cluster rotates is chosen randomly among the sites of the cluster. The probability of rotation per unit of time is chosen similarly to the translational move, that is:

$$P_r(\mathcal{C}) = \frac{1}{\tau_{r0}\, R_\perp} \tag{4}$$

where $R_\perp$ is the value of the cluster radius perpendicular to the effective rotation, and τ_{r0} a typical rotation time for a single molecule. Here too, random mixing rule is used for the displaced fluid.

2.4.4. the algorithm

First, we have total freedom on the initial configuration. The two most popular initial states are : the complete mixed system, where the molecules A, B, C are put randomly on the lattice ; or the multiphase system, where the initial system is divided into two or more homogeneous phases (a unique cluster of A molecules in a homogeneous mixed fluid of B and C molecules, for example). Both initial states correspond clearly to two different initial mixing ways.

The definite algorithm is as follows :

- at a given time, all the possible elementary movements (exchange of two neighboring molecules, translations or rotations of clusters) are considered with their corresponding probabilities per unit of time;
- the sum, Σ, of all probabilities is computed and the list of probabilities is normalized by this sum. That way, the sum of all normalized probabilities takes the value 1 (and this corresponds to time re-scaling) ;
- then, one event is selected in that list according to its probability. This event is realized, and the physical time is implemented by the amount $\log(1/X)/\Sigma$, with X a random number uniformly distributed on $]0, 1]$ (to insure that the increment of time is Poisson-distributed).

Journal of Physics: Conference Series **352** (2012) 012047 doi:10.1088/1742-6596/352/1/012047

The system evolves according to these simple probabilistic rules and is able to find naturally metastable states in the canonical ensemble. As we know that the ouzo phase is metastable, the Kinetic Monte-Carlo method is preferred in the context. A pictorial example is shown on the figures 2, 3 and 4, where three states of a same 2D ternary system on the hexagonal lattice, at three different times, are shown. The system size is 100 × 100 lattice sites, that is a small size which has been chosen only for clear visualization (sizes of 1000 × 1000 × 1000 on the 3D lattice can be reached easily). The values of the exchange energies were respectively: $W_{AB} = 2.5$ which corresponds to a limited solubility of A in B (roughly speaking, exchange energy value < 2 means that the two substances are soluble one in the other, while a value > 2 means insolubility); $W_{AC} = 7$, that is A is insoluble in C ; $W_{BC} = 1.2$ that is B is soluble in C.

It is clear on the figures that the system tends to form clusters of molecules A surrounded by cloud of molecules B in the fluid C. Then, due to attraction of the molecules B for C, two clouds of molecules B are seldom to merge, that stabilizes the already formed structures of A-clusters with B shell. This is consistent with the remark that no drop coalescence is observed in experimental ouzo process [23].

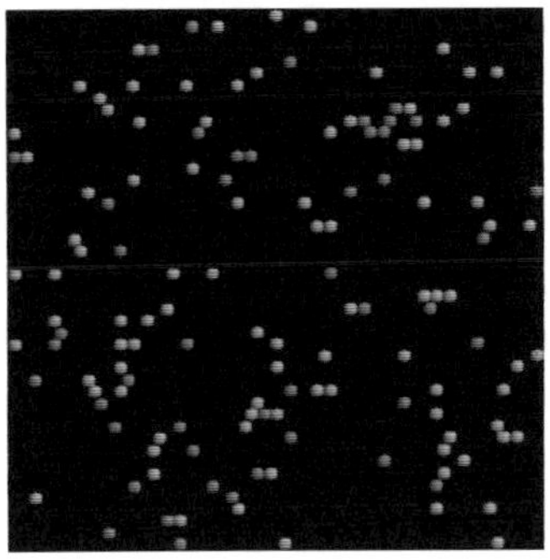

Figure 2. Totally mixed initial state of a 2D ternary system on the hexagonal lattice. The volume fractions are respectively : 2.5% of molecules A (yellow circles) ; 5% of molecules B (green circles) ; 92.5% of molecules C (the black part).

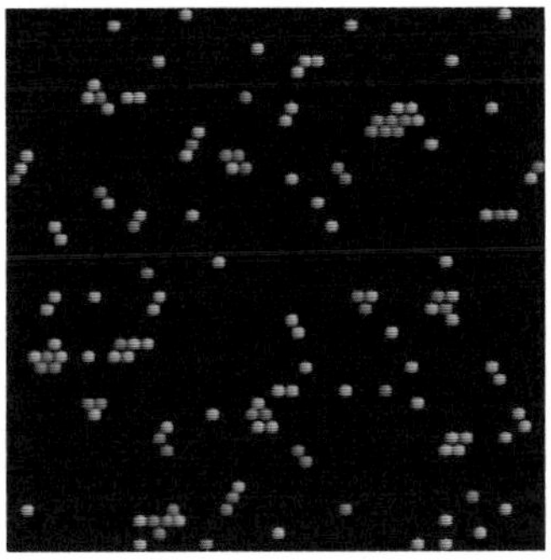

Figure 3. At a later time, the system begins to self-organize in small clusters of molecules A (yellow), because they are essentially insoluble in the fluid C (black). The system evolution is irreversible at this stage, and the total energy is decreasing with the time.

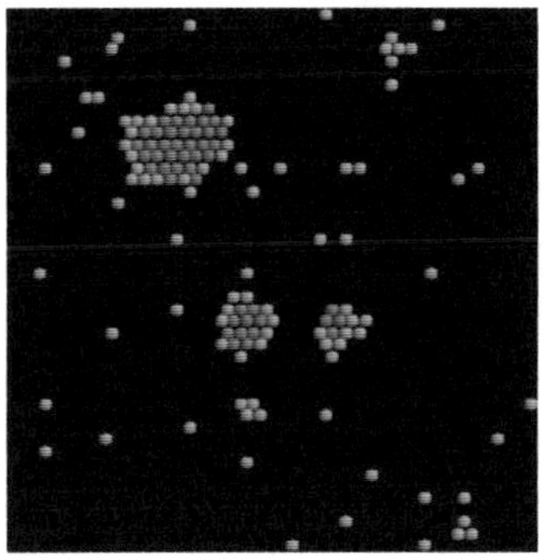

Figure 4. The steady state is made of clusters of molecules A (yellow) surrounded with molecules B (green). That way, the system acquires a small total energy in a metastable phase, even if the strict minimum energy should be for separated phases.

2.5. *The importance of microfluidic devices*

Microfluidic devices [24] have been shown to be able to control finely patterned self-organization [25] or crystallization [26] at the microscopic scale. This is a promise candidate for obtaining high-quality nanoparticles. Also, a number of works using microfluidic, were devoted to the general formation of emulsion [27, 28]. It is then interesting to use this technique to investigate the mechanisms acting in the early stages of the self-emulsification effect. The general concept of such experiments is sketched on the figure 5.

Indeed, as the process is highly irreversible at the beginning, the short time history of the evolution is of prime importance to understand the spontaneous appearance of the metastable

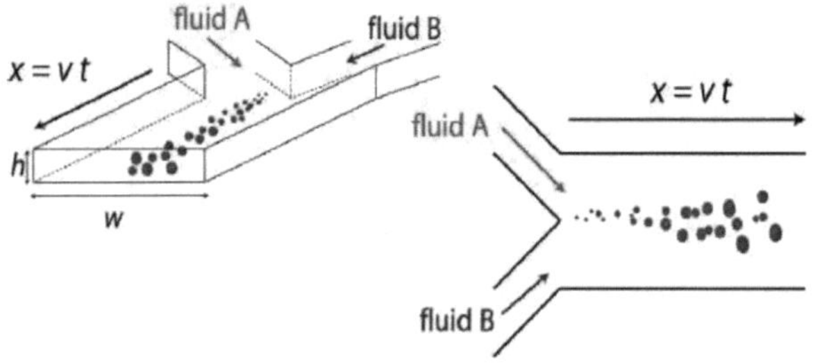

Figure 5. Schematic representation of the microfluidic "Y-junction" used in LOF laboratory, Pessac, France (J.-B. Salmon) to study the ouzo effect. Typical microfluidic geometries are (h and w are defined on the left-hand figure): $h = 10 \sim 50\,\mu$m, $w = 10 \sim 1000\,\mu$m. Water and solvent+solute are introduced through converging channels, meet at the junction and flow out through the exit channel. Interdiffusion takes place in the laminar exit stream. The distance downstream in the channel is related to the time elapsed since the two streams were put into contact: the distance x from the junction corresponds to the outcome of interdiffusion-reaction over a time $t = x/v$ (v is the average velocity in the channel). It is thus possible to analyze in details the very beginning of the ouzo process with specific analytical tools at time scales down to several tens of ms ($v \simeq 1$ cm/s, $x \simeq 100\,\mu$m), and to compare with numerical simulations in the same geometry.

phase. Moreover, the short times are accessible to the Kinetic Monte-Carlo method, then direct comparison between real experiments and numerical experiments with the same geometry and time range, is possible. This work is presently in progress.

3. A short introduction to medication delivery

Now, we would like to finish with possible application of the ouzo effect in an important field: pharmaceutical drug-delivery.

In most of the cases, active ingredient of a drug must reach special inner place in the body of a patient. The target can be diseased tissues – such as malignant cells –, or virus-infected cells. The fastest and most direct way to act, is to use parenteral injection in the circulatory system. This way avoids indeed first-pass metabolism, that is the sudden decrease of drug concentration occurring in the digestive system. However, severe conditions must be fulfilled to insure the intravenous therapy be safe. Here we will focus on two important aspects: the medication has to be a liquid solution or a fine dispersion in a liquid solvent ; the solvent must not react with blood and blood vessel, from the double point of view of the chemistry (chemical inactivity) and physics (isotonic liquid).

3.1. the liposomes

Using a homogeneous liquid solution is the simplest way, but there are immediate drawbacks: the liquid active chemical compound may interact with any cell irrigated with the blood. It results in possible disorder due to uncontrolled toxic interaction.

Alternatively, a very active research domain in the field pertains to preparation of dispersions of nanometric particles in isotonic solvent [29]. The active ingredient is then trapped inside the nanometric particles, and it is released when the target is reached. The most popular particles – discovered fifty years ago –, are the liposomes [30]. A liposome is a closed vesicle of lipid bilayers which encapsulates small amount of aqueous solution in which the active ingredient is dissolved. As water cannot pass through the lipid layers, the aqueous solution is permanently trapped in this small bag, until the vesicle opens. From the point of view of physics, the system is then a mixture of two liquid phases (the blood as the continuous phase, and the aqueous solution with the active drug as the dispersed phase), which are separated one from the other by lipid interface.

The protective layer is treated such that the immune system does not destroy the vesicle for a long time in the bloodstream, typically for a day or more. The stealth character is obtained studding particular chemical compound (such as polyethylene glycol) on the surface of the outer lipid layer, making it inert in the circulatory system and in the cleansing organs. On the other hand, a chemical mechanism breaks up the layer when the liposome is close to its target. This is achieved grafting special peptides on the surface of the liposome. Indeed, the peptides bind specifically to special receptors present on the target cells, then, once bound to a receptor, the lipid layer becomes weak and opens, releasing the tiny amount of active ingredient in a concentrated way at the right place.

Even if lipid vesicles are used nowadays to carry a number of pharmaceutical payloads, their formation remains generally very delicate, while the efficiency depends dramatically on characteristics hard to control, such as: polydispersity, shelf-life, and batch-to-batch reproducibility. Actually, liposomes are not generated spontaneously, and one needs input of energy to produce them. This is generally achieved by high-energy sonication of natural or synthetic phospholipids in the medication solution [31]. However, the quality of the resulting liposomes is poorly controlled using this method. More sophisticated methods were contrived, but they all encounter the problem of the small production rate. At the end, the difficulty for obtaining reliable and efficient manufacturing techniques, results in high production costs which still limit wide applicability of the method.

3.2. the alternative use of spontaneous emulsions

A liposome dispersion is a particular case of nanometric emulsion. However, up to now, ordinary emulsion (with addition of surfactant to stabilize the droplets) could not be used for intravenous injection because the usual surfactants easily destroy the blood cell walls. Then, the ouzo effect appears to be the perfect candidate to manufacture monodisperse droplets of therapeutic drug in aqueous medium compatible with blood. Moreover, the typical sizes for the droplets appearing by the ouzo effect are comparable to the optimum efficient sizes (around 100 to 500 nm) to destroy the cancer cells [32]. Then, as soon as we shall understand how the phase diagram of the ouzo effect is build and is structured according to the solute compound, such a biotechnology will be at our fingertips.

4. Conclusion

We have seen in this review that the ouzo effect is a versatile way of forming nanometric droplets of a liquid encapsulated in a protective shell. Such a process is attractive : it is spontaneous, mild, reproducible and gives rise to controlled monodisperse emulsion. The only serious flaw is that we do not understand yet the mechanisms leading to the self-organized systems. Without such a knowledge, one can only proceed by “trial-and-error” to find the emulsification operating window. In 2011, a three-years CNRS project started in France to elucidate the ouzo effect. This project is expected to give soon the keys to understand, thus to control, the spontaneous nanoprecipitation process.

5. Acknowledgments

This work is part of the CNRS project ANR-10-BLAN-942-04 *LimOuzInE* (2011-2013) dedicated to "understanding and predicting the ouzo limit". The team, networking four laboratories in France, is : François Ganachaud (Institut National des Sciences Appliquées, Lyon), Bernard Cabane (Ecole Supérieure de Physique et de Chimie Industrielles, Paris), Jean-Baptiste Salmon (Laboratory Of the Future, Pessac) and Robert Botet (Laboratoire de Physique des Solides, Orsay).

References

[1] Vitale S A and Katz J L 2003 *Langmuir* **19** 4105.
[2] Horn D and Lüddecke E. 1996 "Preparation and Characterization of Nano-sized Carotenoid Hydrosols" in *Fine Particles Science and TechnologyFrom Micro to Nanoparticles* ed.: E. Pelizzetti, Kluwer, Dordrecht, 761.
[3] Wakerly M G, Pouton C W, Meakin B J and Morton F S 1986 *Am. Chem. Soc. Symp.* **Ser. 311** 242.
[4] Simion F A, Starch M S, Witt P S, Woodford J D and Edgett K L 1998 in *Textbook of Cosmetic Dermatology*, 2nd ed. Baran R, Maibach H I and Martin Dunitz, London, 309.
[5] Pouton C W 2000 *Eur. J. Pharm. Sci.* **11** S93.
[6] Ganachaud F and Katz J L 2005 *Chem. Phys. Chem.* **6** 209.
[7] Stainmesse S, Orecchioni A.-M., Nakache E, Puisieux F and Fessi H 1995 *Colloid Polym. Sci.* **273** 505.
[8] Kim J Y , Lee H K , Baik K J and Kim S C 1997 *J. Appl. Polym. Sci.* **65** 2643.
[9] Aubry J, Ganachaud F, Cohen Addad J-P and Cabane B 2009 *Langmuir* **25** 1970.
[10] Ruschak K J and Miller C A 1972 *Ind. Eng. Chem. Fundam.* **11** 534.
[11] Grillo I 2003 *Cooloid Surf. A: Physicochem. Eng. Aspects* **225** 153.
[12] Carteau D, Pianet I, Brunerie P, Guillemat B and Bassani D M 2006 *Langmuir* **23** 3561.
[13] Yu W, Tabosa do Egito E S, Barratt G, Fessi H, Devissaguet J P and Puisieux F 1993 *Int. J. Pharm.* **89** 139
[14] Bouchemal K, Briançon S, Perrier E and Fessi H 2004 *Int. J. Pharm.* **280** 241.
[15] Gallardo M, Couarraze G, Denizot B, Treupel L, Couvreur P and Puisieux F 1993 *Int. J. Pharm.* **100** 55.
[16] Jones R A L 2002 *Soft Condensed Matter* Oxford University Press.
[17] Rapaport D C 2004 *The Art of Molecular Dynamics Simulation* 2nd revised edition, Cambridge University Press.
[18] Kabrede H and Hentschke R 2006 *Physica A: Statistical Mechanics and its Applications* **361** 485.
[19] Meng B and Weinberg W H 1994 *J. Chem. Phys.* **100** 5280.
[20] Hettema H and Mc Featers J S 1996 *J. Chem. Phys.* **105** 2816.
[21] Berthelot D 1898 *Comptes Rendus* **126** 1703.
[22] Hastings W K 1970 *Biometrika* **57** 97.
[23] Sitnikova N L, Sprik R, Wegdam G and Eiser E 2005 *Langmuir* **21** 7083.
[24] De Mello A and De Mello J 2004 *Lab Chip* **4** 11.
[25] Kenis P J A, Ismagilov R F and Whitesides G M 1999 *Science* **285** 83.
[26] Zheng B, Tice J D, Roach L S and Ismagilov R F 2004 *Angew. Chem. Int.* **43** 2508.
[27] Zhao H, Wang J X, Wang Q A, Chen J F and Yun J 2007 *Ind. Eng. Chem. Res.* **46** 8229.
[28] Link D R, Anna S L, Weitz D A and Stone H A 2004 *Phys. Rev. Lett.* **92** 054503.
[29] Zhang L, Gu F X, Chan J M, Wang A Z, Langer R C and Farokhzad O C 2008 *Clinical Pharmacology and Therapeutics* **83** 761.
[30] Bangham A D, Horne R W, Glauert A M, Dingle J T and Lucy J A 1962*Nature* **196** 952.
[31] Stryer S 1988 *Biochemistry* 3rd ed., New York, W. H. Freeman and co.
[32] Siwak D R, Tari A M and Lopez-Berestein G 2002 *Delivery. Clin. Cancer Res.* **8** 1172.

Biofouling of various metal oxides in marine environment

T. Kougo[1] D.Kuroda[1], N.Wada[1], H.Ikegai[2] and H.Kanematsu[1]
[1]Department of Materials Science and Engineering, Suzuka National College of Tecknology, Shiroko-cho, Suzuka, Mie, 510-0294, Japan.
[2]Department of Chemistry and Biochemistry, Suzuka National College of Tecknology, Shiroko-cho, Suzuka, Mie, 510-0294, Japan.
E-mail: kougo@mse.suzuka-ct.ac.jp

Abstract. Biofouling has induced serious problems in various industrial fields such as marine structures, biomaterials, microbially induced corrosion (MIC) etc. The effects of various metals on biofouling have been investigated so far and the mechanism has been clarified to some extent(1,2), and we proposed that Fe ion attracted lots of bacteria and formed biofilm very easily(3). In this study, we investigated the possibility for biofouling of *Pseudomonas aeruginosa* on various metal oxides such as Fe_2O_3, TiO_2, WO_3, AgO, Cr_2O_3 etc. And in addition of such a model experiment on laboratory scale, they were immersed into actual sea water as well as artificial sea water. As for the preparation of metal oxides, commercial oxide powders were used as starting material and those whose particle sizes were under 100 micrometers were formed into pellets by a press. Some of them were heated to 700 ℃ and sintered for 10 hours at the temperatures. After the calcinations, they were immersed into the culture of P. aeruginosa at 35 ℃ in about one week. After the immersion, they were taken out of the culture and the biofouling behaviors were observed by optical microscopy, low pressure scanning electron microscopy (low pressure SEM) etc. Biofouling is generally classified into several steps. Firstly, conditioning films composed of organic matters were formed on specimens. Then bacterial were attached to the specimen's surfaces, seeking for conditioning films as nutrition. Then bacteria formed biofilm on the specimens. In marine environment, more larger living matters such as shells etc would be attached to biofilms. However, in the culture media, only biofilms were formed.

1. Introduction

Biofouling is defined as the unwanted accumulation of biological material on man-made surfaces[1]. It produces various failures and drawbacks. For example, biofouling onto inside walls of cooling pipes in various factories along sea sides or in nuclear power plant etc., where sea water is mainly used, decreases the cooling capability, which leads to the increase of carbon dioxide in atmosphere finally. Biofouling is classified into microfouling and macrofouling mainly and the mechanism can be described schematically, as shown in Figure 1[1]. Firstly, organic compounds are adsorbed to solid surfaces to form thin organic films called "conditioning films". The carbon compounds attract planktonic bacteria in oligotrophic environments and they attach to solid surfaces. They gather around on solid surfaces and excrete polysaccharide due to quorum sensing at the same time, which lead to

[1] To whom any correspondence should be addressed.

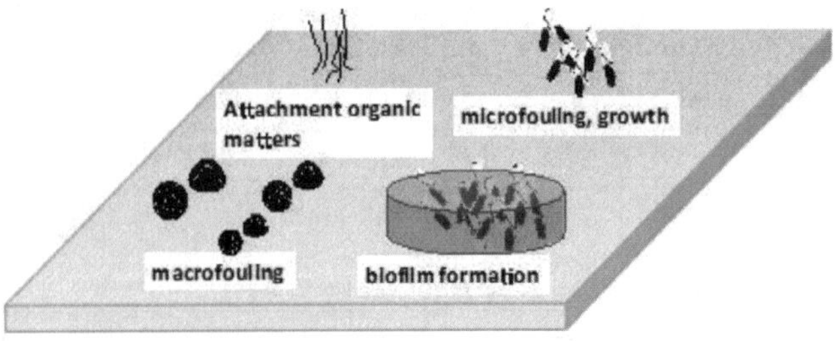

Figure 1 Process of microfouling and macrofouling.

the formation of biofilm. Excretion of EPS is usually induced by quorum sensing[3]. The biofilm formed on solid surfaces attracts larger creature such as oyster and acorn shell, and macrofouling occur continuously. Therefore, biofouling is related directly to biofilm formation. To understand the correlation between metallic materials as caririers and biofouling much more, it would be the best way to investigate the correlation between biofilm formation and metallic materials. Therefore, we have searched for metal elements inhibiting biofilm formation and found some of them so far[4]-[5]. Kanematsu et al[5]reported that some metal ions from the metallic surfaces would inhibit the growth of bacteria attached to the surfaces and the biofilm formation. On the other hand, metal oxides are generally hard to be ionized in aqueous solutions due to the high bonding stability. It would be explained that metal oxides are made by the covalent bonding between metallic and non-metallic components. It would be interesting to clarify how metal ions as cation would affect biofilm formation and biofouling. In addition, it might lead also to the possibility that ceramics materials could be applied to marine structures, cooling pipes, biomaterial etc.

In this study, we focused on some general metal oxides and to investigate their inhibition capabilities of biofilm formation. Various tableting metal oxides were immersed into ordinary bouillon medium and investigated how they would affect bacteria and biofilm formation.

2. Experimental

2.1 Specimens

Various metal oxide powders such as Ag_2O, CeO_2, Cr_2O_3, Fe_2O_3, TiO_2, ZnO and WO_3 (Wako Pure Chemicals, Japan Aerosol Co.) were broken up by agate mortar. Then agglomerated particles were removed by sifter and the particles whose average diameters were below 100 micrometers were collected. Those fine particles were tableted by a tableting machine (JASCO). They were pressured step by step and finally, they were compressed in a single direction at $400kgf/cm^2$ for 20 minutes. Finally, the tablets of 10 mm diameters were formed. They were sintered at 700 ℃ for 10 h in ambient atmosphere.

2.2 Characterization of specimens

All of the specimens were subjected to X-ray diffraction (XRD) analysis to identify and confirm the structures. XRD was carried out by using a general purpose machine for the purpose (RINT-2100, Rigaku). A copper electrode was used as target electrode. The voltage of X-ray tube was 40kV and the current was 20 mA.

2.3 Culture media and immersion tests

Specimens were immersed into ethanol and dried for 4 hours. Then they were immersed into cell suspension of *Pseudomonas aeruginosa* strain PAO1. The suspension was made in the following way. Strain PAO1was cultured in ordinary bouillon (meat extract 10 g, peptone 10 g, NaCl 15 g in distilled water 1L, pH = 7.2) at 35 ℃ for 24 h and then diluted with ordinary bouillon to give a cell density of 10^5 cells/ml. Concentration was checked by CFU (Colony Forming Unit) method. Metal oxide tablets sterilized with ethanol were put in a 24 holes well plate each of which was filled with 2 ml of cell suspension. The plates were kept at 35 ℃ for 7 days in darkness. Then each solution was taken as sample and diluted, and that of 0.1 ml was inoculated into an ordinal agar culture media (1.5 wt% agar was added to ordinary bouillon). It was kept in an incubator at 35 ℃ for 24 h. After the culture procedure, the number of colonies formed on the agar was measured and calculated. As for the biofilm,

Journal of Physics: Conference Series **352** (2012) 012048 doi:10.1088/1742-6596/352/1/012048

bacteria in biofilms were stained with crystal violet and the absorbance intensity was measured for the evaluation how many bacteria existed in them. Concretely, each oxide tablet was immersed into 0.2 wt% crystal violet solution after the culture and rinsed out by phosphoric acid buffer solution. Then the crystal violet attached to the specimen was washed out by ethanol and the absorption strength of the solution was measured by a microplate reader (Model 550, BIO-RAD).

3. Results and discussion

3.1 Identification of oxide specimens by XRD

Those specimens such as WO_3, CeO_2, Fe_2O_3, Cr_2O_3 and ZnO showed quite the same structures with those of starting materials. The results of XRD analysis of these oxides are shown in Figure 2. They indicate clearly that the structures after the sintering at 700 ℃ did not change from the starting points at all. On the other hand, Figure 3 (a) and (b) shows the exceptional results. Figure 3 (a) is the XRD result for Ag_2O specimen. After sintering, silver phase increased in XRD results. It indicates that the silver oxide was reduced by sintering to 700 ℃ to much extent. Figure 3 (b) shows the XRD results for titanium oxide specimen. Before sintering, the main structure of titanium oxide specimen was anatase, while that after sintering was rutile type one. From those results, the specimens except silver oxide were used after sintering. On the other hand, silver oxide was used as non-sintering specimen.

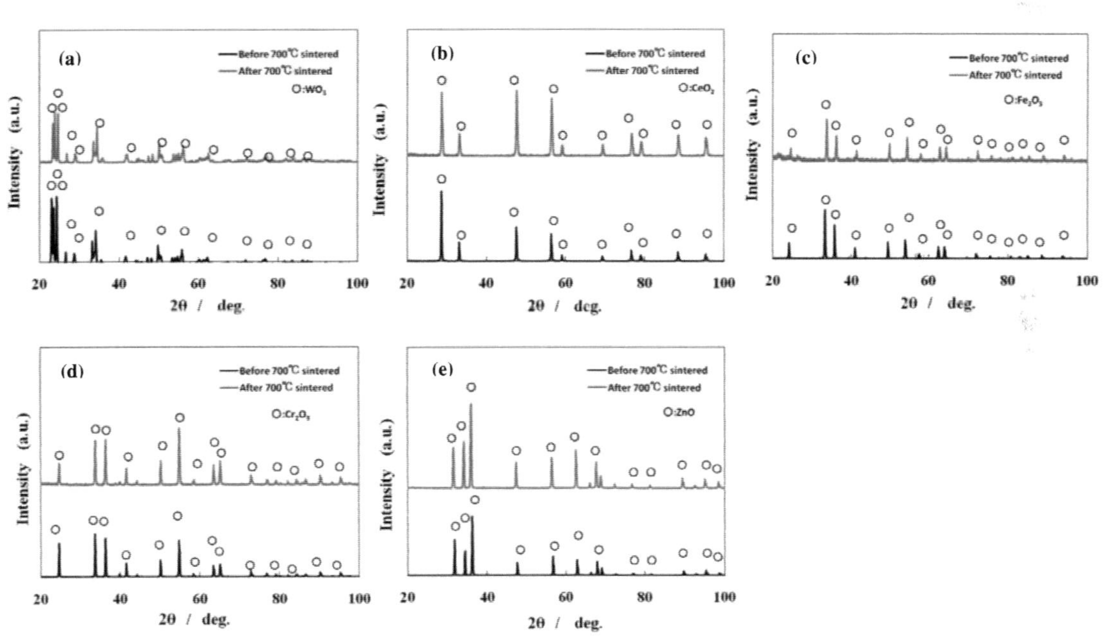

Figure 2 XRD patterns of metal oxide with sintered.
(a)WO_3, (b)CeO_2, (c)Fe_2O_3, (d)Cr_2O_3 and (e)ZnO.

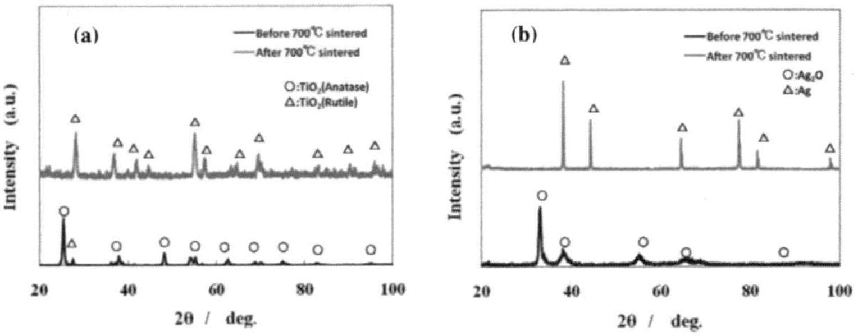

Figure 3 XRD patterns of metal oxide with sintered.
(a) TiO_2 and (b) Ag_2O.

3.2 Immersion tests

All of the oxide specimens were immersed into the bacteria cell suspension. Then, a portion of the suspension was sampled and cultured on agar medium to check bacterial viability. Figure 4 shows the suspension diluted one hundred times. As for WO_3, Fe_2O_3, TiO_2, ZnO and CeO_2, the numbers of colonies were too much to be counted, while those for Cr_2O_3 and Ag_2O were almost zero. It suggests very clearly that the latter two oxides have very strong antibacterial effects for *Pseudomonas aeruginosa*. Figure 5 shows the viability of P.Aruguinosa in the metal-oxide-suspensions as revealed by plate counting. Those results made it possible to differentiate the antibacterial effects among other oxides. In addition of silver oxide and chromium oxide, cerium oxide (CeO_2) also showed the relatively strong antibacterial effect. On the other hand, other oxides such as WO_3, Fe_2O_3, TiO_2 and ZnO did not show any remarkable antibacterial effects. Generally, oxides have antibacterial effects, when photoexcited, since the light-excited oxides could produce radical hydroxide ions. However, the series of experiments were carried out under dark conditions.

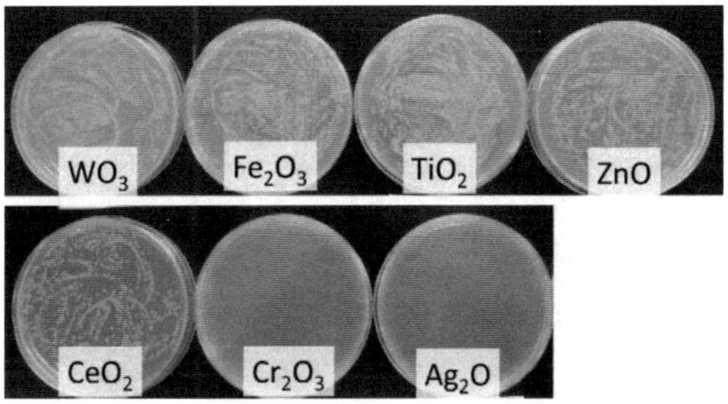

Figure 4 Inoculated solutions diluted a hundred times.

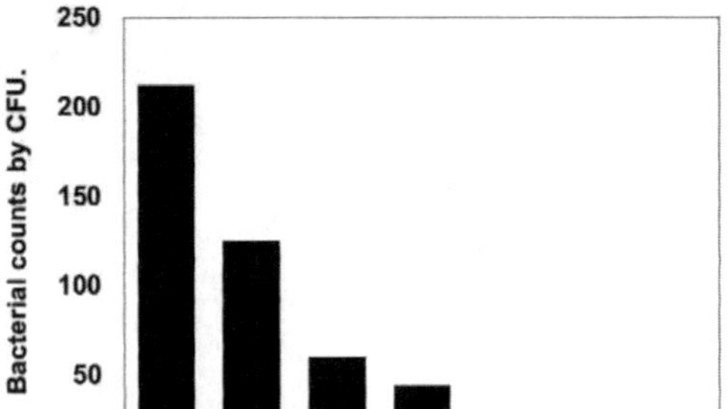

Figure 5 Inoculated solutions diluted a thousand.

Therefore, the antibacterial effects of silver oxide and chromium oxide would be produced due to the dissolutions of chromium and silver as ion, respectively. Originally, both metallic ions have been well-known for their relatively strong antibacterial effects[6]-[7]. In addition to both Cr_2O_3 and Ag_2O, CeO_2 also showed the antibacterial effects. It is also well known that the CeO_2 often shows the effect, particularly as nano particle[8]. Also in this case, dissolved cerium ion would work and show the antibacterial effect. Other oxides did not show so significant antibacterial effects. It could be attributed to that these oxides were relatively stable as oxide and that they did not produce ions in the solutions, even though titanium and zinc ions often show high antibacterial effects[4].

Usually, antibacterial effects are not always same with the inhibition capability of biofilm formation completely[9]. However, the specimens were generally porous different from metallic specimens and crystal violet could attach to the pores much more easily. Therefore, the absorbance intensity might depend on the porosity of specimens to some extent. In addition to that, crystal violet bound to some specimens specifically. For example, tungsten oxide was stained to much extent, since the dimethylamino group of crystal violet binds to the tungsten oxide surface specifically[10]. On the other hand, some researchers found that the carboxylic acid adsorbed to some metallic oxides such as titanium oxide[11]. However, we have to evaluate the biofilm inhibition capability finally and it requires some staining method by pigments. Therefore, the evaluation method by crystal violet did not work in this investigation. This would be an important topic for the future to find them. In this study, we measured the antibacterial effects by dissolved metallic ions from metallic oxides specimens and evaluated the extent of biofilm formation for those oxides specimens indirectly. The evaluation is based on the premise that the antibacterial effect induced by metallic ions in the bacterial suspension solution could inhibit the biofilm formation. The hypothesis would be absolutely right. However, some other factors might be involved with the biofilm formation. It needs further investigations in the future to clarify the inhibition effects of biofilm formation by oxides much more.

4. Conclusions

The inhibition capability of biofilm formation for the seven metallic oxides, WO_3, Fe_2O_3, TiO_2, ZnO, CeO_2, Cr_2O_3 and Ag_2O, were investigated. The evaluation by the adsorption intensity of crystal violet to biofilms on oxide specimens could not be used, since it also depended on the porosity of specimens and also the specific adsorption of composed organic groups to oxide surfaces. Therefore, we evaluated the extent to measure the colony numbers in bacteria (Pseudomonas aeruginosa) suspension solutions. As results, Cr_2O_3, Ag_2O and CeO_2 showed antibacterial effects and therefore, we presumed that these three oxides would have high inhibition capabilities of biofilm formation due to the effect of dissolved ions at this point. Some new effective pigments are now developed to use the evaluation in the future.

5. References

[1] Flemming H.-C. 2009 *Marine and Industrial Biofouling* (Berlin Heidelberg, Germany: Springer Verlag)
[2] Ikigai, H., Kanematsu, H, Kuroda, D. 2011. *Journal of the Japan Institute of Light Metals*, ***61*(4)**, 160-166.
[3] H. M. Gan, L. Buckley, E. Szagedi, A. O. Hudson and M. A. Savka, *J. Bacterio.l*, 2009, **191(8)**, 2551-2560.
[4] Kanematsu, H., Ikegai, H., and Kuroda, D. 2011. *Rust Prev. Cont. Jap.*, **55(10)**, 369-377.
[5] Kanematsu, H., Ikegai, H., and Kuroda, D. 2011 *J. High Temp. Soc. Jap*, ***37*(1)**, 17-24.
[6] N. Kalantari, and S. Ghaffari, *Iran. J. Environ. Health. Sci. Eng.*, 2008, **5(3)**, 173-178.
[7] G.J. Zhao and S.E. Stevens, *Biometals,* 1998, **11**, pp. 27–32.
[8] Shibli, S>M>A, Archana, S.R, Ashraf, *P.M, 2008, Corr.Sci.*, 50.
[9] Kanematsu, H., Ikigai, H., & Yoshitake, M. 2009. *Inter. Mol. Sci.*, ***10*(2)**, 559-571.
[10] Sayama, K., Kasuga, K., Yanagida, M., Sugihara, H., 2009, *Japan Patent Application (Tokukai)* 2009-70648.
[11] O'Regan, B., Graetzel, M. 1991, *Nature*, **353**, 737.

Journal of Physics: Conference Series **352** (2012) 012049 doi:10.1088/1742-6596/352/1/012049

Applicability of Carbonated Electric Arc Furnace Slag to Mortar

S Yokoyama[1], R Arisawa[2], M N N Hisyamudin[3], K Murakami[4], A Maegawa[4] and M. Izaki[1]

[1] Department of Mechanical Engineering, Toyohashi University of Technology, Hibarigaoka, Tempaku-cho, Toyohashi, Aichi, 441-8105, Japan
[2] Graduate Program of Mechanical Engineering for Master's Degree, Toyohashi University of Technology, Hibarigaoka, Tempaku-cho, Toyohashi, Aichi, 441-8105, Japan
[3] Faculty of Mechanical and Manufacturing Engineering, University Tun Hussein Onn Malaysia, 86400 Parit Raja Batu Pahat Johor, Malaysia
[4] Research Section of Material Technology, Mie Prefecture Industrial Research Institute, Takacyaya, Tsu, Mie, 514-0819, Japan

E-mail: yokoyama@me.tut.ac.jp

Abstract. Authors have been studying the absorption of CO_2 in the steelmaking slag. In this study, an application of the electric arc furnace slag after the carbonation to admixture of mortar was investigated with the JIS (A6206-1997) method for ground granulated blast-furnace slag for concrete. The percent flows for the test mortar were smaller than that for the standard mortar. The percent flow of the carbonated slag whose average particle size of more than approximately 4 μm increased with an increase in the average size of the particles. Because the compressive strengths of the test mortar cured for 91 days were almost the same as those cured 28 days, the slag after the carbonation was thought not to have self-hardening property for a medium and long term. The compressive strength for the test mortar was almost unchanged within a range of approximately 2 to 7 μm of the average particle size, and it in this range was highest. The activity indexes for the test mortar prepared with the slag after the carbonation ranged from approximately 40 to 60 %.

1. Introduction

The increase in the atmospheric concentrations of the green house gases has caused the global warming, which has brought about the climate change. The climate change will bring us into misfortunes, such as the catastrophe, the food shortages and ecocide, if present warming trends continue. Steel industries exhaust CO_2, which is one of the green house gases, in reduction iron ores by cokes and operation. Meanwhile, they discharge a lot of slag. The slag contains CaO and MgO that react with CO_2. The behavior is discharged from the same place of emission source of CO_2. Then, we conceived adsorption and fixation of CO_2 in steel slag.

Until now, we have studied the absorption behaviors of CO_2 into a chemical reagent CaO, a waste concrete, metallic powders and slags that were discharged from electric arc furnace (EAF) under dry and wet grinding [1-6]. In the case that the EAF slags were wet ground in CO_2 atmosphere by a vibrating ball mill, most of the slags can absorb more CO_2 than the quantity of the emitted CO_2 that

Journal of Physics: Conference Series **352** (2012) 012049 doi:10.1088/1742-6596/352/1/012049

was estimated from consumed electric power of the vibrating ball mill in operation. A large amount of the residue that is composed of precipitated $CaCO_3$ and finely ground unreacted slag is discharged from the process. A shortage of final disposal place in Japan limits the disposal of the slag on land. An ocean disposal of the slag is also considered, but the sea dumping does not readily advance even if safety of the slag is confirmed. From the viewpoint of effective use of resources, it is not good that the slag after the carbonation was put into the discard.

Blast furnace slag has been already used as cement and fine and coarse aggregates for concrete. Also, EAF oxidizing slag has been used as fine and coarse aggregates for concrete. Ground granulated blast furnace slag has also been used as an admixture for concrete and mortar. In addition to it, the fine inorganic particles, such as fly ash, silica fume, anhydrite and lime stone, have been used as admixtures to strengthen concrete and mortar. The usage of those slags, ashes and silica fume in concrete has been studied [7-26], and the usage of some of them is regulated by Japanese Industrial Standards (JIS) and so force. The fine particle whose size is less than about 5 μm is used as a repairing material of a crack of concrete. Some of the authors have studied on the application of inorganic materials, such as rock powders, sand and cement sludge, aluminum dross ash, oyster shell, glass fiber and fiber reinforced plastic to concrete [27-30].

The slag particles after carbonating treatment became small by wet grinding. Then, as a series of fundamental study of carbon capture storage utilized slag, the possibility of an application of the slag after the carbonation to admixture of mortar was investigated experimentally in this study to utilize them effectively.

2. Experiment

2.1. *Carbonation of slag*

The slags discharged from electric arc furnaces were used in this study. The chemical compositions of them were listed in table 1. The concentrations of metallic constituents were measured with an inductively-coupled plasma atomic emission spectrometry (ICP-AES) after the alkali fusion of the slag. Subsequently, the composition was calculated from the concentration obtained with an ICP-AES, assuming that the detected elements existed as compounds as shown in table 1. The slags discharged from EAFs are classified into two groups. One of them is oxidizing slag discharged after the smelting of oxidizing in an EAF. The slags OA, OB and OC are the oxidizing slag. The other is reducing slag discharged after the smelting process of reduction, and the slags RA and RB are the reducing slag. The slags OA, OB and RA, were discharged from an EAF in which stainless steel was smelted, and the others were discharged after normal steelmaking in an EAF. Generally, the reducing slag contains more CaO and CaS, and less MnO than the oxidizing slag. The normal steel oxidizing slag OC has more FeO than the stainless steel oxidizing slags, OA and OB.

The crystalline structure and the texture of slag are influenced by cooling condition. Figure 1 shows the X-ray diffraction (XRD) patterns of the slags as examples. All slags used in this experiment were not vitrified but crystallized. The detected crystals were also summarized in table 1.

Table 1. Chemical compositions and crystalline phases of used slag before carbonation

Slag type	Chemical composition / mass%									crystals
	CaO	SiO_2	Al_2O_3	MgO	CaS	Cr_2O_3	MnO	TiO_2	FeO	
OA	35.4	36.3	13.5	6.87	0.22	2.07	3.31	1.98	0.23	$2CaO{\cdot}MgO{\cdot}SiO_2$, $2CaO{\cdot}Al_2O_3{\cdot}SiO_2$
OB	33.0	44.1	5.39	7.68	0.33	3.29	4.09	1.02	0.74	$2CaO{\cdot}MgO{\cdot}SiO_2$, $CaO{\cdot}SiO_2$,$2CaO{\cdot}Al_2O_3{\cdot}SiO_2$,$Mg_2Al(OH)_7$
OC	20.5	20.2	20.1	6.21	-	1.72	4.07	-	26.8	$2CaO{\cdot}Al_2O_3{\cdot}SiO_2$, $2CaO{\cdot}SiO_2$,$FeO{\cdot}Al_2O_3$, FeO
RA	47.1	30.5	10.5	5.81	2.53	1.27	0.47	0.44	1.17	$2CaO{\cdot}SiO_2$,$MgO{\cdot}SiO_2$
RB	50.3	21.1	18.7	4.12	4.02	0.19	0.60	-	0.93	$2CaO{\cdot}Al_2O_3{\cdot}SiO_2$, $2CaO{\cdot}SiO_2$,MgO

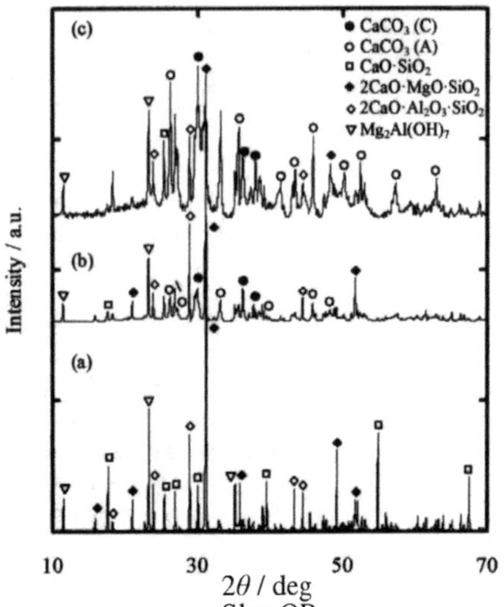

(a): before grinding; (b): after 7.2 ks ground; (c): after 14.4 ks ground.

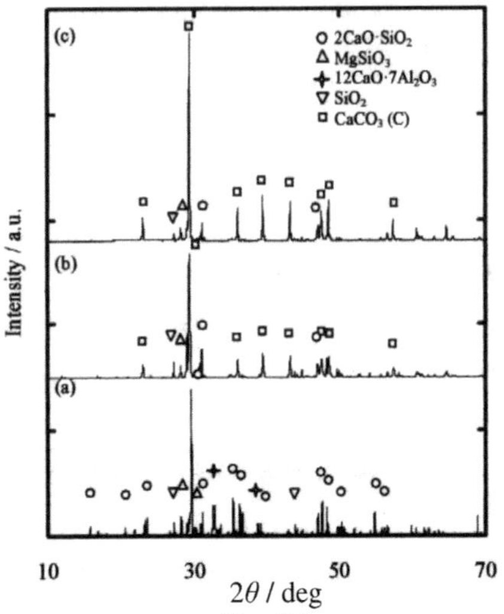

(a): before grinding; (b): after 3.6 ks ground; (c): after 14.4 ks ground.

Figure 1. XRD patterns of slag RA and OB. $CaCO_3$ (C) and (A) were the calcite and aragonite respectively.

Figure 2 shows the schematic of the experimental apparatus for carbonation [3, 4, 6]. This apparatus was used for measuring absorption of CO_2 with the constant pressure method. The slag of 0.4 kg in weight was set in the alumina lined grinding vessel. After the vacuum displacement of air in the container with CO_2, the purified water of 0.4 dm^3 in volume was poured into the container under CO_2 gas flow. Then, the slag was started the wet grinding using 180 alumina balls whose diameter was 20 mm. The pressure of CO_2 in the grinding vessel was kept to be 101.3 kPa by moving the piston of the gas volume meter.

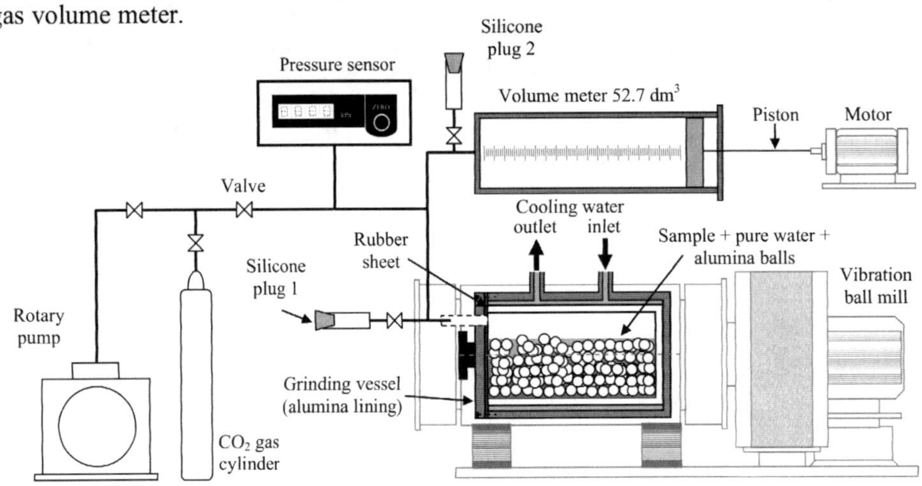

Figure 2. Apparatus used for carbonation.

Figure 3 shows the behavior of CO_2 absorbed [3-6]. Except for the slags OA and OC, generally, the amount of CO_2 absorbed rapidly increased in the early stage of grinding, and, then, it gradually increased. As for the slags OA and OC, the amount of CO_2 absorbed increased almost linearly. In comparison of amount of CO_2 absorbed at grinding time of 14.4 ks, the amount of CO_2 absorbed in the slag RA was highest, and those in the slags RB, OB, OA, OC decreased in this order. The slags used in this mortar test were wet ground for 14.4 ks. The slags OB, RA and RB whose amount of absorbed CO_2 became to be approximately half of that for 14.4 ks of grinding time were also used in this mortar test. The slag after wet grinding was passes through the filter whose pore was 0.45 μm in diameter. Then it was naturally dried. The slag hardened, but the agglomerated slag was so ductile that it was easily fractured by fingers. In addition, the agglomerate came apart apparently when it was dipped into water. The slag agglomerate was sieved, and it whose diameter was less than 850 μm was used for the mortar test.

Table 2 shows the specimens used for the mortar test. The specimen OB0 was unground slag. Figure 1 shows the XRD patterns of the specimen as examples. $CaCO_3$ was detected in the wet ground slag. The intensity of the peaks for the $CaCO_3$ increased with an increase in the grinding time. Conversely, the intensities of the peaks for the other substances that were originally contained in the slag decreased or disappeared as the grinding progressed. The crystalline phases of the slags after the carbonation were summarized in Table 2. The $CaCO_3$ was estimated to form by that the reaction between Ca^{2+} eluted from the slag and CO_3^{2-} dissolved from CO_2 occurred at the interface between water and CO_2. The overall formation reaction of it was expressed by:

$$CaO + CO_2 = CaCO_3 \tag{1}$$

Based on eq. (1), the carbonation ratio, *CR*, in table 2 was calculated by:

$$CR = \frac{Q_{CO_2}}{N_{CaO}} \times 100 \tag{2}$$

where Q_{CO2} is the absorbed CO_2 [mol], and N_{CaO} is the amount of substance of CaO in slag [mol]. In our previous work [6], the relation between the pH and the concentration of Mg dissolved in water during grinding suggested the formation of $MgCO_3{\cdot}3H_2O$ that was not detected by a XRD. Because the quantity of formation of it could not be evaluated, the carbonation ratio was calculated assuming that the CO_2 absorbed reacted with only the CaO in the slag. In addition, the relation between the pH and the concentration of Ca dissolved in water suggested the formation of $CaCO_3{\cdot}6H_2O$. However, the formation of it does not influence the carbonation ratio. Figure 4 shows the specimens after the carbonation. Those particles were plate like.

According to the standard, the specific surface area of the slag particles should be measured with the permeation method. As mentioned before, the slag after carbonation cohered. The gas permeation method like the Blaine method does not give accurate specific surface area of the agglomerated particle. Then, size of the slag particle was measured from the image as shown in figure 4. Table 2 shows the range and average of the particles size. The average size of particle ground for short time was larger than that for long time. The average size of the specimen OC4 was larger than that of the other ground specimens. This thing resulted from that the slag OC was difficult to be broken because the slag OC contained much FeO in itself.

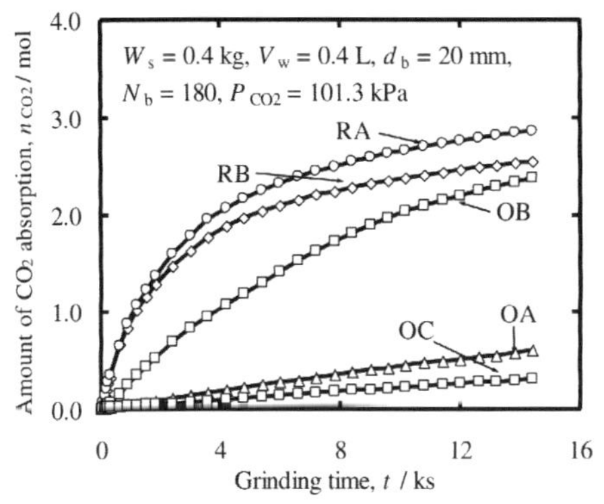

Figure 3. Behavior of absorbed CO_2 in each slag.

Figure 4. Carbonation treated slag particles for 14.4 ks.

Table 2. Slag used for preparation of mortar.

Specimen	Slag	Grinding time /ks	Carbonation ratio, *CR* /%	Particle size /μm		Crystals
				Range	Average	
OA4	OA	14.4	24	1-15	4.4	$2CaO \cdot MgO \cdot SiO_2$, $2CaO \cdot Al_2O_3 \cdot SiO_2$, $CaCO_3$
OB0	OB	0.0	0	100-850	450	$2CaO \cdot MgO \cdot SiO_2$, $CaO \cdot SiO_2$, $2CaO \cdot Al_2O_3 \cdot SiO_2$, $Mg_2Al(OH)_7$
OB2	OB	7.2	41	1-10	5.1	$CaCO_3$, $2CaO \cdot MgO \cdot SiO_2$, $2CaO \cdot Al_2O_3 \cdot SiO_2$
OB4	OB	14.4	80	0.5-7	2.3	$CaCO_3$, $2CaO \cdot MgO \cdot SiO_2$, $2CaO \cdot Al_2O_3 \cdot SiO_2$
OC4	OC	14.4	21	2-40	14.1	$2CaO \cdot Al_2O_3 \cdot SiO_2$, $2CaO \cdot SiO_2$, $FeO \cdot Al_2O_3$, FeO, $CaCO_3$, Al_2O_3
RA1	RA	3.6	50	1-10	6.5	$CaCO_3$, $2CaO \cdot SiO_2$, $MgO \cdot SiO_2$, SiO_2
RA4	RA	14.4	85	0.5-8	3.5	$CaCO_3$, $2CaO \cdot SiO_2$, $MgO \cdot SiO_2$, SiO_2
RB1	RB	3.6	49	0.5-6	4.0	$CaCO_3$, $2CaO \cdot Al_2O_3 \cdot SiO_2$, $2CaO \cdot SiO_2$
RB4	RB	14.4	71	0.5-5	1.8	$CaCO_3$, $Ca_4Si_2O_6(CO_3)(OH,F)_2$

2.2. *Flow and compressive tests*

The preparation of mortar, the flow test and the compressive test was performed based on the JIS A6206-1997 (Ground granulated blast-furnace slag for concrete). This regulation contains some of the matter of the JIS R 5201-1997 (Physical testing method for cement). Here, those are explained briefly.

The standard mortar was prepared with 0.450 ± 0.002 kg of the ordinary portland cement, 1.350 ± 0.005 kg of the standard sand that was natural silica sand whose size distribution was adjusted, and 0.225 ± 0.001 kg of the tap water. The test mortar was prepared with displacing half of the cement to the carbonated slag. Therefore, it was prepared with 0.225 ± 0.002 kg of the portland cement, 0.225 ± 0.002 kg of the slag, 1.350 ± 0.005 kg of the standard sand and 0.225 ± 0.001 kg of the water. These materials were kneaded by the mixer respectively. After that, the mortal was poured in the flow cone. After the flow cone was removed rightly above, it was vibrated fifteen times for fifteen seconds. The flow value was obtained from averaging the maximum diameter of the cone and the diameter that was in the orthogonal direction of the maximum diameter.

The kneaded mortar was also poured in the metallic die. The shape of the mold mortar was a square prism (40×40×160 mm). The mortar was set in the box in which the relative humid and the temperature were controlled to be over 90 % and 293 ± 1 K respectively. Subsequently, it was hold for twenty seven or ninety days in the water whose temperature was 293 ± 1 K. The compressive strength was measured by the compression testing machine (Shimadzu UMH 100). The test piece for measurement of the compressive strength was prepared by cutting the hardened mortar by half. Three of them were used for the compressive test.

3. Results and Discussions

3.1. *Flow value*

The flow value represents the fluidity and the softness of mortar. Figure 5 shows the percent flow of each specimen. Here, the percent flow, F [%], is the ratio of flow values, and it is defined by:

$$F = \frac{l_2}{l_1} \times 100 \tag{3}$$

where l_1 and l_2 are the flow values of the standard and the test mortar respectively. The flow value of the standard mortar was 188. The percent flows of the specimen OB4, RA4 and RB4 could not be measured, because their mortar cracked under the vibration. Generally, the percent flows of the test mortar were lower than that of the standard mortar. It was reported that the flow value for the mortar blended with fine ground particles of the molten slag that was discharged from a garbage incineration plant was almost the same as the standard value [17]. The percent flow of the carbonation treated slag seemed to be inferior to that for the molten slag. In the case that the finely ground mineral powder used in concrete paste, the more amount of water has to be added into paste to enhance and hold the workability [31]. Therefore, addition of much water or an air–entraining or a water reducing admixture and so on is thought to be necessary to improve the flow value of the test mortar.

The percent flows of the specimen of OB0, OB2 and OB4 decreased in this order. Because the carbonation ratios of OB0, and OB2 and OB4 increased in this order, the percent flow decreased with an increase in the carbonation ratio. The low carbonation ratio means the large quantity ratio of the unreacted CaO to the total CaO in slag. The unreacted CaO is thought to contribute to the hardening of the mortar, and the percent flow is estimated to be lowered by the hardening. However, the percent flow of the mortar that contained much unreacted CaO was larger than that contained less unreacted CaO. This is thought to mean the percent flow is not necessarily influenced by the chemical property of the slag.

As for the fluidity of the cement paste that was mixed with fine limestone, it was reported that there was the optimal conditions of the combination of the roundness and the distribution of the particles [32]. Then, it was investigated that the effect of the average size of the slag on the percent flow, and the relation of both was shown in figure 6. As mentioned above, the flow values of the specimen OB4, RA4 and RB4 could not be measured. The average particle sizes of them were below approximately 4 μm. The free water is well-known to contribute the fluidity [31, 32]. These smaller particles hold water in the space among them, and the particles cohered each other [33]. This results in a decrease in the quantity of the free water, and seemed to lower the ratio of flow values. Except for percent flow of the unground specimen OB0, the percent flow for the average size of more than approximately 4 μm showed the upward tendency with an increase in the average size, but the tendency was not clear. Probably, the shape and distribution is thought to affect the ratio of flow value.

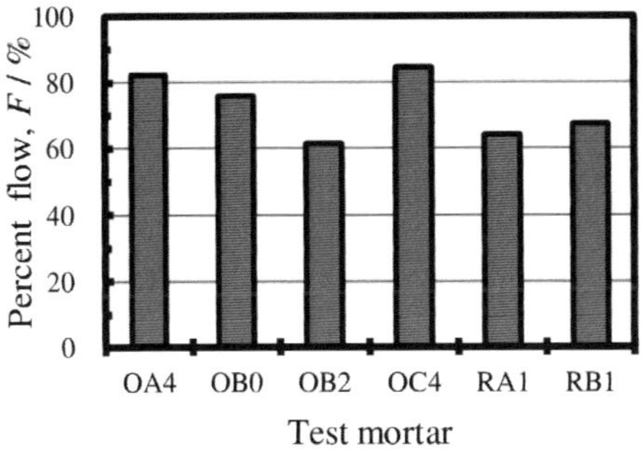

Figure 5. Percent flow of each test mortar.

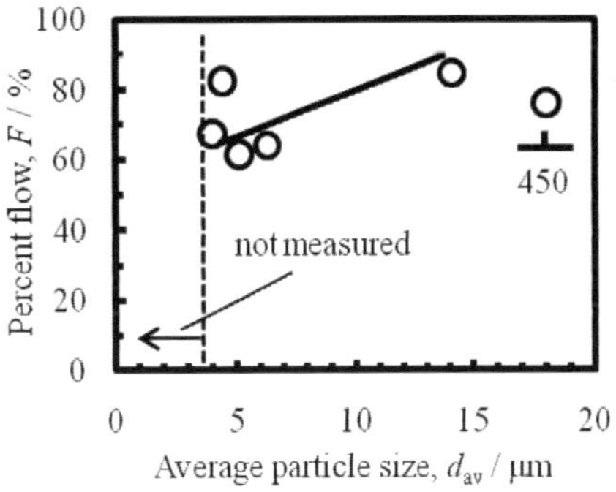

Figure 6. Relation between percent flow and average particle size of slag after carbonation.

3.2. *Compressive strength*

Whereas the flow values of the specimen OB4, RA4 and RB4 could not be measured, it was possible to put them in the mold and make their test pieces for the compressive test. The amplitude and frequency for making the test piece were 0.8 mm and 2800 Hz, and those for measuring the flow value were 10 mm and 1 Hz. The small amplitude and large frequency seemed to make it possible to fill their mortar in the mold well, and cracking of their compacts was thought not to occur at the time of separation of them from the die.

Figure 7 shows the compressive strength of each mortar. The compressive strengths of the test mortars were smaller that that of the standard mortar. The compressive strength of the mortar blended with fine molten slag particles decreased with an increase in the replacement ratio of the slag to cement [17]. The compressive strength of the mortar in which the replacement ratio was 25 % was almost the same as that of the standard mortar in which the ratio was 0 %, but it for 50 % of the ration became smaller than that the standard [25]. In this work, the replacement ratio was 50 %. The smaller compressive strengths for the test mortars seemed to be similar to those for the mortar mixed with the molten slag.

As shown in Figure 7, the compressive strengths of the test mortar cured for 91days were larger than that for 28 days. However, because the 95 % confidence limit of the measured compressive strength was ± 3 MPa, it may be said that only the difference in the compressive strengths of the specimen OB0 was the meaningful. Therefore, the slag OB, which was not carbonated, has self-hardening property, whereas the slags after carbonation seem not to have self hardening property in the term of 28 to 91 days.

The compressive strengths for the specimens OB2 and OB4 were almost the same, ant those were larger than that for the specimen OB0. The compressive strengths for the specimens RA1 and RA4 were almost the same, whereas the compressive strength for RB1 was larger than that for RB4. The carbonation ratios of the specimens OB0, OB2 and OB4 increased in this order. Similarly, the carbonation ratio of the specimen RA1 and RA4, RB1 and RB4 increased in this order respectively. Therefore, the compressive strength for the test mortar was not influenced by the carbonation ratio. The low carbonation ratio means that the much unreacted CaO exists in the slag. This means that the unreacted CaO does not contribute to the hardening of mortar.

As shown in figure 5, the percent flows of the specimens OB4, RA4 and RB4 could not be measured. The compressive strength of the specimens OB4 and RA4 were relatively large, those were almost the same as the specimens OA4, OB2, RA1and RB1 whose percent flows over approximately 60 %. Accordingly, the percnt flow did not influence the compressive strength.

Figure 8 shows the relation between the compressive strength and the average size of the slag. The compressive strength was almost constant over a range of approximately 2 to 7 μm of the average particle size. When the average particle size exceeded about 7 μm, it decreased with an increase in the average size. This result coresponds with the results of the molten slag [17]. The compressive strength for about 1 μm of the average size was small. This seems to result from cohesion of the slag particles [33].

The compressive strengths of these test mortar are lower than those of mortar for concrete, but there are mortar that have such strengths. Fine particles whose average size are a few micron meters have been used as injectable materials to grout crucks. These test mortar seem to be able to be used such repairing materials.

In the JIS, the compressive strength of mortar is evaluated by the activity index. Here, the activity index, A_s, is defined by

$$A_S = \frac{c_2}{c_1} \times 100 \tag{4}$$

where c_1 and c_2 are the compressive strengths of the test pieces of the standard and test mortar cured for the same time respectively. Figure 9 shows the activity index of each specimen. Because the compressive strength of the standard mortar cured for 91days was almost the same as that for 28 days, the relation between the activity index and the test mortar was similar to that between the compressive strength and the test mortar as shown in figure 7. The activity index of the test mortar was within a range of approximately 40 to 60 % except for the specimen OB0 which was unground slag. The activity indexes for the test mortar were smaller than those regulated by the JIS A 6206 and the ASTM C 989 for the ground granulated blast–furnace slag. The slag after carbonation was not used as alternate material of the ground granulated blast–furnace slag.

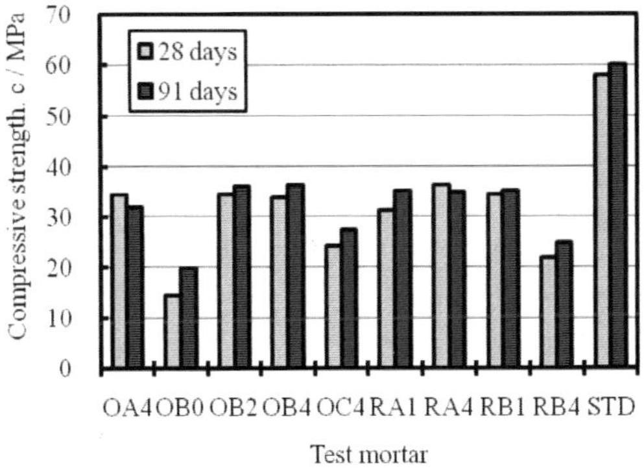

Figure 7. Compressive strength of each mortar.

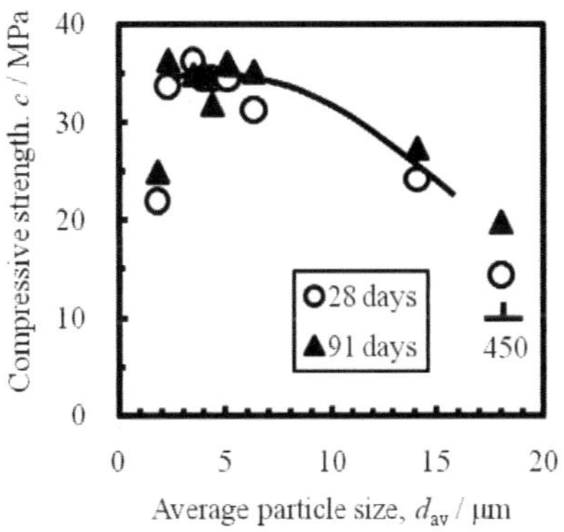

Figure 8. Relation between compressive strength and average particle size of slag after carbonation.

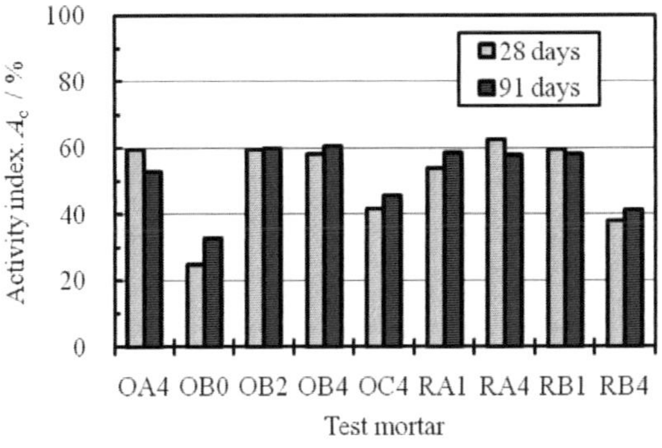

Figure 9. Activity index of each test mortar.

4. Conclusions

In this study, it was tried that the EAF slag after carbonating treatment was applied to the admixture of mortar. A flow value and a compressive strength of test mortar were measured based on the JIS A6206-1997 (Ground granulated blast-furnace slag for concrete). The obtained results could be summarized as follows:

(1) The percent flows for the test mortar were smaller than that for the standard mortar. The average particle size of the carbonated slag influenced the percent flow. The percent flows for the average particle size of less than approximately 4 μm could not be measured. The percent flows for the carbonated slags whose size were larger than approximately 4 μm increased with an increase in the average size of the particles.

(2) It was possible to make the compact for compressive test with mortar whose flow value could not be measured.

(3) The compressive strengths of the test mortar cured for 91 days were almost the same as those cured 28 days. It was thought form this result that the slag after the carbonation did not have self-hardening property for a medium and long term.

(4) The compressive strengths for the test mortar were smaller than that for the standard mortar. The activity indexes for the test mortar prepared with the slag after the carbonation ranged from approximately 40 to 60 %.

(5) The compressive strength became large when the average particle size became large from approximately 1 to 2 μm. The compressive strength for the test mortar was almost unchanged within a range of approximately 2 to 7 μm of the average particle size. When the average particle size exceeded about 7 μm, the compressive strength decreased with an increase in the average size.

References

[1] Hisyamudin MNN, Yokoyama S, Kawakami M and Umemoto M. 2009 Powder Tech **196** 156
[2] Hisyamudin MNN, Yokoyama S, Suzuki T, Motozuka S, Shimomura T, Sasano J, Izaki M. 2010 Adv. Mater. Res. **129-131** 799
[3] Hisyamudin MNN, Yokoyama S, Umemoto M. 2010 Mater. Sci. Forum **654-656** 2927
[4] Yokoyama S, Sato R, Hisyamudin MNN, Umemoto M. 2010 ISIJ Intern. **50(3)** 482
[5] Yokoyama S, Hisyamudin MNN, Hirano S. 2010 Materi Sci Forum **654-656** 2923
[6] Hisyamudin MNN, Yokoyama S. Intern. J Greenhouse Gas Control (in submission)
[7] Honda K, Uechi Y, Morooka H, Yasunaga T. 1992 Cem. Sci. Concr. Tech. **46** 174
[8] Kinoshita M, Saitou K, Arashima T, Yuki Y. 1995 Cem. Sci. Concr. Tech. **49** 192
[9] Fujiwara H, Maruoka M, Fujiwara Y, Hiroshima A. 2006 Cem. Sci. Concr. Tech. **60** 441
[10] Kato H, Taga G, Hiranaka S, Doi H. 2004 Cem. Sci. Concr. Tech. **58** 17
[11] Shoya M, Aba M, Suguta S. 1999 Cem. Sci. Concr. Tech. **53** 448
[12] Watanabe S, Hori A, Naruse H, Nagataki S. 2008 Cem. Sci. Concr. Tech. **62** 419
[13] Tachibana K, Kobayashi T, Ishi M. Cem. 1997 Sci. Concr. Tech. **51** 906
[14] Murai H, Kobayashi T, Nagaoka S. 1997 Cem. Sci. Concr. Tech. **51** 316
[15] Yamamoto T, Hironaga M. 2007 CRIEPI Rep No. N06018
[16] Yamamoto T, Kanazu T. 1998 CRIEPI Rep No. U97061
[17] Kawashima H, Mizuno H, Iisaka T, Sugiyama A. 2004 Cem. Sci. Concr. Tech. **58** 593
[18] Mendi B, Kenai S, Khatib J, Aït-Mokhtar A. 2009 Constr Build Mater **23** 625
[19] Zelić J, Krstulović R, Tkalčec E, Krolo P. 2000 Cem. Concr. Res. **30** 145
[20] Zhang Y, Zhang X. 2008 Constr. Build Mater. **22** 1634
[21] Kumar S, Kumar R, Bandopadhyay A, Alex TC, Kumar BR, Das SK, Mehrotra SP. 2008 Cem. Concr. Comp. **30** 679
[22] Qasrawi H, Shalabi F, Asi I. 2009 Constr. Build Mater. **23** 1118
[23] Nili M, Afroughsabet V. 2010 Constr. Build Mater. **24** 927
[24] E-H. Kardi, R. Duval 2009 Constr. Build Mater. **23** 3388
[25] Saitoh T, Taniyama N, Ozaki H, Matsumoto F, Nara Y. 2002 Summaries of tech papers of annual meeting Archi. Inst. Jap. 715
[26] Mitsui M, Fukuzawa K, Miyasaka Y, Kojima H. 2003 Proceedings Jap. Concr. Inst. **25(1)** 1451
[27] Maegawa A, Yuasa Y, Murakami K. 1997 Rep. Mie. Prefec. Ind. Res. Inst. **21** 30
[28] Maegawa A, Murakami K, Yuasa Y, Tomii K. 2000 Rep. Mie. Prefec. Ind. Res. Inst. **24** 113
[29] Maegawa A, Yuasa Y, Murakami K. 2000 Rep. Mie. Prefec. Ind. Res. Inst. **25** 87
[30] Tanaka M, Murakami K, Maegawa A. 2009 Rep. Mie. Prefec. Ind. Res. Inst. **33** 107
[31] Orchard DF. 1972 *Concrete Technology, vol.1 Properties of material* (London: Applied Science Publishers) 92
[32] Masuda K, Kakinuma Y, Aikawa Y, Sakai E. 2008 Cem. Sci. Concr. Tech. **62** 13
[33] Nawa T, Eguchi H, Ookubo M. 1991 Concr. Struc. Pave. **433/V-15** 139

Journal of Physics: Conference Series **352** (2012) 012050 doi:10.1088/1742-6596/352/1/012050

Elution of zinc in dust discharged from electric arc furnace in carbonic acid solution

S Yokoyama[1], T Sasaki[2], J Sasano[1] and M Izaki[1]

[1]Department of Mechanical Engineering, Toyohashi University of Technology, Hibarigaoka, Tempaku-cho, Toyohashi, Aichi, 441-8105, Japan
[2] Graduate Program of Mechanical Engineering for Master's Degree, Toyohashi University of Technology, Hibarigaoka, Tempaku-cho, Toyohashi, Aichi, 441-8105, Japan

E-mail: yokoyama@me.tut.ac.jp

Abstract. The dust discharged from an electric arc furnace (EAF) is a valuable resource of zinc. As a fundamental study of extraction of zinc, iron and chlorine in the EAF dust, the elution behavior of them in carbonic acid solution was studied. The influence of the weight of the EAF dust on the elution behavior was examined in this study. Experiment was carried out putting the EAF dust from 1 g to 200 g in weight into 1 L of water that was introduced by CO_2. Generally, the pH in the aqueous solution increased with an increase in weight of the additive EAF dust. Maximums of the eluted concentrations of zinc and chloride ion increased with an increase in the weight of the additive EAF dust whereas the extraction ratios of both of them decreased with an increase in the weight of the additive EAF dust. Iron in the EAF dust remained in the dust without elution. The limit of extraction of zinc from the EAF dust to water was given by the solubilities of $ZnFe_2O_4$ and ZnO expressed by eq. (6) and eq. (9) respectively.

1. Introduction

Zinc coated steel is a superior materials economically to prevent rusting and galvanic corrosion. It has been widely used as a material for buildings, vehicles, electronics and so force. Therefore, iron and steel scrap contains zinc in itself. Steelmaking in electric arc furnace (EAF) uses steel scrap as the primary raw material. Because zinc is a volatile metal, zinc was collected as dust from off-gas in an EAF steelmaking process. Ninety two percent of zinc that is used in Japan is imported. Therefore, EAF dust is a valuable resource of zinc. Because it is very important to recover zinc from industrial wastes, numerous researches have been performed until now [1-11]. Zinc is one of the environmentally regulated substances. Then, EAF dust cannot be disposed easily. Whereas zinc has been already recovered from EAF dust, it is desired to develop of the recovery process of zinc with higher efficiency. In this work, as a fundamental study of the recovery of zinc from EAF dust, we studied the elution behaviour of EAF dust, especially zinc, iron and chlorine, in carbonic acid solution.

2. Experiment

2.1. *Dust*

The dust that was discharged from an electronic arc furnace (EAF) was used in this experiment. Table 1 shows the chemical compositions of the dust. The compositions were measured with an EPMA, an

X-ray fluorescence and a combustion analysis. Figure 1 shows the XRD pattern of the dust. The Fe_3O_4, $ZnFe_2O_4$, $CaCO_3$, SiO_2, KCl and C were identified in the used EAF dust. However, the metallic particle of stainless steel was observed in the EAF dust. In addition, as described later, ZnO that was not detected by an XRD existed in the EAF dust. Here, the XRD patterns of Fe_3O_4 and $ZnFe_2O_4$ were almost the same each other. However, there are slight differences in the peak at around 60 degree (2θ) between them. It was judged from the detection of these peaks around 60 degrees and the chemical composition that both Fe_3O_4 and $ZnFe_2O_4$ existed. The EAF dust of 1 g to 200 g in weight was used in the experiment.

Table 1. Chemical compositions of used EAF dust.

Element	Fe	Cr	Ca	Zn	Si	Mn	Na	Mg	Ni	K	Cu	Pb	Al	O	C	Cl
mass%	32.1	9.2	7.8	7.2	6.4	3.6	3.6	1.7	2.8	1.5	0.7	0.4	0.3	15.3	2.6	4.8

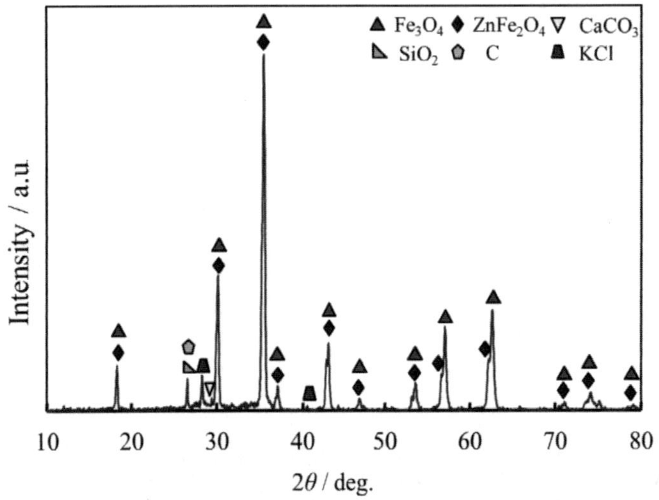

Figure 1. XRD pattern of used EAF dust.

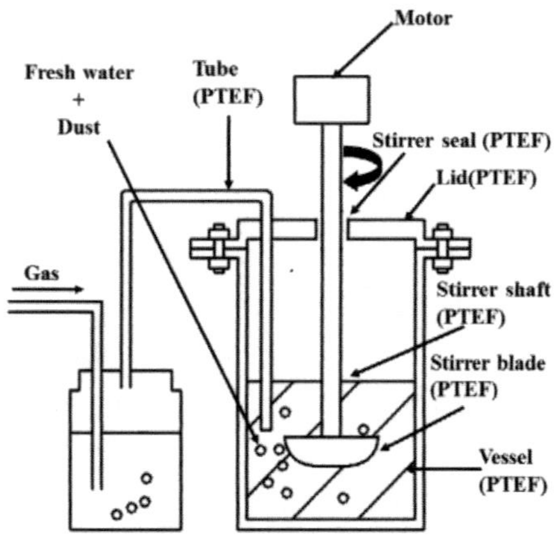

Figure 2. Schematic of used experimental apparatus.

2.2. *Procedure*

Figure 2 shows the schematic of the experimental apparatus. The experimental apparatus and procedure were as similar as those described in the previous paper [12, 13]. However, in order to prevent the aqueous solution with EAF dust from vaporizing, CO2 gas was passed through pure water prior to introduction of gas into the aqueous solution. This treatment diminish made amount of water due to vaporization a one twentieth. The parts that contact with the aqueous solution are made of PTFE. Commercial grade CO_2 was used in the experiment. The apparatus was set in a simple globe box that was made of vinyl sheet to prevent interfusion of suspended dust in the air into the vessel. First, the pure water of 1 L in volume was poured into the vessel. Subsequently, CO_2 gas was introduced into the water at the flow rate of 100 mL/min (NTP) with mixing by the blade at the rotation speed of 200 rpm. After the hydrogen ion potential (pH) became stable, the EAF dust of weight of 0.1-200 g was put in the vessel. During experiment, most EAF dust was suspended in the aqueous solution, whereas a few of the dust was laid at the bottom of the vessel. The pH of the aqueous solution in the container was measured by a pH meter with glass electrode. The aqueous solution that was sucked by a whole pipette was passed through the filter of which pore size was 0.45 μm. The concentrations of iron and zinc in the filtrate were measured with an inductively–coupled plasma atomic emission spectroscopy (ICP-AES), and the chloride ion in the filtrate was measured with a chloride ion meter using a solid membrane.

3. Results and discussions

3.1. *Change in pH with time*

Figure 3 shows the change of pH in the aqueous solution with time. Generally, the pH increased steeply immediately after the addition of the EAF dust. Subsequently, it gradually increased. In addition, the pH increased with an increase in weight of the additive FAF dust.

In this study, CO_2 was introduced into the aqueous solution. Accordingly, following dissolution reaction of CO_2 occurred:

$$CO_2 + H_2O = 2H^+ + CO_3^{2-} \quad (1)$$

$$CO_2 + H_2O = H^+ + HCO_3^- \quad (2)$$

These dissolution reactions of CO_2 into the aqueous solution generate H^+, and decrease the pH. In addition, adsorption of H^+ to the surface hydroxyl, -OH, of the EAF dust is thought to occur [14]. This adsorption reaction is given by

$$\text{-OH}+H^+=\text{-OH}_2^+ \quad (3)$$

This adsorption reaction increases the pH. Furthermore, as mentioned later, the dissolutions of substances of the EAF dust into water decrease the pH. The elution of CO_2 into the aqueous solution, the adsorption of H^+ and the dissolutions of substances of the EAF determined the behavior of pH. The increase in weight of the additive EAF dust implies the increase in the interfacial area of EAF dust and solution. Therefore, pH is influenced by the interfacial area. The pH is mainly determined by adsorption of H^+ to the surface hydroxyl of the FAF dust.

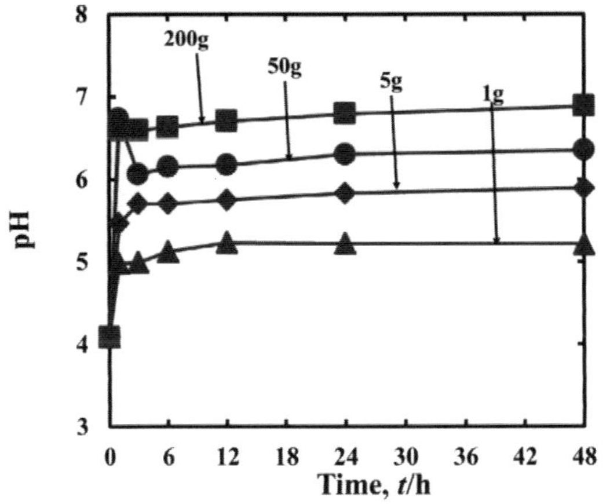

Figure 3. Change in pH in aqueous solution with time.

3.2. *Elution behavior*

3.2.1. *Chlorine*

Figure 4 shows the elution behaviour of chlorine from the EAF dust into the aqueous solution. Generally, the eluted concentrations of chloride ion increased steeply at an early stage of the elution. Subsequently it gradually increased with time. Figure 5 shows the relation between the eluted concentration of chloride ion at the end of experiment and the weight of the additive EAF dust. Whereas the eluted concentration of zinc increased with an increase in the weight of the additive EAF dust, it did not increase proportionally with an increase in the weight of the additive EAF dust.

3.2.2. *Iron*

Figure 6 shows the elution behavior of iron. The eluted concentrations of iron were below approximately 0.27 mg/L. The influence of the weight of the additive EAF dust on the elution behavior of iron was not observed. Maximums of the eluted concentrations of iron for the additive EAF dust of 5 and 50 g in weight were almost the same each other, and were higher than those for the additive EAF dust of 1 and 200 g in weight.

Generally, the solubilities of FeO and $Fe(OH)_2$ are larger than those of Fe_2O_3 and $Fe(OH)_3$ [12]. As mentioned earlier, the used EAF dust contains Fe_3O_4 and metallic stainless steel. Iron ion may be eluted as a Fe^{2+}. Following two things are considered as the reason that the eluted concentration iron is small. One was that the hexavalent chromium ion, which was observed in the aqueous solution, oxidized Fe^{2+}. Another was the formation of $FeCO_3$ due to the reaction between Fe^{2+} and CO_3^{2-}. The relation between maximum of the eluted concentration of iron and the pH was as similar as the solubility of iron carbonate. Further study was needed for clarify the reason.

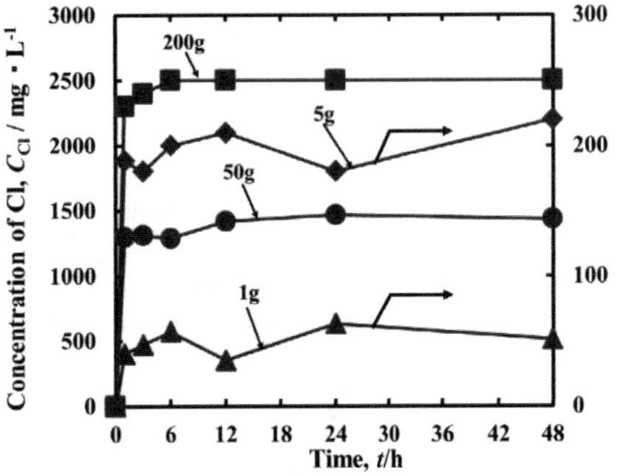

Figure 4. Change in eluted concentration of chloride ion with time.

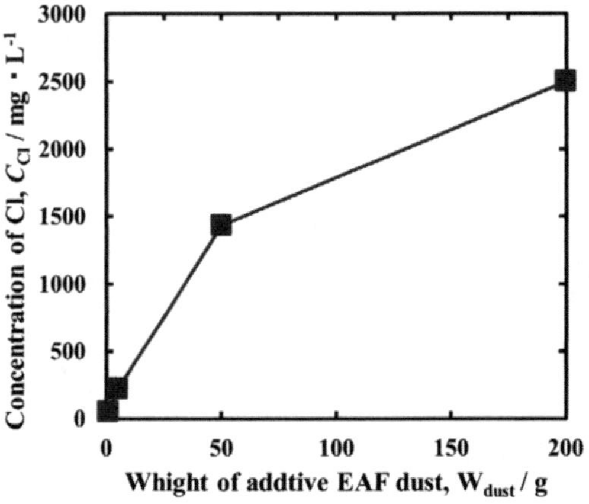

Figure 5. Relation between eluted concentration of chloride ion and weight of additive EAF dust. The concentration of chloride ion is the concentration at the elution time of 48 h.

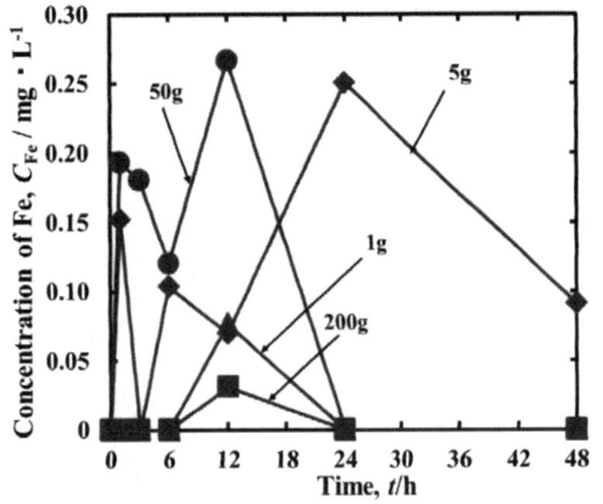

Figure 6. Change in eluted concentration of iron with time.

3.2.3. *Zinc*

Figure 7 shows the elution behavior of zinc. Generally, the eluted concentration of zinc increased steeply at an early stage of the elution. Subsequently, in the case of the additive EAF dust of 50 g and

Journal of Physics: Conference Series 352 (2012) 012050 doi:10.1088/1742-6596/352/1/012050

200 g, the eluted concentrations of zinc decreased with time. In the case of the additive EAF dust of 1 g and 5 g, the eluted concentrations of zinc were unchanged with time.

Figure 8 shows the solubility of ZnO, $Zn(OH)_2$, $ZnCO_3$ and $ZnFe_2O_4$. Here, the partial pressure of CO_2 was 1 atm. The each solubility was calculated with both Debye-Hückel theory and the combination of the following thermodynamic data [15].

$$ZnO+2H^+=Zn^{2+}+H_2O \quad \log K=11.20 \tag{3}$$
$$Zn(OH)_2+2H^+=Zn^{2+}+2H_2O \quad \log K=12.48 \tag{4}$$
$$ZnCO_3+2H^+=Zn^{2+}+CO_2+H_2O \quad \log K= 8.22 \tag{5}$$
$$ZnFe_2O_4+2H^+=Zn^{2+}+2Fe_2O_3+4H_2O \quad \log K=9.67 \tag{6}$$
$$Zn^{2+}+2H_2O =ZnOH^0+2H^+ \quad \log K=-16.80 \tag{7}$$
$$Zn^{2+}+3H_2O =Zn(OH)_3^-+3H^+ \quad \log K=-27.68 \tag{8}$$
$$ZnO+Cl^-+2H^+ =ZnCl^++2H_2O \quad \log K=10.73 \tag{9}$$
$$H_2O=2H^++OH^- \quad \log K=20.78 \tag{10}$$

where K is the equilibrium constant. In addition, the relation between the eluted concentration of Zn and pH, which were obtained experimentally, was also shown in figure 8.

In the case of the weight of the additive EAF dust of 200 g, the eluted concentrations of zinc ranged between the solubility of ZnO given by eq. (3) and $ZnFe_2O_4$ given by eq. (6), and approached the solubility of $ZnFe_2O_4$. This result means that the used EAF dust contains ZnO which is not detected by an XRD. In the case of the weight of the additive EAF dust of 50 g, the eluted concentrations of zinc were below the solubility of $ZnFe_2O_4$ and approached the solubility of $ZnFe_2O_4$. These two results means that $ZnFe_2O_4$ forms by the reaction between Zn^{2+} eluted from ZnO and Fe_3O_4. In the case of the weight of the additive EAF dust of 1 g and 5 g, the eluted concentrations of zinc approached the solubility of ZnO which is given by eq. (9) according as the concentration of chloride ion. Accordingly, eluted concentration of zinc is influenced by concentration of chloride ion. However, the solubility of ZnO given by eq. (9) was as similar as that of $ZnCO_3$ given by eq. (5). Further study is needed for the elution reaction of dust into aqueous solution.

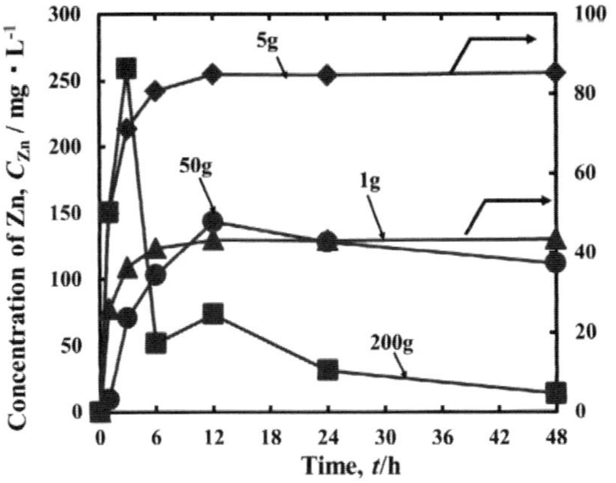

Figure 7. Change in eluted concentration of zinc with time.

3.3. *Extraction ratio*

Table 2 shows the extraction ratio in zinc, iron and chlorine from EAF dust to aqueous solution. Here, the extraction ratio of the substance is the ratio of extracted weight of the substance in aqueous solution and weight of the substance in the additive dust before the extraction. In addition, maximums of the eluted concentrations were used for calculation of the extraction ratio. The extraction ratios of chlorine and zinc decreased with an increase in the weight of the additive EAF dust. The extraction ratio of iron was nearly equal to zero, and was not influenced by the additive weight of the EAF dust.

Accordingly, if a small amount of the EAF dust is put in carbonic acid solution, zinc and chlorine in the EAF dust are extracted from the dust into carbonic acid solution. In addition, it was found that iron in the dust remained in the dust.

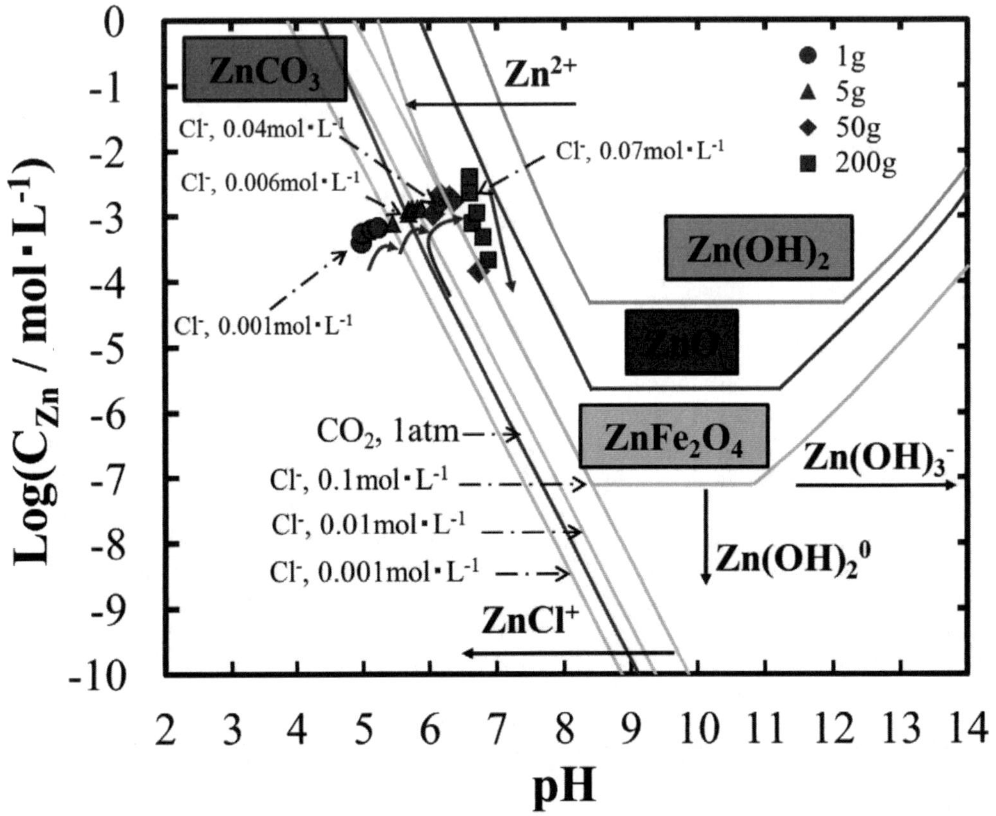

Figure 8. Solubility of ZnO, $Zn(OH)_2$ and $ZnFe_2O_4$.

Table 2. Extraction ratio in zinc, iron and chlorine from EAF dust to aqueous solution. [%]

	Weight of additive EAF dust [g]			
	1	5	50	200
Zn	60	24	4.0	1.8
Cl	100	94	62	26
Fe	0.02	0.02	0.002	0.00005

4. Conclusions

We studied the influence of weight of the additive EAF dust on elution behavior of the EAF dust into water. The obtained main results can be summarized as follows:

(1) The pH in the aqueous solution increased with an increase in weight of the additive EAF dust.

(2) Maximums of the eluted concentrations of zinc and chloride ion increased with an increase in the weight of the additive EAF dust. However, extraction ratios of both of them decreased with an increase in the weight of the additive EAF dust.

(3) Iron in the EAF dust remained in the dust without elution.
(4) The limit of extraction of zinc from the EAF dust to water was given by the solubility of eq. (6) and eq. (9) according as the experimental conditions. The eluted behavior of zinc was influenced by the pH and the concentration of zinc in the solution.

References

[1] Nakayama T and Taniishi H 2011 *Nippon Steel Eng. Co.,Ltd. Tech Rev* **2** 25
[2] Nagai K, Matsumoto Y and Watanabe H 2007 JMMIJ **123** 726
[3] Suzuki M, Niitsuma K and Takeuti N 2004 JMMIJ **120** 284
[4] Liang R, Kikuchi E, Sakamoto H, Fujita T and Numata Y 2002 J Soc Powder Technol J apan **39** 161
[5] Akae N, Uesugi H, Shimada T, Takahashi R and Akiyama T 2002 *Shigen-to-Sozai*, **118** 95
[6] Barakat M A 2003 JOM **55** 26
[7] Sharma K D, Row B R L 1985 *Hydrometallurgy* **13** 377
[8] Jiang J C, Yang X U and Zhao Y C 2010 Adv Mater Res **113 – 116** 2299
[9] Oustadakis P, Tsakiridis PE, Katsiapi A and Agatzini-Leonardou S 2010 *J Hazard Mater 179* 1
[10] Baik D S and Fray D 2000 *J Min Proc and Extrac Metall* **109** 121
[11] Gotfryd L, Chmielarz A and Szołomicki Z 2011 *Physicochem Probl Miner Proc* **47** 149
[12] Yokoyama S, Suzuki A, Izaki M and Umemoto M 2009 *Tetsu-to-Hagane* **95** 434
[13] Yokoyama S, Suzuki A, Nor N H B, Kanematsu H, Ogawa A, Takahashi T, Izaki M and Umemoto M 2010 *ISIJ International* **50** 630
[14] Sato T, Fukushi K and Yoneda T 2007 *J. MMIJ* **123** 132
[15] Lindsay W L 1979 *Chemical equilibria in soil* (Caldwell, Blackburn) p12, 150

Journal of Physics: Conference Series **352** (2012) 012051 doi:10.1088/1742-6596/352/1/012051

Influence of amount of oxidizing slag discharged from stainless steelmaking process of electric arc furnace on elution behavior into fresh water

S Yokoyama[1], T Shimomura[2], M N N Hisyamudin[3], T Takahashi[4] and M. Izaki[1]

[1] Department of Mechanical Engineering, Toyohashi University of Technology, Hibarigaoka, Tempaku-cho, Toyohashi, Aichi, 441-8105, Japan
[2] Graduate Program of Mechanical Engineering for Master's Degree, Toyohashi University of Technology, Hibarigaoka, Tempaku-cho, Toyohashi, Aichi, 441-8105, Japan
[3] Faculty of Mechanical and Manufacturing Engineering, University Tun Hussein Onn Malaysia, 86400 Parit Raja Batu Pahat Johor, Malaysia
[4] Department of Chemical Science and Technology, Miyakonojo National College of Technology, Yoshio-cho, Miyakonojo, Miyazaki, 885-8567, Japan

E-mail: yokoyama@me.tut.ac.jp

Abstract. Fundamental study was carried out for provision for acidification of soil due to acid rain. The influence of weight of the additive slag on elution behavior of the slag into water was studied in this study. Elution experiment was carried out on a basis of JIS K 0058-1. Generally, the pH in the aqueous solution increased with an increase in weight of the additive slag. The pH converged to approximately eight. Calcium, magnesium and manganese, which were essential elements for plants, were eluted from the slag irrespective to elution condition. The eluted concentrations of Ca and Mg increased with an increase in weight of the additive slag. Silicon and zinc were also eluted depending on the conditions. Aluminum that was harmful for plants was not eluted from the used slag.

1. Introduction

A large quantity of slag has been formed in steelmaking process yearly. Because final disposal yard is tight in Japan, re-use of the slag in costal water where the environmental quality standards are not strict have been studied [1-10]. On the other hand, steelmaking slag contains the substances that are essential for plants. The slag is expected to be a supply source of minerals to soil [11-14]. A certain slag has been already used as a fertilizer. In these studies, field-tests of the slag were mainly performed. However, only a few researchers studied fundamentally the elution from the slag [15-17]. In our previous works [18, 19], we studied the elution behaviour of the slag that was discharged from oxidizing process of normal steel in an electric arc furnace (EAF). It was found that the elution concentration of the used slag was less than the environmental quality standards (EQSs). The elution behaviour of calcium and magnesium were expressed by the parabolic law. The elution rate increased with an decrease in the initial pH in the aqueous solution. The slag has buffering action to pH of water and is anticipated to be neutralization effect on acid rain. According to the ministry of the environment in Japan, the average hydrogen ion exponent (pH) of rainfall in Japan is 4.68. Rain of which pH was less than four fell sometimes and somewhere in Japan. The pH of rainfall at Lake Ijira is 4.51, and it is the lowest in Japan [20]. The damage to ecosystem with the acid rain has not been reported until now, but the monitoring has been continued. Addition of the slag to soil may ease the acidification of soil and effusion of minerals from soil due to the acidification. Then, in this work, as a series of fundamental study of elution of the slag into water, we studied the elution behaviour of the small quantity of slag in water.

2. Experiment

2.1. *Slag*

The used slag in this experiment was the oxidizing slag which was discharged from stainless steel making in an electronic arc furnace (EAF). The slag was ground, and then sieved. The slag of which size was 1-2 mm was used for the experiment. Table 1 shows the chemical compositions of the slag. The chemical compositions of the oxides were converted from the composition of the metallic components which were determined by an inductively-coupled plasma atomic emission spectrometry (ICP-AES) after alkali fusion of the slag. As shown in figure 1, the X-ray diffractometer (XRD) revealed that the slag was mainly composed of $2CaO{\cdot}MgO{\cdot}SiO_2$, $2CaO{\cdot}Al_2O_3{\cdot}SiO_2$ and $CaO{\cdot}SiO_2$.

Table 1. Chemical composition of used slag. [mass%]

CaO	SiO_2	Al_2O_3	MgO	CaS	Cr_2O_3	MnO	TiO_2	FeO	NiO	CuO	ZnO
33.0	44.1	5.39	7.68	0.33	3.29	4.09	1.02	0.74	0.06	0.24	0.01

2.2. *Procedure*

Because the experimental apparatus and procedure were described in detail in the previous paper [18], a brief account for them was given here. Elution experiment was carried out on a basis of JIS K 0058-1 (Test method for chemicals in slags- Part 1: Leaching test method). Figure 2 shows the schematic of the experimental apparatus. The parts of the apparatus which were contact with water were made of PTFE. The apparatus was set in a simple globe box that was made of vinyl sheet to prevent interfusion of suspended dust in the vessel. The slag of 0.1-100 g in weight set in the vessel. A pH of the purified water was pre-adjusted to be 6.0 by addition of super precision analysis grade HCl. The purified water of 1 L in volume was softly poured in the vessel. Subsequently, the water in the vessel was mixed at a rotating speed of 100 rpm. The slag was laid silently in the container during mixing. The aqueous solution was sucked by a whole pipette, and it was passed through the filter of which pore size was 0.45 μm. The filtrate was analyzed with an ICP-AES. The pH of the aqueous solution in the container was also measured by a pH meter.

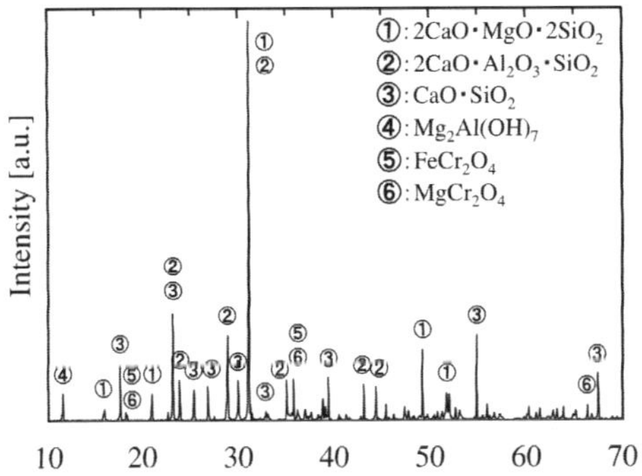

Figure 1. XRD pattern of used slag.

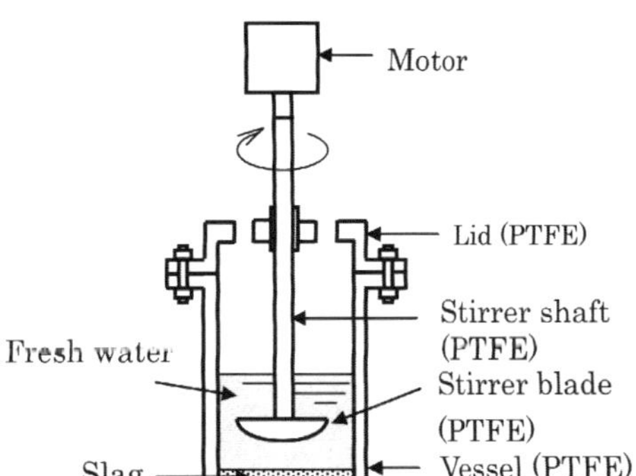

Figure 2. Schematic of experimental apparatus.

3. Results and discussions

3.1. *Change in pH with time*

Figure 3 shows the change of pH in the aqueous solution with time. In the case of the slag weight of 100 g, the pH increased steeply in the early stage of the elution, and, subsequently, it gradually decreased. In the case of the slag weight of 10 g and below, the pH also increased steeply in the early stage of the elution, and, subsequently, it gradually increased. Generally, the pH increased with an increase in the slag weight. The pHs seem to converge to approximately eight.

The behavior of pH may be determined by the following three reactions. One is adsorption and desorption of H^+ to and from surface hydroxyl of the slag and decomposition of the surface hydroxyl [21]. Another is the dissolution of components of the slag into water. Basically, these reactions give the buffering action the slag-water system. The other is formation of H^+ by following dissolution of CO_2 contained in the air.

$$CO_2 + H_2O = 2H^+ + CO_3^{2-} \quad (1)$$

$$CO_2 + H_2O = H^+ + HCO_3^- \quad (2)$$

The increase in the weight of the additive slag implies the increase in the interfacial area of slag-solution. Therefore, pH is influenced by the interfacial area. The pH is mainly determined by adsorption and desorption of H^+ to and from surface hydroxyl of the slag and decomposition of the surface hydroxyl. The threshold value of pH is approximately eight. When the pH in the solution is below eight, H^+ adsorbs on the slag surface. Oppositely, when the pH is above eight, H^+ desorbs into the aqueous solution. This gives the slag – water system buffering action. In north Europe and Canada, acidification of lake becomes problem. Addition of a small amount of the slag into acid lake may improve the pH in acid water in lake.

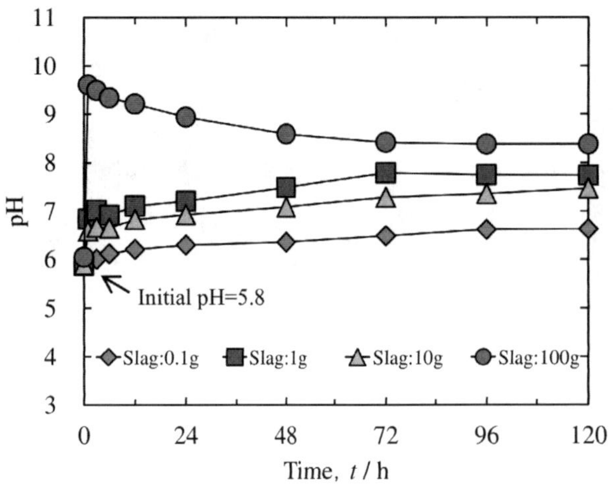

Figure 3. Change in pH with time.

3.2. *Elution behavior*

3.2.1. *Main substances*

Figure 4 and Figure 5 show the elution behavior of Ca and Mg from the slag into the aqueous solution respectively. Generally, the eluted concentrations of Ca and Mg increased steeply at an early stage of the elution. Subsequently they gradually increased with time. In addition, the eluted concentration of Ca and Mg increased with an increase in the weight of the additive slag. In comparison with the eluted concentration of Mg, the eluted concentration of Ca was higher than that of Mg. The increase in the weight of the additive slag implies the increase in the interfacial area between slag and aqueous solution. However, the eluted concentration does not increase tenfold even if the interfacial area becomes large tenfold. As mentioned above, pH in the aqueous solution became large with an increase in weight of the additive slag. In previous work [18,19], it was found that the elution rate decreased with an increase in the pH. Therefore, the eluted concentration did not increase proportionally with the interfacial area.

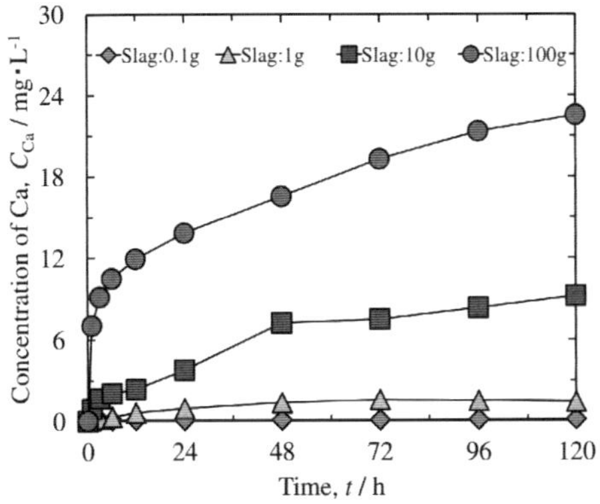

Figure 4. Change in eluted Ca with time.

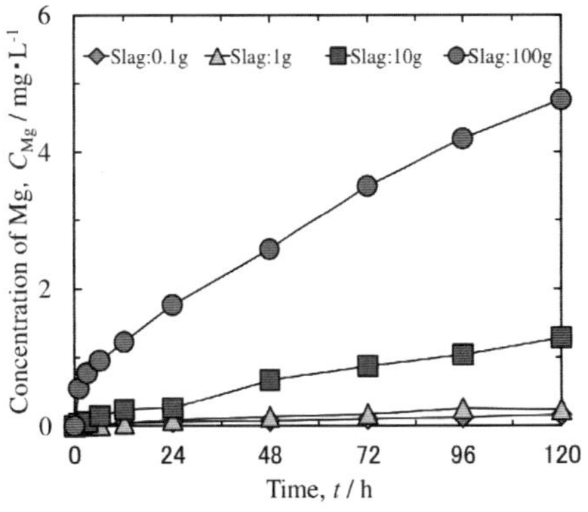

Figure 5. Change in eluted Mg with time.

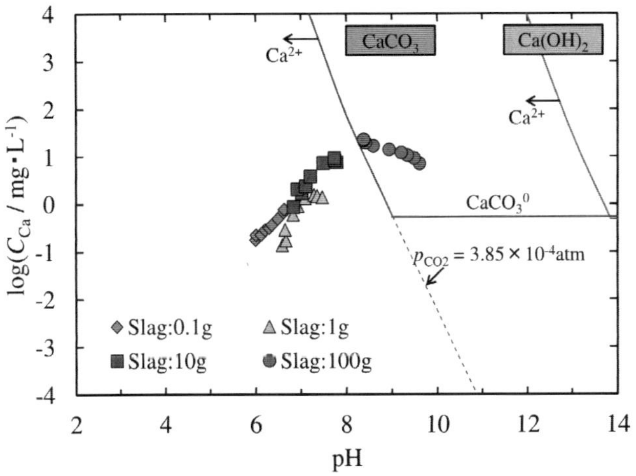

Figure 6. Solubility of $Ca(OH)_2$ and $CaCO_3$.

Figure 6 shows the solubility of $Ca(OH)_2$ and $CaCO_3$. Here, the partial pressure of CO_2 was 3.85×10^{-4} atm in the air. The each solubility was calculated with both Debye-Hückel theory and the combination of the following thermodynamic data [22].

$$CaCO_3+2H^+=Ca^{2+}+CO_2(g)+H_2O \qquad \log K=9.74 \qquad (3)$$

$$Ca(OH)_2+2H^+=Ca^{2+}+2H_2O \qquad \log K=22.8 \qquad (4)$$

$$CaCO_3=CaCO_3^0 \qquad \log K=-5.27 \qquad (5)$$

where K is the equilibrium constant. In addition, the relation between the eluted concentration of Ca and pH, which were obtained experimentally, was also shown in figure 6. In the case of weight of the additive slag of 10 g and below, the eluted concentrations of Ca were lower than the solubility of $CaCO_3$. In the case of weight of the additive slag of 100 g, the eluted concentrations of Ca ranged in the solubility of $CaCO_3$. If Ca in the slag existed as $CaCO_3$, Ca cannot be eluted in aqueous solution. In addition, the XRD revealed that no $CaCO_3$ formed after experiment. Although the eluted Ca concentration existed in the region of formation of $CaCO_3$, it could be said that the reverse reaction of eq. (3) did not proceed. The concentration of Mg was below the solubility of $Mg(OH)_2$ and $MgCO_3$.

Figure 7 shows the elution behavior of Si. In the case of weight of the slag of 100 g, the eluted concentrations of Si increased steeply at an early stage of the elution, and, subsequently, it gradually increased with time. In the case of weight of the slag of 10 g, the eluted concentrations of Si were lower than the detection limit of the ICP-AES at an early stage of the elution, and, then, it increased gradually. It may be said that the elution of Si occurred when the pH is smaller than approximately 7.5, but further study is necessary for understanding the elution behavior of Si. In the case of weight of the additive slag of 1 g and below, the eluted concentrations of Si were lower than the detection limit of the ICP-AES. Generally, the eluted concentration of Si increases with an increase in weight of the additive slag. Here, the eluted Si concentration was below the solubility of H_2SiO_3. Accordingly, it is thought that the elution of Si of the slag proceeds after the formation of H_2SiO_3 on the slag surface [18].

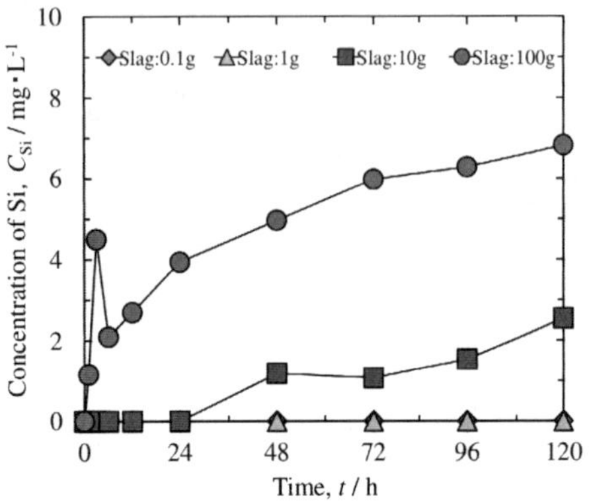

Figure 7. Change in eluted Si with time.

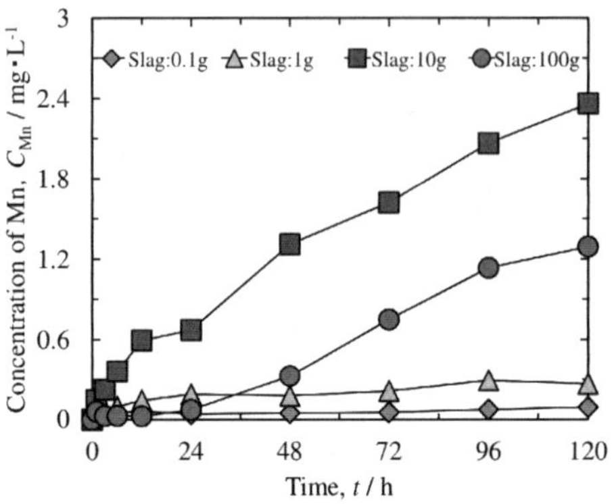

Figure 8. Change in eluted Mn with time.

Figure 8 shows the elution behavior of Mn. Generally, the eluted concentration of Mn increased steeply at an early stage of the elution. Subsequently they gradually increased with time. However, when 100 g of the slag was eluted, Mn was not almost eluted at an early stage of the elution. In addition, the concentration of Mn increased with an increase in the weight of the additive slag except for the result of 100 g of the additive slag.

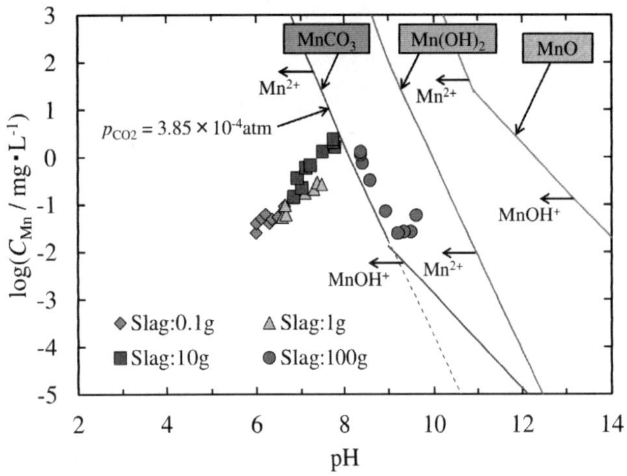

Figure 9. Solubility of MnO, $Mn(OH)_2$ and $MnCO_3$.

Figure 9 shows the solubility of MnO, $Mn(OH)_2$ and $MnCO_3$. Here, as mentioned above, the each solubility was calculated with the combination of the following thermodynamic data [22].

$$MnO+2H^+=Mn^{2+}+H_2O \qquad \log K=18.39 \qquad (6)$$
$$Mn(OH)_2+2H^+=Mn^{2+}+2H_2O \qquad \log K=15.19 \qquad (7)$$
$$MnCO_3+2H^+=Mn^{2+}+CO_2+H_2O \qquad \log K=8.08 \qquad (8)$$
$$Mn^{2+}+H_2O=MnOH^++H^+ \qquad \log K=-10.95 \qquad (9)$$

In addition, the relation between the eluted concentration of Mn and pH, which were obtained experimentally, was also shown in figure 9. In the case of weight of the additive slag of 10 g and below, the eluted concentrations of Mn were lower than the solubility of $MnCO_3$. In the case of weight of the additive slag of 100 g, the eluted concentrations of Mn ranged in the solubility of $MnCO_3$ at an early stage of the elution, and, subsequently, the concentration increased along the solubility of $MnCO_3$. Although the XRD revealed that there was no $MnCO_3$in the slag, $MnCO_3$ seemed to be at the surface of the slag. The elution of Mn proceeded when the pH in aqueous solution became lower that the solubility of $MnCO_3$.

Here, the concentrations of Al and Fe were not detected with the ICP-AES. It is favorable for plant that Al is not eluted from the slag, because Al is harmful for plants. The elution of Fe was preferable because Fe was essential element for plants. However, the elution of Fe did not occur.

Table 2 shows extraction ratio from the slag to the aqueous solution. Here, the extraction ratio was the ratio in weight of a substance contained in the additive slag to a substance extracted in aqueous solution. In addition, the eluted concentrations at the end of the elution experiment were used for calculation. The extraction ratios for Ca, Mg and Mn decreased with an increase in weight of the additive slag, whereas the extraction ratio of Si increased with an increase in weight of the additive slag. These results can be explained qualitatively as follows. The increase in the additive weight of the slag means the increase in the interfacial area. The elution rate increased proportionally with an increase in the weight of additive slag. On the other hand, the elution was also influenced by pH. The smaller the pH, the larger the elution rate. Accordingly, in the case of the elution of Ca, Mg and Mn, the elution of them is considered to be influenced by pH than interfacial area. Oppositely, the elution of Si is considered to be strongly influenced by interfacial area. The extraction ratio for Ca was almost the same as the ratio for Mg. This result implied that elution of Ca and Mg occurred depending on the amounts of them contained in the slag.

Table 2. Extraction ratio from slag to aqueous solution. [%]

	Weight of additive slag [g]			
	0.1	1	10	100
Ca	1.8	0.32	0.21	0.053
Si	0.0	0.0	0.07	0.018
Mg	1.9	0.28	0.15	0.057
Mn	1.6	0.46	0.41	0.022

3.2.2. *Minor substances*

Figure 10 shows the elution behavior of Cr. The eluted concentration of Cr increased at an early stage of the elution. Subsequently, the eluted concentration of Cr decreased and, then, was not detected with the ICP-AES. The environmental quality standard (EQS) for soil pollution in Japan regulates the eluted concentration of hexavalent Cr. The eluted concentration of Cr detected with the ICP-AES was lower than the solubility of $Cr(OH)_3$. Accordingly, the eluted Cr was estimated to be trivalent Cr. In addition, the environmental quality standard refers only the eluted concentration of Cr six hours after the elution test. The eluted concentration of Cr obtained experimentally was lower than EQS.

Figure 11 shows the elution behavior of Zn. In the case of weight of the additive slag above 10 g, after the eluted concentration of Zn increased at an early stage of the elution, it decreased. Subsequently, the eluted concentration of Zn increased and decreased again. In the case of weight of the additive slag of 1 g, after the eluted concentration of Zn increased at an early stage of the elution, it decreased. After that, the eluted concentration of Zn was not detected. In the case of weight of the additive slag of 0.1 g, the eluted concentration of Zn was not detected through the elution.

It is estimated that the decrease in the eluted concentrations of Cr and Zn are caused by coprecipitation and adsorption of them on the slag. However, further study of this behavior of Cr and Zn is needed.

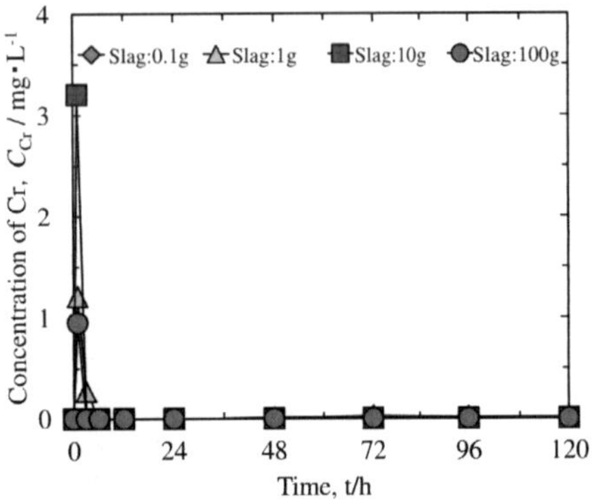

Figure 10. Change in eluted Cr with time.

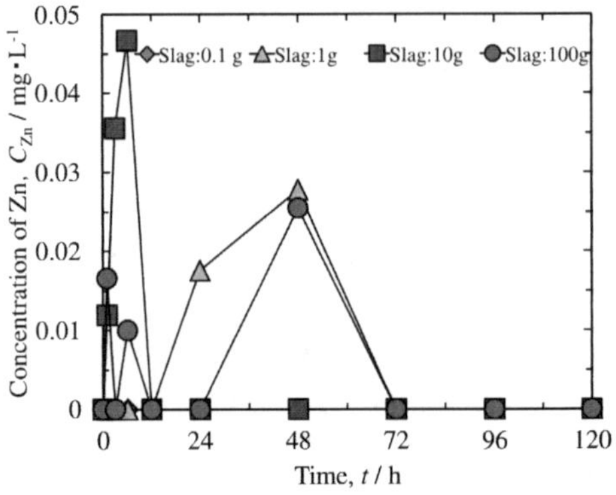

Figure 11. Change in eluted Zn with time.

Table 3. Eluted concentrations of elements essential for plant. [mg/L]

	Weight of additive amount of slag [g]											
	Elution for 6h				Elution for 24h				Elution for 120h			
	0.1g	1g	10g	100g	0.1g	1g	10g	100g	0.1g	1g	10g	100g
Total N	-	-	-	-	-	-	-	-	-	-	-	-
Total P	-	-	-	-	-	-	-	-	-	-	-	-
Total K	-	-	-	-	-	-	-	-	-	-	-	-
Total Ca	0.226	0.290	2.028	10.5	0.316	0.908	3.76	13.8	0.773	1.39	9.21	22.5
Total Mg	0.0317	0.0154	0.144	0.958	0.0570	0.0869	0.261	1.77	0.156	0.234	1.28	4.76
Total S	-	-	-	-	-	-	-	-	-	-	-	-
Total Mn	0.0502	0.0963	0.362	0.0264	0.0403	0.189	0.671	0.0724	0.0918	0.266	2.36	1.29
Total B	-	-	-	-	-	-	-	-	-	-	-	-
Total Fe	0	0	0	0	0	0	0	0	0	0	0	0
Total Cu	-	-	-	-	-	-	-	-	-	-	-	-
Total Zn	0	0	0.0467	0.0100	0	0.0176	0	0	0	0	0	0
Total Cl	-	-	-	-	-	-	-	-	-	-	-	-
Total Mo	-	-	-	-	-	-	-	-	-	-	-	-
Total Si	0	0	0	2.09	0	0	0	3.94	0	0	2.55	6.82

3.3. *Essential elements for plant*

Table 3 shows the essential elements for plants and the eluted concentrations. Here, silicon is not an essential element usually, but is an essential element for gramineous plants. This slag can supply water with Ca, Mg and Mn irrespective to the elution conditions. Depending on elution condition, this slag can supply Zn and Si, too. This slag contains no P, K and so on. However, a certain steelmaking slag contains a lot of them. These slag may supply soil with various minerals. This slag satisfies the environmental quality standard for soil in Japan, but, as mentioned above, the eluted concentration of Cr was high at an early stage of the elution. Therefore, adding of this slag into soil directly seems to have the risk of pollution of soil. However, the eluate from the slag can be used for soil because of monitoring the substances and the concentration of them. This eluate can be used as a fertilizer.

4. Conclusions

We studied the influence of weight of the additive slag on elution behavior of the slag into water. The obtained main results can be summarized as follows:

(1) The pH in the aqueous solution increased with an increase in weight of the additive slag.

Journal of Physics: Conference Series **352** (2012) 012051 doi:10.1088/1742-6596/352/1/012051

(2) Calcium, magnesium and manganese, which were essential elements for plants, were eluted from the slag irrespective to elution condition. Silicon and zinc were eluted depending on the conditions. Irrespective to elution condition, iron was not detected in the aqueous solution, and aluminum, which was harmful for plants, was not detected.
(3) The extraction ratio of calcium was almost the same as the ratio of Mg. The elution of them proceeded depending on the amounts of them in the slag.

References

[1] Nakagawa M 2011 *J. JSCE* **96** 29
[2] Miki S, Kato T, Tsutsumi N *J. Jap. Soc. Water Environ.* **32** 33
[3] Matsunaga H, Tanishige T, Fujii T 2009 *Anual J. Costal Eng. JSCE* **56** 1196
[4] Minato T, Yamazaki S, Sato Y 2009 *J. Jap. Asoc. Costal Zone Studies* **22** 37
[5] Honda H, Suzuki M 2008 *JFE Tech. Rep.* **19** 23
[6] Miyata Y, Sato Y, Shimizu S 2008 *JFE Tech. Rep.* **19** 1
[7] Yamamoto T 2003 *Tetsu-to-Hagane* **89** 494
[8] Ito K, Nishijima W, Syoudo E, Okada M 1997 *J. Jap. Soc. Water Env.* **20** 670
[9] Ito K, Nishijima W, Syoudo E, Okada M 1996 *J. Jap. Soc. Water Env.* **19** 501
[10] Sugie K, Taniguchi A 2008 *Tetsu-to-Hagane* **94** 569
[11] Murakami K, Goto I 2010 *KAGAKU to SEIBUTSU* **48** 608
[12] Kondo M, Kajisa T, Ishiguro S 2006 Trans. J*ap. Soc. Irrigation, Drainage and Rural Eng.* 475
[13] Shinoda E, Murakami K, Goto I 2005 *Nihon Dojou Hiryougaku zasshi* **76** 891
[14] Murakami K, Shinoda E, Nakamura A, Goto I 2005 *Nihon Dojou Hiryougaku zasshi* **76** 59
[15] Miki T, Futatsuka T, Shitogiden K, Nagasaka T, Hino M 2004 *ISIJ Int.* **44** 935
[16] Inoue R, Suito H 2002 *Tetsu-to-Hagane*, **88** 347
[17] Suito H, Inoue R 2002 *ISIJ Int.***42** 921
[18] Yokoyama S, Suzuki A, Izaki M and Umemoto M 2009 *Tetsu-to-Hagane* **95** 434
[19] Yokoyama S, Suzuki A, Nor N H B, Kanematsu H, Ogawa A, Takahashi T, Izaki and Umemoto M 2010 *ISIJ International* **50** 630
[20] Press release from Ministry of Environment, Mar. 27, 2009, http:// www.env.go.jp/press/ press.php?serial=10971
[21] Sato T, Fukushi K and Yoneda T 2007 *J. MMIJ* **123** 132
[22] Lindsay W L 1979 *Chemical equilibria in soil* (Caldwell, Blackburn) p12, 150

Ammonia-oxidizing activity and microbial community structure in acid tea (*Camellia sinensis*) orchard soil

K Okamura[1*], A Takanashi[2], T Yamada[1,2] and A Hiraishi[1,2]

[1] Research Center for Agrotechnology and Biotechnology and

[2] Department of Environmental and Life Sciences, Toyohashi University of Technology, Toyohashi 441-8580, Japan

E-mail: okamura@recab.tut.ac.jp

Abstract. The purpose of this study was to determine the ammonia-oxidizing activity and the phylogentic composition of microorganisms involved in acid tea (*Camellia sinensis*) orchard soil. All soil samples were collected from three sites located in Tahara and Toyohashi, Aichi Prefecture, Japan. The potential nitrification rate (PNR) was measured by the chlorate inhibition method. The soil pH of tea orchards studied ranged from 2.78 to 4.84, differing significantly from sample to sample, whereas that of meadow and unplanted fields ranged from 5.78 to 6.35. The PNR ranged from 0.050 to 0.193 μg NO_2^--Ng^{-1} h^{-1} and were positively correlated with the soil pH ($r^2 = 0.382$, $p<0.001$). Bulk DNA was extracted from a tea orchard soil (pH 4.8; PNR, 0.078 μg NO_2^--Ng^{-1} h^{-1}) and subjected to PCR-aided clone library analyses targeting archaeal and bacterial *amoA* genes. The detected archaeal clones separated from the cluster of the 'Soil clones' and tightly clustered with the clones originating from other acidic soil environments including the Chinese tea orchard soil. These results suggest that the specific archaeal populations dominate as the ammonia oxidizers in acid tea-orchard soils and possibly other acid soils, independent of geographic locations, which results from the adaptation to specific ecological niches.

1. Introduction

Nitrification, the oxidation of ammonia (NH_3) to nitrate (NO_3^-), is a critical step in the nitrogen cycle in nature and has agricultural and environmental consequences for the bioavailability of nitrogen as a plant nutrient, nitrate leaching to groundwater, and the release of greenhouse gases (NO, N_2O) into the atmosphere. The rate-limiting step of nitrification is the conversion of ammonia to nitrite (NO_2^-), which is performed by both ammonia-oxidizing archaea (AOA) within the phylum *Thaumarchaeota* [1], and ammonia-oxidizing bacteria (AOB). Both groups of prokaryotes have been detected in a wide range of soil ecosystems [2, 3].

Journal of Physics: Conference Series **352** (2012) 012052 doi:10.1088/1742-6596/352/1/012052

The potential for archaeal ammonia oxidation has been confirmed in culture enrichments and isolates [4, 5]. In most soils, archaeal *amoA* genes, a gene coding ammonia monooxygenase subunit A, are more abundant than those of bacteria, indicating that AOA have a greater role in soil ammonia oxidation than AOB [3, 6]. In soil ecosystems, nitrifying populations in terms of quantity, quality, and activity vary with vegetation type, location, and environmental conditions [7]. The pH of soil is one of the important factors on the nitrification activity and phylogenetic diversity of ammonia oxidizers [8, 9]. Especially at acidic environments, the absence of nitrification activity of AOB was reported in some highly acid soils [8, 10].

Tea (*Camellia sinensis*) orchard soils provide an excellent environment to examine the relative contribution of AOA and AOB in nitrification processes at lower pH. Tea is an important economic crop and is planted widely in acid red soils of the humid subtropical zones in Japan. High levels of nitrogen fertilizer are applied to tea orchards, altering N cycling and microbial community structure [11]. Tea orchard soils are often acidic in the rage of pH from 3.5 to 5.5, but known to keep high nitrification activity despite the potential high sensitivity of AOB to low pH [12, 13]. It was suggested that AOA populations are responsible for the major part of nitrification in acidic soils, that soil pH and N fertilizer input influences both the community structure of ammonia oxidizers and nitrification potential, and that different AOA and AOB phylotypes occupy distinct pH ranges (i.e., niche differentiation) [14]. The types of soil and vegetation have potentially additional effects on microbial community composition [15], and it is therefore pertinent to compare tea soils with other vegetated systems on the same soil type.

This study determined the activity and phylogenetic compositions of ammonia oxidizers in tea orchard soils from two different sites. The nitrification activities of soils were also analyzed in adjacent meadow and unplanted field with different pH to evaluate the effect of environmental factors (i.e. pH, vegetation type, and soil type) on the nitrification activity of soils.

2. Materials and methods

Study sites and soil sampling. Three soil samples were collected from two sites located in Tahara (Fig. 1A, 1) and one site from Toyohashi (Fig. 1A, 2), Aichi Prefecture, Japan. The selected tea soils from the Tahara area, where the N application rate is high, represent a wide range of orchard ages, fertilizers, and lime applications. In the Toyohashi area, soil samples were taken from a low-N-applied tea orchard. All soil samples were taken from the furrows of the tea plants (Figs. 1B and 1C), because fertilization was mainly performed in these regions. In addition, adjacent meadow and unplanted field soils were sampled to evaluate the effect of vegetation on ammonia- oxidizing communities. These samples were

Fig. 1. Location of the sampling sites (A) and photos of the tea orchard, red circle indicates the position of furrow in tea orchard (B) and an enlarged photo of the point of red circle in Fig. 1B. Red arrow indicate the sampling space (C)

collected from three sampling plots randomly chosen within each tea orchard or reference soils. Six random soil cores (5-cm diameter by 15-cm length) were taken from each sample and mixed. All soil samples were sieved through 2 mm sieving and then used in the following experiments.

Soil chemical analysis. Soil pH was measured using a glass electrode (soil/water, 1 : 2.5, w/v). The ammonia oxidation potential was determined from the decrease in the concentration of nitrite (NO_2^-) after incubation of soils amended with ammonium (NH_4^+) and chlorate as inhibitor of nitrite oxidation to nitrate (NO_3^-) according to Kurola *et al.* [16] with slight modifications. Briefly, 20 g of fresh soil was added to 300-mL conical flask containing 80 mL of citrate-phosphate buffer (pH 3.5) with 1 mM $(NH_4)_2SO_4$. Potassium chlorate with a final concentration of 10 mM was added to inhibit the nitrite oxidation. The suspension was incubated in darkness at 25 °C for 24 h and then treated with 5 mL of 2 M KCl to extract nitrite, followed by colorimetric detection at 540 nm with N-(1-naphthyl) ethylenediamine dihydrochloride. The potential nitrification rate (PNR) was calculated from the linear increase in concentrations of NO_2^-.

Construction of clone libraries and sequence analysis. Bulk DNA from a tea orchard soil (pH 4.8; PNR, 0.078 μg NO_2^--Ng^{-1} h^{-1}) which was pretreated with $CaCO_3$ [17] was extracted using a PowerSoil DNA Isolation kit according to the manufacturer's instructions. Archaeal and bacterial *amoA* genes were amplified by PCR with the soil DNA as the template and pair primer sets of Arch-*amoA*F (5'-STAATGGTCTGGCTTAGACG-3') /Arch-*amoA*R (5'-GCGGCCATCCATCTGTATGT-3') and *amoA*-1F (5'-GGGGTTTCTACTGGTGGT-3')/*amoA*-2R (5'-CCCCTCKGSAAAGCCTTCTTC-3'), respectively. PCR products were purified with the Wizard Gel and PCR Clean-up System and subcloned using a pT7Blue Perfectly Blunt™ Cloning kit according to the manufacturer's instructions. More than 30 positive clones from each sample were sequenced using a BigDye Terminator v3.1 cycle sequencing kit and an Applied Biosystems 3130xl genetic analyzer. Multiple alignment of sequence data, calculation of the corrected evolutionary distance, and construction of a neighbor-joining (NJ) phylogenetic tree were performed using the ARB program package. The TREEFINDER program package was used for tree construction.

3. Results and discussion

The pH of the tea orchard soils ranged from 2.78 to 4.84, differing significantly from sample to sample, whereas those of meadow and unplanted fields ranged from 5.78 to 6.35 (Fig. 2). The PNR ranged from 0.050 to 0.193 μg NO_2^--N g^{-1} h^{-1} and was positively correlated with soil pH (r^2 = 0.382, p<0.001). Within the tea orchard soil, the PNR value ranged from 0.027 to 0.125 and had no correlation with soil pH (Fig. 2A). Three points of the tea orchard with different PNRs had the lowest pH, which might be due to a composite factors such as high N fertilizer application, the resultant acidification by nitrification, plant growth, and phytochemical

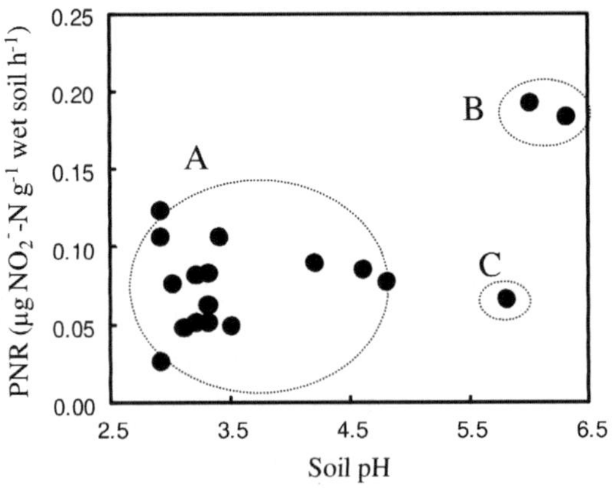

Fig. 2. The correlation between soil pH and PNR (potential nitrification rate). A, tea orchard soil; B, unplanted field soil; C, meadow soil.

inputs (ex. organic acids). These results suggest that PNR was affected not only by soil pH but also by other factors. Yao *et al.* [15] indicated that PNR was positively correlated with total nitrogen (r^2 = 0.71), organic carbon (r^2 = 0.67), and N fertilizer application (r^2 = 0.27) and negatively correlated with soil pH (r^2 = 0.28).

Phylogenetic analyses were performed on *amoA* gene sequences of 22 bacterial and 13 archaeal clones from one site (pH 4.8; PNR, 0.078 μg NO_2^--Ng^{-1} h^{-1}) (Fig. 3). Out of the 22 bacterial *amoA* clones (Fig. 3A), only 3 clones were related to *Nitrosomonas* sequences, and the remaining 19 sequences clustered with the *Nitrosospira* group. Archaeal *amoA* gene sequences fell into two major groups (Fig. 3B). One of the groups comprised 11 sequences that clustered with the 'Acidic soil group' sequences. The other clustered with the sequences from sea, sediment and coral environments, which fall within the 'marine-associated group'. Interestingly, the detected AOA clones separated from the cluster of the 'Soil clones' and tightly clustered with the clones originating from other acidic soil environments including the Chinese tea orchard soil. These results suggest that the specific AOA populations dominate as ammonia oxidizers in acid soils, particularly acid tea orchard soils, as a result of the adaptation to specific ecological niches. These results suggest the adaptation of ammonia-oxidizing AOAs to particular tea orchard soil over geographic locations.

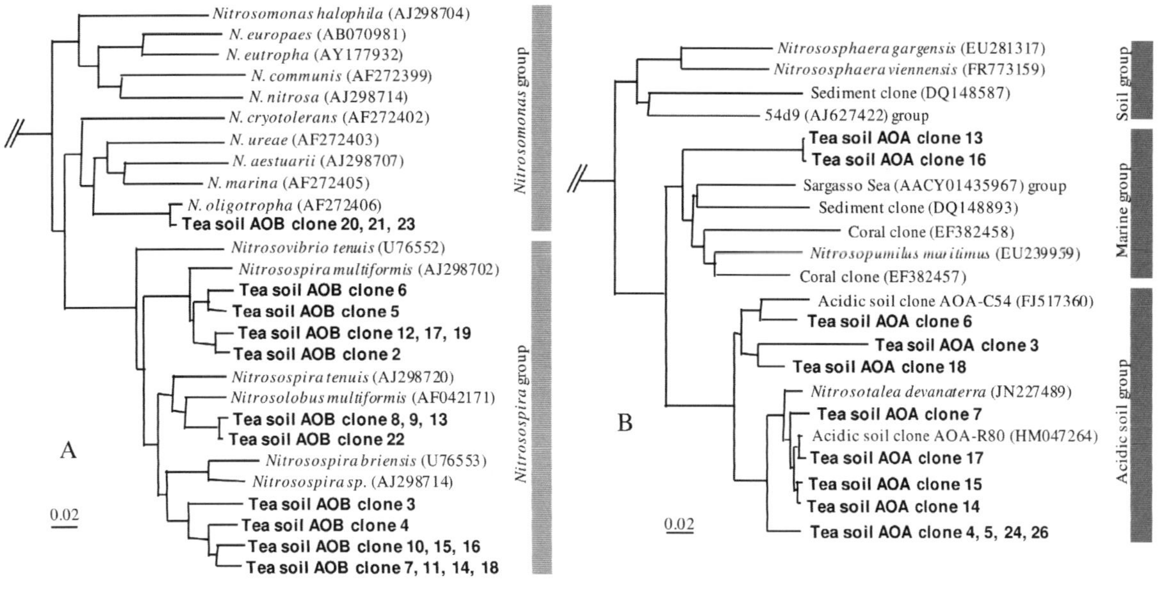

Fig. 3. Phylogenetic trees based on AOB *amoA* gene sequences (A) and AOA *amoA* gene sequences (B) showing relationships of the isolates and environmental clones. The trees were inferred by the NJ method. The *amoA* gene sequences of *Nitrosococcus oceanus* (U96611) and *Nitrosocaldus yellowstonii* (EU239961) were used as an outgroup to root the trees A and B, respectively. The accession numbers for the sequences are given in parentheses behind the species and clone names. The sequences determined in this study are shown by bold letters. Bar = 2% sequence divergence (K_{nuc}).

Acknowledgements

This work was supported by a Grant-in-Aid for Young Scientists ((B) 23710091), the Japanese Ministry of Education, Culture, Sports, Science and Technology.

References

[1] Brochier-Armanet C *et al.* 2008 *Nat. Rev. Microbiol.* **6** 245–252
[2] Di H J *et al.* 2009 *Nat. Geosci.* **2** 621–624
[3] Leininger S *et al.* 2006 *Nature* **442** 806–809
[4] Hatzenpichler R *et al.* 2008 *Proc. Natl. Acad. Sci. USA* **105** 2134–2139
[5] Könneke M *et al.* 2005 *Nature* **437** 543–546
[6] Prosser J I and Nicol G W 2008 *Environ. Microbiol.* **10** 2931–2941
[7] Webster G *et al.* 2005 *Environ. Microbiol.* **7** 676–684
[8] de Boer W and Kowalchuk G A 2001 *Soil Biol. Biochem.* **33** 853–866
[9] Nicol G W *et al.* 2008 *Environ. Microbiol.* **10** 2966–2978
[10] de Boer W *et al.* 1996 *Soil Biol. Biochem.* **28** 203–211
[11] Yao H *et al.* 2000 *Microb. Ecol.* **40** 223–237
[12] Xue D *et al.* 2009 *J. Environ. Sci.* **21** 1225–1229
[13] Xue D, Yao H Y and Huang C Y 2006 *Plant Soil* **288** 319–331
[14] Yao H *et al.* 2011 *Appl. Environ. Microbiol.* **77** 4618–4625
[15] Jones C M and Hallin S 2010 *ISME J.* **4** 633–641
[16] Kurola J et al. 2005 *FEMS Microbiol. Lett.* **250** 33–38
[17] Sagova-Mareckova M *et al.* 2008 *Appl. Environ. Microbiol.* **74** 2902–2907

Journal of Physics: Conference Series **352** (2012) 012053 doi:10.1088/1742-6596/352/1/012053

Extraction and refining of essential oil from Australian tea tree, Melaleuca alterfornia, and the antimicrobial activity in cosmetic products

Q Huynh[1, 2], T D Phan[1], V Q Q Thieu[1, 2], S T Tran[2] and S H Do[2]

[1]Faculty of Chemical Engineering, Ho Chi Minh City University of Technology (HCMUT), 268 Ly Thuong Kiet Street, Ho Chi Minh City, Vietnam

[2]Refinery and Petrochemicals Technology Research Center (RPTC), Ho Chi Minh City University of Technology, 268 Ly Thuong Kiet Street, Ho Chi Minh City, Vietnam

Email: ttsuong87@yahoo.com

Abstract. Tea tree oil (TTO) comes from the leaves of *Melaleuca alterniformia* that belongs to the myrtle family (Myrtaceae). It is one of the most powerful immune system stimulants and sorts out most viral, bacterial and fungal infections in a snap, while it is great to heal wounds and acnes. In Vietnam, Melaleuca trees can grow on acid land that stretches in a large portion of lands in the Mekong Delta region. So, there are some Melaleuca plantations developed under the Vietnamese government plans of increasing plantation forests now. However, TTO contains various amounts of 1,8-cineole that causes skin irritant. So TTO purification is very necessary. In this study, the purification of TTO that meet International Standard ISO 4730 was carried out via two steps. The first step is steam distillation to obtain crude TTO (terpinen-4-ol 35% v/v) and the average productivity is among *2.37% (v/wet-wt)* or *1.23% (v/dry-wt)*. In the second step, the cleaned TTO is collected by vacuum distillation column and extraction yield of the whole process is about 0.3% (w/w). Besides, high concentration essential oil was applied in the cosmetic products to increase its commercial value.

1. Introduction

Melaleuca alternifornia, commonly known as Australian tea tree and Melaleuca tree, grow fast and can be harvested as early as six months. They have been used medicinally for centuries by Australian aboriginal people. Tea tree oil (TTO) that comes from their leaves, contains over 100 components (mostly monoterpenes, sesquiterpenes and terpene alcohols). [4], [5] The studies demonstrated that terpinen-4-ol, a monoterpene is the most abundant (minimum 30%) and responsible for most of the antimicrobial activity, besides TTO also contains various amounts of 1,8-cineole that causes skin irritant. In an agriculture-based country like Vietnam, the acid land is actually not used for cultivation except for Melaleuca trees. So develop techniques to extract and refine essential oil from *Melaleuca alterniformia* leaves and research its cosmetic applications will improve the living standard of families in the area, especially for the farmers living on acid land.

Most essential oils available today are extracted by steam distillation among, although there are various extraction methods such as distillation, CO_2 supercritical extraction, and solvent extraction. It's the oldest form of essential oil extraction, quite simple and the best method for distilling leafy materials. Moreover, this process not only causes minimum changes to the essential oil composition during extraction, but also the steam is readily available, cheap, not hazardous and can be recycled. So steam distillation is used to collect crude TTO (35 – 45% of volume of Terpinen-4-ol) [1].

After steam distillation, the commercial value of TTO is not high, so it should be refined to increase its commercial value and fit with tea tree oil standard. Some refinement methods are vacuum distillation, crystallization, chromatography column… In this experiment, we focus on vacuum distillation to decrease TTO boiling point, thus limiting degradation of heat-sensitive volatiles.

Tea Tree Oil has always been appreciated for its biological activity. So TTO is used as antiseptic, antifungal to treat cuts, burns, abrasions and acne in a range of cosmetic products such as antiseptics, deodorants, shampoo and soaps. In this study, we used TTO as preservatives, germicidal ingredient in two cosmetic products: hand cleaner and body lotion. [3]

2. Materials and methods

The technical process for extracting and refining TTO was illustrated in Figure 1

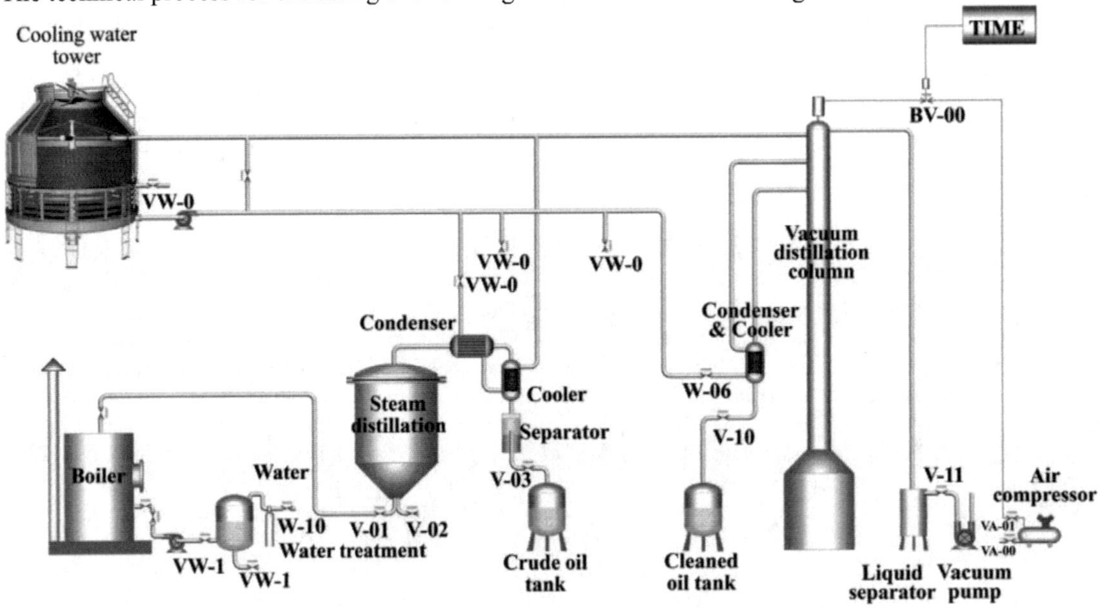

Figure 1: Technical process for TTO extraction and refinement

2.1. Essential oil extraction – steam distillation process [2]

Materials in this experiment is Australian melaleuca trees that were planted in Tan Phuoc District - Tien Giang Province - Vietnam.

TTO were produced by steam distillation from the leaves and terminal branches of Melaleuca tree. Dry materials were obtained after 24 hours of drying, then determined the moisture of these materials (representative sampling). We distilled both dry and wet materials to exam impact of moisture in raw materials to essential oil amount extracted. Steam for distillation is supplied from outside boiler of which capacity is 350 kg water per hour. The main device for this process is distillation vessel of which dimensions are 1110 mm diameter, 1689 mm height. The steam flowrate into this one is adjusted automatically by controller connected to the temperature sensor and the pressure sensor. The operation parameters of distillation vessel are 120 °C and 2 bar in 3 hours.

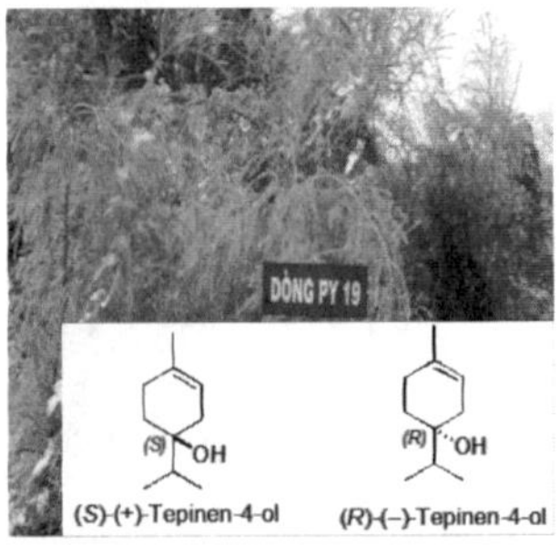

Figure 2: (A) Melaleuca Tree in Vietnam, (B) two isomers (S) and (R) of terpinen-4-ol

Mixture from condenser includes water and essential oil. Because of the difference of their specific gravity, they will become two liquid phases in the separator. The amount of essential oil that dissolved in the distillate water is small (negligible) and therefore recovery of essential oil dissolved in the water is unnecessary.

2.2. Essential oil refinement – vacuum distillation process [2]
Materials for vacuum distillation process are crude TTO (35 – 45% of volume of Terpinen-4-ol) from steam distillation.

The purpose of TTO purification is to create product that has low concentration of 1,8-cineole and high concentration of terpinen-4-ol. In this experiment, we carried out in conditions: 6000 ml of feed volume, 5 mmHg and a reflux ratio of around 1:1 (1 part returned condensate to 1 part condensate take off).

First, feed the crude essential oils into the reboiler of the vacuum distillation column. Then, the feeds are boiled by indirect heating resistor via thermal oil heater. A vacuum generator is connected to the top of distillation tower to create low-pressure in system. In front of the vacuum generator, we put the liquid separator, which is cooled by liquid nitrogen, in order to condense completely essential oil entrained by the vacuum line and to protect this engine.

The reflux ratio was set by adjusting On/Off time of solenoid valve to take TTO out. When the valve closes (timer Off), TTO vapour was condensed at the top of the tower and made a reflux stream back to the tower. Conversely, when the valve opens (timer On), TTO vapour was brought to condenser. Because On/Off cycle is very short (in a few seconds), it can be considered as continuous circulation process. Two thermometers are put on the bottom and top of distillation column to determine each interval exactly.

2.3. Cosmetic applications of TTO
Materials are some cosmetic ingredients from Mai Phuong Cosmetic Company – Vietnam, TTO with high concentration of terpinen-4-ol that is produced at RPTC

First, we created the formulations for two cosmetic products: body lotion and hand cleaner. [6]

Table 1: Formula of cream base

Materials	***Concentration % w/w***
Emulpharma 165	1
Beeswax	4
MC30 (Hydrogenated polyisobutene)	2
1,3 Butylene Glycol	4
Floraeste K-20W	0.5
Aquagel 45	3.5
Salacos 99	2
TC-Carbomer 380	0.4
Water	82.6

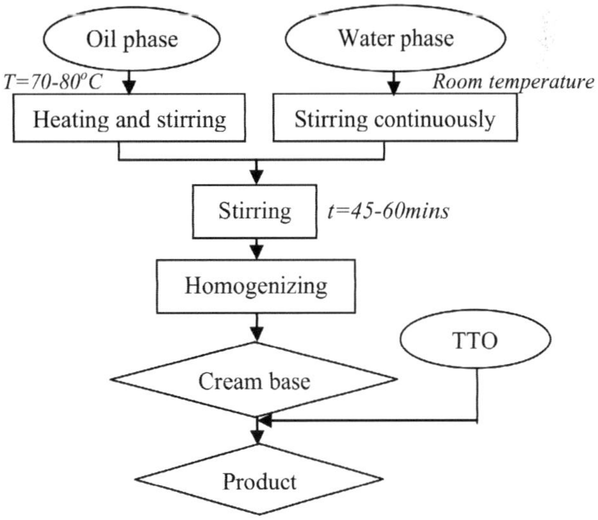

Figure 3: Procedure for body lotion

Table 2: Formula of hand cleaner

Materials	*Concentration % w/w*
Sodium lauryl sulfate (SLS)	20
Cocamido propyl betain (CAPB) 30%	1
Coconut Diethanolamine (CDE)	0.06
Beeswax	1
Ethanol	0.5
Glycerin	3
TC-Carbomer 380	1
Water	73.24

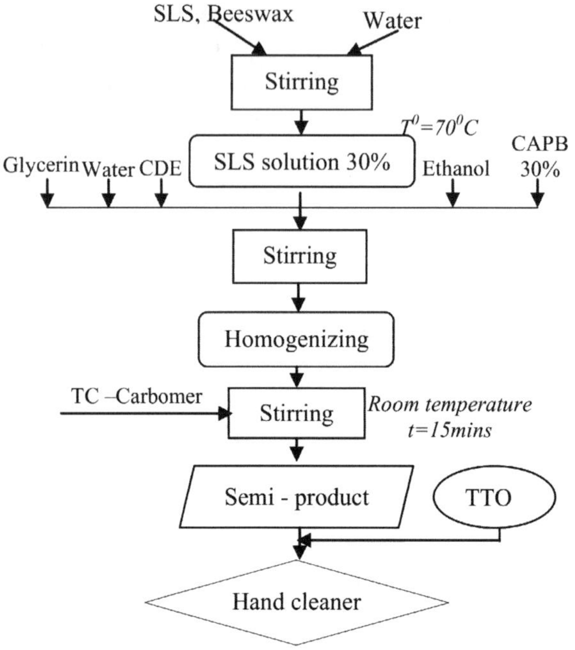

Figure 4: Procedure for hand cleaner

Next, these products were examined anti-microbial (for *E. coli, Staphylococcus aureus*) at the different ratios of the formulation and cleaned TTO. General principle was counting quantity of bacteria before and after using cosmetic products. From there, the suitable content was chosen. This survey was carried out at Bio-Technology Laboratory – Ho Chi Minh City University of Technology.

Besides, customer's evaluations were also collected (based on the satisfaction of 28 female customers and the product characteristics). Finally, in order to ensure product quality and safety for customer, these cosmetic products were inspected by 3113/1999/QĐ-BYT method (according to Vietnamese Standard) at Ho Chi Minh City Pasteur Institute – Vietnam

3. Results and Discussions:

3.1. TTO extraction and refinement

For essential oil extraction, we carried out 19 batches to determine average productivity of TTO. The results were shown in Figure 5.

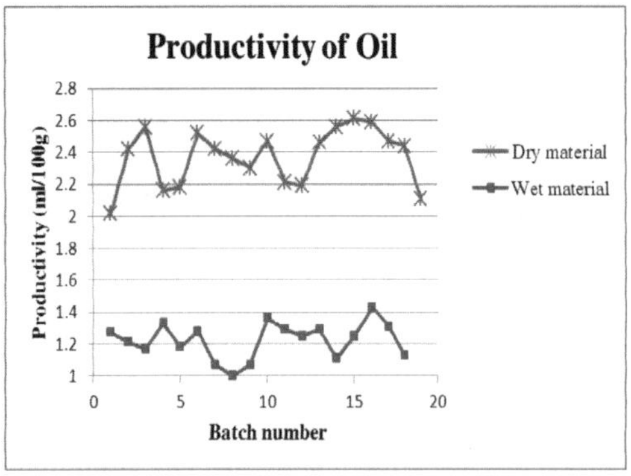

Figure 5: Crude TTO productivity for Dry materials and Wet materials

Table 3: Component of steam-extracted TTO before purification

Component	% Concentration (v/v)
α-pinen	1.96 - 2.72
Terpinen-4-ol	31.8 - 47.02
α-terpinen	8.65 - 9.04
Limonen	3.49 - 4.68
p-cymen	1.56 - 7.215
1,8-cineol	4.44 - 7.141
γ-terpinen	16.747- 19.20
Terpinolen	5.58-13.71
α-terpineol	2.79-3.32

Journal of Physics: Conference Series **352** (2012) 012053 doi:10.1088/1742-6596/352/1/012053

The yield of essential oil is typically 1 to 2% of wet material weight [1]. In this study, the average productivity is among *2.37% (v/wet-wt)* or *1.23% (v/dry-wt)*, when Melaleuca trees were planted in Tien Giang province – Vietnam. The results are also consistent with some studies in other provinces of Vietnam (Long An, Can Tho, An Giang…). The essential oil percentage decreased significantly after several days, so either the wet material or the dry one can be used for essential oil extraction. The GC used in this analysis was an Agilent 6890N with a flame ionization detector. Separation took place in DB-5 column ($0.32mm \times 30m \times 0.25\mu m$), using nitrogen as carrier gas. The GC oven temperature was initially controlled at 50^0C, then it was increased at a rate of 5^0C per min to 80^0C which was held for 1min, finally it was increased at a rate of 20^0C per min to a final temperature 220^0C which was held for 10 min. According to Table 3, two significant components of TTO after steam distillation meet International Standard ISO 4730 (terpinen-4-ol >30% and 1,8-cineol <15). However, we still continued refining crude TTO and compared the benefits between crude TTO production and cleaned TTO production in order to determine the TTO type that we will apply in industrial scale.

For essential oil refinement, the results were described in Figure 6. We can divide this process into 5 intervals. The main component of *Interval 1* which was from the beginning of distillation to 30 min and 0→15% oil volume, was α-terpinen (8-12% v/v). The main components of *Interval 2* which was from end of interval 1 to 210 min and 15→30-35% oil volume, were α-terpinen (20-30% v/v) and γ-terpinen (25-30% v/v). The main component of *Interval 3* which was from 210-330 min and about 30-35%→60-65% oil volume, was γ-terpinen (46-52% v/v). The main component of *Interval 4* which was 330-480 min and 60-65%→75-80% oil volume, was terpinen-4-ol (78-95% v/v). Finally, I*nterval 5* contains amount of TTO that was still in the reboiler of the vacuum distillation column, has 75-80%→95-100% oil volume. This interval has about 50-70% (v/v) of terpinen-4-ol and 10-15% (v/v) α-terpineol.

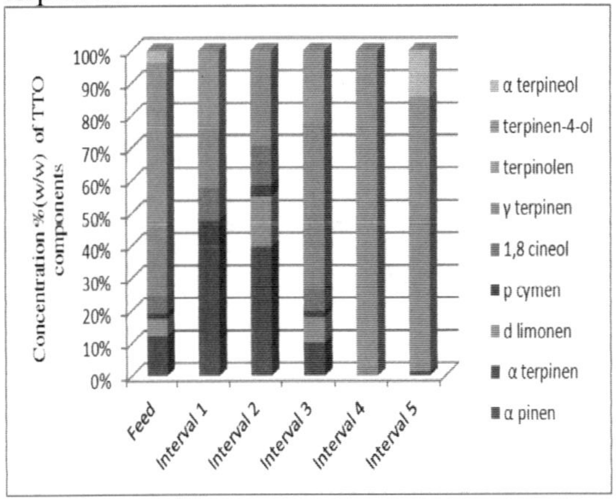

Figure 6: The essential oil components during refinement

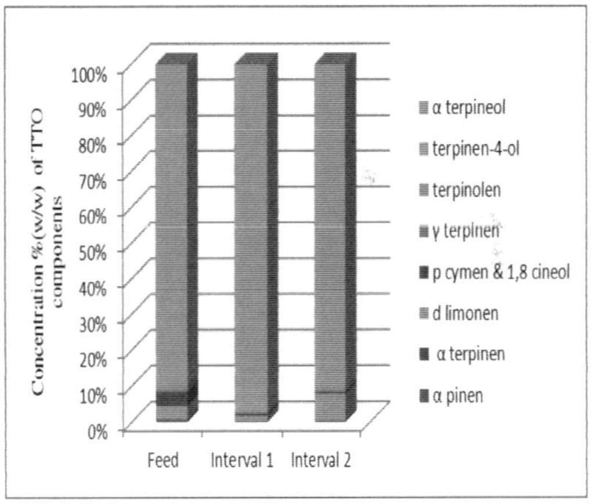

Figure 7: The essential oil components during refinement when mixing interval 5 of many batches

The percentage of terpinen-4-ol in different intervals from refinement process are different. The highest concentration of terpinen-4-ol is in interval 4 and lower in previous ones. The concentration of terpinen-4-ol decreases after interval 4, so we shutdown the operation here. Interval 5 that has significant amounts of terpinen-4-ol, will be mixed with interval 5 in many batches and purified again. According to Figure 7, we divided this process into 2 intervals. The main component of first interval that was from the beginning of distillation to 210 min, 0→45% oil volume, was 90-95% (w/w) of terpinen-4-ol. The main component of final interval was α-terpineol that has 35-40% (w/w). Extraction yield of whole process is about 0.3% (w/w) means we can obtain about 3kg of cleaned TTO (minimum 95% v/v) from 1 ton of Melaleuca leaves.

Table 4 showed that producing TTO can bring high economic efficiency. Besides, ROR of cleaned TTO product is about 1.5 times higher than ROR of crude TTO product, but cleaned TTO purification need more complex technology and more money than crude TTO. Depend on economic abilities and technical conditions, we will choose kind of TTO product.

Table 4: Benefits from TTO

Unit for price: Thousand VND

(1) **Products**	*(2)* **Price (VND/liter)**[a]	(3) **Yield (liter/day)**	(4) **Cost /liter**	(5) **Total costs (VND/day)**	(6) **Revenues (VND/day)**[b]	(7) **Income (VND/day)**[c]	(8) **Return on revenue (ROR) (%)**[d]
Crude TTO	1.000	4,47	620	2.770	4.470	1.700	38
Cleaned TTO	8.000	1,34	3.276	4.390	10.720	6.330	59

[a] The lowest price of TTO in Vietnamese market (2010)

[b] Revenues (VND/day) = (2) * (3) means amount of money received by selling TTO during a certain time period (a day)

[c] Income (VND/day) = (6) - (5) means revenues minus total cost, over a given period of time (a day)

[d] Return on revenues (ROR) % = (7)/(6)*100 . This is a means of measuring the profitability of a business by dividing the income by the amount of revenue

1 USD = 21014VND

3.2. Two cosmetic products

For body lotion, it has quite high anti-microbial ability even in small concentrations, Figure 8. Based on economic and anti-microbial ability, we choose it containing 0.5% (w/w) of cleaned TTO. However, anti-microbial requirements in body lotion product are lower than the cleaning products, especially the smell product is very important. TTO has bit unpleasant odor, so the TTO percentage in body lotion should be only about 0.3% (w/w).

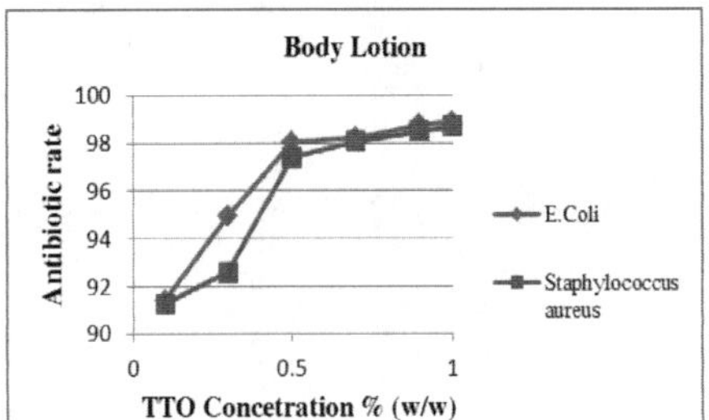

Figure 8: Anti-microbial ability of TTO in Body Lotion

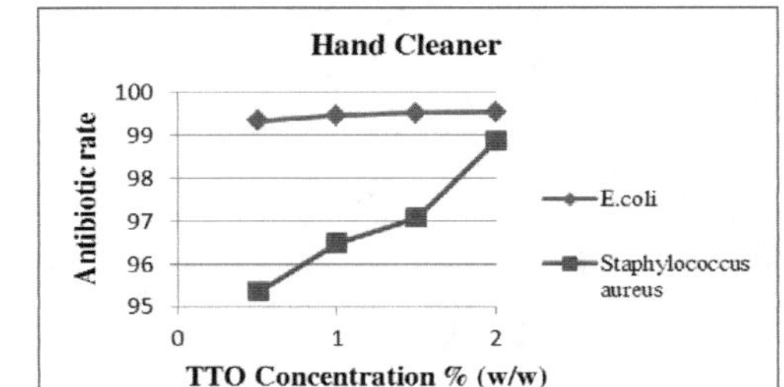

Figure 9: Anti-microbial ability of TTO in Hand Cleaner

Anti-microbial ability of hand cleaner product is as good as body lotion product, Figure 9. The ability to kill *Staphylococcus aureus* is lower than *E. coli*. But, it is quite high generally. Therefore, we chose 0.5% (w/w) of cleaned TTO in hand cleaner.

Besides, these cosmetic products are appreciated by 28 female customers. According to microbial test result of Ho Chi Minh City Pasteur Institute, they were not infected bacteria and satisfied microbial standards for cosmetic field.

Figure 10: Some cosmetic products from TTO produced in RPTC

Four cosmetic products from TTO: crude TTO, cleaned TTO, body lotion and hand cleaner were studied in RPTC, Figure 10. All of them were made to improve commercial value of Melaleuca trees, so living quality of farmers in Mekong Delta region would increase.

4. Conclusion:

Both TTO types satisfy the requirements given in International Standard ISO 4730. The concentrations of crude and cleaned TTO are about 35-48%, 95-100% of volume of terpinen-4-ol, respectively. Among them, cleaned TTO has higher commercial value and its ROR is about 1.5 times higher than ROR of crude TTO. The technologies for these processes are simple with cheap materials and quality of TTO is high. Besides, the antimicrobial activity of TTO in cosmetic products was still kept so the economic value of TTO increased. Therefore, TTO and its cosmetic products actually bring benefits for Vietnamese farmers who live on agriculture in acid land area

* **Acknowledgement:** We would like to thank Dr. Seiji Iwasa for not only giving us the opportunity to work in his laboratories but also being a great mentor and acknowledge Jica's Office, Tien Giang Department of Science Technology for their financial supports.

References

[1] C. F. Carson, K. A. Hammer and T. V. Riley 2006 Melaleuca alternifolia (Tea Tree) Oil: a Review of Antimicrobial and Other Medicinal Properties *Clinical Microbiology Reviews* **19** 50-62
[2] Geofrey R.Davis 1999 *Tea Tree Oil Distillation (*Australia: Harwood Academic) pp 155-168
[3] K.A. Hammer et al 2006 A review of the toxicity of Melaleuca alternifolia (tea tree) oil *Food and Chemical Toxicology* **44** 616–625
[4] Brophy, N. W. Davies, I. A. Southwell, I. A. Stiff and L. R. Williams 1989 Gas chromatographic quality control for oil of Melaleuca terpinen-4-ol type (Australian tea tree) *J. Agric. Food Chem* **37** 1330-1335
[5] Calcabrini, A. Stringaro, L. Toccacieli, S. Meschini, M. Marra, M. Colone, G. Salvatore, F. Mondello, G. Arancia and A. Molinari 2004 Terpinen-4-ol, the main component of Melaleuca alternifolia (tea tree) oil inhibits the in vitro growth of human melanoma cells *J. Investig. Dermato* **122** 349-360
[6] Vuong Ngoc Chinh 2005 *Cosmetic-Fragrance* National University of Vietnam 321

Journal of Physics: Conference Series **352** (2012) 012054 doi:10.1088/1742-6596/352/1/012054

Production of D-lactic acid from sugarcane bagasse using steam-explosion

Chizuru Sasaki, Ryosuke Okumura, Ai Asakawa, Chikako Asada and Yoshitoshi Nakamura

Department of Life System, Institute of Technology and Science, The University of Tokushima, 2-1 Minamijosanjima-cho, Tokushima 770-8506, Japan

E-mail: csasaki@bio.tokushima-u.ac.jp

Abstract. This study investigated the production of D-lactic acid from unutilized sugarcane bagasse using steam explosion pretreatment. The optimal steam pressure for a steaming time of 5 min was determined. By enzymatic saccharification using Meicellase, the highest recovery of glucose from raw bagasse, 73.7%, was obtained at a steam pressure of 20 atm. For residue washed with water after steam explosion, the glucose recovery increased up to 94.9% at a steam pressure of 20 atm. These results showed that washing with water is effective in removing enzymatic reaction inhibitors. After steam pretreatment (steam pressure of 20 atm), D-lactic acid was produced by *Lactobacillus delbrueckii* NBRC 3534 from the enzymatic hydrolyzate of steam-exploded bagasse and washed residue. The conversion rate of D-lactic acid obtained from the initial glucose concentration was 66.6% for the hydrolyzate derived from steam-exploded bagasse and 90.0% for that derived from the washed residue after steam explosion. These results also demonstrated that the hydrolyzate of steam-exploded bagasse (without washing with water) contains fermentation inhibitors and washing with water can remove them.

Introduction

Lactic acid is widely used in food, pharmaceutical, and chemical industries. In particular, lactic acid has been used as a raw material for manufacturing poly(lactic acid), one of the most promising biodegradable polymers. For this reason, many researchers have studied

L-lactic acid production strategies using unutilized cellulosic plant material and starchy biomass such as paper sludge, water wastes, agro wastes, and beer production byproducts [1-4]. Recently, it was discovered that D-type and L-type lactic acids form stable stereocomplexes that have thermal properties different from those of the original poly(lactic acid) made from only L-type or D-type lactic acid [5-7]. However, few studies have focused on strategies for producing D-lactic acid from unutilized plant biomass [8, 9].

The main components of unutilized cellulosic biomass are cellulose, hemicellulose, and lignin. Because cellulose is surrounded by a hard network of lignin, it is necessary to decompose it using an environmentally friendly pretreatment method prior to its conversion into glucose using an enzymatic method.

Steam explosion is a hydrothermal pretreatment method for plant biomass that uses high-pressure and high-temperature steam without the addition of any chemicals and involves chemical effects such as auto hydrolysis, defibriation, and delignificaion. Compared to the generally used chemical treatments, it has several potential advantages: it eliminates the use of toxic substances, such as strong acids or alkalis, chemical tolerance equipment, waste solution processing, and preliminary feedstock size reduction.

Sugarcane bagasse, a byproduct of the sugar industry, is an abundant source of lignocellulose. Although it is used as fuel for boilers, a large quantity of this material is accumulated in sugar processing plants, which leads to environmental problems. Advanced utilization of this material as a carbon resource holds promise for the bioproduction of useful chemicals like lactic acid.

In the present study, optimal steam explosion pretreatment conditions for sugarcane bagasse followed by enzymatic saccharification were investigated. In addition, efficient microbial conversion of hydrolyzates derived from sugarcane bagasse to D-lactic acid by *Lactobacillus delbrueckii* was investigated.

Materials and Methods

Sugarcane bagasse

The raw sugarcane bagasse (bagasse) from Kyuyo Sugar Co. Ltd. (Okinawa, Japan) used in these experiments was cut into 1–3 cm-sized chips. The chemical composition (based on dry material) of bagasse was 45.1% α-cellulose, 22.3% Klason lignin, 27.9% hemicellulose, and 4.7% other.

Steam explosion

Pretreatment of the bagasse was conducted in a steam explosion apparatus (Japan Chemical Engineering and Machinery Co. Ltd., Osaka, Japan). The reactor was charged with 100 g (dry matter) of feedstock per batch. Saturated steam from the boiler was then allowed to enter the reactor to heat the bagasse at a controlled pressure of 15, 20, 25, 30 or 35 atm. Each pressure was maintained for 5 min and then the reactor was suddenly depressurized. The exploded bagasse was recovered in a cyclone and cooled to room temperature.

Components analysis of bagasse and steam-exploded bagasse

Components analysis of untreated bagasse and steam-exploded bagasse was conducted by Wayman's method [10] using distilled water and methanol as extractive solvents. The method is illustrated in Fig. 1. The amount of components, i.e., water-soluble material (WSM), methanol-soluble material, acid-soluble lignin, holocellulose (cellulose and hemicellulose), and Klason lignin, was determined as follows.

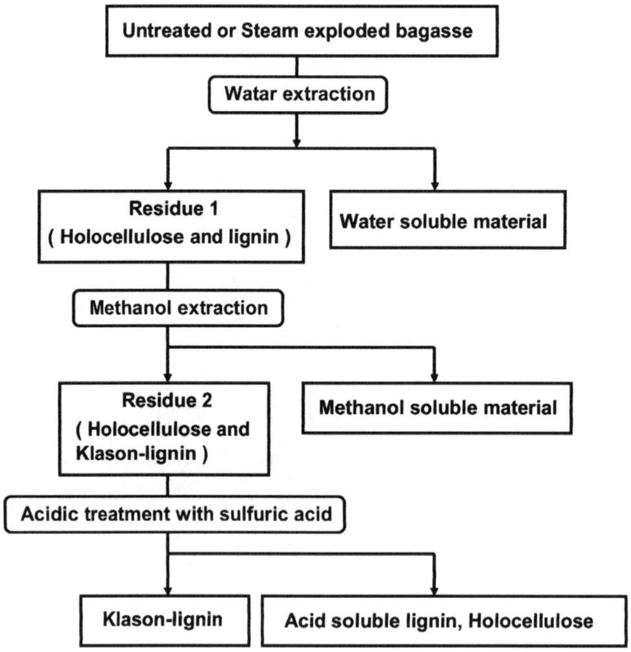

Fig. 1 Component analysis method of untreated bagasse and steam exploded bagasse

Five grams of dry untreated or steam-exploded bagasse was added to 100 mL of distilled water and extracted for 24 h at room temperature. The solid (Residue 1) and liquid materials were separated by filtration, the filtrate was recovered from the liquid, then concentrated, dried, and weighed (water-soluble material). Residue 1 was extracted at 80°C for 24 h in a Soxhlet extractor using 150 mL methanol to dissolve the methanol-soluble material. After concentration and drying of the extract, the methanol-soluble material was weighed. Residue 2 from the methanol extraction consisted of holocellulose and Klason lignin, a high-molecular weight lignin. This residue (1 g) was added to 15 mL of 72% (w/v) sulfuric acid and kept at room temperature for 4 h. The residue was placed in a 100 mL conical flask, washed with 560 mL distilled water, and then autoclaved for 1 h. After the insoluble material was washed with distilled water, it was heat-dried at 105°C to a constant weight and weighed (Klason lignin). The weight of holocellulose was calculated by subtracting the weight of Klason lignin from 1 g of Residue 2.

Enzymatic hydrolysis

Steam-exploded bagasse was enzymatically hydrolyzed using Meicellase (produced by *Trichoderma viride*) provided by Meiji Seika Co., Ltd. Enzymatic hydrolysis was performed using a 0.05 M sodium phosphoric acid buffer (pH 4.5) at 50°C on a rotary shaker at 140 rpm for 48 h. The substrate concentration and enzyme loading were 100 g/L and 50 FPU/g substrate, respectively. The supernatant was centrifuged to remove solid waste and analyzed for glucose. All enzymatic hydrolysis experiments were performed in duplicate.

Microorganism and inoculum cultivation

Lactobacillus delbrueckii NBRC 3534, a homofermentative D-lactic acid producer, a high optical purity D-lactic acid producing strain (content of D-type lactic acid is 98.5%) [8], was used to produce D-lactic acid from the hydrolyzate of steam-exploded bagasse. The microorganism was subcultured every four weeks. The strain was precultured in 100 mL of medium in a 300 mL flask at 37°C for 24 h using a static incubator. The preculture media comprised 0.1 g/L $(NH_4)_2SO_4$, 10 g/L glucose, 0.1 g/L KH_2PO_4, 0.1 g/L $MgSO_4{\cdot}7H_2O$, and 1.0 g/L yeast extract. Later, the cells were harvested by centrifugation (2000×*g*, 15 min), rinsed thoroughly with sterile distilled water, centrifuged again, and then resuspended in sterile distilled water.

Optimization of initial glucose concentration of the medium

To determine optimal initial glucose concentration for D-lactic acid fermentation using *Lactobacillus delbrueckii* NBRC 3534, four glucose concentrations of 21.0, 32.3, 44.5, and 53.6 g/L were examined. The other nutrients added were 5.0 g/L yeast extract, 10.0 g/L peptone, 5.0 g/L sodium acetate, 2.0 g/L ammonium citrate, 2.0 g/L K_2HPO_4, 0.1 g/L $MgSO_4 \cdot 7H_2O$, 0.05 g/L $MnSO_4 \cdot H_2O$, 1.0 g/L Tween 80[12]. The strain was cultured in 100 ml of medium in a 300 ml flask at 37°C for 72 h using a static incubator.

D-lactic acid fermentation using bagasse hydrolyzate

The nutrient medium for D-lactic acid fermentation contained hydrolyzate derived from steam-exploded bagasse (substrate concentration of 10% w/v), nutrient medium, 0.05 M acetate buffer (pH 5.0), and 10% (v/v) preculture solution. The nutrient medium and buffer were autoclaved at 121°C for 20 min. The hydrolyzate was added after sterilizing with a 0.22 μm pore size filter. Cultivation was carried out in a 300 mL flask with 100 mL of the medium using a static incubator at 37°C. Aliquots of the samples were collected and assayed for determining ethanol and residual glucose concentrations.

Determination of residual glucose and produced D-lactic acid

Residual glucose and produced D-lactic acid concentrations in the fermentation medium were determined by the mutarotase GOD method (Glucose C-II test, Wako Pure Chemicals Co., Ltd., Japan) and an enzymatic UV test with D-lactate dehydrogenase (D-LDH) (enzymatic fluid D-lactic acid, R-Biofarm, GmbH, Germany), respectively.

Results and discussion

Composition analysis of steam-exploded bagasse

To break down the lignin network that surrounds cellulose and increase the accessibility of the enzyme to cellulose, steam explosion was used to treat the bagasse. Table 2 shows the chemical composition of untreated bagasse and steam-exploded bagasse (steam pressures of 15, 20, 25, 30, and 35 atm; steaming time 5 min) separated by the extraction method illustrated in Fig. 1. The highest amount of WSM (17.8%), which contains sugars derived from holocellulose, was obtained at a steam pressure of 15 atm. In general, hemicellulose in the holocellulose hydrolyzes under less severe conditions than cellulose [13, 14]; therefore, mainly degradation of hemicellulose occurred. Above 20 atm, the amount of WSM decreased owing to the degradation of pentose, i.e., xylose or arabinose sugars derived from hemicellulose, which changed to insoluble materials such as furfural [15]. With increased production of sugars and further degraded materials, the amount of holocellulose decreased (steam pressures of 15 and 35 atm resulted in 48.6% and 34.4%, respectively). The amount of methanol-soluble material increased with steam pressure. Steam pressures of 15 and 30 atm resulted in 7.9% and 23.7%, and it contained low-molecular lignin [16-18]. Furthermore, the maximum reduction of Klason lignin (reduction rate 19.8%) was observed at a steam pressure of 20 atm. At higher pressure, a gradual increase was observed owing to the combination of lignin with low-molecular lignin and water-soluble materials [16-18].

Table 1 Chemical composition (%) of untreated and steam exploded bagasse at different steam pressures (steaming time 5 min)

Components	Untreated bagasse	Steam pressure (atm) [Steaming temperature]				
		15 [200℃]	20 [213℃]	25 [224℃]	30 [234℃]	35 [243℃]
Water soluble material	1.9	17.8	13.7	11.2	9.7	10.0
Methanol soluble material	1.5	7.9	22.0	22.8	23.7	22.6
Acid soluble lignin	1.8	0.9	0.4	0.4	0.4	0.4
Holocellulose	72.6	53.3	46.2	44.6	41.7	39.0
Klason-lignin	22.2	20.1	17.8	21.0	24.5	28.0
Total	100	100	100	100	100	100

Enzymatic saccharification of steam-exploded bagasse

The effect of steam explosion conditions on the enzymatic saccharification of bagasse was studied using a substrate concentration of 50 g/L and results are presented in Fig. 2(A). Steam pressures of 15, 20, 25, 30, and 35 atm for a steaming time of 5 min were investigated. The maximum amount of glucose produced was observed at a steam pressure of 25 atm and steaming time of 5 min: 364 mg/g dry steam-exploded bagasse, which corresponds to a 73.7% glucose yield from untreated bagasse. For steam pressures of 20, 30, and 35 atm, the glucose yield decreased more than that for steam pressure of 15 atm. Fig. 2(B) shows the enzymatic saccharification of washed steam-exploded bagasse, i.e., WISM, after steam explosion as a substrate (substrate concentration is 50 g/L). Under all conditions, the amount of glucose produced increased. Similar higher glucose concentration was achieved at steam pressures of 20 and 25 atm. The value was 469 mg in 1 g of steam-exploded dry bagasse, corresponding to a 94.9% glucose yield from untreated bagasse. These results indicated that if inhibitors for saccharification, such as low-molecular lignin and water-soluble decomposed saccharide, i.e. furfural and 5-hydroxymethylfurfural (5-HMF), the amount of these materials were 16.7, 3.5 mg in 1 g of steam-exploded dry bagasse at steam pressure 25 atm (data not shown), were washed with water, the enzyme could access cellulose in the treated bagasse. Dunlop et al. [15], Palmqvist et al.[19] reported that during high temperature or acid pretreatment of wood, water-soluble inhibitors such as sugar- and lignin-derived materials are formed. From the viewpoint of low energy input, this study indicated that the optimal condition for steam explosion of bagasse was steam pressure of 20 atm for a steaming time of 5 min.

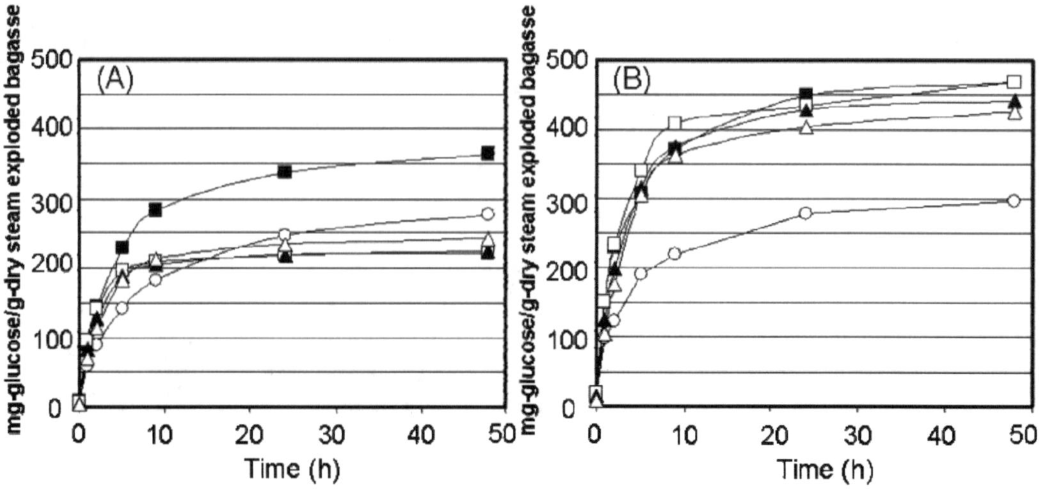

Fig. 2 Enzymatic saccharification of steam exploded bagasse (A) and water insoluble residue after steam explosion (B). Symbols; (○) steam pressure 15 atm, (□) 20 atm, (■) 25 atm, (△) 30 atm and (▲) 35 atm, at steaming time 5 min.

D-lactic acid production from bagasse

The optimal initial glucose concentration for D-lactic acid by *Lactobacillus delbrueckii* NBRC 3534 using was investigated before production of D-lactic acid from bagasse. Maximum D-lactic acid yield (from initial amount of glucose) and the residual ratio of glucose for a fermentation period of 72 h are summarized in Table 2. The maximum yield was observed at an initial glucose concentration of 21.0 g/L, i.e., 78.6%. In the case of an initial glucose concentration greater than 21.0 g/L, the residual ratio of initial glucose increased. For D-lactic acid production from bagasse in this study, an initial glucose concentration was adjusted at 20.0 g/L.

Table 2 D-lactic acid production and amount of residual glucose by *Lactobacillus delbrueckii* NBRC 3534 at different initial glucose concentrations after 72 h cultivation.

Initial glucose Conc. (g/l)	D-lactic acid Conc. (g/l)	Yield (%)	Residual glucose Conc. (g/l)	Residual ratio of initial glucose (%)
21.0	16.5	78.6	0.8	3.8
32.3	17.7	54.8	10.8	33.4
44.5	18.7	42.0	23.0	51.6
53.6	17.7	33.0	33.9	63.2

Fig. 3 shows the time courses of glucose consumption and D-lactic acid production by *Lactobacillus delbrueckii* NBRC 3534 using available glucose, steam-exploded bagasse (steam pressure of 20 atm, steaming time 5 min), and water-insoluble residue after steam explosion (20 atm, 5 min) as substrates. When water-insoluble residue after steam explosion was used as the substrate, D-lactic acid production behavior similar to the results obtained using commercially available glucose was observed. In this case, the maximum amounts of D-lactic acid produced were 16.5 g/L (available glucose) and 17.1 g/L (water-insoluble residue), at a 72 h cultivation time, which corresponds to 81.7% and 90.0% of the conversion rate of the initial glucose concentration, respectively. On the other hand, when steam-exploded bagasse was used as the substrate, i.e., without washing with water after steam explosion treatment, the amount of D-lactic acid produced decreased (conversion rate was 66.5%) and no dramatic glucose consumption was observed. This phenomenon also showed the presence of fermentation inhibition materials in the steam-exploded bagasse, similar to the case of enzymatic saccharification. This study found that a simple method with low environmental loading, such as washing only with water, could remove the inhibitors for D-lactic acid producers and cellulase from steam-exploded bagasse. It was determined that 48.9 g of D-lactic acid can be produced from 100 g of water-insoluble residue after steam explosion (steam pressure of 20 atm and steaming time 5 min) using *Lactobacillus delbrueckii* NBRC 3534. Moreover, from 100 g of bagasse, 42.2 g of D-lactic acid can be produced by this method.

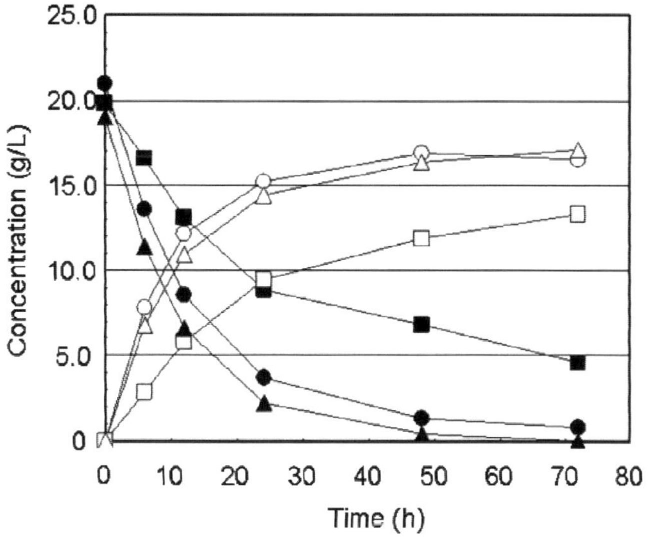

Fig. 3 Residual glucose concentration and D-lactic acid production profiles by *Lactobacillus delbrueckii* NBRC3534. Symbols; (●) Glucose with available glucose, (○) D-lactic acid derived from available glucose, (■) Glucose with steam exploded bagasse, (□) D-lactic acid

derived from steam exploded bagasse, (▲) Glucose with water insoluble residue after steam explosion and (△) D-lactic acid derived from water insoluble residue after steam explosion.

Conclusions

This study showed that high amounts of glucose could be produced from enzymatic hydrolyzate of steam-exploded bagasse. In particular, washing the steam-exploded sample with water effectively removes the inhibitors that interrupt the reaction between cellulose and the enzyme (cellulase). Furthermore, it was determined that higher amounts of D-lactic acid could be produced from a carbon source derived from steam-exploded bagasse washed with water after steam explosion than from that prepared without washing using *Lactobacillus delbrueckii* NBRC 3534. Washing with water is a simple and environmentally friendly method for removing inhibitors from the pretreated residue. This study will contribute to improving large-scale industrial methods for the production of D-lactic acid from unutilized plant materials.

Acknowledgements

A part of this study was funded by Shiseido Female Researcher Science Grant and Grant-in-Aid for Young Scientist (B)(No. 21750159) from the Ministry of Education, Culture, Sports, Science and Technology of Japan.

References

[1] Marques S, Santos J.A.L, Girio F.M, Roseiro J.C. 2008 *Eng. J.* **41** 210
[2] Nakasaki K, Adachi T 2003 *Biotechnol. Bioeng.* **82** 263
[3] John R.P, Nampoothiri M, Pandey A 2006 *Process Biochem.* **41** 759
[4] Shindo S, Tachibana T 2004 *J. I. Brewing* **110** 347
[5] Ikeda Y, Jamshidi K, Tsuji H, Hyon S.H. 1987 *Macromolecules* **20** 904
[6] Tsuji H, Hyon S.H, Ikeda Y 1991 *Macromolecules* **24** 2719
[7] Slivniak R, Domb A.J 2002 *Biomacromolecules* **3** 754
[8] Fukushima K, Sogo K, Miura S, Kimura Y 2004 *Macromol. Biosci.* **4** 1021
[9] Lu Z, Lu M, He F, Yu L 2009 *Bioresource Technol.* **100** 2026
[10] Chua M.G.S, Wayman M 1979 *Can. J. Chem.* **57** 141
[11] Mikami, H. Ishida Y 1983 *Bunseki Kagaku* (in Japanese) **32** E207
[12] Mercier P, Yerushalmi L, Rouleau D, Dochain D, 1992 *J. Chem. Technol. Biot.* **55** 111
[13] Palmqvist E. Hahn-Hagerdal B 2000 *Bioresource Technol* **74** 25
[14] Ando H, Sakaki T, Kokusho T, Shibata M, Uemura Y, Hatate Y 2000 *Ind. Eng. Chem. Res.* **39** 3688

[15] Dunlop A.P 1948 *Ind. Eng. Chem.* **40** 204

[16] Nakamura Y, Sawada T, Inoue E 2001 *J. Chem. Technol. Biot.* **76** 879

[17] Kobayashi F, Take H, Asada C, Nakamura Y 2004 *J. Biosci. Bioeng.* **97** 426

[18] Asada C, Nakamura Y, Kobayashi F 2005 *Biotechnol. Bioprocess E* **10** 346

[19] Palmqvist E, Hahn-Hagerdal B, Galbe M, Zacchi G. 1996 *Enzyme Microb. Tech.* **19** 470

AUTHOR INDEX

Abderrahmane, A 76, 80
Abe, K 284, 307
Abe, M. 110, 190, 231
Akiyama, T 127, 132, 148, 153, 158
Aktsipetrov, O 199
Aoki, K 301
Aoyagi, N 217
Arai, S 252
Arisawa, R 338
Asada, C 374
Asakawa, A 374
Baba, Y 284, 307
Botet, R 323
Do, S H 367
Dou, S X 51
Fujieda, S 137
Fujikado, T 43
Fujimoto, K 158
Fujisawa, I 301
Fukuda, K 137
Fukuda, M 252
Grishin, A M 51, 60
Gross, H J 317
Hamada, S 98
Handa, H. 110, 231
Harrison, R J 105
Hashimoto, Y 307
Hatsukade, Y 239, 246
Hayasaka, Y 204
Hiraishi, A 85, 362
Hirota, S 278
Hisyamudin, M N N 338, 354
Hosokai, Y 70
Huynh, Q 367
Ichiyanagi, Y 70, 143
Ikegai, H 333
Ikigai, H 211
Imanaka, S 268
Inoue, M 175, 252
Ishikawa, R 28, 85, 114, 121
Iwasa, S 85, 114, 121
Iwase, M 127
Izaki, M 338, 347, 354
Jalalian, A 51
Jeyadevan, B 137
Kallumadil, M 231
Kanamori, Y 153
Kanda, H 43
Kanematsu, H 211, 225, 333
Kang, E G 167
Kataoka, M 284, 307
Kato, M 252
Kawaguchi, T 252
Kawahata, M 98
Kawashima, T 183, 204
Kawauchi, S 288
Khartsev, S I 60
Kidowaki, H 148
Kikuchi, Y 288
Kim, Y T 167
Kimura, H 35
Kimura, T 183
Kishida, A 183
Kitamoto, Y 110
Kitamura, Y 246, 301
Ko, P J 1, 18, 28
Kobayashi, T 35, 225
Kogo, T 211
Koide, S 76, 80
Kolmychek, I 199
Konagai, M 28
Kougo, T 333
Kuboyama, H 252
Kunugihara, D 278
Kuroda, D 211, 333
Kurokawa, A 70, 143
Kurokawa, Y 28
Kurosawa, R 239
Li, Z 1
Machida, Y 70
Maegawa, A 338
Maezawa, Y 35
Mamonov, E 199
Masaki, T 110
Masuzawa, T 183
Matsufuji, Y 175
Matsumoto, N 204
Matsushita, N 110
Maydykovsky, A 199
Mihara, Y 278
Morisato, K 225
Moriyama, K 252
Moshchalkov, V 199
Murakami, K 338
Murozono, M 153
Murzina, T 199
Nagai, M 183, 204
Nagao, Y 85
Nagata, A 158
Nakagawa, J 132
Nakagawa, T 190
Nakahara, M 225
Nakajima, A 35
Nakamura, M 127, 132
Nakamura, Y 175, 374
Nakayama, H 278
Neviarouskaya, A 312
Noda, T 35, 43
Nomura, Y 274
Numano, R 85

AUTHOR INDEX

Oguri, M 204
Ohnishi, Y 153
Ohshima, T 80
Ohta, J 35, 43
Ohtani, T 246
Ohtani, Y 153
Okada, H 80, 85
Okada, J 274
Okamoto, R 301
Okamoto, Y 85, 114, 121
Okamura, K 362
Okawa, T 217
Oku, T 90, 127, 132, 148, 153, 158
Okumura, R 374
Ono, Y 153
Ooie, T 307
Orito, N 288
Pan, Y 43
Phan, T D 367
Saito, H 70
Sakai, E 288
Sakai, N 70, 143
Sandhu, A 1, 18, 28, 76, 80, 85, 114, 121
Sasagawa, K 35, 43
Sasaki, C 374
Sasaki, T 347
Sasano, J 347
Sato, K 288
Sato, S 80
Sato, T 217
Sawadsaringkarn, Y 35
Sekino, H 98, 105
Shibasaki, I 76
Shibata, T 183, 204
Shimizu, K 217
Shimomura, T 354
Shinoda, K 137
Silhanek, A 199
Suda, Y 85, 217
Sugiyama, J 268
Suzuki, A 90, 127, 132, 158
Suzuki, S 137, 246
Tada, M 190
Taguchi, K 268, 274, 278
Taira, S 70
Takahashi, H 76, 80
Takahashi, T 354
Takamura, T 18
Takanashi, A 362
Takashina, T 225
Takeda, A 158
Takeuchi, D 301
Takeuchi, H 70, 143
Takikawa, H 85, 217
Tamura, K 274
Tanaka, H 288
Tanaka, M 307
Tanaka, S 239, 246
Tanaka, T 190, 288
Tanizawa, Y 85
Tanoue, H 217
Tero, R 85, 114, 121
Thieu, V Q Q 367
Tokuda, T 35, 43
Tokumitsu, K 127, 132
Totsuka, M 268
Tran, S T 367
Tsetserukou, D 312
Tsuzuki, K 85, 114, 121
Uchida, Y 239
Ue, H 217
Ueda, T 110, 231
Umegaki, G 183
Umekage, S 288
Uruno, H 217
Usui, A 70
Valev, V 199
Verbiest, T 199
Wada, N 70, 143, 333
Yamada, M 127, 132
Yamada, T 362
Yamada, Y 183
Yamamoto, S 252
Yamamura, S 284, 307
Yang, Y 1
Yano, S 70, 143
Yanoh, T 70, 143
Yatsushiro, S 284, 307
Yokoi, Y 105
Yokoyama, S 211, 338, 347, 354
Yoshida, K 211
Yoshida, N 85
Zhu, L 70, 143

Institute of Physics
Dirac House, Temple Back
Bristol BS1 6BE UK

CURRAN ASSOCIATES INC
proceedings.com